MW00611583

Theory of LINEAR PHYSICAL SYSTEMS

Theory of physical systems from the viewpoint of classical dynamics, including Fourier methods

Ernst A. Guillemin

Massachusetts Institute of Technology

Dover Publications, Inc.
Mineola, New York

Bibliographical Note

This Dover edition, first published in 2013 is an unabridged republication of the work originally published by John Wiley & Sons, Inc., New York, in 1963.

International Standard Book Number

ISBN-13: 978-0-486-49774-7
ISBN-10: 0-486-49774-7

Manufactured in the United States by Courier Corporation
49774701
www.doverpublications.com

To Mary Grace and Russ

Preface

This book deals mainly with topics that lie between those treated in *Introductory Circuit Theory* and those treated in *Synthesis of Passive Networks*. These are topics in linear system theory with restriction to physical systems but not confined to passive or bilateral systems.

Briefly, some of these topics have to do with the topological, algebraic and matrix aspects of setting up equilibrium equations involving, besides the usual two-terminal passive *R*'s, *L*'s and *C*'s, elements which are symbolically represented by boxes with three or more terminals projecting. Incidentally, the embedding of such elements tells us how to deal with networks that have been torn apart into smaller sections and hence includes the analysis of networks by the method of tearing.

Matrix aspects touch upon synthesis as well as upon analysis questions, and to establish the pertinent relationships between matrices and the related impedance functions the concept of normal coordinate transformations is essential. This is discussed not only for two-element-kind networks, where its application has long been known, but for *RLC* networks as well. Two different approaches are taken here: one depends upon expressing the network equilibrium on a mixed-variable basis; the other upon applying a suitable complex transformation directly to the matrices on a loop or node basis. The interesting aspect of the latter method is that elements of the transformation matrix turn out to be the residues of the associated impedances obtained by familiar methods of partial-fraction expansion. The astounding result here is that (except for a simply formable normalization factor) the direction cosines of the normal coordinates are the residues of the pertinent impedance functions.

Another topic that belongs in this category is a construction of the impulse response of a network in terms of its poles and zeros in a form that is usually gotten by Laplace methods, which shows that this result does not require Laplace theory for its derivation.

This group of topics (expanding those in *Introductory Circuit Theory*) is rounded out by discussing the synthesis problem starting with parameter matrices for one-, two- and three-element-kind networks, deriving the essential energy relations for the characterization of impedances, and in terms of these deriving the properties of impedances (for example, the positive-real property) and methods of testing a given function for these properties. These subjects are dealt with in a simplified manner compared with that used in *Synthesis of Passive Networks* but sufficiently detailed to be meaningful and useful.

The remaining part of the book (which is the larger part) is devoted to Fourier and related theory, and its application to the solution of linear systems. Most of this section of the book deals only with Fourier *theory*; the part dealing with its application to linear systems is restricted to a single chapter (except for the discussion of numerical methods of evaluation).

Fourier theory is introduced as a method of creating a desired interference pattern from steady sinusoids, by applying this theory first to periodic functions and then extending the method to aperiodic ones. The paramount issue here is the evaluation of error and its dependence upon the spectral width, which results in the fact that one can make the error arbitrarily small, although not zero, with a sufficiently large but still finite spectral width. Through the simple use of Cauchy's integral law one can then extend the procedure to constructing such interference patterns with sinusoids involving complex frequencies; that is to say, through the use of sinusoids having exponentially modulated amplitudes. This amounts to extending the Fourier integral to a double-ended Laplace transform, and one thus recognises that the Fourier transform is the Laplace transform evaluated along the j-axis. Here one must, of course, include impulses in the spectral functions, which is a process that fits in with familiar time-domain procedures and leads directly to the numerical evaluation of inverse transforms by means of impulse-train methods.

A topic that logically fits in at this point is the evaluation of time- and frequency-domain errors, the mean-square type of error and the kind of frequency-domain truncation causing it, the related Gibbs phenomenon, the Cesàro and other types of truncation or summation and their time-domain aspects.

All these things are presented before Fourier theory is applied to network analysis. Thus the limitations due to Fourier theory alone and those that are introduced by network properties are presented separately, avoiding confusion in the student's mind.

It is emphasized that in the use of Fourier methods for finding network response everything in the analysis remains as it is in the familiar classical differential-equation approach except that the excitation function is represented as an interference pattern of sinusoids (involving real or complex frequencies). The solution still consists of a forced part (particular integral) and a force-free part (complementary function). However, if the real parts of corresponding complex frequencies whose interference pattern yields the desired excitation are larger or at least as large as the largest of all the natural frequencies, then the forced part of the solution dominates and the force-free part is negligible. The solution is formed from the particular integral alone. When the particular integral does not dominate one must add another integral yielding the force-free (transient) part. The use of Fourier methods is still possible now but no longer simple. These matters are seldom if ever pointed out in other textbooks on Fourier theory and its application.

Rather, in many of the modern textbooks, Fourier methods are dealt with as a completely new approach which is based upon a set of postulates having little if anything to do with classical dynamics. According to this attitude the linear system has no natural frequencies, no normal coordinates, not even an equilibrium equation. The total solution is given by the forced part of the solution alone—according to dictum, not because of any physical reason. In fact, one is supposed to use only mathematical reasoning. There is a basic inconsistency between such an approach and the one which is based upon classical dynamics. This abstract point of view may be acceptable, however, for certain types of application where the system is a black box and has no normal physical properties.

Fourier or Laplace theory introduced in this way is useful for handling certain problems, for instance, in the discussion of ideal bandpass filtering. But let it not be implied that it is a general approach that is likewise applicable to physical systems, for this is untrue. I believe that this textbook is the first one to point out that such inconsistencies exist in the teaching of Fourier theory. Let us give the mathematically abstract approach a different name like "black-box" Fourier theory so that the student will be forewarned to expect discrepancies between this brand of Fourier theory and the kind that is consistent with and based upon classical dynamics, which has been the basis for all the circuit theory he has been taught so far.

Next in the list of topics presented is that dealing with methods of numerical evaluation which are based upon the impulse-train concept. The sampling of functions, arbitrarily or uniformly, is a closely related subject. Important in this discussion is a related method of error evaluation and error transformation from frequency to time domains. It is extremely interesting to find here that this process is basically the same as the almost forgotten C. L. Fortescue method of "symmetrical coordinates."

Finally, since the topic dealing with real-part sufficiency is closely related to the exact and approximate methods of Fourier transformation it seems logical to close with a discussion of Hilbert transforms. These lead rather logically to Bode's resistance-integral theorem and its applications, which are used to present practical applications or illustrations.

In closing, I wish to express gratefulness to all my associates who, through the years, have helped teach this subject and thereby have made many new and useful contributions that are reflected in the present state of this text material. I would like to single out for special mention Thomas G. Stockham, Jr., who has contributed many interesting discussions and useful points of view.

ERNST A. GUILLEMIN

Wellesley Hills, Mass.
May 1963

Contents

CHAPTER I

Solution of Algebraic Equations When Auxiliary Conditions Are Specified

1. The Mixed Situation

Consider the set of b linear simultaneous equations

$$\begin{aligned} a_{11}x_1 + a_{12}x_2 + \cdots + a_{1b}x_b &= y_1 \\ a_{21}x_1 + a_{22}x_2 + \cdots + a_{2b}x_b &= y_2 \\ \cdot\ \cdot\ \cdot\ \cdot\ \cdot\ \cdot\ \cdot\ \cdot\ \cdot\ \cdot\ \cdot\ \cdot\ \cdot\ \cdot \\ a_{b1}x_1 + a_{b2}x_2 + \cdots + a_{bb}x_b &= y_b \end{aligned} \tag{1}$$

which may be written in matrix form as

$$[A] \cdot x] = y] \tag{2}$$

where

$$[A] = \begin{bmatrix} a_{11} & a_{12} & \cdots & a_{1b} \\ a_{21} & a_{22} & \cdots & a_{2b} \\ \cdot & \cdot & \cdot & \cdot \\ a_{b1} & a_{b2} & \cdots & a_{bb} \end{bmatrix} \tag{3}$$

and

$$x] = \begin{bmatrix} x_1 \\ x_2 \\ \cdot \\ \cdot \\ \cdot \\ x_b \end{bmatrix}; \quad y] = \begin{bmatrix} y_1 \\ y_2 \\ \cdot \\ \cdot \\ \cdot \\ y_b \end{bmatrix} \tag{4}$$

In the usual situation, the quantities $y_1 \cdots y_b$ are given and $x_1 \cdots x_b$ are to be found. If

$$[B] = [A]^{-1} = \begin{bmatrix} b_{11} & b_{12} & \cdots & b_{1b} \\ b_{21} & b_{22} & \cdots & b_{2b} \\ \cdot & \cdot & \cdot & \cdot \\ b_{b1} & b_{b2} & \cdots & b_{bb} \end{bmatrix} \tag{5}$$

denotes the inverse of the matrix 3, then the desired solution in this case is indicated by

$$[B] \cdot y] = x] \tag{6}$$

Denoting the determinants of $[A]$ and $[B]$ by A and B respectively and their cofactors by A_{sk} and B_{sk}, we have in the familiar manner

$$b_{sk} = \frac{A_{ks}}{A} \quad \text{and} \quad a_{sk} = \frac{B_{ks}}{B} \tag{7}$$

The matrix $[A]$ must, of course, be nonsingular in order for this solution to exist; or, what amounts to the same thing, the equations 1 must be independent since the stated condition is satisfied if the determinant A is nonzero, whereupon $B = 1/A$ is likewise nonzero.

A somewhat different situation is the one in which some of the x_k's and some of the y_k's are known and the equations 1 are to be used to find the remaining x_k's and y_k's. This is a sort of *mixed* situation. To be specific, let us suppose that n of the x_k's and l of the y_k's are known, and $l + n = b$. A fitting notation to accommodate this specification is had by partitioning the column matrices $x]$ and $y]$ as indicated by writing

$$x] = \begin{bmatrix} x_l \\ \cdots \\ x_n \end{bmatrix} \quad \text{and} \quad y] = \begin{bmatrix} y_l \\ \cdots \\ y_n \end{bmatrix} \tag{8}$$

in which $x_l]$ involves the quantities $x_1 \cdots x_l$; $x_n]$ the quantities $x_{l+1} \cdots x_b$; and analogous definitions apply to $y_l]$ and $y_n]$. The rows and columns in the matrices $[A]$ and $[B]$ are correspondingly partitioned into groups of l and n, yielding the forms

$$[A] = \begin{bmatrix} a_{ll} & \vdots & a_{ln} \\ \cdots & \cdot & \cdots \\ a_{nl} & \vdots & a_{nn} \end{bmatrix} \tag{9}$$

and

$$[B] = \begin{bmatrix} b_{ll} & \vdots & b_{ln} \\ \cdots & \cdot & \cdots \\ b_{nl} & \vdots & b_{nn} \end{bmatrix} \tag{10}$$

Equations 1 are now written

$$\begin{aligned} a_{ll} \cdot x_l] + a_{ln} \cdot x_n] &= y_l] \\ a_{nl} \cdot x_l] + a_{nn} \cdot x_n] &= y_n] \end{aligned} \tag{11}$$

Known are $x_n]$ and $y_l]$; we are to solve for $x_l]$ and $y_n]$. This is done quite readily by using the first equation to get

$$a_{ll} \cdot x_l] = y_l] - a_{ln} \cdot x_n]$$

or (12)

$$x_l] = a_{ll}^{-1} \cdot \{y_l] - a_{ln} \cdot x_n]\}$$

whereupon substitution into the second equation yields

$$y_n] = a_{nl} \cdot a_{ll}^{-1} \cdot y_l] + (a_{nn} - a_{nl} \cdot a_{ll}^{-1} \cdot a_{ln}) \cdot x_n] \quad (13)$$

In general a_{ll} must here be nonsingular so that the solution 12 for $x_l]$ exists and the second equation in set 11 can be solved for $y_n]$ to yield 13. Rank of the matrix $[A]$ must be at least l, but this condition is, of course, not sufficient to assure that a_{ll} will be nonsingular.

If a_{ll} is singular then both $x_n]$ and $y_l]$ cannot be specified independently. For a solution to be possible at all, the vector defined by column matrix $y_l] - a_{ln} \cdot x_n]$ must belong to the transposed vector set of a_{ll}. When $x_n]$ alone is given, a vector $y_l]$ can always be found to fulfill this condition and a solution is possible although not unique. When $y_l]$ alone is given, a vector $x_n]$ can in general be found to fulfill the pertinent condition only if $l \leq n$ and a_{ln} has rank l. When the rank of a_{ln} is less than l or when $l > n$, and $y_l]$ is specified, equations determining the components of $x_n]$ are not necessarily consistent and a solution in general is not possible. The matrix $[A]$ or its submatrix a_{nn} may be, but does not have to be nonsingular in these considerations.

If we are at liberty to construct $x_n]$ and $y_l]$ from any of the quantities $x_1 \cdots x_b$ and $y_1 \cdots y_b$ respectively, then it is necessary only that $[A]$ have the rank l, for we can first arrange the equations 1 so that the first l of them are independent and then arrange the numbering of the x_k's so that the first l of these are associated with those columns in the first l rows of $[A]$ that are independent. (The fact that the l independent rows of $[A]$ must contain l independent columns is a well-known matrix property that is commonly used as a test for the independence of such a given set of rows.)

An alternate way of getting a solution to this situation is to consider the inverse of Eqs. 11, namely

$$\begin{aligned} b_{ll} \cdot y_l] + b_{ln} \cdot y_n] &= x_l] \\ b_{nl} \cdot y_l] + b_{nn} \cdot y_n] &= x_n] \end{aligned} \quad (14)$$

then use the second of these to get

$$b_{nn} \cdot y_n] = x_n] - b_{nl} \cdot y_l]$$

or (15)

$$y_n] = b_{nn}^{-1} \cdot \{x_n] - b_{nl} \cdot y_l]\}$$

and finally substitute into the first of Eqs. 14 to obtain

$$x_l] = (b_{ll} - b_{ln} \cdot b_{nn}^{-1} \cdot b_{nl}) \cdot y_l] + b_{ln} \cdot b_{nn}^{-1} \cdot x_n] \tag{16}$$

Here the submatrix b_{nn} must be nonsingular. In this regard, remarks analogous to those made above with reference to $[A]$ now apply to $[B]$.

If the given matrix is $[A]$, then the solution given by Eqs. 12 and 13 involves less computation; on the other hand, if $[B]$ is given, Eqs. 15 and 16 yield the desired solution with fewer computations. In either case, solution of the original equations 1 is less tedious when we have a mixed situation than it is when the y's are known and the x's are to be found since inversion of a submatrix is easier than inversion of the entire matrix. Incidentally, comparison of Eqs. 12 and 16 shows that

$$a_{ll}^{-1} = (b_{ll} - b_{ln} \cdot b_{nn}^{-1} \cdot b_{nl}) \tag{17}$$

and comparison of Eqs. 13 and 15 yields

$$b_{nn}^{-1} = (a_{nn} - a_{nl} \cdot a_{ll}^{-1} \cdot a_{ln}) \tag{18}$$

which are a pair of collaterally useful relations.

2. The Auxiliary Conditions Are Linear Constraints

Instead of having $y_l]$ and $x_n]$ known, we might have l linear constraint relations among the y's and n linear constraint relations among the x's specified. We will write these relations in the form

$$\begin{aligned} \alpha_{l+1,1}x_1 + \alpha_{l+1,2}x_2 + \cdots + \alpha_{l+1,b}x_b &= x'_{l+1} \\ \alpha_{l+2,1}x_1 + \alpha_{l+2,2}x_2 + \cdots + \alpha_{l+2,b}x_b &= x'_{l+2} \\ \cdots\cdots\cdots\cdots\cdots\cdots\cdots\cdots \\ \alpha_{b1}x_1 + \alpha_{b2}x_2 + \cdots + \alpha_{bb}x_b &= x'_b \end{aligned} \tag{19}$$

and

$$\begin{aligned} \beta_{11}y_1 + \beta_{12}y_2 + \cdots + \beta_{1b}y_b &= y'_1 \\ \beta_{21}y_1 + \beta_{22}y_2 + \cdots + \beta_{2b}y_b &= y'_2 \\ \cdots\cdots\cdots\cdots\cdots\cdots \\ \beta_{l1}y_1 + \beta_{l2}y_2 + \cdots + \beta_{lb}y_b &= y'_l \end{aligned} \tag{20}$$

in which the coefficients α_{sk} and β_{sk} are assumed here to have only the values $+1$, -1, or zero.

Since these constraints are by assumption independent, we can (through renumbering of the variables, if necessary) assume the pertinent partitioned matrices to have the so-called *canonic* forms

$$[\alpha_{nb}] = [\alpha_{nl} \vdots u_n] \tag{21}$$

and

$$[\beta_{lb}] = [u_l \vdots \beta_{ln}] \tag{22}$$

in which u_n and u_l are unit matrices of order n and l respectively, and α_{nl} and β_{ln} are submatrices involving l columns or n columns as indicated by the subscripts. Furthermore, we can obviously augment these matrices to assume the forms

$$[\alpha] = \left[\begin{array}{c:c} u_l & 0 \\ \hdashline \alpha_{nl} & u_n \end{array}\right] \tag{23}$$

and

$$[\beta] = \left[\begin{array}{c:c} u_l & \beta_{ln} \\ \hdashline 0 & u_n \end{array}\right] \tag{24}$$

and then write for the constraint relations in matrix form

$$[\alpha] \cdot x] = x'] \tag{25}$$

$$[\beta] \cdot y] = y'] \tag{26}$$

Thus Eq. 25 includes Eqs. 19 plus the identities $x_k \equiv x'_k$ for $k = 1, 2, \ldots l$, while 26 is equivalent to the constraints 20 plus identities $y_k \equiv y'_k$ for $k = l + 1, \ldots b$. The zeros appearing in the partitioned matrices 23 and 24 are null submatrices having the appropriate numbers of rows and columns.

Including such identities with the constraint relations obviously imposes no additional conditions upon the resulting solution to our problem while it introduces the algebraic convenience of rendering the constraint relations in the form of a linear transformation of variables (coordinate transformation) that incidentally is reversible since the pertinent matrices 23 and 24 are nonsingular. Thus we can also write

$$[\alpha]^{-1} \cdot x'] = x] \tag{27}$$

and

$$[\beta]^{-1} \cdot y'] = y] \tag{28}$$

in which it is easy to see that

$$[\alpha]^{-1} = \left[\begin{array}{c:c} u_l & 0 \\ \hdashline -\alpha_{nl} & u_n \end{array}\right] \tag{29}$$

and

$$[\beta]^{-1} = \left[\begin{array}{c:c} u_l & -\beta_{ln} \\ \hdashline 0 & u_n \end{array}\right] \tag{30}$$

Equations 27 and 28 substituted into the original Eq. 2 now yield

$$[A] \cdot [\alpha]^{-1} \cdot x'] = [\beta]^{-1} \cdot y']$$

or

$$[\beta] \cdot [A] \cdot [\alpha]^{-1} \cdot x'] = y'] \tag{31}$$

while substitution of 25 and 26 into the inverse Eq. 6 gives

$$[B] \cdot [\beta]^{-1} \cdot y'] = [\alpha]^{-1} \cdot x']$$

or

$$[\alpha] \cdot [B] \cdot [\beta]^{-1} \cdot y'] = x'] \tag{32}$$

which is the inverse of 31.

Recalling the constraint relations 19 and 20 we note that the n quantities $x'_{l+1}, x'_{l+2}, \ldots, x'_b$ and the l quantities $y'_1, y'_2, \ldots, y'_l$ are known. Hence in the equation 31 or 32, n of the $x'_1 \cdots x'_b$ and l of the $y'_1 \cdots y'_b$ are known; and we are to solve for the remaining variables in each group. This is the problem dealt with in the previous article; and so we see that, except for a coordinate transformation, that problem is the same as one in which the given algebraic equations together with a set of b independent linear constraints (some involving the x's and some the y's) are to be satisfied simultaneously.

Substituting from Eqs. 9, 24 and 29 we see that the resultant matrix in Eq. 31 is given by

$$[\beta] \cdot [A] \cdot [\alpha]^{-1} = \left[\begin{array}{c:c} u_l & \beta_{ln} \\ \hdashline 0 & u_n \end{array}\right] \times \left[\begin{array}{c:c} a_{ll} & a_{ln} \\ \hdashline a_{nl} & a_{nn} \end{array}\right] \times \left[\begin{array}{c:c} u_l & 0 \\ \hdashline -\alpha_{nl} & u_n \end{array}\right]$$

$$= \left[\begin{array}{c:c} (a_{ll} + \beta_{ln}a_{nl} - a_{ln}\alpha_{nl} - \beta_{ln}a_{nn}\alpha_{nl}) & (a_{ln} + \beta_{ln}a_{nn}) \\ \hdashline (a_{nl} - a_{nn}\alpha_{nl}) & a_{nn} \end{array}\right] \tag{33}$$

If we denote this as a matrix $[C]$, given in partitioned form by

$$[C] = \left[\begin{array}{c:c} c_{ll} & c_{ln} \\ \hdashline c_{nl} & c_{nn} \end{array}\right] \tag{34}$$

then the matrix equation 31 is equivalent to

$$\begin{aligned} c_{ll} \cdot x'_l] + c_{ln} \cdot x'_n] &= y'_l \\ c_{nl} \cdot x'_l] + c_{nn} \cdot x'_n] &= y'_n \end{aligned} \tag{35}$$

which has precisely the form of Eqs. 11; and the known quantities are $x'_n]$ and $y'_l]$. With proper change in notation, Eqs. 12 and 13 represent the desired solution.

In many of the circuit-theory applications to which we shall apply these methods it turns out that the submatrices a_{ln} and a_{nl} are identically zero; in fact in a great many cases the matrix $[A]$ is diagonal (this is so for passive bilateral networks without mutual-inductive coupling). The matrix $[C]$ then is considerably simplified, specifically, instead of the form 33, we have

$$[C] = \left[\begin{array}{c:c} (a_{ll} - \beta_{ln} a_{nn} \alpha_{nl}) & \beta_{ln} a_{nn} \\ \hdashline -a_{nn} \alpha_{nl} & a_{nn} \end{array}\right] \tag{36}$$

3. Symmetry Conditions

In cases where the matrix $[A]$ of the original equations is symmetrical, it is of interest to know the conditions on the constraint relations 19 and 20 for which this symmetry is not destroyed. The answer to this question is easy to find. The resultant matrix of Eq. 31 or 32 is symmetrical for a symmetrical $[A]$ (or $[B]$) if

$$[\alpha]^{-1} = [\beta]_t$$

or

$$[\alpha] = [\beta]_t^{-1}, \quad [\beta] = [\alpha]_t^{-1} \tag{37}$$

That is to say, $[\alpha]$ should be the inverse transpose (also called the *reciprocal*) of $[\beta]$ or vice versa. In terms of Eqs. 23 and 30, or 24 and 29, this means that

$$\alpha_{nl} = -(\beta_{ln})_t \quad \text{or} \quad \beta_{ln} = -(\alpha_{nl})_t \tag{38}$$

The necessary and sufficient condition on the constraint relations 19 and 20, with matrices 21 and 22, for which symmetry is preserved, is thus evident.

A rather interesting interpretation of this condition is had from the constraint relations in their augmented form 25 and 26. If we form the transpose of 26 we have

$$\underline{y} \cdot [\beta]_t = \underline{y'} \tag{39}$$

where the notation for a row matrix (the transpose of the corresponding column matrix) is evident. In view of 37 this gives

$$\underline{y'} = \underline{y} \cdot [\alpha]^{-1} \tag{40}$$

and if we premultiply both sides of 25 with this row matrix we get

$$\underline{y} \cdot x] = \underline{y'} \cdot x'] \tag{41}$$

or, written out more fully,

$$x_1y_1 + x_2y_2 + \cdots + x_by_b = x_1'y_1' + x_2'y_2' + \cdots + x_b'y_b' \tag{42}$$

In circuit problems the x's and y's are respectively currents and voltages, and the scalar product 41 or 42 represents power. The left-hand side of 42 represents the associated power in terms of the original variables; the right-hand side represents the power in a new coordinate system or in the presence of constraint relations (whichever way we wish to interpret the introduction of the primed variables). The result 42 states that the condition upon the constraints 19 and 20 for which symmetry is preserved yields invariance of the associated power.

Under these conditions the constraints upon the x's are not independent of the constraints upon the y's, for the matrix of the relations 19 is then the negative transpose of the matrix of the relations 20. We shall refer to this relationship as one of *consistency* and say that the matrices $[\alpha]$ and $[\beta]$ fixing the constraints or the coordinate transformation are *consistent* if condition 37 preserving both symmetry and power invariance is fulfilled.

4. Abridgment

Returning to the original equations in their partitioned form as given by Eqs. 11, the situation frequently occurs in which $y_n] \equiv 0$ and we are interested only in the variables in $x_l]$ and their dependence upon the nonzero quantities in $y_l]$. That is to say, we wish to suppress the variables in $x_n]$ and obtain an abridged set of equations relating $x_l]$ to $y_l]$.

This result is readily obtained from Eqs. 11. If we set $y_n]$ in the second of these equations equal to zero and solve for $x_n]$ we have

$$x_n] = -a_{nn}^{-1} \cdot a_{nl} \cdot x_l] \tag{43}$$

whereupon substitution into the first of these equations yields

$$[a_{ll} - a_{ln} \cdot a_{nn}^{-1} \cdot a_{nl}] \cdot x_l] = y_l] \tag{44}$$

which is the desired result.

We can write it in the form

$$[\bar{A}] \cdot x_l] = y_l] \tag{45}$$

with

$$[\bar{A}] = [a_{ll} - a_{ln} \cdot a_{nn}^{-1} \cdot a_{nl}] \tag{46}$$

Here 45 is referred to as the abridged form of Eqs. 1; and formula 46 expresses the matrix of this abridged set of equations in terms of submatrices in the appropriately partitioned form 9 of the matrix 3 pertaining to the original set. The manipulations indicated in 46 are sometimes referred to as the abridgment process. This process is often used in network theory where we are interested only in the relation between voltages and currents at a small selected number of meshes or node pairs (points of access).

Problems

1. The equilibrium equations of a network on the loop basis are:

$$\begin{aligned} i_1 + 2i_2 - 3i_3 + i_4 + i_5 &= e_{s_1} \\ -i_1 - i_2 + 5i_3 - 4i_4 + i_5 &= e_{s_2} \\ -i_1 - i_2 + 6i_3 - 5i_4 + 2i_5 &= e_{s_3} \\ i_1 + i_2 - 4i_3 + 5i_4 - i_5 &= e_{s_4} \\ i_1 + 3i_2 - 2i_3 - 3i_4 + 2i_5 &= e_{s_5} \end{aligned}$$

(*a*) For the excitation voltages:

$$e_{s_1} = 5, \quad e_{s_2} = 1, \quad e_{s_3} = 5, \quad e_{s_4} = 6, \quad e_{s_5} = -1 \text{ volts}$$

solve for the loop currents.

(*b*) The excitation voltages for the first three loops are the same as in part (*a*); currents in loops 4 and 5 are required to have the values $i_{s_4} = 4$, $i_{s_5} = 5$ amperes. Solve for the excitation voltages e_{s_4} and e_{s_5} that will yield such a result, and find the currents i_1, i_2, i_3.

(*c*) The situation is the same as in part (*b*) but $e_{s_1} = e_{s_2} = e_{s_3} = 1$ volt and $i_{s_4} = i_{s_5} = 1$ ampere. Again solve for i_1, i_2, i_3 as well as for the voltages e_{s_4} and e_{s_5}.

(*d*) Repeat part (*c*) with $e_{s_1} = e_{s_3} = e_{s_5} = 1$ volt and $i_{s_2} = i_{s_4} = 1$ ampere. Find i_1, i_3, i_5 and e_{s_2}, e_{s_4}.

2. In the following equations:

$$\begin{bmatrix} 1 & 2 & -3 & 1 & 1 \\ -1 & -1 & 5 & -4 & 1 \\ -1 & -1 & 6 & -5 & 2 \\ -1 & 0 & 8 & -8 & 4 \\ 0 & 1 & 3 & -4 & 3 \end{bmatrix} \times \begin{bmatrix} i_1 \\ i_2 \\ i_3 \\ i_4 \\ i_5 \end{bmatrix} = \begin{bmatrix} e_{s_1} \\ e_{s_2} \\ e_{s_3} \\ e_{s_4} \\ e_{s_5} \end{bmatrix}$$

the excitation voltages have the values

$$e_{s_1} = 2, \quad e_{s_2} = 0, \quad e_{s_3} = 1, \quad e_{s_4} = 3, \quad e_{s_5} = 3 \text{ volts}$$

Solve for the currents. Is this solution unique? Are the currents an independent set? Will the equations have a solution for an arbitrary set of excitation voltages? For any other set? If so, how is such a set defined?

3. (*a*) For the [*A*]-matrix of prob. 2, choose $i_4 = i_5 = 1$ ampere and $e_{s_1} = e_{s_2} = e_{s_3} = 1$ volt. Solve for i_1, i_2, i_3, e_{s_4}, e_{s_5}.

(*b*) Choose $i_2 = i_4 = 1$ and $e_{s_1} = e_{s_3} = e_{s_5} = 1$. Solve for $i_1, i_3, i_5, e_{s_2}, e_{s_4}$. If a solution is not possible, revise the value of e_{s_5} so that unique values for e_{s_2} and e_{s_4} will result, and express i_1 and i_3 in terms of i_5.

4. The first three rows of a symmetrical matrix of order 5 and rank 3 are:

$$\begin{matrix} 1 & 2 & 1 & -1 & 4 \\ 2 & 2 & -1 & 3 & 1 \\ 1 & -1 & 3 & 2 & 1 \end{matrix}$$

Construct the remaining two rows. Are they unique?

5. The matrix equation

$$[A] \cdot x] = y]$$

has a diagonal $[A]$-matrix with the element values 1, 2, 3, 4, 5. Constraints among the x's and among the y's read

$$x_1 - x_2 + x_4 = 1$$
$$x_2 - x_3 + x_5 = 2$$

and

$$y_1 - y_4 = 1$$
$$y_2 + y_4 - y_5 = 2$$
$$y_3 + y_5 = 1$$

Derive and solve the resulting equations for x_1, x_2, x_3.

6. Equilibrium equations for a five-loop network are given by

$$\begin{bmatrix} 14 & -1 & -2 & 1 & 1 \\ -1 & 6 & -1 & 2 & -1 \\ -2 & -1 & 5 & -2 & 1 \\ 1 & 2 & -2 & 5 & -2 \\ 1 & -1 & 1 & -2 & 1 \end{bmatrix} \times \begin{bmatrix} i_1 \\ i_2 \\ i_3 \\ i_4 \\ i_5 \end{bmatrix} = \begin{bmatrix} e_1 \\ e_2 \\ e_3 \\ e_4 \\ e_5 \end{bmatrix}$$

(*a*) Abridge this set so that only the currents and voltages in loops 1, 2, 3 are involved.

(*b*) Alternately, abridge the set so as to suppress the variables i_2 and i_4. In this case e_2 and e_4 will no longer be involved.

7. The $[A]$-matrix in the previous problem is replaced by the following one of rank 2:

$$\begin{bmatrix} 5 & -2 & 4 & 8 & -4 \\ -2 & 4 & -4 & 4 & -4 \\ 4 & -4 & 5 & 1 & 1 \\ 8 & 4 & 1 & 29 & -19 \\ -4 & -4 & 1 & -19 & 13 \end{bmatrix}$$

As in prob. 6, abridge to the variables i_3, i_4, i_5. Again to the variables i_1, i_3, i_5. What conclusions can you draw from these results? Can you prove the pertinent matrix property?

8. Given the third-order matrix

$$\begin{bmatrix} 6 & -1 & -2 \\ -1 & 4 & -1 \\ -1 & -1 & 3 \end{bmatrix}$$

border it with zeros so as to have the form

$$\begin{bmatrix} 6 & -1 & -2 & 0 & 0 \\ -1 & 4 & -1 & 0 & 0 \\ -1 & -1 & 3 & 0 & 0 \\ 0 & 0 & 0 & 0 & 0 \\ 0 & 0 & 0 & 0 & 0 \end{bmatrix}$$

Now add the matrix of prob. 7 and abridge to the variables i_1, i_2, i_3. On the basis of these results, state a general method for augmenting a third-order matrix to fifth order in such a way that subsequent abridgment regains the given matrix. Apply this method to the given third-order matrix above so that the abridgment applies to variables i_1, i_3, i_5.

9. For the graph shown here, construct a tie-set matrix $[\beta]$ for loops consisting of branches 1568, 2348, 578, 12569, 23457 and any others picked at random.

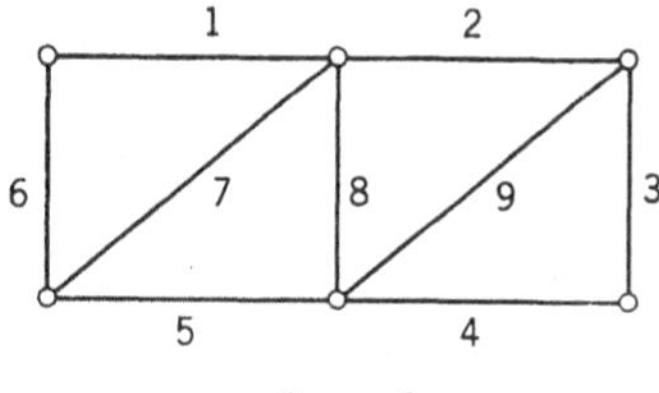

PROB. 9.

Now construct a cut-set matrix $[\alpha]$ for cut sets consisting of branches 567, 1278, 14789, 249, 36789, and any others picked at random.

Observe that all rows in $[\alpha]$ are orthogonal to all rows in $[\beta]$. Show why this must be so.

CHAPTER II

Regarding Equilibrium Equations

1. Direct Algebraic Approach

The given network we consider to consist of linear inductive, resistive and capacitive elements. For the moment we shall assume these to be passive and bilateral, although, as is shown in following discussions, the removal of these restrictions requires no essential changes in the procedures to be described here. In an already familiar manner,* we designate the number of inductive elements by λ, the number of capacitances by σ, and the number of resistances by ρ.

Each element is considered a network branch so far as topological structure is concerned, and is provided with a number for its identification and a reference arrow that serves both for the branch current and the branch voltage. In order that the volt-ampere relation for a branch (like $v = rj$ for a resistive branch) shall not involve a minus sign, it follows that the branch voltage v must be a voltage *drop* in the reference direction common to this voltage and the associated branch current j.

Inductive branches are assumed to be numbered consecutively from 1 to λ, resistive branches from $\lambda + 1$ to $\lambda + \rho$, and capacitive branches from $\lambda + \rho + 1$ to $\lambda + \rho + \sigma = b$. There is no compulsion about this numbering sequence, of course, but it is convenient to choose a standard convention and assume that it will be the rule unless specific mention is made to the contrary.

Each inductive branch may be mutually coupled with all other inductive branches. Hence the pertinent parameter matrix has the form

$$[l] = \begin{bmatrix} l_{11} & l_{12} & \cdots & l_{1\lambda} \\ l_{21} & l_{22} & \cdots & l_{2\lambda} \\ \cdots & \cdots & \cdots & \cdots \\ l_{\lambda 1} & l_{\lambda 2} & \cdots & l_{\lambda\lambda} \end{bmatrix} \tag{1}$$

* See E. A. Guillemin, *Introductory Circuit Theory*, John Wiley and Sons, 1953 (hereafter referred to as *ICT*), Ch. X.

while the branch parameter matrices for resistances and elastances are diagonal, namely

$$[r] = \begin{bmatrix} r_{\lambda+1} & 0 & 0 & \cdots & 0 \\ 0 & r_{\lambda+2} & 0 & \cdots & 0 \\ \cdots & \cdots & \cdots & \cdots & \cdots \\ 0 & 0 & 0 & \cdots & r_{\lambda+\rho} \end{bmatrix} \tag{2}$$

and

$$[s] = \begin{bmatrix} s_{\lambda+\rho+1} & 0 & 0 & \cdots & 0 \\ 0 & s_{\lambda+\rho+2} & 0 & \cdots & 0 \\ \cdots & \cdots & \cdots & \cdots & \cdots \\ 0 & 0 & 0 & \cdots & s_b \end{bmatrix} \tag{3}$$

If p denotes the derivative operator d/dt and p^{-1} denotes indefinite integration, namely,

$$p^{-1} = \int dt \tag{4}$$

then the total volt-ampere relations for the branches are compactly expressed with the help of a branch-operator matrix,

$$[A] = \left[\begin{array}{c:c:c} lp & 0 & 0 \\ \hdashline 0 & r & 0 \\ \hdashline 0 & 0 & sp^{-1} \end{array}\right] \tag{5}$$

in which 1, 2 and 3 are embedded submatrices. Thus the pertinent volt-ampere relations for the branches of our network are given by

$$[A] \cdot j] = v] \tag{6}$$

in which the branch currents and voltages are elements in the column matrices

$$j] = \begin{bmatrix} j_1 \\ j_2 \\ \cdot \\ \cdot \\ \cdot \\ j_b \end{bmatrix} \quad \text{and} \quad v] = \begin{bmatrix} v_1 \\ v_2 \\ \cdot \\ \cdot \\ \cdot \\ v_b \end{bmatrix} \tag{7}$$

Equation 6, like Eq. 1 of Ch. I, is a set of b simultaneous algebraic equations involving the variables $j_1 \cdots j_b$ and $v_1 \cdots v_b$. Our network problem is like the situation described in art. 2 of Ch. I in which constraint relations exist among the j's and the v's. Constraints among the j's are the Kirchhoff current-law relations, and constraints among the v's are the Kirchhoff voltage-law relations. As is well known,* the current-law constraints are n in number (the number of branches in an $(n + 1)$-node tree) and the voltage law yields l independent constraints, making $l + n = b$ constraints altogether, as is assumed in the discussion of art. 2, Ch. I. In fact the constraint relations as expressed by Eqs. 19 and 20 of Ch. I apply here with only minor changes in notation. To be specific, we have in place of the former, the Kirchhoff current-law equations

$$\begin{aligned} \alpha_{l+1,1}j_1 + \alpha_{l+1,2}j_2 + \cdots + \alpha_{l+1,b}j_b &= i_{s_1} \\ \alpha_{l+2,1}j_1 + \alpha_{l+2,2}j_2 + \cdots + \alpha_{l+2,b}j_b &= i_{s_2} \\ \cdots\cdots\cdots\cdots\cdots\cdots\cdots\cdots \\ \alpha_{b1}j_1 \quad + \alpha_{b2}j_2 \quad + \cdots + \alpha_{bb}j_b \quad &= i_{s_n} \end{aligned} \tag{8}$$

and in place of the latter, the Kirchhoff voltage-law equations

$$\begin{aligned} \beta_{11}v_1 + \beta_{12}v_2 + \cdots + \beta_{1b}v_b &= e_{s_1} \\ \beta_{21}v_1 + \beta_{22}v_2 + \cdots + \beta_{2b}v_b &= e_{s_2} \\ \cdots\cdots\cdots\cdots\cdots\cdots\cdots\cdots \\ \beta_{l1}v_1 + \beta_{l2}v_2 + \cdots + \beta_{lb}v_b &= e_{s_l} \end{aligned} \tag{9}$$

Thus the x's and y's are replaced respectively by j's and v's, and instead of using primed quantities for the right-hand members, we recognize their physical identities as current and voltage sources $i_{s_1} \cdots i_{s_n}$ and $e_{s_1} \cdots e_{s_l}$. The matrices $[\alpha_{nb}]$ and $[\beta_{lb}]$ in 8 and 9 we recognize as being the cut-set and tie-set matrices respectively. The rows in the α-matrix are independent cut sets, and those in the β-matrix are independent tie sets or loops.

If these are based upon the selection of a tree in the given network graph, and if the corresponding links are numbered consecutively from 1 through l while the n tree branches are identified by numbers $l + 1$ through b, then it is not hard to see that the α- and β-matrices in Eqs. 8 and 9 assume the canonic forms (Eqs. 21 and 22 of Ch. I) given by

$$[\alpha_{nb}] = [\alpha_{nl} \vdots u_n] \tag{10}$$

$$[\beta_{lb}] = [u_l \vdots \beta_{ln}] \tag{11}$$

* See, for example, *ICT*, Chs. I and II.

As pointed out in Ch. I, nothing is gained or lost algebraically if we add to the current constraints 8 a set of l identities and to the voltage constraints 9 a set of n identities; and if we let the former pertain to the variables $j_1 \cdots j_l$ and the latter to $v_{l+1} \cdots v_b$, then the correspondingly augmented α- and β-matrices read

$$[\alpha] = \begin{bmatrix} u_l & \vdots & 0 \\ \cdots & \cdot & \cdots \\ \alpha_{nl} & \vdots & u_n \end{bmatrix} \tag{12}$$

and

$$[\beta] = \begin{bmatrix} u_l & \vdots & \beta_{ln} \\ \cdots & \cdot & \cdots \\ 0 & \vdots & u_n \end{bmatrix} \tag{13}$$

as given by Eqs. 23 and 24 of Ch. I. Instead of using primed quantities in the identity relations we again are motivated by physical reasoning to use a more suggestive notation. Thus we recognize that the link currents $j_1 \cdots j_l$ are identified as loop currents and that the tree-branch voltages $v_{l+1} \cdots v_b$ are identified as node-pair voltages. The identity relations themselves are written

$$\begin{aligned} j_1 &\equiv i_1 & v_{l+1} &\equiv e_1 \\ j_2 &\equiv i_2 & v_{l+2} &\equiv e_2 \\ &\cdots & &\cdots \\ j_l &\equiv i_l & v_b &\equiv e_n \end{aligned} \tag{14}$$

and this suggests the definition of column matrices

$$i] = \begin{bmatrix} i_1 \\ \cdot \\ \cdot \\ \cdot \\ i_l \\ i_{s_1} \\ \cdot \\ \cdot \\ \cdot \\ i_{s_n} \end{bmatrix} = \begin{bmatrix} i_v \\ \cdots \\ i_s \end{bmatrix}; \quad e] = \begin{bmatrix} e_{s_1} \\ \cdot \\ \cdot \\ \cdot \\ e_{s_l} \\ e_1 \\ \cdot \\ \cdot \\ \cdot \\ e_n \end{bmatrix} = \begin{bmatrix} e_s \\ \cdots \\ e_v \end{bmatrix} \tag{15}$$

where in the partitioned forms, the subscripts s and v designate *source* quantities or *variables* respectively, the term "source quantity," from a purely algebraic standpoint, being merely a convenience in identifying those variables whose values are known or fixed while the ones designated as "variables" are those whose values are unknown and hence are to be determined from the given equation 6. This attitude fits in nicely with the fact that "sources" are more properly regarded as constraints, and the equations in which these quantities appear (Eqs. 8 and 9) are constraint relations.

The Kirchhoff constraints with embedded identities are thus expressed by the compact matrix equations

$$[\alpha] \cdot j] = i] \quad \text{or} \quad [\alpha]^{-1} \cdot i] = j] \tag{16}$$

and

$$[\beta] \cdot v] = e] \quad \text{or} \quad [\beta]^{-1} \cdot e] = v] \tag{17}$$

it being clear, through their manner of formation, that the matrices $[\alpha]$ and $[\beta]$ are nonsingular and hence possess inverses.

The result of introducing these constraint relations into the original Eq. 6 is now readily obtained by substituting for $j]$ from Eq. 16, for $v]$ from Eq. 17, and then premultiplying both sides by $[\beta]$ to get

$$[\beta] \cdot [A] \cdot [\alpha]^{-1} \cdot i] = e] \tag{18}$$

Dimensionally it is clear that the triple product $[\beta] \cdot [A] \cdot [\alpha]^{-1}$, given by Eqs. 33 and 34 of Ch. I and denoted there as the matrix $[C]$, is an impedance. We denote it here as

$$[Z] = \begin{bmatrix} Z_{ll} & \vdots & Z_{ln} \\ \cdots & \cdot & \cdots \\ Z_{nl} & \vdots & Z_{nn} \end{bmatrix} \tag{19}$$

and, with the help of the notation in 15, write Eq. 18 in the form

$$\begin{aligned} Z_{ll} \cdot i_v] + Z_{ln} \cdot i_s] &= e_s] \\ Z_{nl} \cdot i_v] + Z_{nn} \cdot i_s] &= e_v] \end{aligned} \tag{20}$$

which parallels the Eqs. 35 of Ch. I. The impedance matrix $[Z]$ is given by Eq. 33 of Ch. I and hence we have

$$\begin{aligned} Z_{ll} &= a_{ll} + \beta_{ln}a_{nl} - a_{ln}\alpha_{nl} - \beta_{ln}a_{nn}\alpha_{nl} \\ Z_{ln} &= a_{ln} + \beta_{ln}a_{nn} \\ Z_{nl} &= a_{nl} - a_{nn}\alpha_{nl} \\ Z_{nn} &= a_{nn} \end{aligned} \tag{21}$$

in which a_{ll}, a_{ln}, a_{nl} and a_{nn} are submatrices in the partitioned form of the matrix $[A]$.

Note, however, that this partitioning is not that shown in Eq. 5. There the rows and columns are partitioned into groups of λ, ρ and σ, according to the numbers of inductive, resistive and capacitive elements. Here, as in Eq. 9 of Ch. I, the partitioning is done with respect to the numbers of links l and tree branches n. It is clear that if there is no mutual-inductive coupling (which is true in many cases), then the submatrices a_{ln} and a_{nl} are identically zero.

Moreover, since the link currents circulate upon the same contours to which the Kirchhoff voltage constraints apply, and the node-pair voltages are identified with tree branches associated with cut sets to which the Kirchhoff current constraints apply, it follows according to established topological principles* (and further detailed discussion given in the succeeding articles) that the matrices $[\alpha]$ and $[\beta]$ are reciprocal (Eqs. 37 and 38 of Ch. I apply), and so we have

$$\alpha_{nl} = -(\beta_{ln})_t \quad \text{or} \quad \beta_{ln} = -(\alpha_{nl})_t \tag{22}$$

For these reasons the equation 21 yielding significant submatrices in the impedance matrix $[Z]$ assume the simpler forms given by

$$\begin{aligned} Z_{ll} &= a_{ll} + \beta_{ln}a_{nn}(\beta_{ln})_t \\ Z_{ln} &= \beta_{ln}a_{nn} \\ Z_{nl} &= a_{nn}(\beta_{ln})_t \\ Z_{nn} &= a_{nn} \end{aligned} \tag{23}$$

Since a_{nn} is diagonal in any case, and a_{ll} is diagonal also unless mutual-inductive coupling is present (which is not so in many practical situations), and since the matrix β_{ln} contains only plus or minus "ones" or zeros, the results 23 are extremely easy to evaluate in most cases and hence the equations 20 are readily established.

In our network problem it suffices to solve the first of these equations, namely

$$Z_{ll} \cdot i_v] = e_s] - Z_{ln} \cdot i_s] \tag{24}$$

representing l equations with l unknowns, for the loop currents $i_1 \cdots i_l$ contained in $i_v]$. From these and the $i_{s_1} \cdots i_{s_n}$ (which are known) we then get all the branch currents $j]$ from 16 (the matrix $[\alpha]^{-1}$ being given by Eq. 29 of Ch. I).

The result 24 is commonly referred to as the equilibrium equations on a *loop basis*. The second term on the right-hand side of 24 yields the conversion of current sources into equivalent voltage sources. The second of the equations 20 may be used to compute the node-pair voltages in $e_v]$

* See *ICT*, Chs. I and II.

(after the loop currents are known) and from these the branch voltages can be found by using Eq. 17 (and the known form of $[\beta]^{-1}$ given by Eq. 30 of Ch. I). It is probably easier, however, to get the branch voltages directly from the original Eqs. 6 after the branch currents are found.

Entirely analogous results may be obtained by introducing the constraint relations into the inverse of Eq. 6 involving the inverse of the $[A]$-matrix in Eq. 5 which is given by

$$[B] = \begin{bmatrix} \gamma p^{-1} & \vdots & 0 & \vdots & 0 \\ \cdots & & \cdots & & \cdots \\ 0 & \vdots & g & \vdots & 0 \\ \cdots & & \cdots & & \cdots \\ 0 & \vdots & 0 & \vdots & cp \end{bmatrix} \tag{25}$$

in which submatrices $[\gamma]$, $[g]$ and $[c]$ are the reciprocal inductance, the conductance and the capacitance matrices for the network branches. These are respectively the inverses of the matrices $[l]$, $[r]$ and $[s]$ given by Eqs. 1, 2 and 3 above. The derivative operator p and its inverse (Eq. 4) have the same significance as before.

In terms of the matrix 25, the inverse of the equations expressed by 6 is given by

$$[B] \cdot v] = j] \tag{26}$$

To introduce the constraint relations among the j's and v's we now substitute for $v]$ from Eq. 17, for $j]$ from Eq. 16, and then premultiply both sides by $[\alpha]$. This gives

$$[\alpha] \cdot [B] \cdot [\beta]^{-1} \cdot e] = i] \tag{27}$$

As might have been expected, this result is simply the inverse of Eq. 18.

The triple matrix product involved here is dimensionally an admittance (actually $[B]$ is an admittance and $[\alpha]$ and $[\beta]$ are dimensionless). We denote it by

$$[Y] = \begin{bmatrix} Y_{ll} & \vdots & Y_{ln} \\ \cdots & & \cdots \\ Y_{nl} & \vdots & Y_{nn} \end{bmatrix} = [Z]^{-1} \tag{28}$$

and, paralleling the form of Eqs. 20, write 27 more explicitly as

$$\begin{aligned} Y_{ll} \cdot e_s] + Y_{ln} \cdot e_v] &= i_v] \\ Y_{nl} \cdot e_s] + Y_{nn} \cdot e_v] &= i_s] \end{aligned} \tag{29}$$

The submatrices involved here are obtained straightforwardly by using the partitioned forms for $[\alpha]$, $[B]$ and $[\beta]^{-1}$ as given by Eqs. 10, 23 and 30 of Ch. I. We find

$$
\begin{aligned}
Y_{ll} &= b_{ll} \\
Y_{ln} &= b_{ln} - b_{ll}\beta_{ln} \\
Y_{nl} &= b_{nl} + \alpha_{nl}b_{ll} \\
Y_{nn} &= b_{nn} + \alpha_{nl}b_{ln} - b_{nl}\beta_{ln} - \alpha_{nl}b_{ll}\beta_{ln}
\end{aligned} \tag{30}
$$

For the same reasons as pointed out in connection with the $[A]$-matrix, the submatrices b_{ln} and b_{nl} in most cases are identically zero. Recognizing again that the reciprocal relation between $[\alpha]$ and $[\beta]$ yields the property expressed in Eqs. 22, we have for the majority of practical situations the simpler results

$$
\begin{aligned}
Y_{ll} &= b_{ll} \\
Y_{ln} &= b_{ll}(\alpha_{nl})_t \\
Y_{nl} &= \alpha_{nl}b_{ll} \\
Y_{nn} &= b_{nn} + \alpha_{nl}b_{ll}(\alpha_{nl})_t
\end{aligned} \tag{31}
$$

which are analogous to Eqs. 23 on an impedance basis and are applicable under the same conditions. Like the relations 23, those in Eqs. 31 are computationally simple to evaluate, although they apply only when mutual-inductive coupling is absent.

Returning to the equations 29, the solution to our network problem involves only the second of these, namely

$$
Y_{nn} \cdot e_v] = i_s] - Y_{nl} \cdot e_s] \tag{32}
$$

expressing equilibrium on a *node basis* in which the variables or unknowns are the node-pair voltages contained in the column matrix $e_v]$. The second term on the right-hand side of Eq. 32 represents the conversion of voltage sources into equivalent current sources.

After the node-pair voltages are known from the solution of Eq. 32, the branch voltages are readily calculated from Eq. 17, and the branch currents are then most easily obtained from Eq. 26.

As pointed out in Ch. I, the reciprocal relationship between the matrices $[\alpha]$ and $[\beta]$ implies power invariance in the pertinent network. We can get a specific expression of this result by noting first from Eq. 16 that the transposed relation $\underline{i} = \underline{j} \cdot [\alpha]_t$ premultiplied into 17 gives

$$
\underline{j} \cdot [\alpha]_t \cdot [\beta] \cdot v] = \underline{i} \cdot e] \tag{33}
$$

If $[\alpha]$ and $[\beta]$ are reciprocal then $[\alpha]_t \cdot [\beta]$ is a unit matrix of order b. The transposed forms of column matrices $j]$ and $i]$, denoted by $\underline{j}$ and $\underline{i}$

respectively, involve the same elements arranged in rows. With the help of the specific forms given in Eqs. 7 and 15, therefore, we see that Eq. 33 yields

$$j_1v_1 + j_2v_2 + \cdots + j_bv_b = i_1e_{s1} + i_2e_{s2} + \cdots + i_le_{sl} + i_{s1}e_1 + i_{s2}e_2 + \cdots + i_{sn}e_n \tag{34}$$

In the physical interpretation of this result, particularly of the right-hand member, it is essential to recall that the reciprocal relation between $[\alpha]$ and $[\beta]$, which is made use of in the derivation of 34, implies that the loop or link currents $i_1 \cdots i_l$ circulate upon contours for which the Kirchhoff voltage constraints involving $e_{s1} \cdots e_{sl}$ apply, and that the node-pair or tree-branch voltages $e_1 \cdots e_n$ pertain to the same node pairs for which the Kirchhoff current constraints involving $i_{s1} \cdots i_{sl}$ are written. It follows that $i_1 \cdots i_l$ are currents delivered by the voltage sources $e_{s1} \cdots e_{sl}$ respectively and that $e_1 \cdots e_n$ are voltages across the current sources $i_{s1} \cdots i_{sn}$. Hence the right-hand member of Eq. 34 represents the total instantaneous power supplied to the network by the sources. Since the left-hand side of this equation is evidently the total instantaneous power absorbed by the branches of the network, we see that this result states the conservation of power for the complete system consisting of the network plus sources.

2. Introduction of Auxiliary Variables; The Topological Approach

Looking back upon the approach given in the preceding article it is significant to observe that the introduction of new current or voltage variables in place of branch currents or voltages is actually not involved, notwithstanding the fact that we did mention such things as "loop currents" and "node-pair voltages" in connection with the identity relations 14. Since these *are* identities, the $i_1 \cdots i_l$ are precisely the same as $j_1 \cdots j_l$, and $e_1 \cdots e_n$ are precisely the same as $v_{l+1} \cdots v_b$. We could have used primed j's and primed v's to denote these quantities (as was done in Ch. I) and we then would have continued to refer to them as branch currents and branch voltages. We chose not to do this only because such a deliberate way of avoiding mention of the terms "loop current" and "node-pair voltage" would have had the appearance of overdoing things for the sake of making a point.

In yielding to the natural temptation of calling things by their proper names, however, we wish at this time to emphasize for reasons of logic that the method of solution given in the preceding article formally does not involve or depend in its basic concepts upon the introduction of variables other than the branch currents and branch voltages themselves.

In sharp contrast with this view, the method of solution to be presented now uses the introduction of auxiliary variables as the key with which the desired result is achieved. Needless to say, the end result is essentially the same as before; only the reasoning by which we arrive at it is different.

Lest the impatient reader may at this point begin to turn pages and skip over the rest of this article, we remind him of the importance of obtaining a solution to the same problem by a variety of different lines of reasoning, for it is thus that the essential features of that problem become sufficiently illuminated to reveal a true understanding of the subtleties and the detailed aspects upon which all else depends.

We begin again with the volt-ampere relations for the branches as expressed compactly by the matrix 6. Represented here are b equations and $2b$ unknowns, for both currents and voltages are unknown. In order to obtain a solution we need b additional equations involving these currents and voltages.

In our search for these additional equations we turn to the Kirchhoff laws; and we use topological methods to determine how many independent relations we can thus obtain. Reasoning in terms of the tree concept and related topological principles,* we find that in an $(n + 1)$-node network we can write n independent current-law equations and l independent voltage-law equations among the j's and v's, and that $l + n = b$, so that the additional equations thus obtained will just suffice to determine the desired solution. But precisely how do we implement the procedure for carrying out such a solution?

The most obvious way of implementing it is to write down the totality of $2b$ equations with $2b$ unknowns and solve this system simultaneously. This is a horrible thought even to contemplate, and we shall not seriously consider it, but let us nevertheless write down this set of equations to see what it looks like. First of all we will write Eq. 6 in the form

$$[A] \cdot j] - v] = 0 \tag{35}$$

which we can alternately write

$$[A \vdots -u_b] \cdot \begin{bmatrix} j \\ \cdots \\ v \end{bmatrix} = 0 \tag{36}$$

in which the submatrix u_b is a unit matrix of order b. The column matrix involves the $2b$ j's and v's.

* As discussed in detail in *ICT*, Chs. I and II.

In analogous fashion we can write the Kirchhoff law equations in the form

$$\begin{bmatrix} 0 & \vdots & \beta_{lb} \\ \cdots & \cdots & \cdots \\ \alpha_{nb} & \vdots & 0 \end{bmatrix} \times \begin{bmatrix} j \\ \cdots \\ v \end{bmatrix} = \begin{bmatrix} e_s \\ \cdots \\ i_s \end{bmatrix} \tag{37}$$

Like the first matrix in 36, the first one here has b rows and $2b$ columns; the right-hand column matrix contains b elements. We can now combine 36 and 37 in the following single matrix equation:

$$\begin{bmatrix} A & \vdots & -u_b \\ \cdots & \cdots & \cdots \\ 0 & \vdots & \beta_{lb} \\ \cdots & \cdots & \cdots \\ \alpha_{nb} & \vdots & 0 \end{bmatrix} \times \begin{bmatrix} j \\ \cdots \\ v \end{bmatrix} = \begin{bmatrix} 0 \\ \cdots \\ e_s \\ \cdots \\ i_s \end{bmatrix} \tag{38}$$

representing $2b$ equations in $2b$ unknowns.

In our desire to avoid such a large number of equations we ask ourselves whether the introduction of new variables might be helpful. In this connection we consider the possibility of introducing currents which by definition circulate upon closed contours. Topologically we are aware of the fact that link currents can be given this physical interpretation and that all branch currents can uniquely be expressed in terms of only l such circulatory currents properly selected. At the same time we can see that if only circulatory currents are involved then Kirchhoff current-law equations are trivial, for circulatory currents automatically fulfill the physical condition upon which this law is based, namely that what enters a node must also leave it. A circulatory current leaves all points that it enters; it does this by definition.

There are many different sets of circulatory currents that we can choose, and we will later on go into further detail as to how we do this algebraically, but for the moment let us content ourselves with the simplest way of choosing loop currents, which is to identify them with a set of link currents.

We still have to consider the Kirchhoff voltage-law equations together with Eq. 6 relating branch voltages and currents. The pertinent Kirchhoff equations are given by

$$[\beta_{lb}] \cdot v] = [u_l \vdots \beta_{ln}] \cdot v] = e_s] \tag{39}$$

and if we substitute for $v]$ from Eq. 6, these will be expressed in terms of the branch currents, which we must next express in terms of the loop currents. Since these are identified with link currents, and the closed paths defined by $[\beta_{lb}]$ are the same as those traversed by these link currents, a well-known property* of $[\beta_{lb}]$ tells us that

$$[\beta_{lb}]_t \cdot i_v] = \begin{bmatrix} u_l \\ \cdot \ \cdot \ \cdot \\ (\beta_{ln})_t \end{bmatrix} \cdot i_v] = j] \tag{40}$$

which incidentally checks the identification of link currents with loop currents.

Substituting this equation into 6, and the result into 39 gives

$$[\beta_{lb}] \cdot [A] \cdot [\beta_{lb}]_t \cdot i_v] = e_s] \tag{41}$$

which represents l equations in terms of the l loop currents contained in $i_v]$. The triple matrix product involved here, assuming a_{ln} and a_{nl} in the partitioned form of $[A]$ to be identically zero, is given by

$$[u_l \ \vdots \ \beta_{ln}] \times \begin{bmatrix} a_{ll} & \vdots & 0 \\ \cdot \ \cdot & \cdot & \cdot \ \cdot \\ 0 & \vdots & a_{nn} \end{bmatrix} \times \begin{bmatrix} u_l \\ \cdot \ \cdot \ \cdot \\ (\beta_{ln})_t \end{bmatrix}$$

$$= a_{ll} + \beta_{ln} a_{nn} (\beta_{ln})_t = Z_{ll} \tag{42}$$

which checks the first equation in the set 23.

Equation 41 is therefore equivalent to

$$Z_{ll} \cdot i_v] = e_s] \tag{43}$$

and represents the desired equilibrium equations on a loop basis.

Comparison with Eq. 24 shows that the term representing the conversion of current sources into equivalent voltage sources is missing, as should be expected because the Kirchhoff current-law equations were ignored in the derivation of 43. This approach to the problem is, therefore, acceptable if either no current sources are present or these have been converted into equivalent voltage sources according to a well-known procedure before this process of derivation is begun.

A line of reasoning entirely dual to the one just given starts by considering whether there exists a way of defining voltage variables that makes the Kirchhoff voltage-law equations trivial. Here we recognize that node-pair voltages or potential differences between nodes have this property, for the algebraic sum of such potential differences around a closed path that begins and ends at any given node yields the potential difference between that node and itself, which is manifestly zero.

* See *ICT*, Ch. I.

In the choice of appropriate node-pair voltages we again have many possibilities, the one that is analogous to choosing link currents for loop currents being the identification of tree-branch voltages with node-pair voltages. If we make this choice, and write the Kirchhoff current-law equations for cut sets in which the tree branches appear singly (one in each equation) then these equations assume the canonic form

$$[\alpha_{nb}] \cdot j] = [\alpha_{nl} \vdots u_n] \cdot j] = i_s] \tag{44}$$

and a well-known property of this $[\alpha]$-matrix yields the relations for the branch voltages in terms of the node-pair voltages that read

$$[\alpha_{nb}]_t \cdot e_v] = \begin{bmatrix} (\alpha_{nl})_t \\ \cdot \quad \cdot \quad \cdot \\ u_n \end{bmatrix} \cdot e_v] = v] \tag{45}$$

which incidentally checks the identification of tree-branch voltages with node-pair voltages.

Substituting this equation into 26 (the inverse of 6), and the result into 44 gives

$$[\alpha_{nb}] \cdot [B] \cdot [\alpha_{nb}]_t \cdot e_v] = i_s] \tag{46}$$

which represents n equations in terms of the n node-pair voltages contained in $e_v]$. The triple matrix product involved here, assuming b_{ln} and b_{nl} in the partitioned form of $[B]$ to be identically zero, is given by

$$[\alpha_{nl} \vdots u_n] \times \begin{bmatrix} b_{ll} & \vdots & 0 \\ \cdot \quad \cdot & \cdot & \cdot \quad \cdot \\ 0 & \vdots & b_{nn} \end{bmatrix} \times \begin{bmatrix} (\alpha_{nl})_t \\ \cdot \quad \cdot \quad \cdot \\ u_n \end{bmatrix}$$

$$= \alpha_{nl} b_{ll} (\alpha_{nl})_t + b_{nn} = Y_{nn} \tag{47}$$

which agrees with the last equation in the set 31.

Equation 46 is, therefore, equivalent to

$$Y_{nn} \cdot e_v] = i_s] \tag{48}$$

and represents the desired equilibrium equations on a node basis.

Comparison of Eq. 48 with 32 shows that the term representing the conversion of voltage sources into equivalent current sources is missing, as should be expected because the Kirchhoff voltage-law equations were ignored in the derivation of 48. This approach to the problem is, therefore, acceptable if either no voltage sources are present or these have been converted into equivalent current sources according to a well-known procedure before this process of derivation is begun.

Such source conversion, of course, is fraught with the annoyance that the current or voltage in the particular branch associated with the source is not preserved, and one must subsequently correct for this inconsistency by an appropriate calculation which is simple enough but may become a computational nuisance if a large number of sources need to be converted.

The procedure leading to the results given by Eqs. 24 and 32, on the other hand, takes care of all source conversions automatically and requires no subsequent correctional computations. For this reason, the approach to our network problem in the form in which it has just been presented is practically less desirable in situations where sources of both kinds are present.

We can, however, overcome this objection in a very simple manner that incidentally sheds additional light upon the concept of loop currents and node-pair voltages. The underlying thought here is suggested by an attitude provided by the algebraic approach discussed in art. 1 above where, in the construction of column matrices $i]$ and $e]$ as given by Eqs. 15, it is pointed out that the only distinction between source quantities and variables from an algebraic point of view is that values of the former we know and those of the latter we do not know. In every other respect, quantities among the currents or among the voltages are homogeneous.

Let us apply this idea to the topological aspects of voltages and currents by suggesting that loop currents and source currents are topologically alike, or that node-pair voltages and source voltages are topologically alike. The choice of loop currents on a topological basis is ordinarily made by choosing a set of closed contours on which these currents are assumed to circulate. These contours or paths may be chosen quite freely, the only condition being that they form an independent set; and this condition is readily checked by writing down the pertinent tie-set matrix and testing its rows for independence (they are independent if an equal number of columns can be found for which the determinant is nonzero).

If we consider source currents to be circulatory currents also (and they are because they circulate from the source through the network and back to the source) then we should be allowed to choose arbitrary closed paths for these also and write additional rows in the tie-set schedule indicating the pertinent confluent sets of branches traversed by these source currents from one terminal of each source to the other; and the test for independence of such additional paths should follow the same pattern as for the loop currents proper.

It does not seem at first thought that we have the right to choose paths for the source currents because we argue physically that these currents do not follow definite paths but divide at every node and permeate generally through the branches of a network. We should be reminded, however, of the fact that we are constructing this view of the situation only for the help it might be in manipulating our algebraic equations in the process of solving them; and in this regard our interpretations do not need to be true in a strict physical sense; the only requirement that they must meet is that they satisfy all pertinent mathematical expressions of physical equilibrium.

In this regard we have a perfect right to stipulate paths on which source currents circulate. The only mathematical expression of physical equilibrium that currents have to meet is Kirchhoff's current law, and circulatory currents automatically comply with this law no matter what paths we choose for them.

This attitude is generally taken whenever we choose a mathematical artifice as an aid to the solution of a set of equations. We play a game of make believe. Just so long as we do not violate the demands of our equilibrium equations we can let our imagination run wild if we wish. Assuming the existence of circulatory currents in the first place amounts to playing make believe. None of these circulatory currents really exist. Only the branch currents exist; but if the actual current distribution is as though the circulatory currents were present then the question as to whether they really exist or not is indeed trivial.

What we can do, then, is to choose l closed paths for loop-current variables in the usual manner and then choose paths for the source currents that would normally appear in a set of n current-law equations—the current sources across node pairs. The pertinent tie-set schedule which we can write down in a straightforward manner, except that the current sources are not included as network branches, will evidently be a square matrix of order b because $l + n = b$. The paths are independent if the corresponding determinant is nonzero. *The choice of paths and of node pairs for the current sources need fulfill no other condition.*

This schedule is an augmented $[\beta]$-matrix, like the matrix 13 in art. 1 above, only this matrix will not be in canonic form if the choice of paths and node pairs is random. Being a $[\beta]$-matrix, its rows yield Kirchhoff voltage-law equations. In those for the source current loops the right-hand members of these equations are the node-pair voltages that appear across the sources. That is why these sources are not included as network branches when writing down the rows of $[\beta]$, as mentioned above.

Consistent with our present attitude we see that in the voltage-law equations given by the rows of $[\beta]$, no distinction is made in the right-hand members between source voltages and node-pair voltage variables. The former are associated with rows pertinent to loop-current paths, the latter with rows pertinent to source-current paths. Just as source and loop currents are treated alike, so source voltages and node-pair voltages are regarded as homogeneous.

If we use the tree concept and base the construction of an augmented $[\beta]$-matrix (according to these ideas) upon it, then we get the canonic form discussed earlier (Eq. 13). For the first l rows we get the matrix β_{lb} of Eq. 11 by inserting links, one at a time, into the tree and noting the closed paths that result. For the remaining n rows we continue in precisely the same manner, treating the current sources across tree branches as though they were voltage sources in links. Each additional row of the $[\beta]$-matrix thus contains a single $+1$ in the column corresponding to the pertinent tree branch. It is clear that we obtain in this way the augmented $[\beta]$-matrix in Eq. 13 which we refer to as its canonic form because it is based entirely upon the choice of a tree; that is to say, this choice alone determines the entire $[\beta]$-matrix.

The rows of this $[\beta]$-matrix yield contours to which the Kirchhoff voltage-law equations apply, and its columns yield relations for the branch currents in terms of loop and source currents. Hence we have

$$[\beta] \cdot v] = e] \tag{49}$$

and

$$[\beta]_t \cdot i] = j] \tag{50}$$

in which $e]$ and $i]$ are defined by Eqs. 15. Together with the volt-ampere relations for the branches as given by Eq. 6, we thus get the result

$$[\beta] \cdot [A] \cdot [\beta]_t \cdot i] = e] \tag{51}$$

which is the same as Eq. 18 for the consistency relation $[\beta] = [\alpha]_t^{-1}$ implicit in the present derivation. For

$$[\beta] \cdot [A] \cdot [\beta]_t = [Z] = \left[\begin{array}{ccc} Z_{ll} & \vdots & Z_{ln} \\ \cdot & \cdot & \cdot \\ Z_{nl} & \vdots & Z_{nn} \end{array}\right] \tag{52}$$

we get the expressions 23 and the equilibrium equations 24 which include the automatic conversion of current to voltage sources.

When we do not choose a canonic form for the $[\beta]$-matrix, a few additional items need further clarification. For example, suppose in choosing an arbitrary path between a node pair for a given source current we encounter some voltage sources, what is to be done with these? If we think in terms of the Kirchhoff voltage-law equation for this path we see that the corresponding node-pair voltage will appear with some additive constant terms determined by these voltage sources. Again, if voltage sources in series with branches are randomly distributed and the paths for loop currents are arbitrarily chosen then the net source voltage for the pertinent Kirchhoff voltage equation is the algebraic sum of all source voltages encountered on the contour with due regard to their polarities. Finally if node pairs across which current sources are located do not coincide with the choice of node pairs for the circulatory source currents then the appropriate values for the latter are found by placing (simultaneously) short circuits across all chosen node pairs. The resulting short-circuit currents with reversed algebraic signs are the desired values.

In this way of approaching the network problem we are in a sense regarding the network graph as having only loops (b independent ones) and no node pairs because node-pair voltages are grouped with source voltages and hence are tentatively at least regarded as known, while all currents are loop currents and are in the same sense regarded as unknown.

In the complete reverse of this situation all voltages play the role of unknown node-pair voltage variables and all currents are tentatively regarded as known sources. This reverse situation is, therefore, dual to the one just described. Topologically we regard the network graph as having only node pairs (b independent ones) and no loops because loop currents, being known sources, are topologically open circuits, and all voltages, being node-pair voltage variables, appear across open-circuited node-pairs.

Here we approach the algebraic formulation by constructing an augmented [α]-matrix through making forthright choices for cut sets associated with voltage sources as well as for node-pair voltage variables. This is done in the following manner.

The n rows corresponding to cut sets associated with node-pair voltages are constructed in the normal manner. These correspond to Kirchhoff current-law equations for which the right-hand members are current sources bridged across the same node pairs. For the remaining l rows suppose we think of the loop currents, for the moment, as identified with link currents and assume that all voltage sources are located in links. If we regard these as node-pair voltages and the sources as currents then each corresponding cut set consists of the pertinent link by itself. Each of the additional l rows of the augmented [α]-matrix contains a single $+1$ in the column corresponding to the link in question.

If, as has been consistently assumed, we number the links consecutively from 1 to l, and these rows of the [α]-matrix are written down first, then this matrix evidently has the canonic form shown in Eq. 12 provided the normal cut sets correspond to branches of the same tree to which the links apply. The choice of a tree alone is thus seen to fix the entire augmented [α]-matrix in canonic form.

While the rows of this $[\alpha]$-matrix yield contours to which the Kirchhoff current-law equations apply, its columns yield relations for the branch voltages in terms of source and node-pair voltage variables. Hence we have

$$[\alpha] \cdot j] = i] \tag{53}$$

and

$$[\alpha]_t \cdot e] = v] \tag{54}$$

with $e]$ and $i]$ again defined by Eqs. 15. Together with the volt-ampere relations for the branches as given by Eq. 26, the equations 53 and 54 thus give us the result

$$[\alpha] \cdot [B] \cdot [\alpha]_t \cdot e] = i] \tag{55}$$

which is the same as Eq. 27 for the consistency relation $[\alpha] = [\beta]_t^{-1}$ implicit in the present derivation. For

$$[\alpha] \cdot [B] \cdot [\alpha]_t = [Y] = \left[\begin{array}{c:c} Y_{ll} & Y_{ln} \\ \hdashline Y_{nl} & Y_{nn} \end{array}\right] \tag{56}$$

we get the expressions 31 and the equilibrium equations 32 which include the automatic conversion of voltage to current sources.

When we do not choose a canonic form for the augmented $[\alpha]$-matrix, a number of items need further clarification. First of all, regarding normal cut sets, it should be recalled that their forthright choice (like the arbitrary choice of loops for the rows of a $[\beta]$-matrix) is visualized in terms of picking up some of the nodes in one hand and letting these coincide while the remaining ones are similarly assembled in the other hand. The pertinent cut set is given by the branches that become stretched as we pull our hands apart.

We can get the same result by drawing on the network graph a contour enclosing the picked-up nodes together with the subgraph that attaches exclusively thereto, being careful to draw this contour so that a given branch is not crossed more than once if it is crossed at all. Such a contour may be referred to as a "cutting surface" since the branches which it crosses are the pertinent cut set. Drawing cutting surfaces at random and noting corresponding cut sets is dual to choosing loops at random when constructing rows of a $[\beta]$-matrix. The branches in the cut sets similarly determine rows of an $[\alpha]$-matrix, and their independence is readily checked in the normal manner. If, in this process, some current sources are cut, their algebraic sum yields the right-hand members in the Kirchhoff current-law equations which the pertinent rows of the $[\alpha]$-matrix imply.

Now let us consider the pseudo cut sets that determine the additional l rows of the augmented $[\alpha]$-matrix. In the canonic form of this matrix (as described above) these cut sets are single links. The "picked-up nodes" in this case are single nodes. Thus, each link is assumed to contain a voltage source which tentatively is considered as a current source whose value is the link or loop current in question, the voltage of this source being, for the moment, regarded as a node-pair voltage. The node which joins this source and the link in series is the "picked-up node," the "cutting surface" being simply a circle enclosing this node. Since it cuts the pertinent link on the one hand and the source on the other, the left- and right-hand sides of a corresponding Kirchhoff current-law equation are obvious.

Any more general form of pseudo Kirchhoff current-law equation can only be a linear combination of equations of this simple form written for a set of links corresponding to the choice of some tree. Since the right-hand sides of these equations are loop currents (regarded tentatively as known source currents) we see that these equations really are defining equations for more general loop-current variables as linear combinations of link currents. The l additional rows in the $[\alpha]$-matrix are thus seen to be any combinations of links corresponding to some chosen tree. Observe that this result is dual to the formation of the n additional rows of an augmented $[\beta]$-matrix as combinations of tree branches forming paths for current sources across node pairs.

One major point remains to be clarified. When the pseudo Kirchhoff current equations involve single links as they do in the canonic form of the $[\alpha]$-matrix, then the associated node-pair voltage is the single known voltage source in series with the pertinent link. What is this node-pair voltage for a more general pseudo Kirchhoff current equation which is a linear combination of such equations written for single links? Since voltages do not appear in these equations at all, the answer is not immediately obvious, yet we must know the values of these voltages because they appear in the right-hand side of Eq. 32 as voltage sources converted to equivalent current sources.

If we are reminded of the fact that the consistency relation $[\alpha] = [\beta]_t^{-1}$ is implicit in the present formulations, the answer to this question is readily established. Thus if we form a more general matrix $[\bar{\alpha}]$ in which some rows are linear combinations of rows in a canonic matrix $[\alpha]$, we can write

$$[\bar{\alpha}] = [\tau] \cdot [\alpha] \tag{57}$$

in which the matrix $[\tau]$ effects the pertinent row combinations. The corresponding matrix $[\bar{\beta}]$ must be reciprocal (i.e. inverse transpose) to

$[\bar{\alpha}]$, as $[\beta]$ is reciprocal to $[\alpha]$. This gives

$$[\bar{\beta}] = [\tau]_t^{-1} \cdot [\beta] \tag{58}$$

In other words the corresponding row combinations in $[\beta]$ are given by the reciprocal of the matrix $[\tau]$ which characterizes the row combinations in $[\alpha]$.

Since premultiplication of $[\beta]$ means premultiplication on both sides of Eq. 17, we see that the corresponding voltages are transformed according to the relation

$$\bar{e}] = [\tau]_t^{-1} \cdot e] \tag{59}$$

As an illustration, suppose $[\bar{\alpha}]$ is $[\alpha]$ with its second row replaced by the sum of the first two rows. Then

$$[\tau] = \begin{bmatrix} 1 & 0 & 0 & \cdots & 0 \\ 1 & 1 & 0 & \cdots & 0 \\ 0 & 0 & 1 & \cdots & 0 \\ \cdot & \cdot & \cdot & \cdot & \cdot \\ 0 & 0 & 0 & \cdots & 1 \end{bmatrix} \tag{60}$$

and

$$[\tau]_t^{-1} = \begin{bmatrix} 1 & -1 & 0 & \cdots & 0 \\ 0 & 1 & 0 & \cdots & 0 \\ 0 & 0 & 1 & \cdots & 0 \\ \cdot & \cdot & \cdot & \cdot & \cdot \\ 0 & 0 & 0 & \cdots & 1 \end{bmatrix} \tag{61}$$

Hence we see that only the source voltage for loop No. 1 is changed and that its new value is $\bar{e}_{s1} = e_{s1} - e_{s2}$. Source voltage for loop No. 2, as well as for all other loops, is unchanged notwithstanding that row No. 2 in $[\alpha]$ is the modified one. This apparent inconsistency is removed if we recognize the distinction between algebraic and topological definitions for loop currents. The matrix $[\alpha]$ defines loop currents algebraically while the rows of $[\beta]$ stipulate the contours on which they circulate and hence define the loop currents topologically. In the example just given, the algebraic definition for loop current No. 2 is changed but its contour is not, while that for loop current No. 1 is changed even though its algebraic definition remains the same.*

An interesting and collaterally useful property of cut-set and tie-set matrices follows directly from the fact that mesh or loop currents automatically satisfy Kirchhoff's current law and that node potentials or

* Further elaboration of this point is given in *ICT*, Ch. I, art. 10.

node-pair voltages automatically satisfy Kirchhoff's voltage law. That is to say, current-law equations in terms of loop currents or voltage-law equations in terms of node-pair voltages must reduce to trivial identities of the form $0 \equiv 0$.

Consider a matrix $[\alpha]$ in which the rows are any cut sets. Regardless of whether these are independent or not, or of the number of such rows, we can always write $v] = [\alpha]_t e]$ expressing branch voltages in terms of the node-pair voltages implied by these rows of α.

Independently, consider a matrix $[\beta]$ in which the rows are any tie sets. Regardless of whether these are independent or not, or of the number of such rows, we can always write $j] = [\beta]_t i]$ expressing branch currents in terms of the loop currents implied by these rows of β.

Now $[\alpha] \cdot j] = i_s]$ are current-law equations for the chosen cut sets, and $[\alpha] \cdot [\beta]_t \cdot i] = i_s]$ are these equations expressed in terms of loop currents. Similarly, $[\beta] \cdot v] = e_s]$ are voltage-law equations for the chosen tie sets, and $[\beta] \cdot [\alpha]_t \cdot e] = e_s$ are these equations expressed in terms of node-pair voltages. It follows that $[\alpha] \times [\beta_t]$ *or* $[\beta] \times [\alpha_t]$ *must be a null matrix* or that *all rows of the* $[\alpha]$*-matrix must be orthogonal to all rows of the* $[\beta]$*-matrix*; and this is true regardless of whether the rows in $[\alpha]$ or in $[\beta]$ are independent or not. The only condition is that rows of $[\alpha]$ define cut sets and rows of $[\beta]$ define tie sets and that these pertain to the same network graph.

3. Introduction of Auxiliary Variables Through an Algebraic Approach

The discussion in the preceding article shows that the augmented cut-set and tie-set matrices, besides containing the Kirchhoff equations, yield algebraic definitions for the loop-current and node-pair voltage variables. Thus the first l rows in the augmented $[\alpha]$-matrix determine algebraically the loop-current definitions while the last n rows in the augmented $[\beta]$-matrix determine algebraically the node-pair voltage definitions.

The first l rows of the $[\beta]$-matrix determine closed paths or loops for Kirchhoff voltage-law equations; and if $[\alpha]$ and $[\beta]$ are consistent, these are the contours on which the loop currents, defined by $[\alpha]$, circulate. The last n rows of the $[\alpha]$-matrix, on the other hand, designate sets of branches determined by cutting surfaces (cut sets) to which Kirchhoff current-law equations apply; and if $[\alpha]$ and $[\beta]$ are consistent, these cut sets pertain to the node-pair voltages defined by $[\beta]$; that is to say, each of these sets of branches is energized separately by its respective node-pair voltage acting alone just as each tie set is energized separately by its respective loop current acting alone.

While these ideas are, in the analytic considerations of the preceding article, motivated primarily by topological principles, it is significant to observe that such interpretation of loop-current or node-pair voltage variables is by no means essential to solution of the network problem. In fact we gain the greatest generality from an analytical point of view if we proceed on a purely algebraic basis and abandon any attempt at topological interpretation, as indeed we must since, in such a general algebraic formulation, the pertinent voltage and current variables cease to have any topological interpretation at all. The following discussion elaborates these thoughts.

The most general way of defining current variables algebraically is to express them as linear combinations of the branch currents, thus:

$$\begin{aligned} \alpha_{11}j_1 + \alpha_{12}j_2 + \cdots + \alpha_{1b}j_b &= i_1 \\ \alpha_{21}j_1 + \alpha_{22}j_2 + \cdots + \alpha_{2b}j_b &= i_2 \\ \cdot \quad \cdot \quad \cdot \quad \cdot \quad \cdot \quad \cdot \quad \cdot & \quad \cdot \quad \cdot \\ \alpha_{l1}j_1 + \alpha_{l2}j_2 + \cdots + \alpha_{lb}j_b &= i_l \end{aligned} \tag{62}$$

which is abbreviated in the matrix form

$$[\alpha_{lb}] \cdot j] = i_v] \tag{63}$$

and in which the coefficients α_{ks} are any real numbers, not necessarily $+1$, -1 or zero. We will continue to refer to the i_k's as loop currents even though they may no longer possess that topological interpretation.

Whether or not they do have a topological interpretation can be answered only by determining the tie-set schedule which the definitions 62 imply. In this connection it should be observed first of all that these algebraic definitions do not "imply" any tie-set schedule at all unless we invoke the consistency condition for which the Kirchhoff voltage equations (determined by $[\beta_{lb}]$) pertain to the same loops that the loop currents are assumed to circulate upon. When the definitions 62 no longer correspond to currents circulating upon closed contours, the consistency condition (which, as will be recalled, results in a symmetrical set of equilibrium equations) nevertheless implies that the algebraic definitions determine a corresponding tie-set matrix; that is to say, the first l rows of the augmented $[\alpha]$-matrix determine the first l rows of the augmented $[\beta]$-matrix. We shall demonstrate and elaborate upon this determination.

Consider the matrices $[\alpha]$ and $[\beta]$ in the partitioned forms

$$[\alpha] = \begin{bmatrix} \alpha_{ll} & \vdots & \alpha_{ln} \\ \cdots & \cdot & \cdots \\ \alpha_{nl} & \vdots & \alpha_{nn} \end{bmatrix}; \quad [\beta] = \begin{bmatrix} \beta_{ll} & \vdots & \beta_{ln} \\ \cdots & \cdot & \cdots \\ \beta_{nl} & \vdots & \beta_{nn} \end{bmatrix} \tag{64}$$

The consistency condition is expressed by the relation

$$[\alpha] \times [\beta]_t = \left[\begin{array}{c:c} u_l & 0 \\ \hdashline 0 & u_n \end{array}\right] \tag{65}$$

where, as before, u_l and u_n are unit matrices of orders l and n, and the zeros represent appropriate null matrices.

With reference to the left half of the right-hand matrix in 65, and the partitioned forms 64, we have

$$\alpha_{ll}(\beta_{ll})_t + \alpha_{ln}(\beta_{ln})_t = u_l \tag{66}$$

$$\alpha_{nl}(\beta_{ll})_t + \alpha_{nn}(\beta_{ln})_t = 0 \tag{67}$$

Here only the first l rows of $[\beta]$ are involved. It is our object to show that these are uniquely determined by the first l rows of $[\alpha]$ alone; which is the same as saying that the solution of Eqs. 66 and 67 for $(\beta_{ll})_t$ and $(\beta_{ln})_t$ yields the *same* results for *any* appropriate cut sets chosen to determine the last n rows in $[\alpha]$.

To this end we first obtain from 67

$$(\beta_{ln})_t = -(\alpha_{nn})^{-1}\alpha_{nl}(\beta_{ll})_t \tag{68}$$

and substitute into 66 to get

$$[\alpha_{ll} - \alpha_{ln}(\alpha_{nn})^{-1}\alpha_{nl}](\beta_{ll})_t = u_l \tag{69}$$

Since the cut sets are independent, the necessary nonsingular character of α_{nn} can always be secured by appropriate branch numbering. The matrix in the square bracket in 69 is an abridged form of $[\alpha]$. Let us denote it by $\bar{\alpha}_{ll}$. Then Eqs. 68 and 69 yield

$$\beta_{ll} = (\bar{\alpha}_{ll})_t^{-1} \quad \text{and} \quad \beta_{ln} = -(\bar{\alpha}_{ll})_t^{-1}(\alpha_{nl})_t(\alpha_{nn})_t^{-1} \tag{70}$$

Observe that the last n rows in $[\alpha]$ are involved only in the matrix product $(\alpha_{nn})^{-1}\alpha_{nl}$ or its transpose. If this product is invariant to replacing the implied cut sets by any independent linear combination of them (which is all that any other appropriate cut sets can be), then our original contention will be proved: namely, that the matrix $[\beta_{lb}]$ is uniquely determined by $[\alpha_{lb}]$.

Such a linear combination of the last n rows of $[\alpha]$ can be written

$$l_{nn} \times [\alpha_{nb}] = [l_{nn}\alpha_{nl} \vdots l_{nn}\alpha_{nn}] = [\hat{\alpha}_{nb}] \tag{71}$$

in which the matrix l_{nn} is nonsingular. We readily see that

$$(\hat{\alpha}_{nn})^{-1}\hat{\alpha}_{nl} = (\alpha_{nn})^{-1}(l_{nn})^{-1}l_{nn}\alpha_{nl} = (\alpha_{nn})^{-1}\alpha_{nl} \tag{72}$$

which demonstrates the desired invariance.

It is collaterally interesting to recognize the topological significance of the invariant matrix 72. When a cut-set matrix is arbitrarily chosen it is denoted by

$$[\alpha_{nb}] = [\alpha_{nl} \vdots \alpha_{nn}] \tag{73}$$

When its construction is based upon the choice of a tree, on the other hand, we get the *canonic* form which reads

$$[\alpha_{nb}]^c = [(\alpha_{nl})^c \vdots u_n] \tag{74}$$

where it is tacitly assumed that tree branches are the last n of all consecutively numbered branches in the network graph. Since the rows in 74 must be a linear combination of the rows in 73, we have

$$[\alpha_{nb}]^c = l_{nn} \times [\alpha_{nb}] = (\alpha_{nn})^{-1} \times [\alpha_{nl} \vdots \alpha_{nn}] \tag{75}$$

and evidently

$$(\alpha_{nl})^c = (\alpha_{nn})^{-1}\alpha_{nl} \tag{76}$$

The invariant matrix 72 is thus revealed to be the canonic form of the submatrix α_{nl}.

In the computation of β_{ll} and β_{ln} by means of Eqs. 70, we can use any appropriate cut-set matrix $[\alpha_{nb}]$ that happens to be handy (the canonic form will do), secure in the knowledge that it will yield the same result as any other.

If in the determination of $[\beta_{lb}]$ by means of Eqs. 70 from a given set of loop-current definitions 62 it turns out that only coefficients $+1$, -1 or zero are involved, then these currents may be given some sort of topological interpretation, perhaps that of ordinary circulatory currents or of "mesh-pair" currents or perhaps that of some topologically weird entities; but in general this tie-set matrix will hardly be worthy of its name although algebraically the solution to our network problem proceeds as smoothly as it does with real loop currents.

The fact that $[\alpha_{lb}]$ determines $[\beta_{lb}]$ uniquely is the important feature of this procedure. A set of algebraic definitions for loop currents (as in Eq. 62 or 63) uniquely determines their topological definitions. The converse is, however, not true. Topological definitions do not uniquely determine corresponding algebraic definitions. There is not a one-to-one correspondence between $[\alpha_{lb}]$ and $[\beta_{lb}]$. While algebraic definitions include or imply topological definitions, the reverse is not true. Topological definitions are less specific than algebraic definitions. An endless variety of algebraic definitions can all correspond to the same topological loop-current entities. We shall now elaborate this point.

Since in Eq. 64 we can interchange the letters α and β, we can do the same in 66 and 67, and get

$$\beta_{ll}(\alpha_{ll})_t + \beta_{ln}(\alpha_{ln})_t = u_l \tag{77}$$

$$\beta_{nl}(\alpha_{ll})_t + \beta_{nn}(\alpha_{ln})_t = 0 \tag{78}$$

From the last equation we have

$$(\alpha_{ln})_t = -(\beta_{nn})^{-1}\beta_{nl}(\alpha_{ll})_t \tag{79}$$

and substitution into 77 gives

$$[\beta_{ll} - \beta_{ln}(\beta_{nn})^{-1}\beta_{nl}](\alpha_{ll})_t = u_l \tag{80}$$

The necessary nonsingular character of β_{nn} can always be secured by proper branch numbering, and 79 and 80 then yield

$$\alpha_{ll} = (\bar{\beta}_{ll})_t^{-1} \quad \text{and} \quad \alpha_{ln} = -(\bar{\beta}_{ll})_t^{-1}(\beta_{nl})_t(\beta_{nn})_t^{-1} \tag{81}$$

where $\bar{\beta}_{ll}$ denotes the matrix in the square bracket of Eq. 80.

So far, this process is exactly analogous to the foregoing one. In fact, Eqs. 79, 80 and 81 are the same respectively as Eqs. 68, 69 and 70 except for an interchange of the letters α and β. We cannot now, however, claim the invariance of $(\beta_{nn})^{-1}\beta_{nl}$ which depends upon the last n rows of $[\beta]$. These determine a set of algebraic definitions for node-pair voltage variables given by the equations

$$\begin{aligned}
\beta_{l+1,1}v_1 + \beta_{l+1,2}v_2 + \cdots + \beta_{l+1,b}v_b &= e_1 \\
\beta_{l+2,1}v_1 + \beta_{l+2,2}v_2 + \cdots + \beta_{l+2,b}v_b &= e_2 \\
\cdots\cdots\cdots\cdots\cdots\cdots\cdots\cdots \\
\beta_{b1}v_1 \quad + \beta_{b2}v_2 \quad + \cdots + \beta_{bb}v_b &= e_n
\end{aligned} \tag{82}$$

which may be abbreviated in the matrix form

$$[\beta_{nb}] \cdot v] = e_v] \tag{83}$$

Not all possible varieties of these defining equations can be derived from an arbitrary set like 82 merely by making independent linear combinations of the rows in $[\beta_{nb}]$. The canonic form of this matrix, which results simply from identifying the node-pair voltages in $e_v]$ with branch voltages of a chosen tree, reads

$$[\beta_{nb}]^c = [0 \ \vdots \ u_n] \tag{84}$$

Obviously if we merely make linear row combinations in this matrix the null submatrix remains null, and we cannot generate more general forms for the defining equations in this way.

What we can do here is start from a matrix

$$[\beta_{nb}]^a = [(\beta_{nl})^a \ \vdots \ u_n] \tag{85}$$

in which $(\beta_{nl})^a$ is completely arbitrary. Through linear row combinations, which leave the matrix $(\beta_{nn})^{-1}\beta_{nl}$ in 80 invariant, and a proper choice of $(\beta_{nl})^a$ we can generate any desired defining equations such as 82. Hence the use of 85 in Eqs. 80 and 81 should yield a matrix $[\alpha_{lb}]$ from which a specified $[\beta_{lb}]$ will be regained by the foregoing process, leading to Eqs. 69 and 70. That this is indeed true we shall now demonstrate.

Using 85 in Eqs. 80 and 81 we have

$$\alpha_{ll} = (\bar{\beta}_{ll})_t^{-1} = [\beta_{ll} - \beta_{ln}(\beta_{nl})^a]_t^{-1}; \quad \alpha_{ln} = -(\bar{\beta}_{ll})_t^{-1}(\beta_{nl})_t^a \tag{86}$$

Substituting into Eqs. 70 and making use of Eq. 76 yields

$$\beta_{ll} = [\beta_{ll} - \beta_{ln}(\beta_{nl})^a]\{u_l + (\beta_{nl})_t^a(\alpha_{nl})^c\}_t^{-1}; \quad \beta_{ln} = \beta_{ll}(\alpha_{nl})_t^c \tag{87}$$

or

$$\beta_{ll} = \beta_{ll}[u_l - (\beta_{ll})^{-1}\beta_{ln}(\beta_{nl})^a] \times [u_l + (\alpha_{nl})_t^c(\beta_{nl})^a]^{-1} \tag{88}$$

If we now recognize by analogy to Eqs. 73 through 76 that

$$[\beta_{lb}] = [\beta_{ll} : \beta_{ln}]; \quad [\beta_{lb}]^c = [u_l : (\beta_{ln})^c] = (\beta_{ll})^{-1}[\beta_{lb}] \tag{89}$$

giving

$$(\beta_{ln})^c = (\beta_{ll})^{-1}\beta_{ln} \tag{90}$$

and that

$$(\beta_{ln})^c = -(\alpha_{nl})_t^c \tag{91}$$

(Eq. 67 or 77 for the canonic case where $\beta_{ll} = \alpha_{ll} = u_l$ and $\alpha_{nn} = \beta_{nn} = u_n$), then it is clear that 88 is an identity and that the matrix $(\beta_{nl})^a$ is indeed arbitrary.

It is this arbitrariness that yields an infinite variety of matrices $[\alpha_{lb}]$ from a given matrix $[\beta_{lb}]$ by means of Eqs. 86. In particular, the choice of a null matrix for $(\beta_{nl})^a$, corresponding to the canonic form 84, yields

$$\alpha_{ll} = (\beta_{ll})_t^{-1} \quad \text{and} \quad \alpha_{ln} \equiv 0 \tag{92}$$

for which the defining equations 62 express the loop currents in terms of only l of the branch currents forming an independent subset.

The duals of all of these results and relationships follow at once. From the right half of the right-hand matrix in Eq. 65 we get the duals of relations 77 and 78 which read

$$\alpha_{ll}(\beta_{nl})_t + \alpha_{ln}(\beta_{nn})_t = 0 \tag{93}$$

$$\alpha_{nl}(\beta_{nl})_t + \alpha_{nn}(\beta_{nn})_t = u_n \tag{94}$$

Interchanging the letters α and β we alternately have the duals of 66 and 67, namely

$$\beta_{ll}(\alpha_{nl})_t + \beta_{ln}(\alpha_{nn})_t = 0 \tag{95}$$

$$\beta_{nl}(\alpha_{nl})_t + \beta_{nn}(\alpha_{nn})_t = u_n \tag{96}$$

From these we obtain uniquely the last n rows in $[\alpha]$ (the cut-set matrix defining node-pair voltages topologically) from the last n rows in $[\beta]$ which define these voltages algebraically as in Eq. 82. This result, which is dual to that given by Eqs. 70, reads

$$\alpha_{nn} = (\bar{\beta}_{nn})_t^{-1} \quad \text{and} \quad \alpha_{nl} = -(\bar{\beta}_{nn})_t^{-1}(\beta_{ln})_t(\beta_{ll})_t^{-1} \tag{97}$$

with

$$\bar{\beta}_{nn} = [\beta_{nn} - \beta_{nl}(\beta_{ll})^{-1}\beta_{ln}] \tag{98}$$

Here the matrix $(\beta_{ll})^{-1}\beta_{ln}$, which Eq. 90 shows to be the canonic form of β_{ln}, is invariant, and hence Eqs. 97 and 98 lead to a unique matrix $[\alpha_{nb}]$.

The reverse process, namely that of determining the algebraic from the topological definitions of the node-pair voltages, like the dual problem involving currents, is not unique. The duals of 84 and 85 are the canonic and the arbitrary forms of $[a_{lb}]$, namely

$$[\alpha_{lb}]^c = [u_l \vdots 0] \tag{99}$$

$$[\alpha_{lb}]^a = [u_l \vdots (\alpha_{ln})^a] \tag{100}$$

Use of the latter in Eqs. 93 and 94 leads to the dual of relations 86, thus:

$$\beta_{nn} = (\bar{\alpha}_{nn})_t^{-1} = [\alpha_{nn} - \alpha_{nl}(\alpha_{ln})^a]_t^{-1}; \quad \beta_{nl} = -(\bar{\alpha}_{nn})_t^{-1}(\alpha_{ln})_t^a \tag{101}$$

If, for the arbitrary matrix $(\alpha_{ln})^a$ we choose a null matrix as in the canonic form 99 for $[\alpha_{lb}]$, we find

$$\beta_{nn} = (\alpha_{nn})_t^{-1} \quad \text{and} \quad \beta_{nl} \equiv 0 \tag{102}$$

for which the defining equations 82 express the node-pair voltages in terms of only n of the branch voltages forming an independent subset.

These defining equations (like Eqs. 62 for loop currents) in general involve coefficients β_{ks} which can be any real numbers, not necessarily $+1$, -1 or zero. We continue to refer to the e_k's as node-pair voltages even though they may no longer possess that topological interpretation.

Whether or not they do have a topological interpretation can be answered by determining from Eqs. 97 and 98 the cut-set matrix which the definitions 82 imply. If this matrix thus found contains only $+1$, -1 or zero elements, then the definitions 82 do have some sort of topological interpretation, not necessarily that of ordinary node-pair voltages. In general this cut-set matrix will no longer define "cut sets" in the usual sense, although algebraically the solution to our network problem proceeds as smoothly as it does with real node-pair voltage definitions.

If we are not interested in having the equilibrium equations on node or loop basis be symmetrical, then we can choose any cut-set matrix $[\alpha_{nb}]$ we wish and associate it with $[\alpha_{lb}]$ in the defining equations 63 to form an augmented $[\alpha]$-matrix, the only condition being that the resulting $[\alpha]$ be nonsingular, which is almost certain to be the case if the defining equations 62 are arbitrary and independent. With the loop-current variables and source currents thus regarded as a homogeneous set, the relations between these and the branch currents are reversible. One can solve for the branch currents in terms of loop currents with this augmented set of equations while the definitions 62 alone, having a nonsquare matrix, are not reversible.

Similarly, if symmetry is no object, we can choose any tie-set matrix $[\beta_{lb}]$ we wish and associate it with $[\beta_{nb}]$ in Eq. 83 to form an augmented β-matrix which is almost certain to be nonsingular if the defining equations 82 are arbitrary and independent. With node-pair voltage variables and source voltages regarded as a homogeneous set, the relations between these and the branch voltages are reversible. One can solve for the branch voltages in terms of node-pair voltages with this augmented set of equations while the definitions 82 alone, having a nonsquare matrix, are not reversible.

4. Parameter Matrices

Consider equilibrium equations 24 on a loop basis with the symmetrical impedance matrix Z_{ll} given by the first of the relations 23 as

$$Z_{ll} = a_{ll} + \beta_{ln}a_{nn}(\beta_{ln})_t \tag{103}$$

or alternately by Eq. 41 in the form

$$Z_{ll} = [\beta_{lb}] \cdot [A] \cdot [\beta_{lb}]_t \tag{104}$$

We are now interested in separating out the effects of the inductance, the resistance and the elastance elements. Hence we consider the partitioned form of $[A]$ given by Eq. 5 and similarly partition the columns of $[\beta_{lb}]$ into groups of λ, ρ and σ, as indicated by writing

$$[\beta_{lb}] = [\beta_{l\lambda} \vdots \beta_{l\rho} \vdots \beta_{l\sigma}] \tag{105}$$

Then we get

$$[\beta_{lb}] \cdot [A] \cdot [\beta_{lb}]_t = [L]p + [R] + [S]p^{-1} \tag{106}$$

with

$$\begin{aligned} [L] &= \beta_{l\lambda} \cdot [l] \cdot (\beta_{l\lambda})_t \\ [R] &= \beta_{l\rho} \cdot [r] \cdot (\beta_{l\rho})_t \\ [S] &= \beta_{l\sigma} \cdot [s] \cdot (\beta_{l\sigma})_t \end{aligned} \tag{107}$$

and the branch parameter matrices $[l]$, $[r]$, $[s]$ as defined in Eqs. 1, 2, 3.

$[L]$, $[R]$ and $[S]$ in 107 are the parameter matrices on a loop basis for the choice of loops determined by the tie-set matrix $[\beta_{lb}]$.

The equilibrium equations 24 (assuming only voltage sources or that the right-hand excitation matrix includes both terms in Eq. 24) become

$$([L]p + [R] + [S]p^{-1}) \cdot i_v] = e_s] \tag{108}$$

These equations and the parameter matrices 107 appearing therein are, of course, symmetrical because of the implied consistency condition. If this condition is not met, the parameter matrices $[L]$, $[R]$, $[S]$, derived in an entirely analogous fashion, turn out to be nonsymmetrical, notwithstanding that the branch parameters are the same bilateral ones contained in the matrices $[l]$, $[r]$, $[s]$ in Eqs. 107. Bilaterality or nonbilaterality of the network elements has nothing per se to do with symmetry or nonsymmetry of resulting loop-parameter matrices.

On a node basis consider equilibrium equations 32 with the symmetrical admittance matrix Y_{nn} given by the last of the relations 31 as

$$Y_{nn} = b_{nn} + \alpha_{nl} b_{ll} (\alpha_{nl})_t \tag{109}$$

or alternately by Eq. 46 in the form

$$Y_{nn} = [\alpha_{nb}] \cdot [B] \cdot [\alpha_{nb}]_t \tag{110}$$

Again we are interested in separating out the effects of inductance, resistance and capacitance elements. Hence we consider the partitioned form of $[B]$ given in Eq. 25 and partition the columns of $[\alpha_{nb}]$ accordingly, as indicated by writing this matrix in the form

$$[\alpha_{nb}] = [\alpha_{n\lambda} \vdots \alpha_{n\rho} \vdots \alpha_{n\sigma}] \tag{111}$$

We then find

$$[\alpha_{nb}] \cdot [B] \cdot [\alpha_{nb}]_t = [C]p + [G] + [\Gamma]p^{-1} \tag{112}$$

where

$$\begin{aligned} [C] &= \alpha_{n\sigma} \cdot [c] \cdot (\alpha_{n\sigma})_t \\ [G] &= \alpha_{n\rho} \cdot [g] \cdot (\alpha_{n\rho})_t \\ [\Gamma] &= \alpha_{n\lambda} \cdot [\gamma] \cdot (\alpha_{n\lambda})_t \end{aligned} \tag{113}$$

and the branch parameter matrices $[c]$, $[g]$, $[\gamma]$ now are respectively the inverses of $[s]$, $[r]$ and $[l]$.

$[C]$, $[G]$ and $[\Gamma]$ in Eqs. 113 are the parameter matrices on a node basis for the choice of node pairs determined by the cut-set matrix $[\alpha_{nb}]$.

The equilibrium equations 32 (assuming only current sources or that the right-hand excitation matrix includes both terms in 32) become

$$([C]p + [G] + [\Gamma]p^{-1}) \cdot e_v] = i_s] \tag{114}$$

Again the parameter matrices and hence the resulting equations are symmetrical because of the implied consistency condition. If this condition is not fulfilled, the parameter matrices, obtained in the same fashion, are nonsymmetrical. Bilaterality of the network elements alone does not imply symmetry of parameter matrices except on a branch basis where the matrices are $[l]$, $[r]$ and $[s]$ given by Eqs. 1, 2, 3.

If we assume for the node basis that $[\alpha_{nb}]$ is based upon a starlike tree (branches emanating from a common datum node) then the node-pair voltage variables are a set of node-to-datum potentials. As is well known, the matrices $[C]$ and $[G]$ can then easily be constructed by inspection, and so can $[\Gamma]$ if no mutual inductive coupling is present. Thus any non-diagonal element G_{sk} in $[G]$, for example, is the negative of the conductance in a branch joining nodes s and k, while each diagonal element G_{kk} equals the sum of conductances in all branches meeting in node k. The algebraic sum of elements in any row of $[G]$, say the kth row, evidently equals the conductance of the kth tree branch.

If we remove all the tree branches, then the sum of elements in any row or column equals zero. In this case the matrix $[G]$ has a zero determinant (because the sum of all rows is zero and this is a linear dependence condition among the rows). Similar remarks apply to $[C]$ and to $[\Gamma]$ and so we can say the same about the admittance matrix Y_{nn} and its determinant.

Physically, removal of the starlike tree cuts the remainder of the network off from the datum node. We speak in this case of a "floating" matrix Y_{nn}. All of the node potentials now are no longer independent. We can arbitrarily assume any value for one of them and compute the others in terms of it. Hence the equilibrium equations are solvable notwithstanding that their determinant is zero.

More generally we can say that the floating property is produced upon removal of all tree branches regardless of whether the tree is starlike or not, provided that such removal increases the number of separate parts composing the total graph. Specifically, the rank of the resulting matrix Y_{nn} equals $(n - \Delta s)$ where Δs denotes the increase in the number of separate parts. In any case, the "floating" matrix and the physical situation upon which it depends represent a special case of the more general one in which the tree is not removed. The usual procedure, not the one leading to a "floating" matrix, is more general.

An analogous result can, of course, be had on a loop basis. Assuming again a starlike tree, the resulting impedance matrix Z_{ll} has the property described above for the "floating" admittance matrix Y_{nn} if we short-circuit all links. The matrix Z_{ll} can then be spoken of as "anchored," being dual to the "floating" matrix Y_{nn}. This "anchored" property of the matrix Z_{ll} is produced upon short-circuiting or shrinking all links regardless of whether the tree is starlike or not, provided the decrease $l = b - n$ in the number of branches exceeds the decrease Δn in the number of nodes, so that the resulting number of topologically independent loops [which is $(b - l) - (n - \Delta n) = \Delta n$] is less than the original number l. The rank of the resulting matrix Z_{ll} equals Δn which may be less than but cannot exceed the original rank l.

5. Equilibrium Equations in Mixed Form

In the equilibrium equations 108 and 114 on loop and node bases, both derivative and integral operators appear. There are situations in network analysis where it is desirable to avoid having both operators as well as the constant term appear in these equations. If we look carefully to see why both operators are involved on either the loop or node basis, it will become clear how we can express network equilibrium in a way that involves only constant terms and first-derivative terms.

Thus, if we choose only currents as variables, as we do on a loop basis, which involves an equilibrium of voltages, then both operators p and p^{-1} must be involved because inductive voltages are proportional to current derivatives while capacitive voltages, being proportional to charge, are functions of current integrals. On a node basis where the equilibrium is expressed in terms of current and we choose voltages as variables we find a similar situation in that capacitive currents are proportional to voltage derivatives while inductive currents are functions of flux linkage or the time integral of voltage.

Instead of consistently using currents or voltages as variables, we can use currents in inductive branches and voltages in capacitive branches and either one or the other in the resistive branches. With such a mixed set of variables, either a voltage equilibrium or a current equilibrium involves only constant and derivative terms provided that on a voltage basis we have only voltage sources and on a current basis only current sources, for otherwise we will again have trouble from the source-conversion terms in our equilibrium equations.

We can overcome this last difficulty if we not only use a mixed set of current and voltage variables but also use both current and voltage equilibrium equations in a manner that we shall now develop. The price we pay for thus avoiding integral terms in the equilibrium equations is an increase in the total number of variables involved, although the situation does not become as bad in this regard as that represented by Eqs. 38. In fact it is less than half as bad as this, since the total number of variables turns out at most to be equal to the total number of branches b and in many practical cases equals only the number of inductive plus the number of capacitive branches.

We begin by defining the branches of our network as consisting of series combinations of R and L, or parallel combinations of G and C. Correspondingly we define the branch operators

$$z_{ks} = l_{ks}p + r_k \tag{115}$$

and

$$y_k = c_k p + g_k \tag{116}$$

Pure resistive branches involve degenerate forms of either of these; therefore no loss in generality is involved by this special designation while an important practical situation is more readily accommodated.

We now depart slightly from the standard branch numbering practice adopted before, in that we number all RL branches consecutively from 1 to λ and all GC branches from $\lambda + 1$ to $\lambda + \sigma = b$. As just pointed out, pure resistive branches (if there are any) are grouped either with the inductances or with the capacitances and hence the group of ρ resistive branches appearing in the preceding discussions does not exist here.

Voltage drops in the RL branches are elements in a column matrix $v_\lambda]$, and currents in the GC branches are those in a column matrix $j_\sigma]$. Then, denoting a square matrix of order λ with elements z_{ks} as z_λ, and a diagonal matrix of order σ with elements y_k as y_σ, we have

$$v_\lambda] = z_\lambda \cdot j_\lambda] \tag{117}$$

and

$$j_\sigma] = y_\sigma \cdot v_\sigma] \tag{118}$$

in which $j_\lambda]$ is a column matrix representing currents in the RL branches and $v_\sigma]$ is one representing voltage drops across the GC branches. Together Eqs. 117 and 118 are the total volt-ampere relations for the branches.

Columns of the tie-set matrix β_{lb} and of the cut-set matrix α_{nb} are partitioned into groups of λ and σ, and the resulting submatrices identified by subscripts that indicate their row and column structure in the usual manner. Voltage sources acting around loops and current sources acting across node pairs are elements of column matrices $e_s]$ and $i_s]$ respectively. The Kirchhoff voltage and current laws are then expressed by the equations

$$\beta_{l\lambda} \cdot v_\lambda] + \beta_{l\sigma} \cdot v_\sigma] = e_s] \tag{119}$$

and

$$\alpha_{n\lambda} \cdot j_\lambda] + \alpha_{n\sigma} \cdot j_\sigma] = i_s] \tag{120}$$

Loop currents are assumed to circulate on the contours defined by the tie-set matrix $[\beta_{lb}]$ and node-pair voltages are consistent with the cut sets defined by $[\alpha_{nb}]$. Denoting these currents and voltages as elements in the column matrices $i_v]$ and $e_v]$ (as in the preceding discussions) we have

$$j_\lambda] = (\beta_{l\lambda})_t \cdot i_v] \tag{121}$$

and

$$v_\sigma] = (\alpha_{n\sigma})_t \cdot e_v] \tag{122}$$

Substituting Eqs. 117 and 118 into 119 and 120, and then substituting for $j_\lambda]$ and $v_\sigma]$ from Eqs. 121 and 122, we get the following equilibrium equations on a mixed basis:

$$\beta_{l\lambda} \cdot z_\lambda \cdot (\beta_{l\lambda})_t \cdot i_v] + \beta_{l\sigma} \cdot (\alpha_{n\sigma})_t \cdot e_v] = e_s] \tag{123}$$

$$\alpha_{n\lambda} \cdot (\beta_{l\lambda})_t \cdot i_v] + \alpha_{n\sigma} \cdot y_\sigma \cdot (\alpha_{n\sigma})_t \cdot e_v] = i_s] \tag{124}$$

As pointed out in the closing paragraphs of art. 2 above, the rows of tie-set and cut-set matrices are orthogonal. Hence we have either

$$\beta_{l\lambda} \cdot (\alpha_{n\lambda})_t + \beta_{l\sigma} \cdot (\alpha_{n\sigma})_t = 0 \tag{125}$$

or

$$\alpha_{n\lambda} \cdot (\beta_{l\lambda})_t + \alpha_{n\sigma} \cdot (\beta_{l\sigma})_t = 0 \tag{126}$$

If we let

$$\gamma_{ln} = \beta_{l\sigma} \cdot (\alpha_{n\sigma})_t = -\beta_{l\lambda} \cdot (\alpha_{n\lambda})_t \tag{127}$$

and write γ_{nl} for the transpose of γ_{ln}, that is,

$$\gamma_{nl} = (\gamma_{ln})_t = -\alpha_{n\lambda} \cdot (\beta_{l\lambda})_t \tag{128}$$

then the equilibrium equations 123 and 124 may be written

$$\beta_{l\lambda} \cdot Z_\lambda \cdot (\beta_{l\lambda})_t \cdot i_v] + \gamma_{ln} \cdot e_v] = e_s] \tag{129}$$

$$-\gamma_{nl} \cdot i_v] + \alpha_{n\sigma} \cdot y_\sigma \cdot (\alpha_{n\sigma})_t \cdot e_v] = i_s] \tag{130}$$

Parameter matrices similar to those defined in Eqs. 107 and 113 may be introduced here by the analogous equations

$$[L] = \beta_{l\lambda} \cdot l \cdot (\beta_{l\lambda})_t; \quad [R] = \beta_{l\lambda} \cdot r \cdot (\beta_{l\lambda})_t \tag{131}$$

$$[C] = \alpha_{n\sigma} \cdot c \cdot (\alpha_{n\sigma})_t; \quad [G] = \alpha_{n\sigma} \cdot g \cdot (\alpha_{n\sigma})_t \tag{132}$$

whereupon we can write Eqs. 129 and 130 more explicitly as

$$\left\{ \begin{bmatrix} L & \vdots & 0 \\ \cdots & \cdot & \cdots \\ 0 & \vdots & C \end{bmatrix} p + \begin{bmatrix} R & \vdots & \gamma_{ln} \\ \cdots & \cdot & \cdots \\ -\gamma_{nl} & \vdots & G \end{bmatrix} \right\} \times \begin{bmatrix} i_v \\ \cdots \\ e_v \end{bmatrix} = \begin{bmatrix} e_s \\ \cdots \\ i_s \end{bmatrix} \tag{133}$$

These are the desired equations in mixed form.

The submatrix γ_{ln} and its transpose are here associated with the $[R]$- and $[G]$-matrices simply because the equations can be written more compactly by doing this, although γ_{ln} actually has nothing in common with resistive elements since it is determined solely by the topology of the network.

We can get a better appreciation of how the form of Eqs. 133 comes about if we recognize that the manipulations in the present derivation amount to regarding the total network as consisting of an interconnection of an *RL* network with a *GC* network, and that γ_{ln} tells us how these two networks are interconnected. That is to say, γ_{ln} is a "connection matrix" that describes how the *RL* portion of our network is joined to the *GC* portion.

In order to understand the mechanism of interconnection more clearly, it is helpful to observe that we can achieve the essential features exhibited by Eqs. 133 without introducing loop currents or node-pair voltages as variables. Thus if we substitute Eqs. 117 and 118 into 119 and 120, and stop there, we have a pair of matrix equations that can be written

$$\left\{\begin{bmatrix} \beta_{l\lambda} \times l & \vdots & 0 \\ \cdots & & \cdots \\ 0 & \vdots & \alpha_{n\sigma} \times c \end{bmatrix} p + \begin{bmatrix} \beta_{l\lambda} \times r & \vdots & \beta_{l\sigma} \\ \cdots & & \cdots \\ \alpha_{n\lambda} & \vdots & \alpha_{n\sigma} \times g \end{bmatrix}\right\} \times \begin{bmatrix} j_\lambda \\ \cdots \\ v_\sigma \end{bmatrix} = \begin{bmatrix} e_s \\ \cdots \\ i_s \end{bmatrix} \tag{134}$$

Here the variables are the currents in the RL branches and the voltages across the GC branches, which together also are b in number even though λ may not equal l or σ equal n. The only significant difference between Eqs. 134 and Eqs. 133 is that in 134 the first matrix is not symmetrical while in 133 it is. Thus in Eqs. 131 and 132 the branch-parameter matrices are pre- and post-multiplied by pairs of transposed matrices, yielding a symmetrical result. In Eqs. 134 the post-multiplication is left off.

If we assume a special situation in which the GC portion of the network forms a tree in the total network graph, then the RL portion represents a set of links; and if we choose link currents as loop currents and tree-branch voltages as node-pair voltages, then (since l now equals λ and n equals σ) $\beta_{l\lambda}$ and $\alpha_{n\sigma}$ are unit matrices of order l and n respectively. Moreover, the consistency condition makes $\alpha_{n\lambda}$ equal to $-(\beta_{l\sigma})_t$; $j_\lambda]$ becomes $i_v]$, and $v_\sigma]$ is $e_v]$; and Eqs. 133 and 134 are alike. They then have the form

$$\left\{\begin{bmatrix} lp & \vdots & 0 \\ \cdots & & \cdots \\ 0 & \vdots & cp \end{bmatrix} + \begin{bmatrix} r & \vdots & \beta_{ln} \\ \cdots & & \cdots \\ -(\beta_{ln})_t & \vdots & g \end{bmatrix}\right\} \times \begin{bmatrix} i_v \\ \cdots \\ e_v \end{bmatrix} = \begin{bmatrix} e_s \\ \cdots \\ i_s \end{bmatrix} \tag{135}$$

in which the first matrix is diagonal if no mutual-inductive coupling is present.

The "connection matrix" in this case is simply β_{ln} which indicates the interconnection of tree branches forming paths for the link currents. If we visualize the GC portion in a separate box having n terminal pairs, then β_{ln} tells us where to attach these terminal pairs to the nodes of the tree. The RL portion, also in a separate box, has l terminal pairs with a series RL branch across each. There is no interconnection of these RL branches except through mutual-inductive coupling (if present). In a rather obvious fashion, the β_{ln}-matrix tells us how to connect terminal pairs of the RL box to terminal pairs of the GC box.

Returning to the more general situation to which Eqs. 133 pertain, we may visualize the RL portion as a box with l terminal pairs whose open-circuit impedance matrix is $[Lp + R]$, and the GC portion as a box with

n terminal pairs whose short-circuit admittance matrix is $[Cp + G]$. The first term in Eq. 123 and the second term in Eq. 124 involve these separate boxes. In the second term of Eq. 123 the part $(\alpha_{n\sigma})_t e_v]$ represents voltage drops in GC branches, and the factor $\beta_{l\sigma}$ picks out series combinations of these branches which, in series with pertinent terminal pairs of the RL box, form closed loops on which the sources in $e_s]$ act. Similarly, in the first term of Eq. 124 the part $(\beta_{l\lambda})_t i_v]$ represents currents in RL branches, and the factor $\alpha_{n\lambda}$ picks out the parallel combinations of these branches which, in parallel with pertinent terminal pairs of the GC box, form cut sets on which the sources in $i_s]$ act. Thus the factors in γ_{ln} determine the manner in which the RL and GC portions are interconnected, although in a random situation this may be difficult to puzzle out if only the resulting equations are given and the network topology is unknown, even if it is known that the equations stem from a physical network in the first place.

One can, of course, concoct boxes full of ideal transformers, provided with terminal pairs that can be joined in series or in parallel with appropriate terminal pairs of the RL and GC boxes so as to carry out the interconnections implied by the second term in 123 and the first term in 124; but such a mental exercise would have little practical value except perhaps to clarify further the interpretation of the "connection matrix" γ_{ln}. This we shall not indulge in here.

6. Accommodation of Multi-terminal Elements

Procedures discussed in the preceding articles are confined to network configurations in which the branches are two-terminal devices such as the ordinary passive and bilateral resistances, inductances and capacitances. When active elements, such as vacuum tubes or transistors, are included, this simple topological picture no longer applies, since these elements are characterized schematically as boxes from which three or more terminals emanate. We shall show that only a minor addition to the existing analysis procedure is needed to accommodate this more general type of network.

It is well known that a box with $p + 1$ terminals is uniquely characterized at these terminals by p voltages and p currents. Assuming that the device is resistive in character, the pertinent volt-ampere relations may be written in the form

$$\begin{aligned} v_1 &= r_{11}j_1 + \cdots + r_{1p}j_p \\ &\cdot\ \cdot\ \cdot\ \cdot\ \cdot\ \cdot\ \cdot\ \cdot\ \cdot \\ v_p &= r_{p1}j_1 + \cdots + r_{pp}j_p \end{aligned} \tag{136}$$

or inversely as

$$\begin{aligned} j_1 &= g_{11}v_1 + \cdots + g_{1p}v_p \\ &\cdot\ \cdot\ \cdot\ \cdot\ \cdot\ \cdot\ \cdot\ \cdot\ \cdot \\ j_p &= g_{p1}v_1 + \cdots + g_{pp}v_p \end{aligned} \tag{137}$$

Topologically the device is regarded as having the structure of a tree consisting of p branches. The $p + 1$ nodes are the terminals of the device and the currents and voltages in Eqs. 136 and 137 are tree-branch currents and voltage drops.

If the tree is assumed to be starlike in form, then the voltages are a node-to-datum set with the common node of the star as a datum, and the currents are terminal currents entering the p tips of the star and returning through the common datum node. The branch currents and voltages for any other assumed tree configuration are readily expressible in terms of those pertinent to a starlike tree in a familiar manner. Therefore, no matter how the volt-ampere relations for a given $(p + 1)$-terminal device may initially be specified, we can always assume that relations like those in Eqs. 136 or 137, pertinent to any tree configuration we care to assume, are known, since we can readily put the given characterization of the multi-terminal device into this form.

It is significant that the matrices in Eqs. 136 or 137 are not necessarily symmetrical and are not diagonal in contrast with matrices pertinent to a tree configuration consisting of passive bilateral branches. The significant point here is that we are permitted topologically to replace the active and/or nonbilateral device by the same configuration that characterizes the subgraph of a passive bilateral box with the same number of terminals.

Topologically any tree structure is an adequate representation for a multi-terminal element, and analytically the relations 136 or 137 are adequate. A network with multi-terminal elements embedded in it is, therefore, representable topologically by a graph containing only two-terminal branches; and the total volt-ampere relations for these branches are readily formulated from given two-terminal parameter values and equations characterizing the multi-terminal devices, the only difference being that the resulting branch resistance or conductance matrix is not completely diagonal because square submatrices like those in 136 and 137 are embedded in it (one for each multi-terminal element). All other steps relating to the selection of variables and formulation of equilibrium equations remain unchanged.

Problems

1. With reference to the adjacent graph, choose branches 5678 as a tree. Define loop currents as link currents and node-pair voltages as tree-branch voltages.

(*a*) Find algebraic relations expressing branch currents in terms of loop and source currents, and invert these so as to express the loop and source currents in terms of branch currents. Write these equations in matrix form.

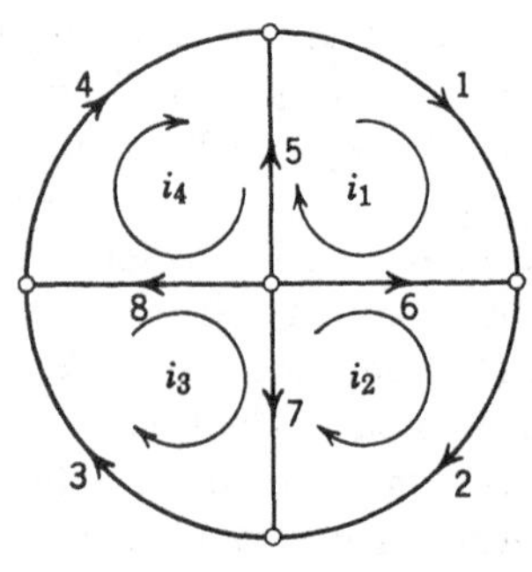

PROB. 1.

(*b*) Find algebraic relations expressing branch voltages in terms of node-pair and source voltages, and invert these so as to express node-pair and source voltages in terms of branch voltages. Write these equations in matrix form.

2. For the situation in prob. 1, assume that all branches are 1-ohm resistances. For the sources let

$$i_{s1} = 1, \quad i_{s2} = -1, \quad i_{s3} = 2, \quad i_{s4} = -4 \text{ amperes}$$

$$e_{s1} = 1, \quad e_{s2} = 2, \quad e_{s3} = -1, \quad e_{s4} = -2 \text{ volts}$$

(*a*) Write equilibrium equations on loop basis. Solve for loop currents and from these obtain values for all branch currents.

(*b*) Write equilibrium equations on node basis. Solve for node-pair voltages and from these obtain values for all branch voltages.

3. With reference to the graph of prob. 1, path choices are made for loop and source currents as indicated in the following sketches:

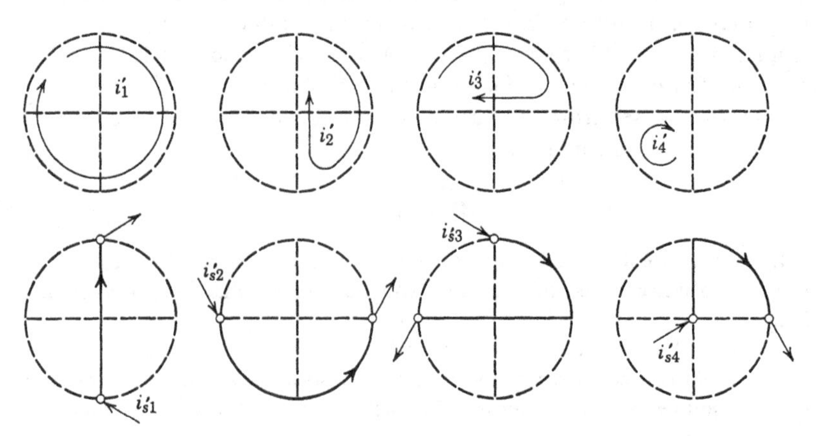

PROB. 3.

For these choices construct the compatible $[\alpha]$- and $[\beta]$-matrices (the augmented cut-set and tie-set matrices).

Voltage sources are e'_{s1}, e'_{s2}, e'_{s3} and e'_{s4} acting in the loops upon which the above loop currents circulate. Node-pair voltages are e'_1, e'_2, e'_3 and e'_4 at the terminals of the indicated current sources. Find the transformations

$$\begin{bmatrix} e'_s \\ \cdot\ \cdot \\ e'_v \end{bmatrix} = \tau \begin{bmatrix} e_s \\ \cdot\ \cdot \\ e_v \end{bmatrix} \quad \text{and} \quad \begin{bmatrix} i'_v \\ \cdot\ \cdot \\ i'_s \end{bmatrix} = \delta \begin{bmatrix} i_v \\ \cdot\ \cdot \\ i_s \end{bmatrix}$$

expressing sources and variables as defined in the above sketches in terms of those pertinent to probs. 1 and 2. What relationship exists between the transformation matrices τ and δ?

4. Choose the same loop-current paths as in prob. 3 but let the source-current paths be determined by the tree chosen in prob. 1. Again construct the compatible $[\alpha]$- and $[\beta]$-matrices as well as the pertinent transformation matrices τ and δ as defined in prob. 3. The primed quantities represent sources and variables as defined in this problem. Show that the same relation between τ and δ holds as in prob. 3.

5. For the original source values and branch resistances given in prob. 2, write loop equilibrium equations involving the loop currents defined in prob. 3. Prove that their solution yields the same branch-current values found in prob. 2. Write node equilibrium equations involving the node pairs and current sources defined in prob. 3. Prove that their solution yields the same branch voltages found in prob. 2.

6. The algebraic definitions for loop currents involve one or the other of the two matrices $[\alpha_{lb}]$ given below:

$$(a) \quad [\alpha_{lb}] = \begin{bmatrix} 1 & -1 & 1 & 0 & -1 & 1 & -1 & 0 \\ 0 & 1 & -1 & 0 & 0 & -1 & 0 & -1 \\ -1 & 1 & -1 & 1 & 1 & -1 & 1 & 0 \\ 0 & 0 & 0 & 0 & 0 & -1 & 0 & -1 \end{bmatrix}$$

$$(b) \quad [\alpha_{lb}] = \begin{bmatrix} -1 & 1 & 0 & 1 & 0 & 0 & 0 & 0 \\ 1 & 0 & 0 & -1 & 0 & 0 & 0 & 0 \\ 1 & -1 & 0 & 0 & 0 & 0 & 0 & 0 \\ 1 & -1 & 1 & -1 & 0 & 0 & 0 & 0 \end{bmatrix}$$

Determine the corresponding tie-set matrices yielding the topological definitions for these loop currents. The pertinent graph is that in prob. 1.

Obtain these results first by using the canonic form for $[\alpha_{nb}]$ and then another form chosen at random.

7. Algebraic definitions for node-pair voltages involve one or the other of the two matrices $[\beta_{nb}]$ given below:

$$(a) \quad [\beta_{nb}] = \begin{bmatrix} 0 & 0 & 0 & 0 & 1 & 0 & -1 & 0 \\ 0 & -1 & -1 & 0 & 0 & 0 & 0 & 0 \\ 1 & 0 & 0 & 0 & 0 & -1 & 0 & 1 \\ 1 & 0 & 0 & 0 & 1 & 0 & 0 & 0 \end{bmatrix}$$

$$(b) \quad [\beta_{nb}] = \begin{bmatrix} 0 & 0 & 0 & 0 & 1 & 0 & -1 & 0 \\ 0 & 0 & 0 & 0 & 0 & 1 & 0 & -1 \\ 0 & 0 & 0 & 0 & -1 & 0 & 0 & 1 \\ 0 & 0 & 0 & 0 & 0 & 1 & 0 & 0 \end{bmatrix}$$

Determine the corresponding cut-set matrices $[\alpha_{nb}]$, first by associating the canonic form for $[\beta_{lb}]$ and then another form chosen at random.

8. For the graph of prob. 1 the following tie-set matrix is given:

$$[\beta_{lb}] = \begin{bmatrix} 1 & 1 & 1 & 1 & 0 & 0 & 0 & 0 \\ 1 & 1 & 0 & 0 & 1 & 0 & -1 & 0 \\ 1 & 0 & 0 & 1 & 0 & -1 & 0 & 1 \\ 0 & 0 & 1 & 0 & 0 & 0 & 1 & -1 \end{bmatrix}$$

together with either matrix $[\beta_{nb}]$ given in prob. 7. The pertinent graph is again that in prob. 1. For each case (*a*) and (*b*), find the corresponding algebraic definitions for loop currents. The answers are given by the matrices $[\alpha_{lb}]$ in prob. 6; see if you can check these results.

9. Consider the cut-set matrix for a given graph written in terms of node-pair voltages. The columns yield relations for branch voltages in terms of these node-pair voltages. n independent columns yield equations that can be solved for the node-pair voltages in terms of branch voltages; and since these solutions, on physical grounds, involve only coefficients that are $+1$, -1 or zero, we may conclude that if the determinant of the n independent equations has a value N, then all of its first-order cofactors can have only the value $+N$, $-N$ or zero.

If we delete one or more rows in the cut-set matrix and consider the number n correspondingly reduced (the graph may lose some branches in this process), the same conclusions still apply; and if we continue until $n = 2$ so that the pertinent determinant has two rows and columns and the cofactor only one, then, clearly, N must be $+1$, -1 or zero for the cofactor and hence also for the determinant. But since this determinant is a cofactor for the case $n = 3$, the determinant for $n = 3$ must also equal $+1$, -1 or zero.

In this manner prove that any cut-set matrix has the property that all of its determinants and subdeterminants can equal only $+1$, -1 or zero; and by duality prove the same for any tie-set matrix.

Incidentally, if the property is proved for a matrix in canonic form, it is true generally, since linear row combinations have no effect upon the value of a determinant.

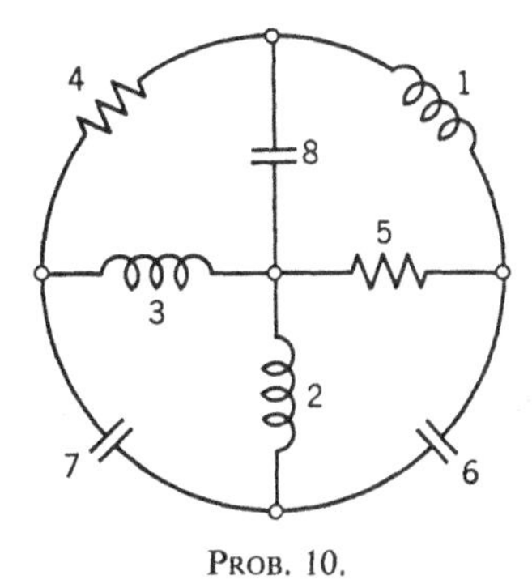

PROB. 10.

10. The network in the adjacent sketch has parameter matrices

$$[l] = \begin{bmatrix} 1 & -1 & 1 \\ -1 & 2 & -1 \\ 1 & -1 & 2 \end{bmatrix}$$

$$[r] = \begin{bmatrix} 2 & 0 \\ 0 & 4 \end{bmatrix}; \quad [s] = \begin{bmatrix} 1 & 0 & 0 \\ 0 & 2 & 0 \\ 0 & 0 & 3 \end{bmatrix}$$

Branch arrows are the same as in the graph of prob. 1. Choose branches 1, 2, 5, 7 as a tree and identify these branch voltages respectively with node-pair voltages e_1, e_2, e_3, e_4. Identify the currents in links 3, 4, 6, 8 respectively with loop currents i_1, i_2, i_3, i_4. Let current sources across node pairs and voltage sources in series with branches be as indicated in the sketches below:

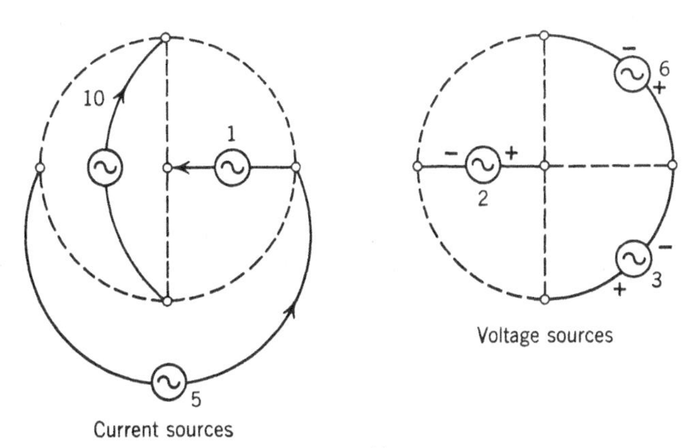

PROB. 10.

Obtain the parameter matrices and equilibrium equations on loop and node bases.

11. In the circuit of prob. 10 with sources as defined there, let loop currents be defined algebraically by the equations

$$\begin{aligned} j_1 - j_2 + j_3 &= i_1 \\ j_1 + 2j_2 + 3j_3 &= i_2 \\ 2j_1 - 3j_2 + j_3 &= i_3 \\ j_4 &= i_4 \end{aligned}$$

Find corresponding symmetrical loop-parameter matrices and equilibrium equations.

12. In the circuit of prob. 10 with sources as defined there, let node-pair voltages be defined algebraically by the equations

$$\begin{aligned} v_1 + v_2 - v_3 + v_4 &= e_1 \\ v_2 - v_3 + v_4 &= e_2 \\ -v_3 + v_4 &= e_3 \\ v_4 &= e_4 \end{aligned}$$

Find corresponding symmetrical node parameter matrices and equilibrium equations.

13. For the graph in prob. 1 and the pertinent matrices $[\alpha_{nb}]$ and $[\beta_{nb}]$, assume that branches 1, 2, 3, 4 are self-inductances having these values while branches 5, 6, 7, 8 are elastances with these corresponding values. Assume sources as in prob. 2 and obtain equilibrium equations in matrix form on a mixed basis.

14. A four-terminal and a three-terminal device, as shown in the following sketches:

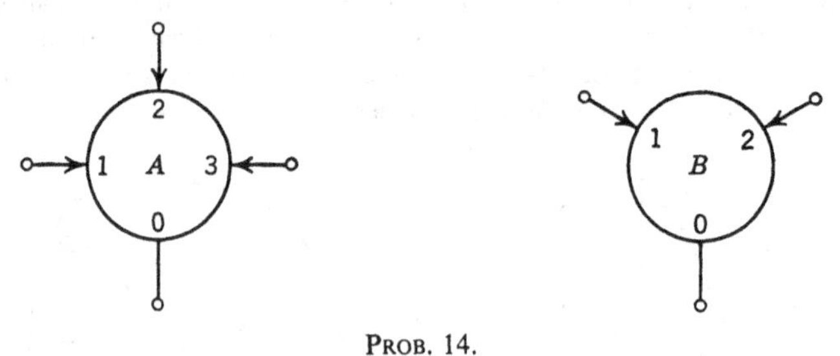

PROB. 14.

are characterized by the following volt-ampere relations in which currents are those entering respective terminals and voltages are drops from these terminals to the corresponding datum node 0:

$$\begin{aligned} v_1 &= j_1 + j_2 + j_3 \qquad & j_1 &= v_1 + \tfrac{1}{2}v_2 \\ v_2 &= j_2 + j_3 \\ v_3 &= -j_1 + j_3 \qquad & j_2 &= -2v_1 + v_2 \end{aligned}$$

These devices are embedded in a network having the graph shown in sketch (*a*) below which is to be replaced by the topologically equivalent graph in sketch (*b*).

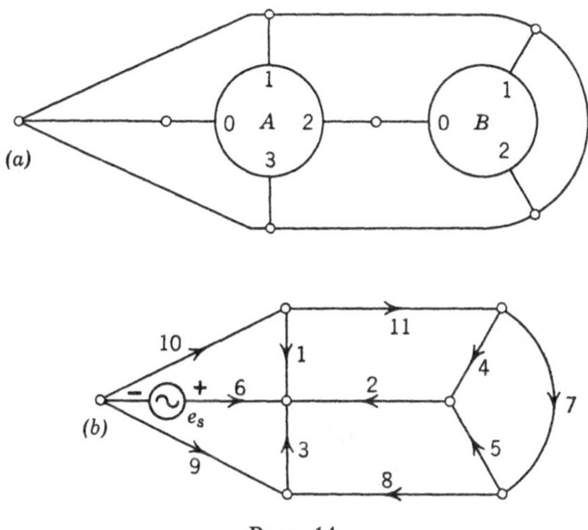

PROB. 14.

Assume that branches 6 through 11 are 1-ohm resistances and that a unit voltage source is located in branch 6 as shown.

In the equivalent graph choose branches 1 through 6 as a tree, branches 7 through 11 as links. Identifying tree branches with node-pair voltages and link currents with loop currents, determine the equilibrium equations for this network on both loop and node bases.

CHAPTER III

Calculation and Characterization of Response

1. Piecewise Analysis of Networks

When the response of a network involving a very large number of branches is to be evaluated, it is necessary to solve a correspondingly large number of simultaneous algebraic equations—a chore that is not too pleasant to contemplate, especially if machine aids are not available. In any case, it seems worthwhile to consider the possibility of separating a given large network into several smaller pieces by plucking judiciously chosen branches out of nodes (like pulling petals from a daisy), analyzing the pieces separately, and then forming the total solution by the proper combination of these separate solutions. Thoughts accompanying the discussion in art. 6, Ch. II suggest a simple way of accomplishing this end.

Suppose that a portion of a large network is detachable therefrom by plucking branches out of $p + 1$ nodes which we refer to as *terminal nodes* for the detached portion. For this portion we write equilibrium equations, for example, on a node-to-datum basis choosing one terminal node as datum, numbering the rest of the terminal nodes consecutively from 1 to p, and continuing the numbering with the remaining nodes inside the detached portion. Excitation is assumed in the form of p current sources acting from datum to the first p nodes.

We now suppress all but the first p node voltages in a familiar manner* and have a set of equations like 137, Ch. II left in which currents $j_1 \cdots j_p$ are identified with the known sources, and voltages $v_1 \cdots v_p$ with the p node-to-datum voltages. We can now re-embed this network portion in the large network and write equilibrium equations for the total network as described in art. 6, Ch. II.

We can in this manner excise several portions, analyze each separately as just described, and after re-embedding them in the total network, treat each like a multi-terminal device.

The method obviously is not restricted to pure resistance networks since all of the above discussion applies formally to impedances or admittances as well as to resistances.

* See art. 4, Ch. I or *ICT*, pp. 531–532.

Another more simple-minded way of apparently achieving a similar result as with this piecewise method of analysis is merely to suppress nodes by means of repeated star-mesh transformations. Computationally, however, this process can become rather involved because the total number of branches between nodes, although initially not very great, may become excessively large, for each star-mesh transformation increases the number of branches since, in the resulting "mesh," each node to which a ray of the pertinent star attaches is connected with every other such node by a branch. The method based upon the idea of multi-terminal devices as described above is not plagued with such a "chain reaction" regarding the creation of branches between nodes.

2. Determinant Evaluation in Terms of Link Sets or Tree Sets

In the process of solving the equilibrium equations on a loop or node basis it becomes necessary to evaluate the determinant of this set of simultaneous equations. In terms of topological reasoning, one can determine the value of this determinant directly from the network graph and the impedances or admittances of its branches. Although there is no computational advantage in this method of evaluation (in fact it is usually more tedious than need be) it directs our attention upon network properties that are collaterally both interesting and useful.

In the following detailed discussion of this method we will assume a pure resistive network because we can then focus all of our attention upon the topological reasoning involved. It is obvious that the results apply as well to branches characterized by impedances or admittances.

Suppose we begin with a set of loop equilibrium equations for which the matrix is given by

$$[R] = [\beta_{lb}] \cdot [r] \cdot [\beta_{lb}]_t \tag{1}$$

in which $[\beta_{lb}]$ is the tie-set matrix having l rows and b columns and $[r]$ is the diagonal branch-resistance matrix of order b. We are interested in the value of the determinant R of the matrix $[R]$. It is evidently a function of the branch resistances r_k which are the diagonal elements in the matrix $[r]$.

With reference to the pertinent network graph, we regard all branches and hence all branch resistances as separated into two groups: (1) those belonging to a chosen tree, and (2) those belonging to the corresponding links determined by this tree. The branches in group 1 we shall refer to as a *tree* and those in group 2 as a *link set*.

For any given graph there is usually more than one way of choosing a tree. The number of possible ways in which a tree can be chosen is, however, always finite and enumerable, although this number may often be extremely large and the process of enumeration long and tedious.

Part (*a*) of Fig. 1 shows a simple five-branch graph in which eight possible trees, shown in part (*b*) of this same figure, may be drawn.

The purely combinatorial problem of determining the number of possible trees for a given graph and of drawing all of their configurations we shall not discuss here. As a by-product of our main topic, however, we will find an interesting algebraic method for determining the number of possible trees for any network graph. Knowing their number, it is possible to draw them by trial.

For a given network graph we thus have a definite number of trees and corresponding link sets. We shall now show that the determinant R is

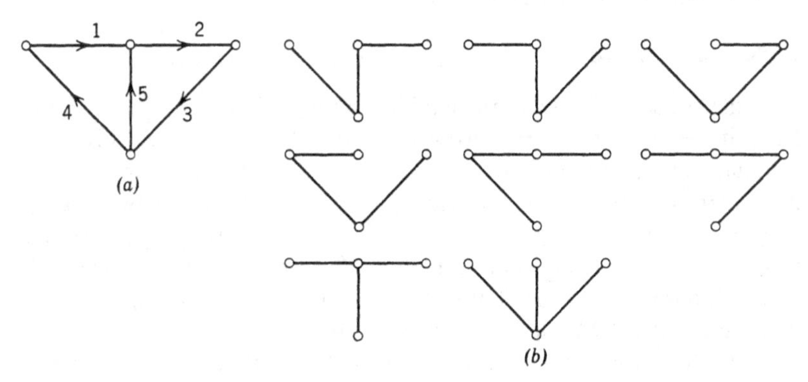

FIG. 1. A 4-node graph (*a*) and its enumerable trees (*b*).

given by a sum of as many terms as there are link sets (or enumerable trees), and that each term is a product of resistances in the branches of a distinct link set. If for the moment we use the term "link set" to mean the product of resistances in a link set, then we can state more simply that the determinant R equals the sum of all link sets.

We will prove this statement by showing: (*a*) that an expression for the determinant can contain no terms other than link sets; and (*b*) that no link sets can be missing in the complete expression for the determinant.

First we need to remind ourselves of some rather obvious properties of network determinants. The conventional determinant evaluation leads to a sum of terms, each of which is a product of l elements (the order of the determinant R being l). According to Eq. 1 each element is a *linear* combination of branch resistances. Therefore, *no term in the determinant evaluation can involve the product of more than* l *branch resistances.* Equation 1 makes it clear, moreover, that if all branch resistances on any tie set are zero, then the corresponding row and column of the determinant are composed wholly of zeros, and the value of this determinant is therefore zero. Observe, however, that the determinant cannot be zero unless *all* resistances on a tie set are zero.

In order to see this, consider a tie-set schedule A based upon l independent loops. In the pertinent graph we can always choose a tree in such a way that a tie-set schedule B based upon it has one row that is identical with any chosen row in schedule A. Moreover, any one of the nonzero elements in this row can represent a link and the remaining ones tree branches. In schedule B, any columns pertaining to tree branches may be deleted without destroying the independence of rows. Since the rows in schedule A are independent linear combinations of the rows in schedule B, deletion of these same tree-branch columns in schedule A likewise leaves its rows independent. Therefore, an abridged version of schedule A corresponding to short-circuiting all but one branch on any chosen loop or tie set still has l independent rows, and hence its Gramian determinant cannot be zero.

From our basic knowledge about trees and links we can say that a link set has the fundamental property that if all of its l branches are opened, no closed paths remain in the network. Furthermore, it is clear that if a set of l branches has this property, it is a link set and if it does not have this property then it is not a link set. In fact we may adopt this statement as a definition for what we mean by a link set, since it is neither stronger nor weaker than the definition of what we mean by a set of links in the first place.

Every link set must contain at least one of the branches in an *arbitrarily* chosen tie set (closed path), for if this were not so, then opening all branches in such a pseudo link set would still leave a closed path in the network, which is impossible (by definition) if the set is truly a link set. No other set of l (or fewer) branches can have this property, since opening all of its branches leaves at least one closed path in the network.

It follows, therefore, that setting *all* resistances in *any* tie set equal to zero is sufficient to cause *all* link sets to vanish; and since setting all resistances in any tie set equal to zero is also sufficient to cause the determinant to vanish, it follows that an expression for the determinant can contain no terms other than link sets because any other term containing l (or fewer) resistances would remain nonzero even though all resistances on a properly chosen tie set were zero. Hence statement (*a*) above is proved.

Regarding statement (*b*), we observe first that each link set contains a unique combination of branches; therefore *all* of the resistances in any one link set do not also appear in any of the others; and so it is possible to have all resistances in one link set nonzero even though one or more resistances in the others are zero. In other words, it is possible for all but one of the link sets to vanish and yet have *all* resistances in the remaining one be nonzero.

As long as one link set has all nonzero resistances, there can be no tie sets completely zero (because it is impossible to find a tie set that does not contain at least one of the nonzero resistances) and hence the determinant cannot be zero. If, in the expression for the determinant, one link set were missing it would be possible to cause the determinant to be zero even though all resistances in the missing link set were nonzero, which means that the determinant could become zero even though there were *no* completely zero tie sets. This is impossible, and so statement (*b*) is proved.

In terms of this method of determinant evaluation we can easily determine the number of trees that may be drawn for any given network graph.* Thus, if we assume that each branch in the network is a 1-ohm resistance, then each link set yielding a term in the expression for the determinant has value 1; and hence the value of the determinant equals the number of link sets, which equals the number of trees. Since the determinant in this case is relatively easy to evaluate, it is correspondingly simple to find the number of trees.

In the example of Fig. 1, we have for the tie-set matrix (choosing j_1 and j_2 as loop currents)

$$[\beta_{lb}] = \begin{bmatrix} 1 & 0 & 0 & 1 & -1 \\ 0 & 1 & 1 & 0 & 1 \end{bmatrix} \tag{2}$$

Hence on a one-ohm-per-branch basis

$$[R] = [\beta_{lb}] \cdot [\beta_{lb}]_t = \begin{bmatrix} 3 & -1 \\ -1 & 3 \end{bmatrix} \tag{3}$$

from which $R = 8$, which checks the number of trees in part (*b*) of Fig. 1.

On a node basis we have in place of Eq. 1 for the short-circuit conductance matrix

$$[G] = [\alpha_{nb}] \cdot [g] \cdot [\alpha_{nb}]_t \tag{4}$$

in which $[\alpha_{nb}]$ is the pertinent cut-set matrix and $[g]$ is the diagonal branch-conductance matrix as described above. The value of the determinant G is given by a sum of all tree sets, where by the term "tree set" in this statement we mean the product of conductances in the branches of a tree.

The proof here is the complete dual of the argument just presented for the evaluation of R. We shall let the reader spell out the details as an exercise.

* First pointed out by Professor Samuel J. Mason of the Electrical Engineering Department, M.I.T.

3. Response Functions Directly Related to Branch Conductances

The topic of the preceding article and its logical continuation have as an objective the development of a procedure whereby one can write down the expression for a driving-point or transfer impedance (or admittance) directly in terms of the branch impedances (or admittances) in the given network, and thus circumvent the necessity, in many cases, for writing equilibrium equations or inverting their matrix. The basic thought here is one of practicality, namely that we hope by such a tactic to avoid much of the usual computational drudgery that accompanies the construction of a solution to a network problem.

In this regard the method which extends the process discussed in the previous article, fails to accomplish its desired objective because the computational tedium associated with it is as great if not greater than that which attends a conventional process of analysis. A major reason for this fact stems from the difficulty involved in the process of enumerating trees and from the fact that their number grows astronomically with the complexity of a network. Computing aids are of little help here either since no simple program has yet been devised for enumeration of trees for a given graph. Moreover, if machine computation is available, there is little incentive to look far beyond well-known existing methods of network analysis.

So-called "short methods" or "direct" methods, therefore, are actually more useful for the additional light that they shed upon the entire network picture than they are in providing shortcuts in the numerical computation of network response. In the immediately following discussion we present such an unconventional approach which incidentally has a decided computational advantage in situations where only a single driving-point or transfer impedance is of interest. Again we shall for the sake of simplicity assume a pure resistive network.

We begin by devising simple ways of relating topology and branch conductance values to the node conductance matrix $[G]$ and vice versa. The number of related node-pair voltages may be equal to or less than the number of branches in a tree appropriate to the given network graph, since one need not consider all of the independent node pairs to be points of access. If we do consider all of them to be accessible node pairs then a node conductance matrix of order n pertains to a network having a total of $n + 1$ nodes (for which the number of tree branches equals n). We shall for the time being confine our attention to this special situation.

Equation 4 then is a formal expression for the node conductance matrix $[G]$ in terms of branch conductances which are the diagonal elements in $[g]$. Since $[\alpha_{nb}]$ can be chosen in a large number of ways for the same network, characterization of the precise form and properties of $[G]$ is not simple. However, for a fixed geometrical tree configuration in a full graph, $[\alpha_{nb}]$ is essentially fixed in form except for a rearrangement of rows (the arrangement of columns being immaterial since it affects only the identities of the diagonal elements in $[g]$).

In this connection it should be clearly understood that the number of distinct *geometrical* tree configurations is far less than the number of trees as distinguished in the previous article and illustrated there in Fig. 1. While the number of possible trees associated with the graph shown there is *eight*, the number of distinct geometrical tree configurations is only *two*: a "linear" tree in which successive branches have one node in common, and a "starlike" tree in which all branches have the same node in common. The number of geometrical tree configurations for n tree branches equals the number of distinct geometrical patterns constructible with n matchsticks.

For a fixed tree geometry, $[G]$ can vary only in a rearrangement of its rows and columns; and a change in the reference direction for a tree-branch or node-pair voltage merely causes all element values in a corresponding row and column to change. We will regard such changes in $[G]$ as not altering its fundamental form and hence consider the number of distinct fundamental forms of $[G]$ as being equal to the number of distinct geometrical tree configurations constructible for a given order n.

Out of the variety of possible tree configurations, two particular ones are of primary interest: the *starlike* tree and the *linear* tree. The first of these implies a node-to-datum set of node-pair voltages and gives rise to the so-called "dominant" matrix $[G]$ in which all nondiagonal terms are negative, the diagonal ones positive, and sums of elements in any row or column are non-negative.* As mentioned above, sign changes resulting from the multiplication of corresponding rows and columns by -1 are disregarded since they can easily be recognized and corrected.

Figure 2 shows a graph for $n = 4$ in which the branches of the starlike tree (drawn with solid lines) are numbered 11, 22, 33, 44, having the common (datum) node 0. The tips of the star are nodes 1, 2, 3, 4. The links (drawn dotted) are numbered 12, 13, 14, etc. according to the nodes which they connect. This is a so-called "full" graph, in that branches connect each node with all other nodes, the total number of branches being $n(n + 1)/2$, the same as the total number of distinct elements in the

* In a wider sense the term "dominant" is used to designate a matrix in which the diagonal terms are positive and as large or larger than the sums of the absolute values of the nondiagonal terms in the same row. We find our less general definition more useful because it exactly fits the familiar node-to-datum situation.

symmetrical matrix $[G]$ of order n. Conductance values of the branches will be denoted by $\overset{*}{g}_{ik}$ corresponding to the branch numbering in this figure, the asterisk above the letter g signifying that these are conductances in a starlike tree.

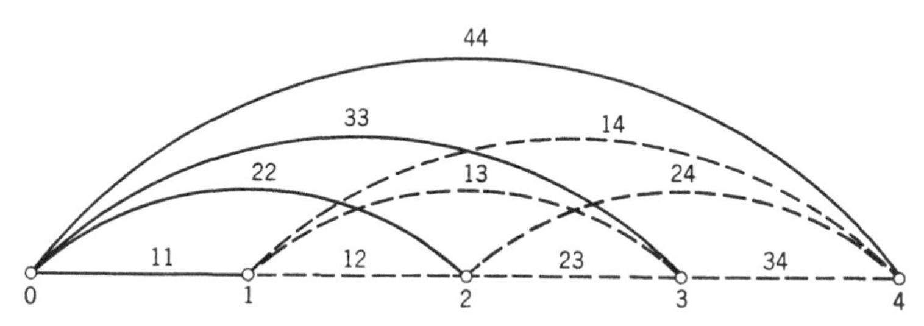

FIG. 2. A 5-node full graph with starlike tree.

Branch numbering for the same graph with a linear tree is shown in Fig. 3. Tree branches are numbered 11, 22, 33, 44, the same as for the starlike tree. Numbering of the links is chosen so that the relations between branch conductances and elements in the corresponding matrix

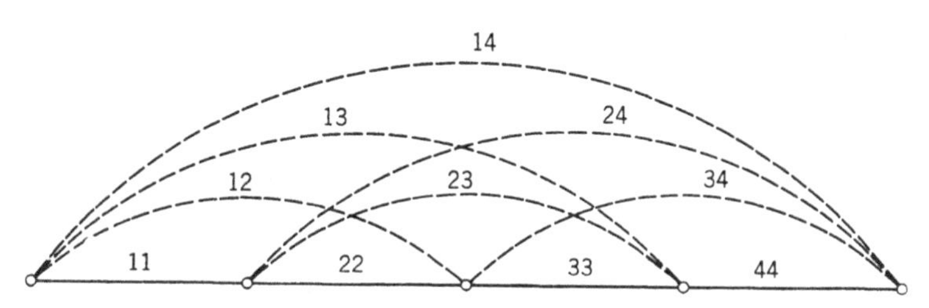

FIG. 3. A 5-node full graph with linear tree.

$[G]$ assume a simple and effective form, as is shown presently. Conductance values here are denoted by g_{ik}, with no additional distinguishing mark.

Equilibrium equations for the network with starlike tree are written

$$[\overset{*}{G}] \cdot \overset{*}{e}] = \overset{*}{i}_s] \tag{5}$$

and for the same network with a linear tree they are given by

$$[G] \cdot e] = i_s] \tag{6}$$

Node-pair voltages in one of these is expressed in terms of those in the other by a linear transformation that reads

$$e] = [T] \cdot \overset{*}{e}] \tag{7}$$

where by inspection of Figs. 2 and 3 we see that the transformation matrix is

$$[T] = \begin{bmatrix} 1 & 0 & 0 & \cdots & 0 & 0 \\ -1 & 1 & 0 & \cdots & 0 & 0 \\ 0 & -1 & 1 & \cdots & 0 & 0 \\ \cdot & \cdot & \cdot & \cdot & \cdot & \cdot \\ 0 & 0 & 0 & \cdots & -1 & 1 \end{bmatrix} \tag{8}$$

with the inverse

$$[T]^{-1} = \begin{bmatrix} 1 & 0 & 0 & \cdots & 0 & 0 \\ 1 & 1 & 0 & \cdots & 0 & 0 \\ 1 & 1 & 1 & \cdots & 0 & 0 \\ \cdot & \cdot & \cdot & \cdot & \cdot & \cdot \\ 1 & 1 & 1 & \cdots & 1 & 1 \end{bmatrix} \tag{9}$$

Substituting 7 into 6 and premultiplying by the transpose of $[T]$ gives

$$[T]_t \cdot [G] \cdot [T] \cdot \overset{*}{e}] = [T]_t \cdot i_s] = \overset{*}{i_s}] \tag{10}$$

in which the identification with $\overset{*}{i_s}]$ on the right-hand side is required by the familiar condition of power invariance. Comparison with Eq. 5 now shows that

$$[\overset{*}{G}] = [T]_t \cdot [G] \cdot [T] \tag{11}$$

In the node-to-datum arrangement with a starlike tree, the algebraic relations between branch conductances and elements $\overset{*}{G}_{ik}$ in the dominant matrix $[\overset{*}{G}]$ are rather obvious. They are given by

$$\overset{*}{g}_{ik} = -\overset{*}{G}_{ik} \qquad \text{for } i \neq k \tag{12}$$

and

$$\overset{*}{g}_{kk} = \sum_{\nu=1}^{n} \overset{*}{G}_{k\nu} \quad \text{or} \quad \overset{*}{G}_{kk} = \sum_{\nu=1}^{n} g_{k\nu} \tag{13}$$

Through use of 8, 9 and 11 we find that

$$[T] \cdot [G] \cdot [T] = [T] \cdot [T]_t^{-1} \cdot [\overset{*}{G}]$$

$$= \begin{bmatrix} 1 & 1 & 1 & 1 & \cdots & 1 & 1 \\ -1 & 0 & 0 & 0 & \cdots & 0 & 0 \\ 0 & -1 & 0 & 0 & \cdots & 0 & 0 \\ 0 & 0 & -1 & 0 & \cdots & 0 & 0 \\ \cdot & \cdot & \cdot & \cdot & \cdot & \cdot & \cdot \\ 0 & 0 & 0 & 0 & \cdots & -1 & 0 \end{bmatrix} \cdot [\overset{*}{G}] \tag{14}$$

whereupon Eqs. 12 and 13 yield

$$[T]\cdot[G]\cdot[T] = \begin{bmatrix} \overset{*}{g}_{11} & \overset{*}{g}_{22} & \overset{*}{g}_{33} & \cdots & \overset{*}{g}_{nn} \\ & \overset{*}{g}_{12} & \overset{*}{g}_{13} & \cdots & \overset{*}{g}_{1n} \\ & & \overset{*}{g}_{23} & \cdots & \overset{*}{g}_{2n} \\ & & & \cdot\ \cdot\ \cdot & \cdot\ \cdot \\ & & & & \overset{*}{g}_{n-1,n} \end{bmatrix} \tag{15}$$

in which the region below the principal diagonal contains elements in which we have no interest. From a comparison of Figs. 2 and 3, noting particularly the difference in notation for the branch conductances, we see that if the right-hand side of 15 is written in terms of the branch conductances of a linear tree, this result reads

$$[T]\cdot[G]\cdot[T] = \begin{bmatrix} g_{11} & g_{12} & g_{13} & \cdots & g_{1n} \\ & g_{22} & g_{23} & \cdots & g_{2n} \\ & & g_{33} & \cdots & g_{3n} \\ & & & \cdot\ \cdot & \cdot\ \cdot \\ & & & & g_{nn} \end{bmatrix} \tag{16}$$

representing a relationship between branch conductances and the node-conductance matrix appropriate to a linear tree that is comparable in simplicity with relations 12 and 13 pertinent to a conductance matrix based upon a starlike tree or node-to-datum variables.

If we tilt the matrix in Eq. 16 so that its principal diagonal becomes horizontal, then the identification of elements in this matrix with pertinent branches in the pyramidal form of graph shown in Fig. 3 becomes strikingly evident, as does also the fact that the number of distinct elements in a conductance matrix $[G]$ of order n exactly equals the number of branches in a full graph with n tree branches.

Recognition of this fact makes it clear that we can always obtain branch conductance values appropriate to a given cut-set matrix $[\alpha_{nb}]$ and node conductance matrix $[G]$, by writing the matrix of Eq. 4 in the equivalent algebraic form

$$G_{ik} = \sum_{\nu=1}^{b} \alpha_{i\nu}\alpha_{k\nu}g_{\nu} \qquad \text{for } k \geq i = 1, 2, \ldots n \tag{17}$$

in which α_{sk} are elements of the cut-set matrix and g_ν are elements in the diagonal branch-parameter matrix $[g]$ in Eq. 4. Solution of this set of $n(n + 1)/2$ simultaneous equations for the same number of unknown g_ν's is straightforwardly possible and leads (after appropriate manipulation) to the same result as that given by Eq. 16.

This result affords an equally simple numerical procedure for computing G_{ik}'s from g_{ik}'s or vice versa since the operation on rows and columns demanded by the transformation matrix $[T]$ or $[T]^{-1}$ is easily carried out. Suppose we denote the matrix 16 by $[\bar{g}]$ and assume as an example

$$[\bar{g}] = \begin{bmatrix} 1 & 2 & 3 & 2 & 1 \\ & 1 & 2 & 2 & 1 \\ & & 1 & 2 & 1 \\ & & & 1 & 1 \\ & & & & 1 \end{bmatrix} \tag{18}$$

Elements below the principal diagonal are of no interest and are not involved in the manipulations.

Noting $[T]^{-1}$ in Eq. 9 we have by inspection

$$[T]^{-1} \cdot [\bar{g}] = \begin{bmatrix} 1 & 2 & 3 & 2 & 1 \\ & 3 & 5 & 4 & 2 \\ & & 6 & 6 & 3 \\ & & & 7 & 4 \\ & & & & 5 \end{bmatrix} \tag{19}$$

Here we write down the first row as in 18, then add to this one the second row in 18 to form the second row in 19, then add to this one the third row in 18 to form the third in 19, and so forth.

Having completed formation of 19 we construct the following matrix by columns:

$$[T]^{-1} \cdot [\bar{g}] \cdot [T]^{-1} = \begin{bmatrix} 9 & 8 & 6 & 3 & 1 \\ & 14 & 11 & 6 & 2 \\ & & 15 & 9 & 3 \\ & & & 11 & 4 \\ & & & & 5 \end{bmatrix} = [G] \tag{20}$$

where we begin by writing down the last column in 19, then add to this one the fourth column in 19 to form the fourth in 20, and so forth as was done with the rows.

Proceeding in the opposite direction we obtain from 20

$$[T]\cdot[G] = \begin{bmatrix} 9 & 8 & 6 & 3 & 1 \\ & 6 & 5 & 3 & 1 \\ & & 4 & 3 & 1 \\ & & & 2 & 1 \\ & & & & 1 \end{bmatrix} \tag{21}$$

Here the first row is the same as in 20. The second row in 21 is the second minus the first in 20 (ignoring the absent term); the third row in 21 is the third minus the second in 20; the fourth is the fourth minus the third in 20, and so forth.

Finally we form

$$[T]\cdot[G]\cdot[T] = \begin{bmatrix} 1 & 2 & 3 & 2 & 1 \\ & 1 & 2 & 2 & 1 \\ & & 1 & 2 & 1 \\ & & & 1 & 1 \\ & & & & 1 \end{bmatrix} = [\bar{g}] \tag{22}$$

by carrying out analogous operations on columns, and thus regain the matrix 18 that we started out with.

This process is about as easy to perform, and the conditions leading to positive elements in $[\bar{g}]$ (which is of interest in a synthesis problem) are almost as easily recognizable as are those pertaining to a dominant $[G]$ matrix appropriate to a graph with starlike tree. These conditions on a $[G]$-matrix appropriate to a linear tree will be implied by designating $[G]$ to be a "uniformly tapered" matrix, the reason for the choice of this term being evident in the numerical example just given.

Having formed the matrix $[G]$ for a given graph and its branch-conductance values, we are next interested in evaluating elements in the inverse of $[G]$—the open-circuit driving-point and transfer impedances—for those are the desired response functions of our network. The following manipulations are aimed at devising a computational scheme whereby these impedances may be obtained directly from the branch-conductance values of the given network with a minimum number of additions, multiplications and divisions. In this derivation the relationships pertinent to a $[G]$-matrix appropriate to the linear tree will play a significant role.

We begin by writing for the symmetrical matrix $[G]$ the representation

$$[G] = [A]\cdot[A]_t \tag{23}$$

and assume for $[A]$ the triangular form

$$[A] = \begin{bmatrix} a_{11} & 0 & 0 & \cdots & 0 \\ a_{21} & a_{22} & 0 & \cdots & 0 \\ a_{31} & a_{32} & a_{33} & \cdots & 0 \\ \cdot & \cdot & \cdot & \cdot & \cdot \\ a_{n1} & a_{n2} & a_{n3} & \cdots & a_{nn} \end{bmatrix} \tag{24}$$

In view of Eq. 16, noting the form of $[T]$ in Eq. 8, we construct the products

$$[T]\cdot[A] = \begin{bmatrix} a_{11} & 0 & 0 & \cdots & 0 \\ (a_{21} - a_{11}) & a_{22} & 0 & \cdots & 0 \\ (a_{31} - a_{21}) & (a_{32} - a_{22}) & a_{33} & \cdots & 0 \\ \cdot & \cdot & \cdot & \cdot & \cdot \\ (a_{n1} - a_{n-1,1}) & (a_{n2} - a_{n-1,2}) & (a_{n3} - a_{n-1,3}) & \cdots & a_{nn} \end{bmatrix} \tag{25}$$

and

$$[A]_t\cdot[T] = \begin{bmatrix} (a_{11} - a_{21}) & (a_{21} - a_{31}) & (a_{31} - a_{41}) & \cdots & (a_{n-1,1} - a_{n1}) & a_{n1} \\ -a_{22} & (a_{22} - a_{32}) & (a_{32} - a_{42}) & \cdots & (a_{n-1,2} - a_{n2}) & a_{n2} \\ 0 & -a_{33} & (a_{33} - a_{43}) & \cdots & (a_{n-1,3} - a_{n3}) & a_{n3} \\ \cdot & \cdot & \cdot & \cdot & \cdot & \cdot \\ 0 & 0 & 0 & \cdots & -a_{nn} & a_{nn} \end{bmatrix} \tag{26}$$

The second of these matrices is almost the negative transpose of the first. In fact, if we add to the first matrix a last row with the elements $-a_{n1}, -a_{n2}, \ldots -a_{nn}$, and then ignore the first row, its negative transpose is the second matrix. This fact suggests that we consider the matrix 25 with the stated additional row, namely the matrix with $n + 1$ rows and n columns given by

$$[H] = \begin{bmatrix} a_{11} & 0 & 0 & \cdots & 0 \\ a_{21} - a_{11} & a_{22} & 0 & \cdots & 0 \\ a_{31} - a_{21} & a_{32} - a_{22} & a_{33} & \cdots & 0 \\ \cdot & \cdot & \cdot & \cdot & \cdot \\ a_{n1} - a_{n-1,1} & a_{n2} - a_{n-1,2} & a_{n3} - a_{n-1,3} & \cdots & a_{nn} \\ -a_{n1} & -a_{n2} & -a_{n3} & \cdots & -a_{nn} \end{bmatrix} \tag{27}$$

In this matrix all columns add to zero; and if the vector set defined by rows is denoted by $h_0, h_1, h_2, \ldots h_n$, then these vectors form the sides of a closed polygon in n-dimensional space. The branch conductance values in the matrix 16 are given by the scalar products

$$g_{ik} = -h_{i-1} \cdot h_k \qquad \text{for } i \leq k \leq n \quad \text{and } i = 1, 2, \ldots n \tag{28}$$

More specifically, if we designate the elements in H as indicated by writing

$$[H] = \begin{bmatrix} h_{01} & 0 & 0 & \cdot & 0 \\ h_{11} & h_{12} & 0 & \cdot & 0 \\ h_{21} & h_{22} & h_{23} & \cdot & \cdot \\ \cdot & \cdot & \cdot & \cdot & \cdot \\ \cdot & \cdot & \cdot & \cdot & h_{n-1,n} \\ h_{n1} & h_{n2} & h_{n3} & \cdot & h_{nn} \end{bmatrix} \tag{29}$$

then we obtain for the g_{ik} the expressions

$$g_{ik} = -\sum_{\nu=1}^{i} h_{i-1,\nu} h_{k\nu} \qquad \text{for } k \geq i \tag{30}$$

Since we are interested in computing elements in the matrix $[H]$ from g_{ik} values in the given network, we manipulate 30 as follows. First we split off the last term in the sum and have

$$g_{ik} = -\sum_{\nu=1}^{i-1} h_{i-1,\nu}\, h_{k\nu} - h_{i-1,i}\, h_{ki} \tag{31}$$

or

$$-h_{i-1,i}\, h_{ki} = g_{ik} + \sum_{\nu=1}^{i-1} h_{i-1,\nu}\, h_{k\nu} \tag{32}$$

For a particular column of $[H]$ (fixed value of i) it will be noticed that the right-hand side of this equation involves only coefficients h in preceding columns (up to and including column $i - 1$). The formula 32 would thus be suited for the sequential computation of the h-coefficients if it were not for the factor $h_{i-1,i}$ on the left. This awkwardness may be removed by introducing quantities

$$p_{ki} = -h_{i-1,i}\, h_{ki} = g_{ik} + \sum_{\nu=1}^{i-1} h_{i-1,\nu}\, h_{k\nu} \tag{33}$$

Since the columns of $[H]$ add to zero, we have

$$\sum_{k=i-1}^{n} h_{ki} = 0 \quad \text{or} \quad h_{i-1,i} = -\sum_{k=i}^{n} h_{ki} \tag{34}$$

and hence

$$\sum_{k=i}^{n} p_{ki} = -h_{i-1,i} \sum_{k=i}^{n} h_{ki} = h_{i-1,i}^{2} \tag{35}$$

From Eq. 33

$$h_{ki} = \frac{p_{ki}}{-h_{i-1,i}} \tag{36}$$

Using this relation and 35 to form

$$h_{i-1,\nu}\, h_{k\nu} = \frac{p_{i-1,\nu}\, p_{k\nu}}{h_{\nu-1,\nu}^{2}} = \frac{p_{i-1,\nu}\, p_{k\nu}}{\sum_{k=\nu}^{n} p_{k\nu}} \tag{37}$$

we get, through substituting back into Eq. 33, the result

$$p_{ki} = g_{ik} + \sum_{\nu=1}^{i-1} \left(\frac{p_{i-1,\nu}\, p_{k\nu}}{\sum_{k=\nu}^{n} p_{k\nu}} \right) \qquad (k \geq i = 1, 2, \ldots n) \tag{38}$$

Note that for $i = 1$ the sum in this expression drops out and we have simply $p_{k1} = g_{1k}$. Elements in the matrix

$$[P] = \begin{bmatrix} p_{11} & 0 & 0 & \cdot & 0 \\ p_{21} & p_{22} & 0 & \cdot & 0 \\ p_{31} & p_{32} & p_{33} & \cdot & 0 \\ \cdot & \cdot & \cdot & \cdot & \cdot \\ \cdot & \cdot & \cdot & \cdot & \cdot \\ p_{n1} & p_{n2} & p_{n3} & \cdot & p_{nn} \end{bmatrix} \tag{39}$$

may be calculated sequentially by formula 38 starting with the first column, then continuing with the second, and so forth, since for any fixed value of the index i, this formula involves only elements in columns 1 to $i - 1$. The sum in the denominator of the summand in 38 is simply the sum of all elements in the νth column. Hence in the computational procedure, each time the elements in an additional column are calculated, their sum may also be recorded below it, so that its value is readily available for computation of the next column.

Elements of the matrix $[A]$, Eq. 24, may now be expressed directly in terms of the p_{ki}. From the form of matrix $[H]$ in Eq. 27 and the notation in 29 we have first of all

$$a_{ik} = \sum_{\nu=k-1}^{i-1} h_{\nu k} = h_{k-1,k} + \sum_{\nu=k}^{i-1} h_{\nu k} \tag{40}$$

and with the help of 34 this gives

$$a_{ik} = \sum_{\nu=k}^{i-1} h_{\nu k} - \sum_{\nu=k}^{n} h_{\nu k} = -\sum_{\nu=i}^{n} h_{\nu k} \tag{41}$$

Now substituting for $h_{\nu k}$ from Eq. 36 and for $h_{k-1,k}$ from Eq. 35 we find

$$a_{ik} = \frac{\sum_{\nu=i}^{n} p_{\nu k}}{\sqrt{\sum_{\nu=k}^{n} p_{\nu k}}} \qquad \text{for } i \geq k = 1, 2, \ldots n \tag{42}$$

Appearance of the radical in this expression does not contradict the well-known requirement that response in a lumped network be a rational function of the branch conductances since the impedances which we shall presently compute are quadratic functions of the a_{ik}, as is also evident from Eq. 23. For this reason it is advisable not to compute the a_{ik}'s until an evaluation of the desired response function in terms of these coefficients is made and the radicals are eliminated.

If the inverse of the matrix $[A]$ is denoted by

$$[A]^{-1} = [B] = \begin{bmatrix} b_{11} & 0 & 0 & 0 & \cdots & 0 \\ b_{21} & b_{22} & 0 & 0 & \cdots & 0 \\ b_{31} & b_{32} & b_{33} & 0 & \cdots & 0 \\ \cdot & \cdot & \cdot & \cdot & \cdot & \cdot \\ b_{n1} & b_{n2} & b_{n3} & \cdot & \cdot & b_{nn} \end{bmatrix} \tag{43}$$

then we have the following relations between the a_{ik} and b_{ik}:

$$a_{11}b_{11} = 1 \tag{44}$$

$$\left.\begin{aligned} a_{11}b_{21} + a_{21}b_{22} &= 0 \\ a_{22}b_{22} &= 1 \end{aligned}\right\} \tag{45}$$

$$\left.\begin{aligned} a_{11}b_{31} + a_{21}b_{32} + a_{31}b_{33} &= 0 \\ a_{22}b_{32} + a_{32}b_{33} &= 0 \\ a_{33}b_{33} &= 1 \end{aligned}\right\} \tag{46}$$

and so forth. Or, starting at the opposite end

$$a_{nn}b_{nn} = 1 \tag{47}$$

$$\left.\begin{aligned} a_{n-1,n-1}b_{n,n-1} + a_{n,n-1}b_{nn} &= 0 \\ a_{n-1,n-1}b_{n-1,n-1} &= 1 \end{aligned}\right\} \tag{48}$$

$$\left.\begin{aligned} a_{n-2,n-2}b_{n,n-2} + a_{n-1,n-2}b_{n,n-1} + a_{n,n-2}b_{nn} &= 0 \\ a_{n-2,n-2}b_{n-1,n-2} + a_{n-1,n-2}b_{n-1,n-1} &= 0 \\ a_{n-2,n-2}b_{n-2,n-2} &= 1 \end{aligned}\right\} \tag{49}$$

and so forth. The open-circuit resistance matrix, which is the inverse of

[G], is given (according to Eqs. 23 and 43) by

$$[R] = [G]^{-1} = [B]_t \cdot [B] \tag{50}$$

The elements of [R] in which we are particularly interested are

$$r_{nn} = b_{nn}^2 \tag{51}$$

$$r_{n-1,n} = b_{n,n-1}b_{nn} \tag{52}$$

and

$$r_{n-2,n} = b_{n,n-2}b_{nn} \tag{53}$$

The first of these is a driving-point function; the other two are transfer functions. Because of the implied linear tree upon which the terminal pairs are based, we see more particularly that 52 is the open-circuit transfer impedance of a grounded two-terminal pair (three-terminal network) while 53 is the open-circuit transfer impedance of an arbitrary two-terminal-pair network since the two terminal pairs involved are not adjacent and hence do not have a terminal in common.

From 47, 51 and 42 we have for the driving point impedance (which, incidentally, through appropriate branch numbering may be the impedance across any chosen node pair in the given network) the surprisingly simple result

$$r_{nn} = \frac{1}{a_{nn}^2} = \frac{1}{p_{nn}} \tag{54}$$

For the transfer function 52 we find straightforwardly

$$\frac{r_{n-1,n}}{r_{nn}} = -\frac{a_{n,n-1}}{a_{n-1,n-1}} = \frac{-p_{n,n-1}}{(p_{n-1,n-1} + p_{n,n-1})} \tag{55}$$

and for the one given by Eq. 53 we get

$$\frac{r_{n-2,n}}{r_{nn}} = \frac{a_{n,n-1}a_{n-1,n-2} - a_{n,n-2}a_{n-1,n-1}}{a_{n-1,n-1}a_{n-2,n-2}}$$

$$= \frac{p_{n,n-1}p_{n-1,n-2} - p_{n,n-2}p_{n-1,n-1}}{(p_{n-1,n-1} + p_{n,n-1})(p_{n-2,n-2} + p_{n-1,n-2} + p_{n,n-2})} \tag{56}$$

Regarding numerical computation of these quantities, the determination of all elements in the matrix [P], Eq. 39, by the formula 38 is found to involve

$$(n-1)\cdot 1 + (n-2)\cdot 2 + (n-3)\cdot 3 + \cdots + 1\cdot(n-1)$$

$$= \sum_{x=1,2,\ldots}^{n-1} \frac{x(x+1)}{2} = \frac{n(n-1)(n+1)}{6} \tag{57}$$

multiplications, the same number of divisions, and

$$\frac{n(n-1)(n+1)}{6}+\frac{n(n-1)}{2}=\frac{n(n-1)(n+4)}{6} \tag{58}$$

additions, or a total of

$$\frac{n(n-1)(n+2)}{2} \tag{59}$$

operations altogether.

The number of additional operations involved in the computation of driving-point or transfer functions 51, 52 or 53 is small in any case and is evident from the relations 54, 55 and 56 in which it should be noticed that

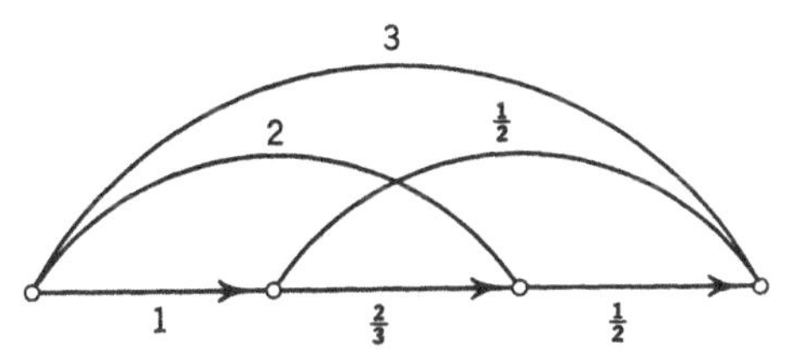

FIG. 4. Network graph to which the calculations in Eqs. 60 and 61 apply.

the sums appearing in the denominators of the last two of these are already available and do not require further addition. Thus computation of r_{nn} requires one additional division, $r_{n-1,n}$ requires additionally one multiplication and one division, and $r_{n-2,n}$ requires additionally four multiplications, a subtraction, and one division.

This method for the calculation of network response appears to be computationally more economical than any other known method,* especially the Kirchhoff combinatorial method mentioned earlier. The common denominator (or determinant) in the expressions obtained by that method has as many terms as the graph has enumerable trees (each term being the product of n branch resistances). Thus in a full graph with $n = 3$ there are 16 trees and this denominator alone involves 15 additions plus 32 multiplications, while our formula 59 yields 15 for all operations (additions, multiplications and divisions).

As an example of this method of calculating response, consider the graph of Fig. 4 for which $n = 3$. Numbers on the branches are conductance values in mhos and the branch numbering is understood to follow the pattern set in Fig. 3. Let the problem be to find the input impedance across the $\frac{1}{2}$-mho tree branch number 3 (which by rearrangement could

* Assuming that only one driving-point or transfer impedance is desired. If *all* response functions are needed (complete inversion of the [G]-matrix) then this method is no better than conventional ones.

be any other branch as well) and the open-circuit transfer impedance between this branch and the other two tree branches (which can also be any other two branches). Construction of the $[P]$-matrix according to formula 38 takes the form indicated in the following schedule:

$$\begin{array}{llll} 1 & & & \\ 2 & \frac{2}{3}+\frac{1}{3}=1 & & \\ 3 & \frac{1}{2}+\frac{1}{2}=1 & \frac{1}{2}+1+\frac{1}{2}=2 & \\ \hline 6=a_{11}^2 & & 2=a_{22}^2 & 2=a_{33}^2 \end{array} \tag{60}$$

from which Eqs. 54, 55 and 56 yield

$$\begin{aligned} r_{33} &= \tfrac{1}{2}\ \text{ohm} \\ r_{23} &= -\tfrac{1}{4}\ \text{ohm} \\ r_{13} &= -\tfrac{1}{24}\ \text{ohm} \end{aligned} \tag{61}$$

The minus signs in the last two arise from the implied reference directions for the tree branches as shown in Fig. 4.

4. Definition of Network Response Functions

Although by implication we already are somewhat familiar with this topic, it is worthwhile formalizing the definition for response functions, the more so since there seems to be some variance in the way in which different writers view this matter, so that the attitude taken in this text ought to be stated clearly at some appropriate point.

Generally speaking, a response function physically is the response of a network per unit of excitation; that is to say, it is the ratio of output to input. If one of these quantities is a voltage and the other a current, then the response function is an impedance or an admittance; while if input and output are both currents or both voltages, then the response function is dimensionless.

It is more important, however, to recognize when the response function is an impedance and when it is an admittance. For example, if the excitation is a voltage and the response a current, then the response function must be an *admittance*, not perchance the pertinent impedance as is (for some strange reason) a common viewpoint. When the excitation is a current and the response a voltage, then the proper response function is an impedance.

Thus we might say that the response function is that quantity which multiplied by the excitation yields the desired response. The operation which yields answers is *multiplication*, not division, as the conventional form of Ohm's law stating $I = E/R$ suggests. Ohm's law in this form is misleading; it should rather be written either as $I = GE$ when E is given and I is sought, or as $E = RI$ when I is given and E is the desired response.

When response functions are consistently defined in this way, then the *poles* of a response function always are the natural frequencies characterizing the response—never the zeros. This state of things adapts very well to the use of complex integration in connection with Fourier and Laplace methods to be discussed later since it is the poles of the integrand and the residues in these poles that we are concerned with in the evaluation process. Poles (not zeros) are the life-giving elements out of which rational functions (like the response functions of lumped networks) are built. To be sure, rational functions have zeros as well as poles, but zeros are points of regular behavior while poles are the singularities, and these are the characterizing elements; a function without singularities must reduce to a constant, which is a rather drab affair.

Response functions fall into two categories: driving-point functions and transfer functions; and these must clearly be distinguished because they have fundamentally different characteristics even though they both look more or less alike since formally they are given by quotients of polynomials. This is not the proper place to go into all the details concerning subtle differences between these two kinds of response functions,* but it is significant even here to know that a driving-point function can be turned upside down but a transfer function cannot, and to know the reason why this is so.

While a transfer function can be a dimensionless ratio, a driving-point function can only be an impedance or an admittance, for one of the quantities (excitation or response) must be a current and the other a voltage—they cannot both be currents or both voltages. Moreover, since both quantities are pertinent to the same terminal pair, the impedance or admittance relating them is the same whether physically it is the current at this point that produces a voltage or the other way about. That is why the upside down of a driving-point impedance is the corresponding driving-point admittance. Both are legitimate response functions.

This is, however, not the case with a transfer function regardless of whether it is dimensionless or not. The reason is easy to see from the fact that turning the function upside down implies a change in the pertinent terminal constraints from short-circuit to open-circuit or vice versa. The details involved here are best seen by considering a two terminal-pair network N characterized in the usual manner† by the open-circuit

* For such a detailed discussion see E. A. Guillemin, *Synthesis of Passive Networks*, John Wiley and Sons, 1957 (hereafter referred to as *SPN*), Chs. I and II.

† See *ICT*, pp. 527–533.

impedance matrix

$$[z] = \begin{bmatrix} z_{11} & z_{12} \\ z_{21} & z_{22} \end{bmatrix} \tag{62}$$

with $z_{12} = z_{21}$ or by its inverse, the short-circuit admittance matrix

$$[y] = [z]^{-1} = \begin{bmatrix} y_{11} & y_{12} \\ y_{21} & y_{22} \end{bmatrix} \tag{63}$$

with $y_{12} = y_{21}$.

Let us begin by considering terminal pair 2 as the output and terminal pair 1 as the input. For an output voltage and an input current we have the response function

$$\frac{E_2}{I_1} = z_{12} \tag{64}$$

For a dimensionless voltage ratio, noting that $E_1 = z_{11}I_1$, we have

$$\frac{E_2}{E_1} = \frac{z_{12}}{z_{11}} \tag{65}$$

For an output current and an input voltage the response function is

$$\frac{I_2}{E_1} = y_{12} \tag{66}$$

and for a dimensionless current ratio, noting that $I_1 = y_{11}E_1$, we get

$$\frac{I_2}{I_1} = \frac{y_{12}}{y_{11}} \tag{67}$$

Turning any of these functions upside down implies that terminal pair 1 becomes the output and terminal pair 2 the input. The right-hand quantities in these equations are then no longer appropriate. The transfer functions that are physically the reciprocals of 64 through 67 are respectively given by

$$\frac{I_1}{E_2} = y_{12} \tag{68}$$

$$\frac{E_1}{E_2} = \frac{z_{12}}{z_{22}} \tag{69}$$

$$\frac{E_1}{I_2} = z_{12} \tag{70}$$

and

$$\frac{I_1}{I_2} = \frac{y_{12}}{y_{22}} \tag{71}$$

None of these are reciprocals of the functions in Eqs. 64 through 67. On account of the reciprocity theorem, 66 and 68 are alike and so are the functions 64 and 70, but this fact has nothing per se to do with our present argument, which firstly points out that the function pairs 64 and 68, 65 and 69, 66 and 70, 67 and 71 are not reciprocal; and secondly that, in all eight of these functions, not a single one has any physical significance as a response function when turned upside down. Thus, for example, $1/z_{12}$ cannot be an output-to-input ratio; neither can z_{11}/z_{12}, nor $1/y_{12}$, or any of the others. Upside down, these functions are ratios of input-to-output which is just the upside down of a transfer function, but not actually a transfer function.

This state of affairs ties in collaterally with the known fact* that although a driving-point function (in a passive network) must have *both* zeros and poles in the left half of the s-plane, a transfer function need only have its poles thus restricted; its zeros may lie anywhere in the complex frequency plane. Since poles of a response function (as already pointed out) are natural frequencies of the response, the latter would grow exponentially if any poles were in the right-half plane (an impossibility in a passive network). The reciprocal of a rational function having right half-plane zeros yields one having right half-plane poles; therefore, we see again that transfer functions in general cannot be turned upside down, but that driving-point functions may be.

Finally we would like to point out at this time (although detailed discussion explaining and elaborating these ideas is given in later chapters of this text) that response functions may also be expressed in the time domain as well as in the frequency domain to which the rational functions that we have just talked about are pertinent. In the time domain the response function is, of course, a function of the time t. It is still defined as the response of the network per unit of excitation, the latter being a unit impulse.† In this sense it is again a ratio of output-to-input; however, the output for any given input time function is not obtained through simple multiplication.

Nevertheless, we are aware of the fact that the impulse may be considered as a building block from which any input function can be constructed through multiplication and summation (integration) alone; and, so long as the network is linear, the response for any input can straightforwardly be construction by addition or integration knowing only the unit impulse response. The process, which is the time-domain equivalent of multiplication, is called *convolution* and will be amply explained in Ch. XIII.

Turning a time-domain response function upside down makes no sense, either with driving-point or transfer functions, and so there is no need to distinguish between these two categories, as is appropriate with frequency-domain functions. On the other hand, there is a time-domain concept

* See *SPN*, Chs. I and II.

† See *ICT*, pp. 190–203.

that bears a fairly close relation to turning a time-domain response function upside down, or perhaps "inside out" is a better term in this case. Thus the response function describes the output of a network when the input is a unit impulse; an impulse is applied at the input terminals and some time function emerges at the output terminals.

We are tempted to ask: if this time function is applied at the output terminals, is it converted back into an impulse at the input terminals? This process might be described as "inverting" the response. The answer, of course, is that the same network with its terminal pairs interchanged will not reconvert the impulse response into an impulse, which amounts to undoing the process of transmission of the applied signal (the impulse) through the network. In fact the reciprocity theorem tells us that if we interchange input and output terminal pairs, the response function is the same.

On the other hand, one can construct a second network that converts the impulse response of a given network into an impulse. In old-fashioned telephone engineer's parlance, this second network is simply the so-called *attenuation* and *phase equalizer* for the first one, the two networks in tandem yielding nothing more than a delay for the input impulse.

This situation is more easily expressed in terms of the logarithmic form of response function known as the *propagation function* for the given transmission network. Thus the natural logarithm of the frequency domain transfer function for a transmission line is commonly referred to as a propagation function; its real and imaginary parts are the *gain* in *nepers* and the *phase lead* in *radians*. Logarithm of the reciprocal transfer function, or the negative of the propagation function, has *loss* or *attenuation* for its real part and phase *lag* for its imaginary part.

Two propagation functions are said to be *complementary* when the sum of their real parts is a constant and the sum of their imaginary parts is a constant times the frequency variable ω. This designation is almost but not quite parallel to that used in connection with impedances or admittances. These, it will be recalled, are said to be complementary if their sum is a constant (usually normalized at unity). In the case of propagation functions the analogous designation would be useless since the complementary network could not be constructed. It can, however, if we allow the sum of propagation functions to have an imaginary part proportional to frequency with an adequate proportionality constant. Thus, in a physical network we can never cancel the delay of a signal (caused by phase lag) because no network can produce a negative delay. The best we can do is to make the delay (phase slope) constant over the essential frequency range.*

In these terms the conversion of a time-domain signal, like the impulse

* The details of this process are discussed later on.

response of a given network, back into an impulse by cascading a second network with the first, becomes straightforwardly obvious. We merely need to make the propagation function of the second network the complement of that of the first one. The ability to do this and the method of accomplishing it have been known in the telephone art for many years.

Recently this process has been rediscovered in connection with a scheme for more successfully transmitting signals through noise. The signals arise in the form of quasi impulses, but instead of transmitting them in that form they are converted by a dispersive network into a function which is spread out in time and correspondingly reduced in amplitude. At the receiving end the complementary network converts the desired signals back into quasi impulses but does not act similarly upon the superimposed noise introduced by the transmission link or noisy channel because the signal and noise spectra during transmission are completely uncorrelated (have nothing in common since one is random while the other has a fancy pedigree bestowed upon it by the first network).

The scheme is spoken of as a method of "coincidence reception" and the two networks involved are referred to as a pair of "matched filters." Regarded in the time domain, the process is conceptually rather sophisticated and indeed was thought to be a novel idea, while in the frequency domain the ability to operate upon signals in this way is strictly "old hat," the scheme having been originally proposed as a method of "scrambling" and "unscrambling" radio signals in order to achieve privacy in communication systems.

Problems

1. For the following network graph use piecewise analysis to find the input current i_s that yields an output voltage $e = 1$ volt. All branches are 1-ohm resistances.

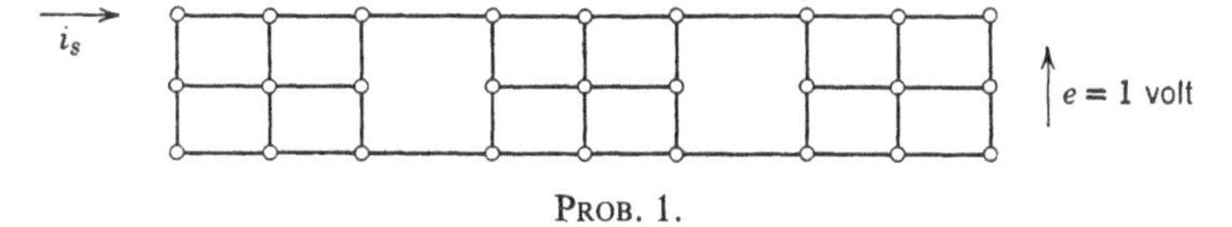

PROB. 1.

2. By constructing the dual of the argument in art. 2, prove, on a node basis, that the value of the determinant G pertinent to Eq. 4 is given by the sum of all tree sets.

3. For the graph of Fig. 1*a*, assume that each branch is a resistance equal to its respective branch number. On a loop basis, evaluate the determinant R as a sum of all link sets and check your result by forming the matrix $[R]$ according to Eq. 1 and evaluating its determinant in the conventional manner. Compare the total computational work involved in the two methods.

4. For the symmetrical matrix

$$[G] = \begin{bmatrix} 4 & 1 & 2 & 1 \\ 1 & 2 & 0 & 1 \\ 2 & 0 & 3 & 2 \\ 1 & 1 & 2 & 1 \end{bmatrix} = [A] \times [A]_t$$

construct $[A]$ in the triangular form 24. From its inverse construct the inverse of $[G]$. Alternately, find the inverse of $[G]$ by systematic elimination and compare the total computational labor with that involved in the first method.

5. For the graph of Fig. 1*a*, assume each branch to be a conductance equal to the respective branch number in mhos. Also assume a 1-volt source in series with branch 1. Find the resulting current in branch 3 by the method summarized in calculations 60 for the network of Fig. 4.

6. For the graph shown here, the conductances of branches 1, 2, 3, 4 are 2 mhos, those of branches 5, 6, 7, 8 are 1 mho. A 1-volt source is located in branch 1.

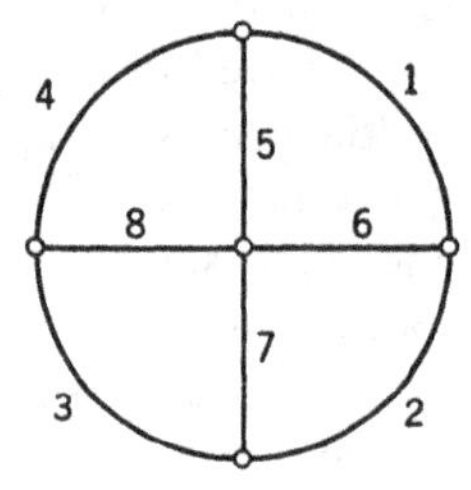

PROB. 6.

Find the voltages across branches 3, 7 and 8 by constructing the appropriate matrix $[P]$, Eq. 39, and then using formula 56, renumbering the branches if necessary and making use of possible simplifications resulting from symmetry.

7.

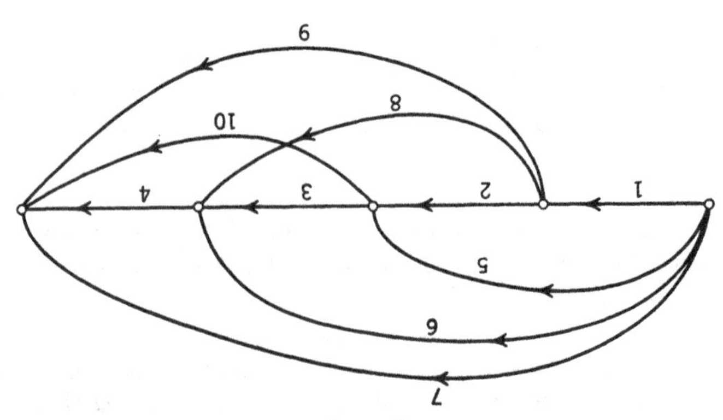

PROB. 7.

Consider the following trees relative to the above graph:

(*a*) 1, 2, 3, 4; (*b*) 1, 5, 6, 7; (*c*) 2, 3, 5, 10

Write down the transformation matrix $[T]$ that relates, as in Eq. 7, the tree-branch voltages in tree (*a*) to those in tree (*b*); those in (*a*) to those in (*c*); and those in (*b*) to those in (*c*).

For tree (*b*) write down the $[G]$-matrix assuming all branches are 1-ohm conductances. By means of this $[G]$-matrix and the above transformation matrices evaluate $[G]$-matrices appropriate to trees (*a*) and (*c*).

CHAPTER IV

Synthesis of Networks from Given Parameter Matrices

1. Realizability Conditions for an $(n + 1)$-Node Network

We shall first discuss how a single-element-kind network can be realized when its parameter matrix is given, and for simplicity assume that the network is purely resistive. Given a short-circuit conductance matrix $[G]$ of order n, the problem is to construct a corresponding n terminal-pair network.

As in art. 3 of Ch. III, we will assume initially that the pertinent network has $n + 1$ nodes. Although this assumption leads to realizability conditions that are more restrictive than they need be, it turns out, as we shall see, that these restrictions are necessary if we extend our interest to the realization of two- and three-element-kind networks.

If a given node conductance matrix $[G]$ is known to be appropriate to a starlike tree, then the necessary and sufficient realizability conditions as an $(n + 1)$-node network are simply that it be a dominant matrix (as described in art. 3 of Ch. III); and if it is known to be based upon a linear tree, then the uniformly tapered property (also described in art. 3 of Ch. III) is necessary and sufficient for its realization. For any given order n, there is a finite number of distinct geometrical tree configurations; and the given $[G]$-matrix must be based upon one of these and satisfy pertinent realizability conditions if a corresponding network with all positive elements is to exist.

If we know the geometrical tree configuration upon which the matrix $[G]$ is based, then we can readily write the appropriate transformation matrix $[T]$ (Eq. 8, Ch. III) connecting the node-pair voltages in that tree to those in a starlike or a linear tree and, by means of a congruent transformation (like that expressed by Eq. 11, Ch. III), transform $[G]$ to a form appropriate to either of these basic tree configurations, whereupon the question regarding its realizability is readily answered, and, if answered in the affirmative, the corresponding network is constructable straightforwardly.

The problem of testing and realizing a given $[G]$-matrix by an $(n + 1)$-node network, therefore, will be solved if we can devise a method for discovering the geometrical tree configuration upon which that $[G]$-matrix is based. This we shall now proceed to do; and we will see, incidentally, that the pertinent tree configuration is unique if $[G]$ is nondegenerate in the sense that all of its elements are nonzero. The method to be described is based upon the recognition that there exists a one-to-one correspondence (properly interpreted, of course) between the pertinent tree configuration and the algebraic sign distribution among elements in the $[G]$-matrix. This algebraic sign pattern enables us to recognize the geometrical configuration of the tree upon which $[G]$ is based, and $[G]$ must be realizable with that tree configuration if it is realizable at all.

Obviously, if $[G]$ contains zeros, there is an ambiguity in the algebraic sign pattern, and the possibility exists that more than one tree may be appropriate. Although one can still apply the method to obtain systematically all appropriate trees in a situation of this sort, the process loses its compactness and we shall, therefore, assume for the time being that $[G]$ has no zero elements.

The easiest way to see that a definite algebraic sign pattern among the elements in $[G]$ is linked with a given geometrical tree configuration, is through physical rather than analytical reasoning. Since $[G]$ is a *short-circuit* driving-point and transfer matrix, we visualize all n terminal pairs across tree branches provided with short-circuiting links and remind ourselves that the values of currents in these links, per volt of ideal voltage source in one of them, equal numerically the elements in $[G]$ inclusive of their algebraic signs relative to chosen reference arrows on the tree branches.

Thus, with the unit voltage source in the short-circuiting link across tree branch No. 1, the currents in the various links numerically equal elements in row 1 of $[G]$. With the unit voltage source in the link across tree branch No. 2, the currents in the n short-circuiting links in value equal the elements in row 2 of $[G]$, and so forth. The voltage rise of the source is made to coincide in direction with the reference arrow on the tree branch in parallel with it, and any resulting current is positive if its direction in the pertinent short-circuiting link agrees with the reference arrow on the tree branch alongside of it, and it is a negative current if its direction is opposite to this reference arrow.

As an example, Fig. 1 shows a tree for $n = 5$ with short-circuiting links across its branches and a voltage source applied to terminal pair No. 1. If we visualize the presence of the rest of the branches associated with this tree in a full graph, it is clear by inspection that the various short-circuit currents have directions as indicated by arrows in the short-circuiting links; and thus we see that all elements in the first row of a [G]-matrix appropriate to this tree (with the assumed reference arrows) are positive.

If we shift the voltage source into the short-circuiting link across branch 2 with its polarity in the same relation to the reference arrow on that branch as is shown for branch No. 1 in Fig. 1, then it is equally simple to see by inspection that the resulting currents for branches 1 and 2 are in positive directions while those in all other short-circuiting links are negative. In the second row of [G], therefore, the first two elements are

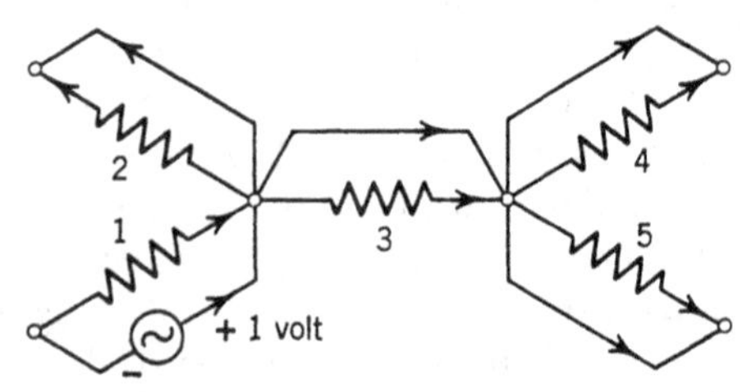

FIG. 1. Example to illustrate relation between algebraic signs in the [G]-matrix and reference arrows on the tree branches.

positive and the rest are negative. Thus it is a simple matter to establish all algebraic signs for the elements of [G]; and one becomes convinced, incidentally, that this sign pattern has nothing to do with element values but is uniquely fixed by the geometrical tree configuration except for interchanging rows and columns (renumbering of tree branches), and the multiplication of rows and columns by minus signs (changing of reference arrows on the tree branches), which, as we shall see presently, are trivial operations so far as recognition of the pertinent tree configuration is concerned.

In order to facilitate the correlation of algebraic sign patterns with geometrical tree configurations, we observe that we can dispense with drawing the tree branches in sketches like the one in Fig. 1. It suffices to draw lines for branches as we are accustomed to do in network graphs and to regard these as the short-circuiting links, their reference arrows being included in the usual manner. A voltage source as well as all other branches in the full graph can easily be imagined to be in their proper places, and directions of pertinent currents can, with the help of the following interpretation, be deduced by inspection.

In Fig. 2 (p. 84) are all six tree configurations for $n = 5$. We can imagine the branches as being water pipes. If in the starlike tree (a) we squirt water into pipe No. 1 in the reference direction it will obviously flow in positive reference directions in all of the four other pipes. Hence in the first row of [G], all elements are plus. If we squirt water in the reference direction through pipe No. 2, it flows positively through pipe No. 1 and negatively through the other three. Thus, in the second row of [G], the first two elements are plus and the rest are minus. This homely analogy makes the process of establishing sign patterns for all trees fast and effortless. The resulting sign patterns thus obtained are indicated as follows by what we might call "sign matrices:"

$$s_a = \begin{bmatrix} + & + & + & + & + \\ & + & - & - & - \\ & & + & - & - \\ & & & + & - \\ & & & & + \end{bmatrix}; \quad s_b = \begin{bmatrix} + & + & + & + & + \\ & + & - & - & - \\ & & + & - & - \\ & & & + & + \\ & & & & + \end{bmatrix};$$

$$s_c = \begin{bmatrix} + & + & + & + & + \\ & + & - & - & - \\ & & + & + & + \\ & & & + & + \\ & & & & + \end{bmatrix} \qquad (1)$$

$$s_d = \begin{bmatrix} + & + & + & + & + \\ & + & + & + & + \\ & & + & - & - \\ & & & + & + \\ & & & & + \end{bmatrix}; \quad s_e = \begin{bmatrix} + & + & + & + & + \\ & + & - & - & - \\ & & + & + & + \\ & & & + & - \\ & & & & + \end{bmatrix};$$

$$s_f = \begin{bmatrix} + & + & + & + & + \\ & + & + & + & + \\ & & + & + & + \\ & & & + & + \\ & & & & + \end{bmatrix} \qquad (2)$$

Since these matrices are symmetrical it is needful only to record signs above the principal diagonal.

Signs on the principal diagonal are always plus for obvious reasons. Observe further that we have chosen reference arrows for the tree branches in such a way that the first row is always a row of plus signs. In an arbitrarily given matrix [G] this state of affairs need not be fulfilled, but we can always multiply rows (and corresponding columns) by minus signs to fulfill this condition and thus convert the given matrix to a sort of normal or basic form with regard to its algebraic sign pattern. This step eliminates once and for all those trivial variants of the given $[G]$-matrix which stem from sign changes of its elements in rows (and in corresponding columns). We may regard the process of making all signs in the first row positive as one of reducing the reference arrows on the branches of the implied tree to a common basic pattern.

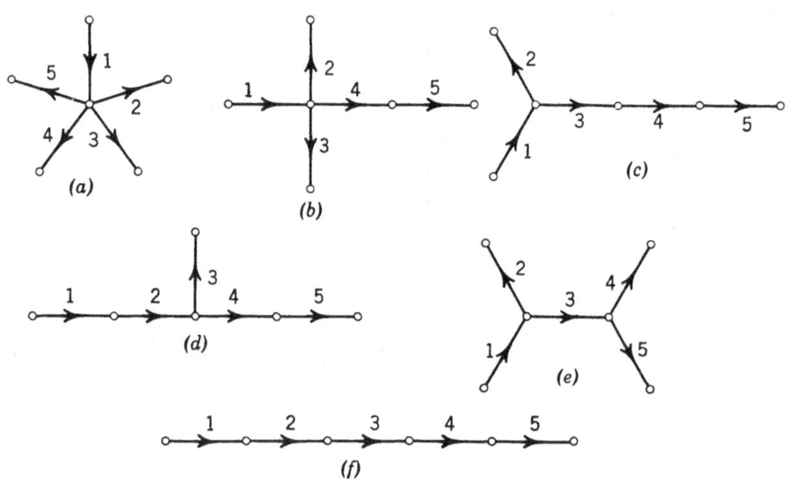

FIG. 2. Detailed example to show how signs in the $[G]$-matrix are correlated with reference arrows in the various possible tree configurations.

Except for row and column interchanges (renumbering of tree branches) each sign matrix uniquely specifies a geometrical tree pattern and vice versa; and we shall presently describe a simple way of constructing the tree from a given sign matrix. In the meantime it is interesting to observe that *the linear tree is the only one for which all signs in the* $[G]$*-matrix are positive.* If a given $[G]$-matrix has all positive elements (when its sign pattern is normalized as just described) then it must be realizable with a linear tree if it is realizable at all; that is to say, it must be a uniformly tapered matrix or else it has no realization in an $(n + 1)$-node network.

In the given $[G]$-matrix with all-positive elements the arrangement of rows and columns may not be such as to give $[G]$ a uniformly tapered form even though it is basically appropriate to a linear tree, since the branch numbering in that tree may not be consecutive. However, one can readily perceive the needed rearrangement of elements in $[G]$ by means of the following procedure.

From the uniformly tapered character of this matrix, or directly from the implied network topology (as given in Fig. 3, Ch. III) and the physical significance of elements in $[G]$, we can state that if $G_{ik} > G_{is}$, then branch (ii) in the linear tree is closer to branch (kk) than to branch (ss) assuming both of these to be on the same side of branch (ii). For example, if $[G]$ is pertinent to a linear tree with branches in consecutive order from left to right, then tree branches (22), (33), . . . are all on the same side of branch (11), and $G_{12} > G_{13} > G_{14} \cdots$ correctly implies that branch (11) is closer to branch (22) than to branch (33), and that it is closer to (33) than to (44), etc. In the second row of $[G]$ we have $G_{23} > G_{24} > G_{25} \cdots$ consistent with the fact that branch (22) is closer to (33) than to (44), etc. In row three we have $G_{34} > G_{35} \cdots$ for branches to the right of (33) and $G_{32} > G_{31}$ for branches to the left of (33), in agreement with the fact that branch (33) is closer on the left to (22) than to (11) and on the right it is closer to (44) than to (55), etc.

Let us refer to these circumstances as *proximity conditions*, and include as one of these the fact that G_{kk} is larger than any element in row or column k. Since elements on the principal diagonal remain there in spite of row and column interchanges (preserving symmetry, of course), this condition holds whether a $[G]$-matrix (appropriate to a linear tree) is pertinent to tree branches in consecutive order or not.

We can now formulate a systematic procedure for finding the tree-branch order appropriate to a given $[G]$-matrix that has all positive elements but for which the tree-branch order is scrambled so that the rows and columns do not appear to fulfill conditions of uniform taper. The procedure is best described in terms of a specific example. Let us consider the following $[G]$-matrix together with tentative branch orders obtained by applying the proximity conditions to respective rows:

$$\begin{bmatrix} 31 & 8 & 25 & 11 & 14 & 23 \\ 8 & 12 & 9 & 4 & 11 & 5 \\ 25 & 9 & 31 & 10 & 17 & 19 \\ 11 & 4 & 10 & 16 & 6 & 14 \\ 14 & 11 & 17 & 6 & 21 & 10 \\ 23 & 5 & 19 & 14 & 10 & 27 \end{bmatrix}$$

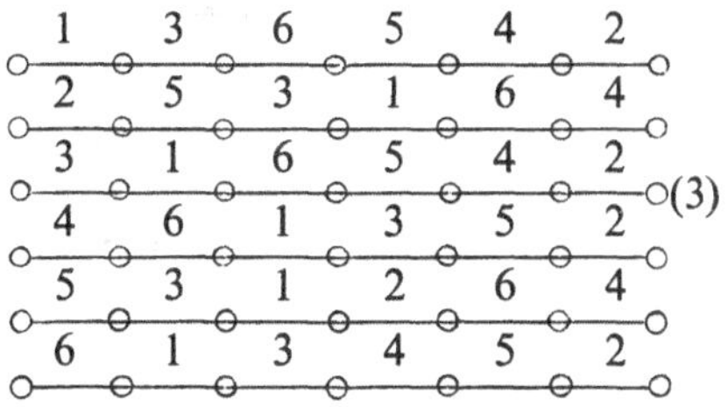

1 3 6 5 4 2

2 5 3 1 6 4

3 1 6 5 4 2 (3)

4 6 1 3 5 2

5 3 1 2 6 4

6 1 3 4 5 2

Since the diagonal element in each row is the largest, the first branch in each tentative tree (let us refer to this as the *lead* branch) pertains to the corresponding row in the $[G]$-matrix. In each tree one tentatively assumes all other branches to be on the same side and orders them according to the proximity conditions. Branches may be transposed from the right to the left of each lead branch so long as the proximity conditions remain fulfilled. In the first tree, for example, we may transpose branches 6 and 4 to the left of the lead branch, keeping 6 closer to 1 than 4. The branch order for this tree then becomes 461352 which is the second tree reversed. Tree 1 can thus be made to agree with tree 2 by transposition. Observe that the reverse is not possible. That is to say, tree 2 cannot be made to agree with tree 1 by transposition.

Since the correct tree is that one to which all tentative trees are reducible by transposition, tree 1 is eliminated, but tree 2 is a possibility. Hence we proceed next to see whether the others are reducible to it. In tree 3 we need merely to transpose branches 5 and 2, keeping 5 nearer to 3 than 2. Tree 4 is tree 2 in reversed order. In tree 5 we can transpose branch 2; and in tree 6 only branch 4 needs to be transposed. We may conclude that tree 2 or tree 4 exhibits the branch order pertinent to the given $[G]$-matrix.

We can easily see that this order must be unique by noting that if a $[G]$-matrix is in uniformly tapered form, *any* rearrangement of its rows and columns destroys that form; the row-and-column arrangement yielding uniform taper is unique. Hence the pertinent tree-branch order is unique; and since any other order is related to this one by a nonsingular transformation, there cannot possibly be two distinct tree-branch orders associated with a realizable $[G]$-matrix whether its rows and columns are scrambled or not. If none of the tentative trees is one to which all others are reducible by transpositions that do not violate the proximity conditions, then the given $[G]$-matrix is not realizable.

The tree branch order 253164 pertinent to the above matrix means that tree-branch voltage v_1 should be renumbered v_4, v_2 should be renumbered v_1, v_3 should remain v_3, v_4 should become v_6, etc. If we express this revision by writing $v_k] = [T]\, v_k']$ in which the revised quantities are indicated with a prime, then

$$[T] = \begin{bmatrix} 0 & 0 & 0 & 1 & 0 & 0 \\ 1 & 0 & 0 & 0 & 0 & 0 \\ 0 & 0 & 1 & 0 & 0 & 0 \\ 0 & 0 & 0 & 0 & 0 & 1 \\ 0 & 1 & 0 & 0 & 0 & 0 \\ 0 & 0 & 0 & 0 & 1 & 0 \end{bmatrix} \tag{4}$$

The correspondingly revised node conductance matrix is obtained through postmultiplication of the given matrix $[G]$ by $[T]$ and premultiplication by its transpose as in Eq. 11, Ch. III. In the present example this process yields

$$\begin{bmatrix} 12 & 11 & 9 & 8 & 5 & 4 \\ 11 & 21 & 17 & 14 & 10 & 6 \\ 9 & 17 & 31 & 25 & 19 & 10 \\ 8 & 14 & 25 & 31 & 23 & 11 \\ 5 & 10 & 19 & 23 & 27 & 14 \\ 4 & 6 & 10 & 11 & 14 & 16 \end{bmatrix} \tag{5}$$

which is the desired normal form.

2. How to Grow a Tree from a Sign Matrix

The procedure for constructing a tree from a given sign matrix is similar in some respects to a method already developed for determining the graph pertinent to a given cut-set matrix.* There, however, the growth pattern for the tree must first be established, while in the present situation one can proceed at once with the growth process for the pertinent tree since any given sign matrix, in contrast with a cut-set matrix, may always be regarded as having an appropriately "ordered form" to begin with.

Since the sign matrices 1 and 2 for the tree configurations in Fig. 2 are too simple to represent worthwhile examples, and moreover are based upon a particular branch numbering sequence that need not be fulfilled in an arbitrarily given situation, we choose to illustrate the method of tree construction proposed here by the following more elaborate sign matrix

$$[S] = \begin{array}{c} \begin{matrix} 1 & 2 & 3 & 4 & 5 & 6 & 7 & 8 & 9 \end{matrix} \\ \begin{bmatrix} + & + & + & + & + & + & + & + & + \\ & + & + & + & + & + & + & - & + \\ & & + & - & - & + & - & + & + \\ & & & + & + & - & + & + & + \\ & & & & + & - & - & + & + \\ & & & & & + & - & + & + \\ & & & & & & + & + & + \\ & & & & & & & + & - \\ & & & & & & & & + \end{bmatrix} \begin{matrix} 1 \\ 2 \\ 3 \\ 4 \\ 5 \\ 6 \\ 7 \\ 8 \\ 9 \end{matrix} \end{array} \tag{6}$$

Numbering of the rows and columns is done to facilitate identification with correspondingly numbered tree branches.

* "How to Grow Your Own Trees from Given Cut-Set or Tie-Set Matrices," *IRE Trans. Circuit Theory*, Vol. CT-6, May 1959, pp. 110–126.

Construction of the tree is accomplished by starting with the last branch, No. 9, and successively adding branches 8, 7, 6, . . . , being guided as to their relative positions by the confluence or counterfluence of reference arrows as demanded by the signs in the respective rows and columns of the matrix [S]. Thus branches 8 and 9 must be counterfluent, while branch 7 is confluent with both 8 and 9. This state of affairs can be met

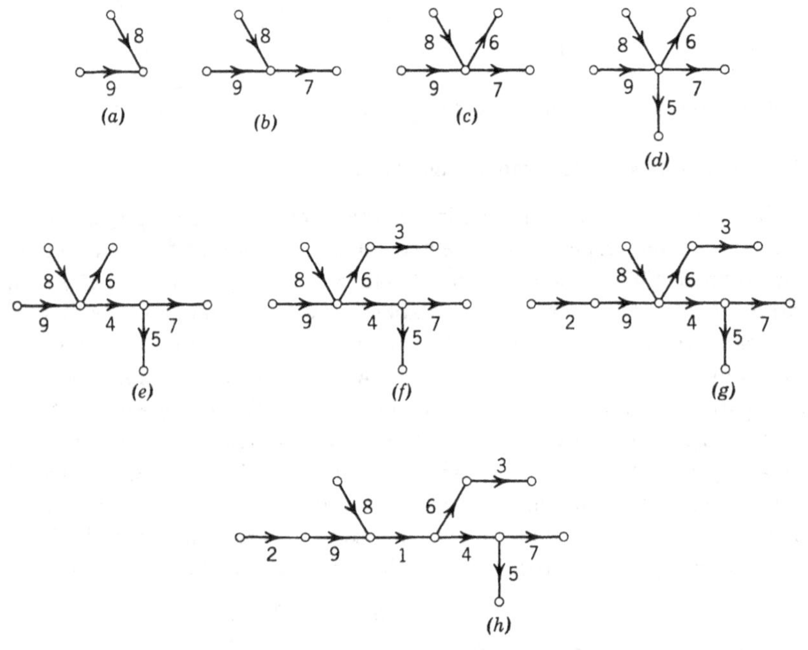

FIG. 3. Showing growth of the tree, branch by branch, pertinent to the sign matrix 6.

only by having branches 7, 8 and 9 meet in a common point as shown in sketch (*b*) of Fig. 3, in which the various sketches (*a*) through (*h*) show the growth of the tree, branch by branch, the position of each added branch being uniquely determined (except for trivial variants) by the signs in the pertinent row of *S*.

Construction of the tree is thus straightforward and always possible unless one encounters a contradiction, in which case no tree exists and the given [*G*]-matrix has no realization in an $(n + 1)$-node network.

3. $(n + 1)$-Node Realization Procedure for an nth-Order [*G*]-Matrix

Once the tree for a given [*G*]-matrix is found, it is a simple matter to write down the transformation matrix [*T*] which, in a congruent transformation, carries [*G*] over into a form appropriate either to a starlike tree (the dominant form) or to a linear tree (the uniformly tapered form). In either case the conditions for its realization present no further problem. Finally, the terminal pairs appropriate to the given matrix can be determined from the geometrical picture upon which construction of the matrix $[T]$ is based.

When trivial variants in the tree structure exist, the following reasoning removes any ambiguities. For example, in Fig. 3 there are two trivial variants. Branches 3 and 6 can be interchanged and so can branches 2 and 9. If we leave them as they are, then in the $[G]$-matrix appropriate to this tree, the familiar physical interpretation of what the elements in $[G]$ stand for tells us that element (13) must be less than element (16); and element (12) must be less than element (19). These conditions can readily be checked; and if fulfilled, the $[G]$-matrix must be realizable with the tree as it stands if it is realizable at all. If either condition is not fulfilled, then we must consider the appropriate variant in the tree structure or else make the pertinent row-and-column interchange in $[G]$.

Observe that the existence of a tree appropriate to the given $[G]$-matrix is not sufficient to assure its realization. Observe also that although we can construct many different $[T]$-matrices connecting the tree of the given $[G]$-matrix with a starlike or a linear tree, it is sufficient to try only one, for if this one fails to yield a realizable $[G]$-matrix (either dominant or uniformly tapered) then no other transformation $[T]$ can do so, since a contrary assumption leads to a contradiction. Hence the realization procedure discussed here does not require repeated trials; one straightforward attack tells the whole story.

4. Collateral Remarks Relative to Realization of the Matrix $[G]$

When the given $[G]$-matrix contains zero elements* then the sign matrix correspondingly contains blank spaces which may be interpreted either as plus or minus signs. If, in the construction of the tree in the above described manner, either sign is admissible for the insertion of a particular tree branch, then a variant in the tree's tentative configuration is possible. If such a variant is nontrivial, and if nontrivial variants occur again in subsequent steps of the tree-growth process, then more than one geometrical tree pattern can be associated with the given $[G]$-matrix. However, it need not follow that the realizability conditions appropriate to these various trees are fulfilled for the given element values in $[G]$.

Although several possibilities need now to be investigated, their number

* Omitting the consideration of situations in which the network consists of more than one separate part, the test for which is described in the above mentioned reference "How to Grow Your Own Trees . . ."

is greatly reduced from the totality of algebraic sign arrangements possible on a purely combinatorial basis by reason of the step-by-step nature of the tree-growing process which enables one by inspection to rule out the majority of trials and hence keep the procedure well within reasonable bounds even in the consideration of degenerate cases.

Collaterally it may be interesting to point out that the present discussion offers an alternate method for the construction of trees (and hence graphs) from given cut-set matrices. Forming the Gramian from the rows of $[\alpha_{nb}]$ yields (according to Eq. 4 of Ch. III) a $[G]$-matrix appropriate to a network with all 1-ohm branches. From its sign matrix the pertinent tree can be constructed (if one exists); and situations for which other methods fail (see in particular the one pertinent to the graph in Fig. 8, p. 126, in the reference cited above entitled "How to Grow Your Own Trees . . .") do not necessarily lead to a degenerate $[G]$-matrix and hence are strictly routine when handled by the method presented here.

At this point we need to remind ourselves that if a given $[G]$-matrix is not realizable by an $(n + 1)$-node network, it may still be realizable by a $2n$-node network in which the terminal pairs to which $[G]$ pertains need not be adjacent, as they manifestly must be in the $(n + 1)$-node graph where each tree branch presents a terminal pair. In a $2n$-node full graph one can always choose a linear tree, and if we identify alternate branches as terminal pairs we have a topological situation that is completely general so far as the realizability of $[G]$ is concerned. In other words, a linear tree in a $2n$-node full graph with every other tree branch yielding an accessible terminal pair is the most general topological structure upon which the realization of $[G]$ can be based. Additional nodes beyond the number $2n$ can do no good since any $[G]$ realizable with more than $2n$ nodes is also realizable with $2n$ nodes because the familiar star-mesh transformation process eliminates the extra nodes and leaves all elements positive.

Before the realization techniques discussed here can be applied to this more general situation, however, one must augment the given n-by-n $[G]$-matrix to one having $2n - 1$ rows and columns. This augmentation process must, of course, fulfill the condition that subsequent abridgment to the accessible terminal pairs regains the originally given n-by-n $[G]$-matrix.

Such an augmentation process can be formulated in simple terms and, as might be expected, it is not unique. Herein lies the greater realization capability of the $2n$-node network over the $(n + 1)$-node network, for one can use the freedom inherent in the nonuniqueness of the augmentation process to help fulfill realization conditions for the implied $2n$-node graph with linear tree. These things are discussed in art. 6 below.

5. Synthesis Procedure When Several Node-Parameter Matrices Are Given

At this point it seems appropriate to discuss the realization of networks containing two or all three kinds of elements R, L and C when their parameter matrices are known. These we assume to be of order n appropriate to equilibrium equations on a node basis for node-pair voltages identified with the branch voltages of a chosen tree in a network graph involving $n + 1$ nodes.

In the synthesis of RC networks both matrices $[G]$ and $[C]$ must be based upon the same geometrical tree configuration (a condition that is easy to meet in the process of matrix construction). In the synthesis of RLC networks where three parameter matrices are involved, all three must separately be $(n + 1)$-node realizable with the same geometrical tree configuration. Realization of the total network is accomplished in two steps. First each matrix is realized by a single-element-kind network for the implied common tree configuration, and then the several networks are joined by connecting corresponding tree branches in parallel (placing the trees in parallel by soldering together their corresponding nodes).

The fact that this procedure yields a network with the desired short-circuit admittance matrix (Eq. 47, Ch. II)

$$Y_{nn} = [C]s + [G] + [\Gamma]^{-1}s \qquad (7)$$

with parameter matrices given by Eqs. 113, Ch. II, is easy to see. The network has n terminal pairs and is characterized by a matrix $[Y]$ equal to the sum of three component $[Y]$ matrices, namely the matrices $[C]s$, $[G]$, and $[\Gamma]^{-1}s$. If three separate n terminal-pair networks having these component $[Y]$-matrices are found, then their parallel combination has a resultant $[Y]$-matrix equal to the sum of the component matrices provided certain conditions assuring no circulating currents resulting from the parallel connection* are fulfilled. As the following discussion shows, these conditions are automatically fulfilled if the $(n + 1)$-node realizations for the separate matrices $[C]$, $[G]$, and $[\Gamma]$ are based upon the same geometrical tree configuration.

The schematic diagram in Fig. 4 illustrates the principle involved here in the parallel connection of two multi-terminal-pair networks. Straightforward extension of the well-known tests (*SPN*, pp. 191–194) for two terminal-pair networks amounts to applying a voltage source E_s to one of the paralleled pairs of terminals as shown in the figure, separately shorting all other terminal pairs, and then testing for zero voltage between

* See *SPN*, pp. 191–194.

those shorted pairs like a-a', b-b' and c-c' that are to be connected in parallel. This test must be repeated n times (for n terminal-pair networks) with the source E_s across each terminal pair.

If the terminal pairs are across corresponding branches of identical tree configurations then their mutual adjacency (which insures a common terminal between pairs) is readily seen to assure that all tests yield zero voltage as required.

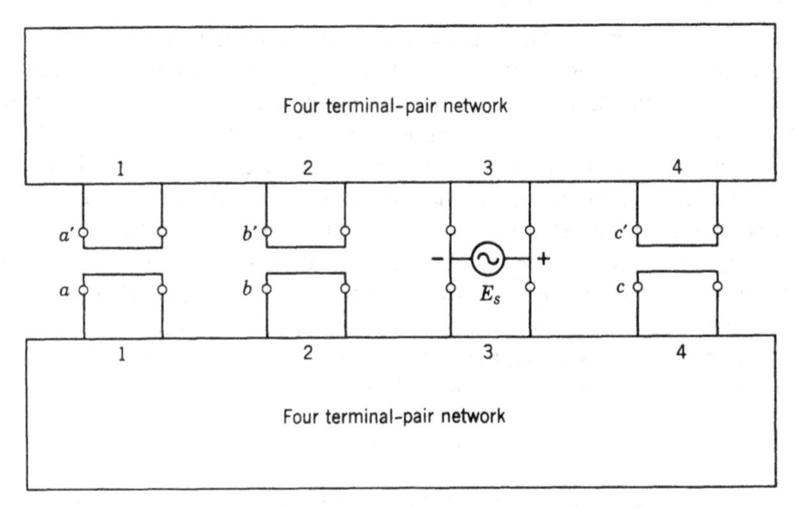

FIG. 4. Illustrating conditions under which parallel connection of component networks is valid.

Therefore, when we approach the synthesis problem from a parameter matrix point of view (rather than one which proceeds from a given impedance or admittance function), all we need for realization of the resulting network is a technique for constructing a single-element-kind network from a given parameter matrix.

6. Realization of an nth-Order $[G]$-Matrix by a $2n$-Node Network

In order to apply the $(n + 1)$-node technique given in arts. 1, 2 and 3 to this situation, one must first augment the given nth-order $[G]$-matrix to one of order $2n - 1$; and this augmentation must fulfill the obvious requirement that subsequent abridgment regains the original matrix. Our first topic, therefore, is concerned with this augmentation process.

Suppose we consider a matrix $[B]$ of order m with its rows and columns partitioned into groups of r and s. The sum $r + s$, of course, equals m.

In this partitioned form we can write

$$[B] = \begin{bmatrix} b_{rr} & \vdots & b_{rs} \\ \cdots & \cdots & \cdots \\ b_{sr} & \vdots & b_{ss} \end{bmatrix} \tag{8}$$

In a set of equilibrium equations having the matrix $[B]$ we assume that the first r variables pertain to accessible terminal pairs. The remaining variables we wish to suppress or eliminate. As shown in art. 4, Ch. I, the abridged matrix pertinent to the desired r points of access is given by

$$[B]_{\text{abr}} = b_{rr} - b_{rs} \times b_{ss}^{-1} \times b_{sr} \tag{9}$$

This same result applies if the r retained variables are not necessarily those designated by the suffixes $1 \cdots r$ but are *any* r of the m original variables. Instead of forming b_{rr} by deleting all rows and columns designated by suffixes larger than r, we delete all rows and columns pertinent to the variables we wish to suppress. The submatrix b_{ss} then consists of the elements located upon the intersections of these deleted rows and columns; the elements in b_{rs} are the remaining ones in the deleted columns while those in b_{sr} are the remaining ones in the deleted rows, the relative positions of all elements being unaltered.

The important point to recognize now is that if the matrix $[B]$ has rank s, then $[B]_{\text{abr}}$ given by Eq. 9 is identically zero. This fact may be seen as follows.

Suppose we construct $[B]$ by writing down s independent rows, each containing $(r + s)$ elements and regarding these rows as represented by the partitioned matrix

$$[b_{sr} \ \vdots \ b_{ss}] \tag{10}$$

Here we may assume that the square submatrix b_{ss} is nonsingular. If we now construct r additional rows that are linear combinations of the rows in 10, these can be written

$$[l_{rs}] \cdot [b_{sr} \cdot b_{ss}] \tag{11}$$

in which $[l_{rs}]$ is a matrix (with r rows and s columns) effecting these row combinations. The matrix

$$[B] = \begin{bmatrix} l_{rs} \times b_{sr} & \vdots & l_{rs} \times b_{ss} \\ \cdots & \cdots & \cdots \\ b_{sr} & \vdots & b_{ss} \end{bmatrix} \tag{12}$$

clearly has rank s and its abridgment according to formula 9 yields a null result. Since any matrix of order $r + s$ and rank s can have the representation 12 (possibly after making appropriate row and column interchanges), the abridgment always yields a null result if only one selects the canceled rows and columns so that b_{ss} is nonsingular. The rows and columns selected to form b_{rr} in 9 are, therefore, not necessarily the first r.

The desired augmentation of a given $[G]$-matrix, as mentioned above, may now be accomplished in the following manner. To be specific let us consider the third-order matrix

$$[G] = \begin{bmatrix} G_{11} & G_{12} & G_{13} \\ G_{21} & G_{22} & G_{23} \\ G_{31} & G_{32} & G_{33} \end{bmatrix} \tag{13}$$

If we wish to augment to the order 4, we first insert a null row and column so as to have

$$[G]_{\text{exp}} = \begin{bmatrix} G_{11} & 0 & G_{12} & G_{13} \\ 0 & 0 & 0 & 0 \\ G_{21} & 0 & G_{22} & G_{23} \\ G_{31} & 0 & G_{32} & G_{33} \end{bmatrix} \tag{14}$$

which we shall refer to as the *expanded* form of $[G]$. The expanded form can alternatively have zeros in any row and column other than the second, but in the present discussions we will always choose for this purpose the *even*-numbered rows and columns since we wish to have these pertain to the branches in a linear tree that do not represent accessible terminal pairs and hence are associated with the variables to be eliminated.

The desired augmented $[G]$-matrix is now formed by adding to $[G]_{\text{exp}}$ any fourth-order matrix $[B]$ of rank 1, thus

$$[G]_{\text{aug}} = [G]_{\text{exp}} + [B] \tag{15}$$

Abridgment of $[G]_{\text{aug}}$ according to the process indicated in Eq. 9, in which the deleted row and column is the second, clearly yields the original matrix in Eq. 13.

If we wish to augment $[G]$ in Eq. 13 to the order 5, we form the expanded matrix

$$[G]_{\text{exp}} = \begin{bmatrix} G_{11} & 0 & G_{12} & 0 & G_{13} \\ 0 & 0 & 0 & 0 & 0 \\ G_{21} & 0 & G_{22} & 0 & G_{23} \\ 0 & 0 & 0 & 0 & 0 \\ G_{31} & 0 & G_{32} & 0 & G_{33} \end{bmatrix} \tag{16}$$

and then get the desired $[G]_{aug}$ from Eq. 15 in which $[B]$ is any fifth-order matrix of rank 2. Abridgment of $[G]_{aug}$ thus found, according to formula 9 with 2 and 4 as deleted rows and columns, again yields the original matrix in Eq. 13.

If $[G]_{aug}$ for the expanded form 14 is realizable as a 5-node network with a linear tree, then this network yields the matrix 13 with its implied terminal pairs 1, 2, 3 identified respectively with the tree branches 1, 3 and 4. If $[G]_{aug}$ for the expanded form 16 is realizable as a 6-node network with a linear tree, then the given matrix 13 is realized with its implied terminal pairs 1, 2, 3 identified with tree branches 1, 3, 5 respectively. For $[G]$ of order 3, this latter network is the most general that need be considered; and the freedom inherent in the construction of the fifth-order matrix $[B]$ of rank 2 provides all the leeway that exists for the realization of $[G]$, beyond the rather restrictive conditions imposed for its realizability as a 4-node network [the $(n + 1)$-node realizability conditions]. The question to be answered next, therefore, is concerned with how to construct the matrix $[B]$ so as to make the most of this leeway.

By the process just described, the given n-by-n $[G]$-matrix can be augmented to an order m which may be $n + 1$ or $n + 2, \ldots$ up to $2n - 1$. It can, of course, be augmented still further, but for reasons already mentioned, this would not yield a greater realizability potential. Realization of the given $[G]$-matrix is successful if $[G]_{aug}$ yields an $(m + 1)$-node network with a linear tree and all positive branch conductances.

As shown in the derivation of Eq. 16 in Ch. III, the branch conductances are obtained from a given $[G]$-matrix appropriate to a linear tree by the relation

$$[\bar{g}] = [T] \cdot [G] \cdot [T] \tag{17}$$

in which $[T]$ is given there by Eq. 8. If we identify $[G]$ in this relation with $[G]_{aug}$ in Eq. 15, then it is clear that we can write for the corresponding branch conductances

$$[\bar{g}]_{aug} = [T] \cdot [G]_{aug} \cdot [T] \tag{18}$$

Furthermore, if we define the additional branch-conductance matrices

$$[\bar{g}]_{exp} = [T] \cdot [G]_{exp} \cdot [T] \tag{19}$$

and

$$[\bar{g}] = [T] \cdot [B] \cdot [T] \tag{20}$$

we have

$$[\bar{g}]_{aug} = [\bar{g}]_{exp} + [\bar{g}] \tag{21}$$

In other words, the additive property applies here. We can separately compute a set of branch conductances pertinent to the expanded $[G]$-matrix as elements of a matrix $[\bar{g}]_{exp}$ and a set of branch conductances pertinent to the auxiliary matrix $[B]$ as elements of a matrix $[\bar{g}]$, and then form the resulting branch conductances simply by adding respective members of these two sets together.

Elements of the matrix $[\bar{g}]_{\text{exp}}$ pertinent to $[G]_{\text{exp}}$ are not all positive; neither are those of the matrix $[\bar{g}]$ pertinent to $[B]$. The crux of the whole problem is to construct $[B]$ in such a way that the positive elements in $[\bar{g}]$ swamp or at least cancel the corresponding negative ones in $[\bar{g}]_{\text{exp}}$, and vice versa, so that the branch-conductance matrix $[\bar{g}]_{\text{aug}}$ has no negative elements.

A trial-and-error procedure is a possible approach to this problem. Assume a matrix $[B]$ of proper order and rank, compute $[\bar{g}]$, add it to $[\bar{g}]_{\text{exp}}$, and see if all resulting elements are positive or zero; if not, revise the structure of $[B]$, etc. It is, however, better to devise a method for the construction of $[\bar{g}]$-matrices directly and thus eliminate the $[B]$-matrix from the procedure altogether. This scheme can be accomplished as follows.

Consider the representation for elements of the branch-conductance matrix $[\bar{g}]$, Eq. 17, as given by Eq. 28 of Ch. III, which we repeat here:

$$g_{ik} = -h_{i-1} \cdot h_k, \qquad \text{for } i \le k \le m \text{ and } i = 1, 2, \ldots m \tag{22}$$

in which the h_k are members in the vector set of matrix $[H]$ given by Eq. 27 or 29 of Ch. III. As shown there by Eq. 30, the scalar product in this expression may be replaced by the more specific algebraic form

$$g_{ik} = -\sum_{\nu=1}^{i} h_{i-1,\nu}\, h_{k\nu}, \qquad \text{for } k \ge i \text{ and } i = 1, 2, \ldots m \tag{23}$$

The rank of $[H]$ is the rank of the matrix $[B]$ in the representation for $[\bar{g}]$ given by Eq. 20. Hence, if we want $[\bar{g}]$ to be representative of a matrix $[B]$ of rank 1, we choose an $[H]$-matrix having a single nonzero column. Let us simplify the notation in this case by denoting elements in this column (the first column of $[H]$) by $h_0, h_1, \ldots h_m$. Then we have

$$g_{11} = -h_0h_1, \quad g_{12} = -h_0h_2, \quad \ldots \quad g_{1m} = -h_0h_m \tag{24}$$

$$g_{22} = -h_1h_2, \quad g_{23} = -h_1h_3, \quad \ldots \quad g_{2m} = -h_1h_m \tag{25}$$

$$g_{33} = -h_2h_3, \quad g_{34} = -h_2h_4, \quad \ldots \quad g_{3m} = -h_2h_m \tag{26}$$

and so forth.

From 24 and 25 we see that

$$g_{22} = \frac{g_{11}g_{12}}{-h_0^2}, \quad g_{23} = \frac{g_{11}g_{13}}{-h_0^2}, \quad \cdots \quad g_{2m} = \frac{g_{11}g_{1m}}{-h_0^2} \tag{27}$$

from 24 and 26, that

$$g_{33} = \frac{g_{12}g_{13}}{-h_0^2}, \quad g_{34} = \frac{g_{12}g_{14}}{-h_0^2}, \quad \cdots \quad g_{3m} = \frac{g_{12}g_{1m}}{-h_0^2} \tag{28}$$

and so forth, while from Eq. 24 alone we get

$$(g_{11} + g_{12} + \cdots + g_{1m}) = -h_0(h_1 + h_2 + \cdots + h_m) \tag{29}$$

But since columns in the $[H]$-matrix must sum to zero, we have

$$h_1 + h_2 + \cdots + h_m = -h_0 \tag{30}$$

and so Eq. 29 yields

$$h_0^2 = g_{11} + g_{12} + \cdots + g_{1m} \tag{31}$$

This result, together with Eqs. 27, 28, and their continuation show that if we want to construct a $[\bar{g}]$-matrix pertinent to a $[B]$-matrix of rank 1, we may do so by writing down any values we please for the elements $g_{11}, g_{12}, \ldots g_{1m}$ in the first row of $[\bar{g}]$, and then calculate elements in the remaining rows by the simple relation

$$g_{ik} = \frac{-g_{1,i-1}g_{1k}}{\sum_{k=1}^{m} g_{1k}} \qquad \begin{matrix} \text{for } k \geq i \\ \text{and } i = 2, 3, \ldots m \end{matrix} \tag{32}$$

A network having these branch-conductance values has the peculiar property that if we pick *any* linear tree and assign terminal pairs to all but one of its branches (this one can be *any* tree branch) then the resulting conductance matrix is identically zero!

The elements $g_{11}, g_{12}, \ldots g_{1m}$ which we choose arbitrarily (except that their sum must be nonzero) are conductances of branches forming a starlike tree (see Fig. 3 of Ch. III). There are many different starlike trees in a full graph; correspondingly there are many sets of m elements in the matrix $[\bar{g}]$ that can be chosen arbitrarily and the rest computed from these by formula 32 (suitably modified as to indexes). The elements in the first row of $[\bar{g}]$ are just a possible set. For the present we shall not elaborate this point further.

More significant is the fact that we have here a method for the construction of a branch conductance matrix $[\bar{g}]$ appropriate to a $[B]$-matrix of any desired order and rank. If $[B_1]$ and $[B_2]$ are like-order matrices, each of rank 1, then (except in degenerate cases) their sum $[B_1] + [B_2]$ has rank 2; and if we add three such matrices we get one of rank 3; and so forth. Since, as pointed out above, the additive property applies to the relation between a matrix $[B]$ and the corresponding branch conductance matrix $[\bar{g}]$, we can form two matrices $[\bar{g}_1]$ and $[\bar{g}_2]$ according to the method using formula 32, and add them to get a $[\bar{g}]$-matrix appropriate to a matrix $[B]$ of rank 2. In the network having branch conductances given by this $[\bar{g}]$-matrix we can pick *any* linear tree, assign terminal pairs to all but two of its branches (any two), and get a resulting conductance matrix $[G] \equiv 0$.

The extension of this procedure to the construction of a $[\bar{g}]$-matrix appropriate to a matrix $[B]$ of any desired order and rank is thus clear. A few simple numerical examples might be interesting at this point.

Consider the branch conductance matrix

$$[\bar{g}_1] = \begin{bmatrix} 1 & -2 & 2 & -3 \\ & -1 & 1 & -\frac{3}{2} \\ & & -2 & 3 \\ & & & -3 \end{bmatrix} \tag{33}$$

which is constructed by writing down a first row at random (its sum in this case is -2) and then rapidly computing the remaining elements by formula 32. Thus $g_{22} = -[1 \cdot (-2)]/(-2) = -1$; $g_{23} = -[1 \cdot (2)]/(-2) = 1$; $g_{24} = -[1 \cdot (-3)]/(-2) = -\frac{3}{2}$; etc. The corresponding $[B]$-matrix (if we are interested in it) may readily be calculated by the method discussed in art. 3, Ch. III. Specifically, Eq. 19 there gives

$$[T]^{-1} \cdot [\bar{g}_1] = \begin{bmatrix} 1 & -2 & 2 & -3 \\ & -3 & 3 & -\frac{9}{2} \\ & & 1 & -\frac{3}{2} \\ & & & -\frac{9}{2} \end{bmatrix} \tag{34}$$

and according to Eq. 20 of Ch. III we get for the matrix $[B]$ corresponding to the branch conductance matrix 33

$$[B_1] = [T]^{-1} \cdot [\bar{g}_1] \cdot [T]^{-1} = \begin{bmatrix} -2 & -3 & -1 & -3 \\ -3 & -\frac{9}{2} & -\frac{3}{2} & -\frac{9}{2} \\ -1 & -\frac{3}{2} & -\frac{1}{2} & -\frac{3}{2} \\ -3 & -\frac{9}{2} & -\frac{3}{2} & -\frac{9}{2} \end{bmatrix} \tag{35}$$

which clearly has rank 1.

In like manner we construct

$$[\bar{g}_2] = \begin{bmatrix} 1 & -2 & 1 & 2 \\ & 1 & -\frac{1}{2} & -1 \\ & & 1 & 2 \\ & & & -1 \end{bmatrix} \tag{36}$$

for which the sum of elements in the first row is 2. Correspondingly we get by the same process as is indicated in Eqs. 33 and 34

$$[B_2] = \begin{bmatrix} 2 & 1 & 3 & 2 \\ 1 & \frac{1}{2} & \frac{3}{2} & 1 \\ 3 & \frac{3}{2} & \frac{9}{2} & 3 \\ 2 & 1 & 3 & 2 \end{bmatrix} \tag{37}$$

having rank 1. Adding $[B_1]$ and $[B_2]$ we have

$$[B] = [B_1] + [B_2] = \begin{bmatrix} 0 & -2 & 2 & -1 \\ -2 & -4 & 0 & -\frac{7}{2} \\ 2 & 0 & 4 & \frac{3}{2} \\ -1 & -\frac{7}{2} & \frac{3}{2} & -\frac{5}{2} \end{bmatrix} \tag{38}$$

Now let us abridge this matrix so as to eliminate variables 3 and 4 (rows and columns 3 and 4 are deleted). Then we have for formula 9

$$\begin{aligned} b_{rs} \cdot b_{ss}^{-1} \cdot b_{sr} &= \begin{bmatrix} 2 & -1 \\ 0 & -\frac{7}{2} \end{bmatrix} \times \begin{bmatrix} 4 & \frac{3}{2} \\ \frac{3}{2} & -\frac{5}{2} \end{bmatrix}^{-1} \times \begin{bmatrix} 2 & 0 \\ -1 & -\frac{7}{2} \end{bmatrix} \\ &= -\tfrac{2}{49} \times \begin{bmatrix} 2 & -1 \\ 0 & -\frac{7}{2} \end{bmatrix} \times \begin{bmatrix} 5 & 3 \\ 3 & -8 \end{bmatrix} \times \begin{bmatrix} 2 & 0 \\ -1 & -\frac{7}{2} \end{bmatrix} \\ &= \begin{bmatrix} 0 & -2 \\ -2 & -4 \end{bmatrix} \end{aligned} \tag{39}$$

and since in this case

$$b_{rr} = \begin{bmatrix} 0 & -2 \\ -2 & -4 \end{bmatrix} \tag{40}$$

we see that the abridgment yields a null matrix as it should.

Alternately, let us abridge matrix 38 so as to eliminate variables 2 and 4. We then have

$$\begin{aligned} b_{rs} \cdot b_{ss}^{-1} \cdot b_{sr} &= \begin{bmatrix} -2 & -1 \\ 0 & \frac{3}{2} \end{bmatrix} \times \begin{bmatrix} -4 & -\frac{7}{2} \\ -\frac{7}{2} & -\frac{5}{2} \end{bmatrix}^{-1} \times \begin{bmatrix} -2 & 0 \\ -1 & \frac{3}{2} \end{bmatrix} \\ &= \tfrac{4}{9} \times \begin{bmatrix} -2 & -1 \\ 0 & \frac{3}{2} \end{bmatrix} \times \begin{bmatrix} \frac{5}{2} & -\frac{7}{2} \\ -\frac{7}{2} & 4 \end{bmatrix} \times \begin{bmatrix} -2 & 0 \\ -1 & \frac{3}{2} \end{bmatrix} = \begin{bmatrix} 0 & 2 \\ 2 & 4 \end{bmatrix} \end{aligned} \tag{41}$$

which is b_{rr} in this case. Hence the abridgment again yields an identically zero result as it should.

We are now in a position to formulate a procedure for the possible realization of a given $[G]$-matrix of order n which is not $(n + 1)$-node realizable, as follows: Insert in the given matrix a null row and column to form a tentative $[G]_{exp}$ and from this calculate a branch conductance matrix $[\bar{g}]_{exp}$ having some positive and some negative elements. Construct a branch conductance matrix $[\bar{g}]$ by the method of formula 32, choosing positive elements in the first row to cancel negative ones in the first row of $[\bar{g}]_{exp}$ as well as any other negative ones in this matrix, and at the same time allowing positive ones in $[\bar{g}]_{exp}$ to absorb resulting negative elements in $[\bar{g}]$. Some trial-and-error manipulation will soon reveal whether or not all resultant elements in the sum $[\bar{g}]_{exp} + [\bar{g}]$ can be made non-negative. If they can, we have a solution; if not, we can revise $[G]_{exp}$ by placing the null row and column in different positions, or we can next form a matrix $[G]_{exp}$ by inserting two nonadjacent null rows and columns in the given $[G]$-matrix.

We now construct two matrices $[\bar{g}_1]$ and $[\bar{g}_2]$ by the method of formula 32 and hence a matrix $[\bar{g}] = [\bar{g}_1] + [\bar{g}_2]$ so that all elements in the sum $[\bar{g}]_{exp} + [\bar{g}]$ are non-negative. Since we now have more free choices of elements, our chances for obtaining a solution are better than before.

Ultimately we can form a matrix $[G]_{exp}$ by inserting $n - 1$ null rows and columns in $[G]$ so that all even numbered rows and columns in the expanded matrix are null. A branch conductance matrix $[\bar{g}] = [\bar{g}_1] + [\bar{g}_2] + \cdots + [\bar{g}_{n-1}]$, in which the $[g_k]$ are constructed by the method of formula 32, affords the maximum number of arbitrary choices available in the process of obtaining a resultant matrix $[\bar{g}]_{exp} + [\bar{g}]$ having no negative elements.

The fact that $[\bar{g}]$ is a linear combination of the component $[\bar{g}_k]$-matrices is a distinct advantage in utilizing the available free choices most effectively. Thus we can first construct $[\bar{g}_1]$ so that $[\bar{g}]_{exp} + [\bar{g}_1]$ has elements as nearly non-negative as may be had by the free choices available in this construction process. Next we construct $[\bar{g}_2]$, utilizing the additional free choices (the same in number as before) as in the previous step so that the sum $[\bar{g}]_{exp} + [\bar{g}_1] + [\bar{g}_2]$ has elements as nearly non-negative as may be. Continuing in this way, each step follows essentially the same pattern with the same objective and carries the result closer to the desired goal.

Additional study, however, needs to be directed toward developing a systematic procedure that will clearly indicate whether or not a solution exists in a given situation.

Meanwhile the present method easily yields numerous solutions in situations which ordinarily are regarded as being rather difficult to solve.

As an example, consider the matrix*

$$[G] = \begin{bmatrix} 9 & 5 & -1 \\ 5 & 9 & 5 \\ -1 & 5 & 9 \end{bmatrix} \tag{42}$$

which by the method of art. 2, is seen to be appropriate to a starlike tree but does not fulfill the dominance conditions. The expanded form

$$[G]_{\text{exp}} = \begin{bmatrix} 9 & 0 & 5 & -1 \\ 0 & 0 & 0 & 0 \\ 5 & 0 & 9 & 5 \\ -1 & 0 & 5 & 9 \end{bmatrix} \tag{43}$$

by the transformation 17, yields the branch conductance matrix

$$[\bar{g}]_{\text{exp}} = \begin{bmatrix} 9 & -5 & 6 & -1 \\ & 5 & -6 & 1 \\ & & 4 & 5 \\ & & & 4 \end{bmatrix} \tag{44}$$

For the matrix $[\bar{g}]$ suppose we write

$$[\bar{g}] = \begin{bmatrix} x_1 & 5 & x_3 & x_4 \\ & \dfrac{-5x_1}{a} & \dfrac{-x_1x_3}{a} & \dfrac{-x_1x_4}{a} \\ & & \dfrac{-5x_3}{a} & \dfrac{-5x_4}{a} \\ & & & \dfrac{-x_3x_4}{a} \end{bmatrix} \tag{45}$$

where we have left three of the free choices arbitrary and have fixed the fourth so as to cancel one of the negative elements in the first row of $[\bar{g}]_{\text{exp}}$. The sum of the first-row elements is denoted by

$$a = x_1 + x_3 + x_4 + 5 \tag{46}$$

* Taken from the paper, "Synthesis Applications of Paramount and Dominant Matrices," by Paul Slepian and Louis Weinberg, *Proc. Nat. Electron. Conf.*, Vol. 14, 1958, pp. 611–630.

If we choose a negative value for x_1 and positive values for x_3 and x_4 as well as for the sum a, then all second-row elements in the matrix 45 are positive, and if we specify that

$$\frac{-x_1 x_3}{a} = 6 \tag{47}$$

then our only remaining concern is with the elements in Eqs. 36, 37 and 47 for which the conditions read

$$\frac{5x_3}{a} \leq 4, \quad \frac{5x_4}{a} \leq 5, \quad \frac{x_3 x_4}{a} \leq 4 \tag{48}$$

or

$$\frac{x_3}{a} \leq \frac{4}{5}, \quad \frac{x_4}{a} \leq 1, \quad \frac{x_3}{a} \leq \frac{4}{x_4} \tag{49}$$

Use of Eq. 47 yields

$$-x_1 \geq \tfrac{15}{2}, \quad a \geq x_4, \quad -x_1 \geq \frac{3x_4}{2} \tag{50}$$

If we choose $-x_1 = \frac{15}{2}$ then 46 and the central inequality in 50 give

$$a = -\tfrac{15}{2} + x_3 + x_4 + 5 \geq x_4 \tag{51}$$

from which we have

$$x_3 \geq \tfrac{5}{2} \tag{52}$$

Several possible solutions are now readily obtained. If we let

$$x_3 = \tfrac{5}{2} \quad \text{and} \quad a = x_4 \tag{53}$$

we get for the matrix 45

$$[\bar{g}] = \begin{bmatrix} -\frac{15}{2} & 5 & \frac{5}{2} & \frac{25}{8} \\ & 12 & 6 & \frac{15}{2} \\ & & -4 & -5 \\ & & & -\frac{5}{2} \end{bmatrix} \tag{54}$$

Alternately, the choices $x_3 = 6$, $x_4 = 4$ lead to $a = \frac{15}{2}$ and we have

$$[\bar{g}] = \begin{bmatrix} -\frac{15}{2} & 5 & 6 & 4 \\ & 5 & 6 & 4 \\ & & -4 & -\frac{8}{3} \\ & & & -\frac{16}{5} \end{bmatrix} \tag{55}$$

which evidently is also acceptable.

If we let $-x_1 = 8$, Eq. 46 yields

$$a = -3 + x_3 + x_4 \geq x_4 \tag{56}$$

or

$$x_3 \geq 3 \tag{57}$$

Here the choice $x_3 = 3$ and $a = x_4$ results in still another solution, namely the one for which

$$[g] = \begin{bmatrix} -8 & 5 & 3 & 4 \\ & 10 & 6 & 8 \\ & & -\frac{15}{4} & -5 \\ & & & -3 \end{bmatrix} \tag{58}$$

7. Realization of an Open-Circuit Resistance Matrix

The problem involved here is the dual of that discussed in art. 3. There we were concerned with the realization of a short-circuit conductance matrix appropriate to equilibrium equations on a node basis. Now we want to develop a network realization procedure given an open-circuit resistance matrix appropriate to equilibrium equations on a loop basis.

Terminal pairs or points of entry are of the "pliers type;" that is to say, they are created by cutting a set of l links in the network graph. In contrast with the situation on a node basis, we cannot assume that we are always dealing with a full graph. Thus if n denotes the number of tree branches in a full graph involving $n + 1$ nodes, then for $n = 1, 2, 3, 4, 5 \ldots$ we have respectively $l = 0, 1, 3, 6, 10 \ldots.$ For l equal to some integer other than these particular ones, the network cannot be represented by a full graph.

Again the solution to our problem depends upon our ability to construct the tree aproppriate to a given resistance matrix. As in the analogous problem on a node basis, no solution exists unless the given matrix is appropriate to some tree configuration; and once that tree structure is determined the total network and its branch resistance values are computed without difficulty. Existence of the tree, however, does not assure that the branch resistances are all positive and hence (as on the conductance basis) it is a necessary though not sufficient condition for the existence of a solution.

When a tree exists, the total network graph (which is found by inserting the links) is unique only if a full graph is involved; that is, when $l = n(n-1)/2$ for integer n values. Otherwise variants in the graph structure exist, although these are not very significant.

Again we shall be able to construct the tree (if it exists) on the basis of the algebraic sign pattern in the given matrix; however, the procedure for accomplishing this end is now considerably more roundabout. The method consists of two steps: First we develop a scheme whereby, from the sign matrix appropriate to the given resistance matrix, we construct a sign matrix appropriate to the corresponding conductance matrix; and from this one we then find the tree by the method given in art. 2 above.

The rows in the open-circuit resistance matrix pertain to loops formed by inserting the links, one at a time, into the tree structure. If two loops i and k have one or more tree branches in common, then the element R_{ik} in the matrix $[R]$ is nonzero; and its sign is plus if the pertinent link currents (according to their reference directions) traverse these common tree branches in the same direction; it is minus if these link-current directions are opposite.

If a third link current circulating on loop s also traverses one or more of these same tree branches, and if the elements R_{ik} and R_{is} are positive, then R_{sk} must likewise be positive. That is to say, if three link currents i, k and s traverse paths that have one or more tree branches in common, and if their reference directions in these common branches coincide, then all three elements R_{ik}, R_{is} and R_{sk} are positive.

Observe, in this connection, that the reference directions must always agree for two of the three currents; and if one of them has a contrary direction, then two of the three elements R_{ik}, R_{is}, R_{sk} (namely the two that pertain to the contrary link current) are negative. Consistent positiveness among the signs of these elements can in this case be achieved by changing all signs in the row and column of $[R]$ pertaining to the contrary current. We will speak of this algebraic sign relationship among the elements R_{ik}, R_{is}, R_{sk} as being *positive* or *potentially positive*.

This sign relationship cannot hold for a situation in which the link currents i, k and s traverse paths that do *not* have a common tree branch, because the common tree-branch condition for the three current paths is both necessary and sufficient for the sign relationship to be positive or potentially positive. Conversely, this sign relationship is a necessary and sufficient condition to establish that the three pertinent currents traverse a common tree branch.

The same reasoning evidently applies to more than three link currents and corresponding elements in the matrix $[R]$; and so we can say that if and only if the algebraic sign of two or more elements in $[R]$ are positive or potentially positive can the pertinent link currents traverse a common tree branch. In the implied network graph, the corresponding links together with the pertinent tree branch form a cut set. We have thus found a way of discovering cut sets solely on the basis of algebraic sign relationships among the elements of the matrix $[R]$.

The number of cut sets equals the number of tree branches. The number of links, which equals the order of the given matrix $[R]$, determines only a lower limit for the number of tree branches, since, in the formula $l = n(n - 1)/2$ pertaining to a full graph, n must be large enough to yield a value for l that is as large or larger than the order of $[R]$. Hence in constructing a tree appropriate to $[R]$ we have theoretically an infinite number of possibilities. We will limit our detailed discussion essentially to the tree with fewest branches since this leads to a full or almost full graph and hence is closest to being the dual of the synthesis procedure based upon a given $[G]$-matrix.

In the expression

$$[R] = [\beta_{lb}] \cdot [r] \cdot [\beta_{lb}]_t \tag{59}$$

which we get from Eqs. 94, Ch. II for a purely resistive network, the tie-set matrix has the form (Eq. 11, Ch. II)

$$[\beta_{lb}] = [u_l : \beta_{ln}] \tag{60}$$

The rows of β_{ln} indicate confluent tree branches traversed by the respective link currents corresponding to these rows, and the columns indicate links belonging to cut sets for the respective tree branches to which these columns pertain. Thus, since $[\alpha]$ and $[\beta]$, Eqs. 12 and 13 of Ch. II, fulfill the consistency condition $[\alpha] = [\beta]_t^{-1}$, we have $\beta_{ln} = -(\alpha_{nl})_t$, as has been pointed out previously; and so β_{ln} is actually a cut-set and a tie-set matrix at the same time. By rows, it is a tie-set matrix (omitting the links in the tie sets); and by columns, it is a cut-set matrix (omitting the tree branches in the cut sets).

Taking cognizance of these facts, the algebraic sign relationship among elements of the matrix $[R]$ enable us to construct the matrix β_{ln}, as will now be shown in detail by means of several examples.

Consider the following sign matrix pertinent to a given $[R]$ of order 10:

$$\begin{array}{cccccccccc|c}
1 & 2 & 3 & 4 & 5 & 6 & 7 & 8 & 9 & 10 & \\
\hline
+ & + & + & + & + & 0 & + & + & + & + & 1 \\
 & + & + & + & 0 & + & + & + & + & + & 2 \\
 & & + & + & - & 0 & + & + & + & + & 3 \\
 & & & + & 0 & - & + & + & + & + & 4 \\
 & & & & + & 0 & + & - & + & - & 5 \\
 & & & & & + & - & + & + & - & 6 \\
 & & & & & & + & + & + & + & 7 \\
 & & & & & & & + & + & + & 8 \\
 & & & & & & & & + & + & 9 \\
 & & & & & & & & & + & 10
\end{array} \tag{61}$$

Numbering of the rows and columns from 1 to 10 is again done for convenience. Incidentally we note the presence of a number of zero elements. These are rather common in open-circuit resistance matrices as opposed to short-circuit conductance matrices. In a full graph, the latter has no zero elements, while the former does. In a less-than-full graph, the resistance matrix may have many zeros.

Scanning along the first row of the matrix 61 we see by inspection that elements at the intersections of rows and columns 1, 2, 3, 4, 7, 8, 9, 10 are all positive; hence this group of links forms a possible cut set. Observe that we have omitted 5 and 6 for obvious reasons; thus 5 is not coupled to 4 and 6 is not coupled to 1. However, again scanning along row 1 we can pick out links 1, 5, 7, 9 as forming another possible cut set. Note here that 2 and 4 are not included because they are not coupled to 5. Branch 3 has a sign inconsistency with 5, as do 8 and 10, while 6 is omitted because its coupling with 5 is zero.

There are no other groups including branch 1, so we now scan along the second row and pick out branches 2, 6, 8, 9. Here 3 has no coupling with 6; R_{46} has the wrong sign; R_{67} and $R_{6,10}$ also have wrong signs. We can pick other possible groups out of row 2, like 2, 7, 8, 9, 10 for instance, but this is a subgroup in the very first one we picked out above; and unless we are interested in constructing a tree with more than the smallest number of branches, we skip such subgroups, as we have skipped similar subgroups in connection with row 1 above.

In the third row of matrix 61 we now pick out branches 3, 5, 8, 10, for which all pertinent R_{sk} are positive after we multiply row and column 5 by -1. As they stand, these elements are *potentially* positive. Row 3 reveals no other groups, so we next consider row 4 and find the group 4, 6, 7, 10 for which the R_{sk} are potentially positive.

The remaining rows are seen to contain only subgroups in the ones already found; and we note, moreover, that the five groups we have picked out imply a tree with $n = 5$, which, in a full graph, yields $l = 5 \cdot 4/2 = 10$. Since this equals the number of links (the order of the matrix 61), we can stop at this point and proceed with the construction of β_{ln}.

The available information enables us to write matrix 62 (see facing page). In the first of these two forms we have merely indicated nonzero elements in the matrix β_{ln} by placing x's in columns according to the cut sets selected from the sign matrix 61. Next these nonzero elements are converted into plus or minus signs, noting that the Gramian formed from the rows must yield the sign matrix 61. Thus, starting tentatively with plus signs in the first row we discover the signs in the first two columns. Then we assume tentatively that the other element in row 2 is also plus,

	11	12	13	14	15
1	x	x			
2	x		x		
3	x			x	
4	x				x
5		x		x	
6			x		x
7	x	x			x
8	x		x	x	
9	x	x	x		
10	x			x	x

$$\rightarrow \begin{array}{c} \begin{array}{ccccc} 11 & 12 & 13 & 14 & 15 \end{array} \\ \left[\begin{array}{ccccc} + & + & 0 & 0 & 0 \\ + & 0 & + & 0 & 0 \\ + & 0 & 0 & + & 0 \\ + & 0 & 0 & 0 & + \\ 0 & + & 0 & - & 0 \\ 0 & 0 & + & 0 & - \\ + & + & 0 & 0 & + \\ + & 0 & + & + & 0 \\ + & + & + & 0 & 0 \\ + & 0 & 0 & + & + \end{array}\right] \end{array} = \beta_{ln} \tag{62}$$

and determine signs in all nonzero elements of the third column. Continuation of this process readily yields all signs. The vacant spaces are filled in with zeros, and the result is a matrix β_{ln} consistent with the sign matrix 61 (differing from the usual convention in that $+1$ or -1 elements are abbreviated by the signs alone).

Except for a reversal of sign (which is unimportant here), the transpose of 62 is the cut-set matrix α_{nl}. Hence the Gramian formed from the columns of 62 yields the following sign matrix on a node basis:

$$\begin{array}{c} \begin{array}{ccccc} 11 & 12 & 13 & 14 & 15 \end{array} \\ \left[\begin{array}{ccccc} + & + & + & + & + \\ & + & + & - & + \\ & & + & + & - \\ & & & + & + \\ & & & & + \end{array}\right] \end{array} \begin{array}{c} \\ 11 \\ 12 \\ 13 \\ 14 \\ 15 \end{array} \tag{63}$$

and by the process of art. 2, the tree is readily constructed as shown in

Fig. 5. The complete graph, in which the links are dotted, is drawn in Fig. 6.

Branch resistance values are easily determined from a given numerical matrix [R], beginning with those tree branches that singly represent nondiagonal elements in [R]. For example, the resistance of branch 11 must equal the element R_{12} or R_{34} or R_{23} or R_{14}. The resistance of branch 14 must equal $-R_{35}$ or $-R_{58}$ or $-R_{5,10}$, and so forth. Thus no special formulas are needed to fill in the branch resistance values once the graph corresponding to [R] is determined; and it is quite clear that the element values in this given matrix must fulfill some rather tight conditions if the resulting network is to exist. By and large, these conditions are more demanding than those pertinent to a given [G]-matrix on the node basis.

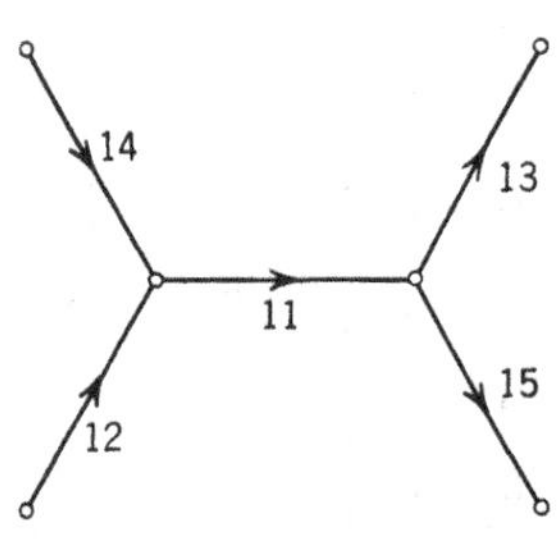

FIG. 5. Tree corresponding to sign matrix 63 constructed according to the process of art. 2.

In this connection, it is alternately possible, of course, to attempt the realization of a given [R]-matrix by computing its inverse and then applying the method of arts. 2 and 3 pertinent to the realization of [G]-matrices. In fact such a procedure, if successful,

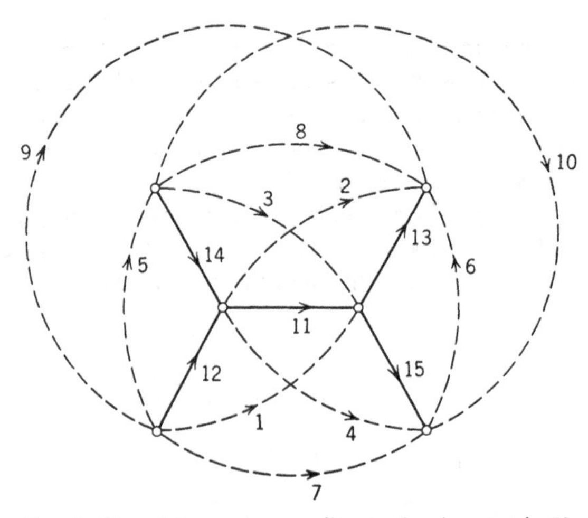

FIG. 6. Complete graph according to the sign matrix 63.

realizes [R] in a full graph, and hence has greater realization potentialities than the dual procedure discussed here.

If we are interested in the synthesis of a purely resistive network, or any other one-element-kind network, this alternate scheme is undoubtedly the thing to use. But if the network is of the two- or three-element-kind variety and we are given two or all three of the open-circuit matrices [R], [L] and [S], then this alternate scheme is not applicable, for there is no way that we can interconnect the separate single-element-kind networks (analogous to the procedure discussed in art. 5) so as to obtain the desired result, while with the dual realization method there also exists a dual to the procedure in art. 5, as we shall show presently.

Meanwhile we will discuss additional examples of the open-circuit matrix realization method in order to illustrate its features and peculiarities more adequately. Let us consider next the sign matrix for a given [R] of order 10 having the following form:

$$
\begin{array}{cccccccccc}
1 & 2 & 3 & 4 & 5 & 6 & 7 & 8 & 9 & 10
\end{array}
$$
$$
\left[\begin{array}{cccccccccc}
+ & 0 & - & - & 0 & + & - & 0 & - & + \\
 & + & 0 & - & - & + & + & - & 0 & - \\
 & & + & 0 & - & - & + & + & - & 0 \\
 & & & + & 0 & 0 & - & + & + & - \\
 & & & & + & - & 0 & - & + & + \\
 & & & & & + & - & 0 & 0 & - \\
 & & & & & & + & - & 0 & 0 \\
 & & & & & & & + & - & 0 \\
 & & & & & & & & + & - \\
 & & & & & & & & & +
\end{array}\right]
\begin{array}{c}
1 \\ 2 \\ 3 \\ 4 \\ 5 \\ 6 \\ 7 \\ 8 \\ 9 \\ 10
\end{array}
\qquad (64)
$$

Starting with the first row, we pick out links for which the R_{sk} elements are all positive or potentially positive. The first such group is 1, 3, 6, 7, having a potentially positive set of nondiagonal elements. It is not hard to recognize this relationship by inspection, as well as to see also that no other links are included in this group. Again in row 1, we can select the group 1, 4, 9, 10 in the same manner; and it is now clear that there are no other groups containing link No. 1. In the second row the first group we discover is 2, 4, 7, 8; and a second one is seen to be 2, 5, 6, 10. In the third row we have the group 3, 5, 8, 9. The remaining rows yield only subgroups. Hence these five are again used to determine a tentative tie-set schedule, namely

	11	12	13	14	15
1	x	x			
2			x	x	
3	x				x
4		x	x		
5				x	x
6	x			x	
7	x		x		
8			x		x
9		x			x
10		x		x	

(65)

in which the columns represent cut sets formed by the stated groups of links.

Now to convert the x's in this schedule into + or − signs, we can begin by arbitrarily choosing a plus sign for the topmost element in each column. This we can do since *all* signs in any column (which are rows of a cut-set matrix) can be changed at will anyway. Forming the Gramian by rows in 65 must yield the signs in 64. Carrying out this operation with the first row fixes all signs in the first two columns in 65, with the second row it fixes all elements in the third and fourth columns, and with the third row it fixes all remaining signs in the fifth column. Forming the Gramian products with all remaining rows yields signs that are consistent with the remaining ones in matrix 64 (as they must if the implied tree structure exists).

We thus are able to replace 65 by the tie-set matrix 66 (see facing page). Forming the Gramian by columns we now get the sign matrix 67, as illustrated on the opposite page, for the corresponding node basis. Here we observe that if we multiply rows and columns 2 and 4 by minus signs, we convert matrix 67 into one in which all signs except those on the principal diagonal are minus, which is characteristic of a dominant

$$\beta_{ln} = \begin{array}{c} \begin{array}{ccccc} 11 & 12 & 13 & 14 & 15 \end{array} \\ \left[\begin{array}{ccccc} + & + & 0 & 0 & 0 \\ 0 & 0 & + & + & 0 \\ - & 0 & 0 & 0 & + \\ 0 & - & - & 0 & 0 \\ 0 & 0 & 0 & - & - \\ + & 0 & 0 & + & 0 \\ - & 0 & + & 0 & 0 \\ 0 & 0 & - & 0 & + \\ 0 & - & 0 & 0 & - \\ 0 & + & 0 & - & 0 \end{array}\right] \end{array} \tag{66}$$

$$\left[\begin{array}{ccccc} + & + & - & + & - \\ & + & + & - & + \\ & & + & + & - \\ & & & + & + \\ & & & & + \end{array}\right] \tag{67}$$

matrix, and hence the tree is starlike. The rest of the solution is now straightforward.

As a third example we consider the sign matrix of an open-circuit resistance matrix given by

$$\begin{array}{c} \begin{array}{cccccccccc} 1 & 2 & 3 & 4 & 5 & 6 & 7 & 8 & 9 & 10 \end{array} \\ \left[\begin{array}{cccccccccc} + & + & + & + & + & + & + & 0 & 0 & 0 \\ & + & + & + & + & + & + & + & + & 0 \\ & & + & + & + & + & + & + & + & + \\ & & & + & + & + & + & + & + & + \\ & & & & + & + & + & + & + & 0 \\ & & & & & + & + & + & + & + \\ & & & & & & + & + & + & + \\ & & & & & & & + & + & + \\ & & & & & & & & + & + \\ & & & & & & & & & + \end{array}\right] \begin{array}{l} 1 \\ 2 \\ 3 \\ 4 \\ 5 \\ 6 \\ 7 \\ 8 \\ 9 \\ 10 \end{array} \end{array} \tag{68}$$

As we shall see, this example illustrates some essential considerations not brought out by the previous ones. Picking out groups of links for which the pertinent R_{sk} elements are all positive, we start with the group 1, 2, 3, 4, 5, 6, 7, and next recognize the group 2, 3, 4, 5, 6, 7, 8, 9. In the third row we can pick the links 3, 4, 6, 7, 8, 9, 10, skipping 5 because it is not coupled with 10.

Next, in the fourth row we are inclined to pick the links 4, 6, 7, 8, 9, 10, again skipping 5 for the same reason as in the previous group. However, we now observe that the group we have just selected is a subgroup under the one picked out of the previous row. Indeed, it becomes clear that any other groups that we can, from here on, pick out, are subgroups. There is nothing wrong with picking subgroups. We can select lots of them; but if we again are interested in finding the tree which has fewest branches, we can select only certain subgroups and we have apparently no guiding principle to indicate how to proceed.

There is, however, such a principle. In the formation of a tentative schedule like 65 in the previous example we can invoke an "exclusion principle" (it was not needed in the other examples) which recognizes that this schedule can have no repetitive rows since these indicate paths for loop currents (or Kirchhoff voltage-law equations) and these must form a distinct set. Therefore, we must exclude from the above formation of subgroups, all such that would yield identical rows in the tentative schedule like 65.

Of course, if we do not limit the number of columns in this schedule then we can admit all the subgroups we want without having to invoke this exclusion principle; but we are aiming for the minimum 5-branch tree, and so we are limited to five columns.

With these ideas in mind we arrive at the tentative tie-set schedule (69) shown on the facing page. Here the column headed 12 is the first one picked out above, column 13 is the second, and column 14 is the third. The column headed 15 is the fourth group selected above with links 6 and 8 discarded. In this form it is still a legitimate subgroup, as is also the one in column 11 which is a subgroup under the one in column 12. It is not hard to see that the selection of subgroups in columns 11 and 15 is unique if we wish to avoid repetitive patterns in the rows. The arrangement of the groups in columns is, of course, arbitrary but we have chosen an arrangement that makes the distinction between successive row patterns clearly evident in a systematic manner.

Since the signs in matrix 68 are all positive, the tie-set matrix β_{ln} is the schedule 69 with plus signs instead of x's and zeros otherwise. Hence the sign matrix on a node basis is seen to consist of all plus signs and no zeros which is characteristic of a linear tree. The reader can finish the rest of the problem as an exercise.

	11	12	13	14	15
1	x	x			
2	x	x	x		
3	x	x	x	x	
4	x	x	x	x	x
5		x	x		
6		x	x	x	
7		x	x	x	x
8			x	x	
9			x	x	x
10				x	x

(69)

8. Synthesis Procedure When Several Loop-Parameter Matrices Are Given

The method here is completely dual to that given in art. 5 for the synthesis of two- and three-element-kind networks from given node-parameter matrices. There each matrix $[G]$, $[C]$ and $[\Gamma]$ is realized as a separate one-element-kind network with the same basic tree configuration, whereupon these networks are connected in parallel by superimposing corresponding nodes or what amounts to the same thing, by connecting corresponding branches in parallel.

On an open-circuit impedance basis where the given matrices are $[R]$, $[L]$ and $[S]$, these are also tentatively realized separately as one-element-kind networks with the same basic tree configuration by the method given in the preceding article, whereupon they are combined by connecting corresponding branches in *series*. That is to say, resulting branch k is a series combination of branch k in the resistive network, branch k in the inductive network, and branch k in the capacitive network, for all branches $k = 1, 2, \dots b$. The resulting terminal pairs are of the pliers type and are located in the links, the same as for each separate one-element-kind network. It is clear by inspection that this process yields the desired result.

It is also rather obvious that it is of no avail to realize each matrix $[R]$, $[L]$ or $[S]$ separately by finding its inverse and then using the node method discussed above for realizing $[G]$-matrices. So far as the individual realizations of the three open-circuit matrices are concerned, the results obtained in this way are good, but one cannot connect corresponding

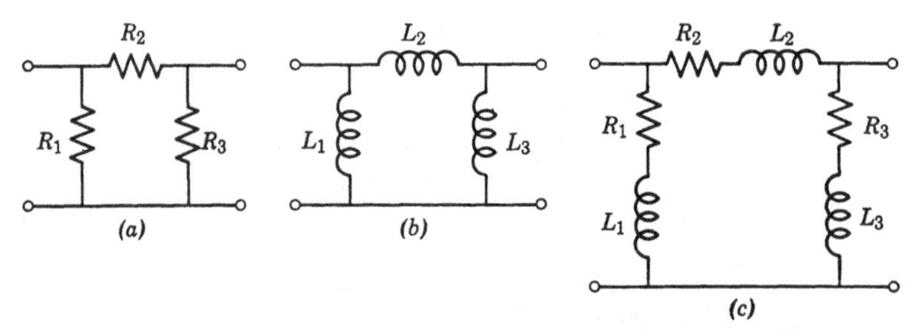

FIG. 7. Open-circuit input impedances of component networks (*a*) and (*b*) are correct, but network (*c*) does not yield an open-circuit input impedance which is the sum of these components.

terminal pairs in series and obtain the proper series combination. Neither can one obtain the desired result by constructing a resultant network in which each branch is a series combination of corresponding branches in the individual one-element-kind networks after the fashion described above where the individual networks are realized as open-circuit impedance matrices.

In order to appreciate the significance of these remarks consider the simple example in Fig. 7. For the individual networks in parts (*a*) and (*b*) we have the open-circuit input impedances

$$R_{11} = \frac{R_1(R_2 + R_3)}{R_1 + R_2 + R_3}, \quad L_{11}s = \frac{L_1(L_2 + L_3)s}{L_1 + L_2 + L_3} \tag{70}$$

in which s is the complex frequency variable. Their sum, which is the open-circuit input impedance of the desired network is

$$z_{11} = R_{11} + L_{11}s = \frac{R_1(R_2 + R_3)}{R_1 + R_2 + R_3} + \frac{L_1(L_2 + L_3)s}{L_1 + L_2 + L_3} \tag{71}$$

while the open-circuit input impedance of circuit (*c*) in Fig. 7 is given by

$$z_{11} = \frac{(R_1 + L_1s)(R_2 + R_3 + L_2s + L_3s)}{R_1 + R_2 + R_3 + (L_1 + L_2 + L_3)s} \tag{72}$$

which is not at all what is wanted.

On the other hand, if we had drawn T-circuits in Fig. 7 instead of π-circuits, the pertinent result would have been correct, since these are forms corresponding to the open-circuit impedance realization method.

9. An Alternate Method of Synthesis for *RLC* Networks from Given Parameter Matrices

The synthesis procedure to be discussed here is tailored to fit given parameter matrices $[R]$, $[L]$, $[G]$ and $[C]$ pertinent to equilibrium expressed on a mixed basis as presented in art. 5, Ch. II. In this connection our intuition immediately suggests that the matrices $[R]$ and $[L]$ be separately realized by the open-circuit impedance method given in art. 7 and combined to form an RL box with λ terminal pairs; further, that the matrices $[G]$ and $[C]$ be separately realized by the short-circuit admittance method given in arts. 2 and 3 and combined to form a GC box with σ terminal pairs; and finally, that these two boxes be properly interconnected to form the desired network. Let us consider the details of this general scheme more carefully.

Any physical network composed of R's, L's and C's can certainly be regarded as having the form of a GC portion connected to an RL portion. Moreover we may think of each portion as being in a separate box having the same number of terminal pairs. In the RL box, which is characterized by loop currents, these terminal pairs are pliers-type entries, and in the GC box, which is characterized by node-pair voltages, they are soldering-iron-type entries. Terminal pairs of the RL box are connected to respective terminal pairs of the GC box. The only question that needs to be answered is how we accommodate this viewpoint in the form of the analysis given in art. 5 of Ch. II.

In this regard we shall have to relinquish some of the generality inherent in that approach, but in turn we gain tremendously in simplification both from an analysis as well as from a synthesis standpoint; and it is not even certain that the loss in generality (which is the price we pay for the simplification gained) ultimately costs us anything with regard to results achievable in a synthesis problem, although this question will require further careful study.

The RL portion of the total graph, considered separately, contains b_L branches of which n_L characterize any tree pertinent to this subgraph and $l_L = b_L - n_L$ are the number of links associated with any tree, or the number of independent loops in the RL subgraph.

The GC portion of the total graph, considered separately, contains b_C branches of which n_C characterize any tree, and this subgraph has $l_C = b_C - n_C$ independent loops or links associated with any tree.

Sketches of the *RL* and *GC* boxes with their terminal pairs and voltage-current designations are shown in Fig. 8. For the *RL* box, the terminal voltages $v_{L1}, v_{L2}, v_{L3} \cdots$ are voltage drops around paths traversed by the

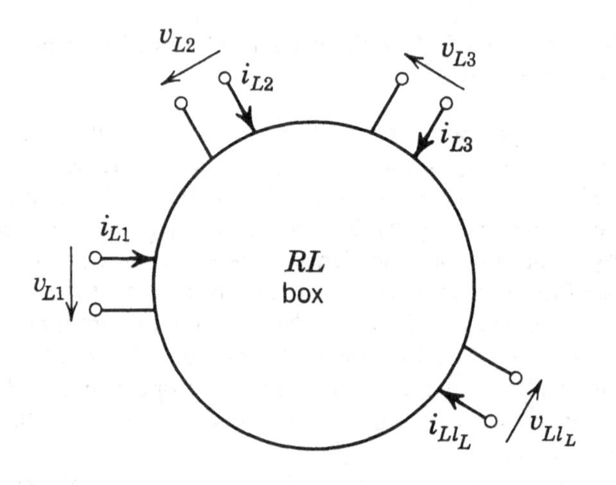

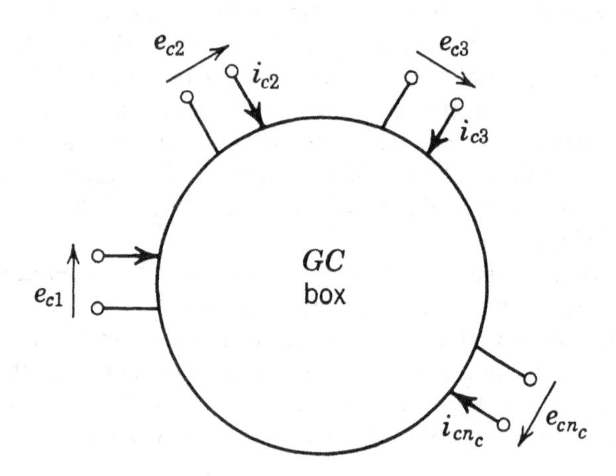

FIG. 8. Total network separated into an *RL* and a *GC* box.

loop currents $i_{L1}, i_{L2}, i_{L3} \cdots$ respectively. For the *GC* box, the terminal voltages $e_{C1}, e_{C2}, e_{C3} \cdots$ are node-pair voltages and hence are voltage rises in their respective reference directions while the terminal currents $i_{C1}, i_{C2}, i_{C3} \cdots$ respond to these node-pair voltages in the normal manner.

The number of terminal pairs in the *RL* box equals the number of l_L of independent loops in the RL subgraph. These terminal pairs are pliers-type entries in a set of links.

The number of terminal pairs in the GC box equals the number n_C of tree branches in the GC subgraph. These terminal pairs are soldering-iron-type entries across a set of tree branches.

Since the total network is formed by connecting terminal pairs in the RL box to respective ones in the GC box, it is necessary that $l_L = n_C$.

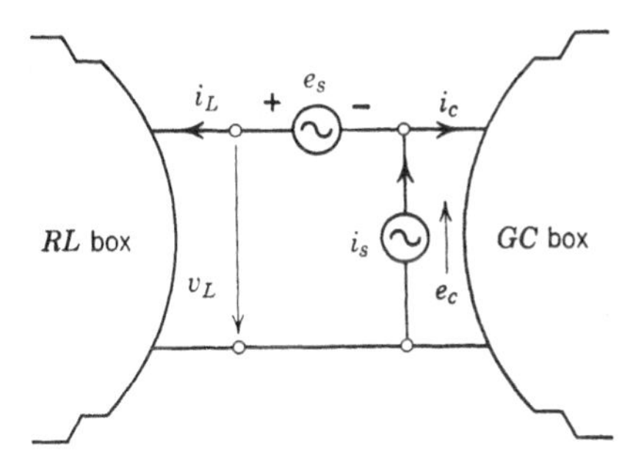

FIG. 9. Situation at a typical interconnection of the RL and GC boxes in Fig. 8.

The RL box is characterized on a loop basis by parameter matrices $[L_L]$ and $[R_L]$. In terms of these the RL box has volt-ampere relations at its terminal pairs given by

$$[L_Lp + R_L] \cdot i_{Lv}] = v_L] \tag{73}$$

in which $i_{Lv}]$ is a column matrix containing the loop or link current variables and $v_L]$ is a column matrix containing the voltage drops around loops.

The GC box is characterized on a node basis by parameter matrices $[C_C]$ and $[G_C]$. In terms of these the GC box has volt-ampere relations at its terminal pairs given by

$$[C_Cp + G_C] \cdot e_{Cv}] = i_C] \tag{74}$$

in which $e_{Cv}]$ is a column matrix containing the node-pair voltage variables and $i_C]$ is a column matrix containing the current responses at these terminal pairs.

The situation at a typical terminal pair after the two boxes are connected is shown in Fig. 9. e_s and i_s are applied voltage and current sources. Voltage- and current-law equations for this situation are respectively given by the equations

$$v_L - e_C = e_s$$

and

$$i_L + i_C = i_s \tag{75}$$

With the help of Eqs. 70 and 71, the equilibrium relations for the total network are therefore seen to be expressed by

$$\begin{aligned}[L_L p + R_L] \cdot i_{Lv}] - e_{Cv}] &= e_s] \\ i_{Lv}] + [C_C p + G_C] \cdot e_{Cv}] &= i_s]\end{aligned} \tag{76}$$

which we can rewrite in either of the following two alternative forms:

$$\begin{bmatrix} L_L p + R_L & \vdots & -u \\ \cdots & \cdots & \cdots \\ u & \vdots & C_C p + G_C \end{bmatrix} \times \begin{bmatrix} i_{Lv} \\ \cdots \\ e_{Cv} \end{bmatrix} = \begin{bmatrix} e_s \\ \cdots \\ i_s \end{bmatrix} \tag{77}$$

or

$$\begin{bmatrix} L_L & \vdots & 0 \\ \cdots & \cdots & \cdots \\ 0 & \vdots & C_C \end{bmatrix} p + \begin{bmatrix} R_L & \vdots & -u \\ \cdots & \cdots & \cdots \\ u & \vdots & G_C \end{bmatrix} \times \begin{bmatrix} i_{Lv} \\ \cdots \\ e_{Cv} \end{bmatrix} = \begin{bmatrix} e_s \\ \cdots \\ i_s \end{bmatrix} \tag{78}$$

in which u is a unit matrix of order $l_L = n_C$.

The last form is comparable and very similar to Eqs. 133 in art. 5 of Ch. II. Equations 78 are considerably simpler in two important respects. First of all the number of variables is considerably smaller since $i_v]$ in Eq. 133 involves loop currents in the total network graph while $i_{Lv}]$ represents loop currents in the *RL* portion alone, and since $e_v]$ in Eq. 133 contains node-pair voltages for the total network whereas $e_{Cv}]$ is restricted to node pairs in the *GC* portion. Thus the total number of variables involved in Eqs. 133 of Ch. II equals b, the number of branches in the total graph, while the number of variables in Eqs. 78 equals $l_L + n_C = 2l_L = 2n_C$, which on an average will be approximately half of b, and thus the present mixed basis on this score compares equally with either of the standard node or loop methods of expressing equilibrium.

Secondly, Eqs. 78 are simpler than Eqs. 133 of Ch. II in that the "connection matrix" γ_{ln} now is simply a unit matrix; interconnection of the two boxes presents no problem.

What is the price we pay for all these advantages? The present situation is not completely general because the *RL* box cannot have more loops than terminal pairs and the *GC* box cannot have more node pairs or tree branches than terminal pairs. A completely general network when separated into an *RL* and a *GC* portion does not necessarily conform to this specification if the topological structure of each is arbitrary and the junctions between them are random.

This state of affairs is, however, consistent with that encountered in *RLC* synthesis on a straight-loop basis or node basis where the $[R]$-, $[L]$-, $[S]$-matrices of order l must separately be l-loop realizable and the $[G]$-, $[C]$-, $[\Gamma]$-matrices of order n must separately be $(n + 1)$-node realizable. In other words, the networks obtained by these procedures are restricted on a loop basis to have no more links than terminal pairs and on a node basis to have no more tree branches than terminal pairs.

Nevertheless, it is at this stage not possible to say whether these relatively simple methods of network realization are limited in the results which can be achieved through their use, since topological randomness and generality with regard to response functions obtainable are by no means to be classed together. It would be equally difficult to prove or to disprove that such is the case.

Problems

1. Show that

$$\overset{*}{[G]} = \begin{bmatrix} 11 & -2 & -3 & -5 \\ -2 & 8 & -2 & 0 \\ -3 & -2 & 8 & -1 \\ -5 & 0 & -1 & 6 \end{bmatrix}$$

and

$$[G] = \begin{bmatrix} 7 & 6 & 2 & 0 \\ 6 & 16 & 10 & 5 \\ 2 & 10 & 12 & 5 \\ 0 & 5 & 5 & 6 \end{bmatrix}$$

are both appropriate to the same network, one for a starlike and the other for a linear tree. Find the network graph and indicate all element values in mhos.

2. The following $[G]$-matrix is appropriate to a 7-node network with linear tree. Determine the consecutive tree-branch numbering and realize the network.

$$[G] = \begin{bmatrix} 31 & 8 & 25 & 11 & 14 & 23 \\ 8 & 12 & 9 & 4 & 11 & 5 \\ 25 & 9 & 31 & 10 & 17 & 19 \\ 11 & 4 & 10 & 16 & 6 & 14 \\ 14 & 11 & 17 & 6 & 21 & 10 \\ 23 & 5 & 19 & 14 & 10 & 27 \end{bmatrix}$$

3. Construct the tree pertinent to the following sign matrix on a node basis:

$$S = \begin{bmatrix} + & - & - & - & - & + & + & + & + & - \\ & + & + & + & + & + & + & - & - & + \\ & & + & - & + & - & - & + & - & - \\ & & & + & + & - & - & + & + & + \\ & & & & + & - & - & + & - & + \\ & & & & & + & + & + & + & - \\ & & & & & & + & + & + & - \\ & & & & & & & + & - & + \\ & & & & & & & & + & + \\ & & & & & & & & & + \end{bmatrix}$$

4. Given the following cut-set matrix

$$\alpha = \begin{bmatrix} 1 & 0 & 0 & 0 & 0 & 0 & 0 & 0 & 1 & 0 & 0 \\ -1 & 1 & 0 & 0 & 0 & 0 & 1 & 0 & 0 & 0 & 0 \\ 0 & 0 & 1 & 0 & 0 & 0 & 1 & 0 & 0 & 1 & 0 \\ 0 & 0 & 0 & 1 & 0 & 0 & 0 & 1 & 0 & 1 & 0 \\ 0 & 0 & 0 & 0 & 1 & -1 & 0 & 1 & 0 & 0 & 0 \\ 0 & 0 & 0 & 0 & 0 & 1 & 0 & 0 & 0 & 0 & 1 \end{bmatrix}$$

Find the pertinent graph and draw it with straight lines of equal length. Number all branches and attach reference arrows. Check your result.

5. From the following cut-set matrices construct the pertinent graphs. Number all branches and attach reference arrows.

$$\alpha_a = \begin{bmatrix} 1 & 0 & 0 & 0 & 1 & 1 & 1 & 0 & 0 & 0 \\ 0 & 1 & 0 & 0 & 1 & 1 & 1 & 1 & 1 & 0 \\ 0 & 0 & 1 & 0 & 0 & 1 & 1 & 1 & 1 & -1 \\ 0 & 0 & 0 & 1 & 0 & 1 & 0 & 0 & 1 & -1 \end{bmatrix}$$

$$\alpha_b = \begin{bmatrix} 1 & 1 & 1 & 0 & 0 & 1 & 0 & 0 & 0 & 0 \\ 0 & 1 & 1 & 1 & 1 & 1 & 0 & 1 & 0 & 0 \\ 0 & 0 & 1 & 1 & 0 & 0 & 1 & 0 & 0 & -1 \\ 0 & 0 & 0 & 0 & 0 & 1 & 0 & 1 & 1 & 1 \end{bmatrix}$$

6. Realize the following conductance matrix

$$[G] = \begin{bmatrix} \frac{19}{4} & -\frac{7}{4} & -\frac{15}{8} \\ -\frac{7}{4} & \frac{23}{4} & \frac{15}{8} \\ -\frac{15}{8} & \frac{15}{8} & \frac{71}{16} \end{bmatrix}$$

7. A branch-conductance matrix $[\bar{g}]$ is to be expressed as

$$[\bar{g}] = [\bar{g}_a] + [\bar{g}_b]$$

in which $[\bar{g}_b]$ generates a $[G]$-matrix of rank 1; that is to say, its elements are determined by formula 32; and $[\bar{g}_a]$ generates one having the expanded form shown in Eq. 43.

If g_{ik}^b denotes an element of $[\bar{g}_b]$, show that for the fourth order we have

$$g_{12}^b = \frac{g_{12} + g_{22}}{1 - \left(\frac{g_{11}^b}{a}\right)}, \quad g_{13}^b = \frac{g_{13} + g_{23}}{1 - \left(\frac{g_{11}^b}{a}\right)}, \quad g_{14}^b = \frac{g_{14} + g_{24}}{1 - \left(\frac{g_{11}^b}{a}\right)}$$

where

$$\frac{g_{11}^b}{a} = -\frac{g_{22} + g_{23} + g_{24}}{g_{12} + g_{13} + g_{14}}$$

and

$$a = g_{11}^b + g_{12}^b + g_{13}^b + g_{14}^b$$

Use these results and the graph resulting from the solution of prob. 1 with linear tree to determine $[\bar{g}]$, and thus construct the $[G]$-matrix of prob. 6.

8. Realize the following open-circuit resistance matrix:

$$[R] = \begin{bmatrix} 3 & -1 & 0 & -1 & 1 & -1 \\ -1 & 3 & -1 & 0 & 1 & 1 \\ 0 & -1 & 3 & -1 & -1 & 1 \\ -1 & 0 & -1 & 3 & -1 & -1 \\ 1 & 1 & -1 & -1 & 3 & 0 \\ -1 & 1 & 1 & -1 & 0 & 3 \end{bmatrix}$$

CHAPTER V

Power and Energy Relations

1. The Energy Functions Associated with Linear Networks

According to the direct algebraic approach discussed in art. 1 of Ch. II, the equilibrium equations for any linear system are expressed by

$$[Z] \cdot i] = e] \tag{1}$$

in which the column matrices $i]$ and $e]$ include both variables and sources, as given there by Eqs. 15; and the impedance matrix $[Z]$ is described in Eqs. 19 and 21. Premultiplying Eq. 1 by the transpose of $i]$ gives

$$\underline{i} \cdot [Z] \cdot i] = \underline{i} \cdot e] = \underline{i_v} \cdot e_s] + \underline{i_s} \cdot e_v] \tag{2}$$

If, in the derivation of 1, the cut-set and tie-set matrices are chosen to be consistent, then the voltage sources in $e_s]$ and the loop currents in $i_v]$ pertain to the same set of links so that i_1 is the current delivered by e_{s1}, i_2 is the current delivered by e_{s2}, etc. Similarly, the current sources in $i_s]$ and the node-pair voltages in $e_v]$ correspond to the same terminal pairs so that e_1 is the terminal voltage of current source i_{s1}, e_2 is that of the source i_{s2}, etc. The right-hand side of Eq. 2 then equals the total instantaneous power delivered by the sources.

It is easy to see that this equals the instantaneous power absorbed by the network. Making use of Eq. 16 in Ch. II the left-hand side of Eq. 2 becomes

$$\underline{i} \cdot [Z] \cdot i] = \underline{j} \cdot [\alpha]_t \cdot [Z] \cdot [\alpha] \cdot j] \tag{3}$$

Noting from Eq. 18 of Ch. II that

$$[Z] = [\beta] \cdot [A] \cdot [\alpha]^{-1} = [\alpha]_t^{-1} \cdot [A] \cdot [\alpha]^{-1} \tag{4}$$

we have

$$\underline{i} \cdot [Z] \cdot i] = \underline{j} \cdot [A] \cdot j] \tag{5}$$

whereupon Eq. 6 of Ch. II yields

$$\underline{i} \cdot [Z] \cdot i] = \underline{j} \cdot v] = j_1 v_1 + j_2 v_2 + \cdots + j_b v_b \tag{6}$$

which clearly is the desired result.

In order to interpret the absorbed power in greater detail we use the partitioned forms for the matrices $[Z]$ and $i]$ (Eqs. 15 and 19, Ch. II) and have

$$P = [i_v \vdots i_s] \times \begin{bmatrix} Z_{ll} & \vdots & Z_{ln} \\ \cdot & \cdot & \cdot \\ Z_{nl} & \vdots & Z_{nn} \end{bmatrix} \times \begin{bmatrix} i_v \\ \cdots \\ i_s \end{bmatrix} \tag{7}$$

or

$$P = \underline{i_v} \cdot Z_{ll} \cdot i_v] + \underline{i_s} \cdot Z_{nl} \cdot i_v] + \underline{i_v} \cdot Z_{ln} \cdot i_s] + \underline{i_s} \cdot Z_{nn} \cdot i_s] \tag{8}$$

If only voltage sources are present (which is commonly the case when equilibrium is expressed on the loop basis) this rather formidable-looking expression reduces to its first term. According to Eqs. 104 and 106 of art. 4, Ch. II,

$$Z_{ll} = [L]p + [R] + [S]p^{-1} \tag{9}$$

and the expression for absorbed power then becomes

$$P = \underline{i_v} \cdot [L] \cdot i_v'] + \underline{i_v} \cdot [R] \cdot i_v] + \underline{q_v'} \cdot [S] \cdot q_v] \tag{10}$$

where we have introduced the loop-charge matrix

$$q_v] = \int i_v] \, dt \tag{11}$$

and a prime is used to denote differentiation with respect to time.

The result expressed by Eq. 10 can be given an interesting physical interpretation if we let

$$2T = \underline{i_v} \cdot [L] \cdot i_v] = \sum_{s,k=1}^{l} L_{sk} i_s i_k \tag{12}$$

$$2F = \underline{i_v} \cdot [R] \cdot i_v] = \sum_{s,k=1}^{l} R_{sk} i_s i_k \tag{13}$$

$$2V = \underline{q_v} \cdot [S] \cdot q_v] = \sum_{s,k=1}^{l} S_{sk} q_s q_k \tag{14}$$

in which the coefficients L_{sk}, R_{sk} and S_{sk} are elements in the loop parameter matrices $[L]$, $[R]$ and $[S]$ respectively. The double sums in these expressions, which are the algebraic equivalents of the matrix products, involve summation over both indexes s and k independently from 1 to l. Thus, for example, the sum in Eq. 12, written out, has the appearance

$$\begin{aligned} 2T = {} & L_{11} i_1 i_1 + L_{12} i_1 i_2 + \cdots + L_{1l} i_1 i_l \\ & + L_{21} i_2 i_1 + L_{22} i_2 i_2 + \cdots + L_{2l} i_2 i_l \\ & + \cdot \quad \cdot \quad \cdot \quad \cdot \quad \cdot \quad \cdot \quad \cdot \quad \cdot \quad \cdot \quad \cdot \quad \cdot \\ & + L_{l1} i_l i_1 + L_{l2} i_l i_2 + \cdots + L_{ll} i_l i_l \end{aligned} \tag{15}$$

from which it is not difficult to see that

$$\frac{dT}{dt} = \sum_{s,k=1}^{l} L_{sk} i_s \frac{di_k}{dt} = \underline{i_v} \cdot [L] \cdot i_v'] \tag{16}$$

Similarly we have

$$\frac{dV}{dt} = \sum_{s,k=1}^{l} S_{sk} q_s \frac{dq_k}{dt} = \underline{q_v'} \cdot [S] \cdot q_v] \tag{17}$$

In the last form we have interchanged the indexes s and k which we can do because $S_{sk} = S_{ks}$. Thus it is obvious that

$$\sum_{s,k=1}^{l} S_{sk} q_s \frac{dq_k}{dt} = \sum_{s,k=1}^{l} S_{sk} \frac{dq_s}{dt} q_k \tag{18}$$

The equivalence of the double sums in 16 and 17 is, therefore, easily recognized by writing out these expressions in detail (which is what the reader should do at this point). Making use of 13, 16 and 17 in 10 we see that the last relation is equivalent to

$$P = 2F + \frac{d}{dt}(T + V) \tag{19}$$

It is quite obvious that $2F$ in Eq. 13 equals the instantaneous rate of energy dissipation in the resistive elements, while T and V in 12 and 14 respectively represent the instantaneous stored energy in the magnetic fields of the inductances and in the electric fields associated with the capacitances. Equation 10 or its equivalent 19, therefore, is an expression of the conservation of energy that is more detailed in character than that given by Eqs. 2 and 6. These merely show that the power supplied by the sources, the right-hand side of Eq. 2, is absorbed by the branches of the network (this power being represented by the right-hand side of Eq. 6). Equation 19, on the other hand, separates the absorbed power into that which is dissipated and that which appears as the time-rate-of-change of energy stored in the associated electric and magnetic fields.

The quantities T, F and V defined by equations 12, 13 and 14 are spoken of as *energy functions* associated with the network. The algebraic expressions for them are called *quadratic forms* since they are homogeneous and quadratic in the pertinent variables.

The matrices $[L]$, $[R]$ and $[S]$ characterizing the coefficients in these expressions are referred to as the matrices of these quadratic forms. The properties of the matrix of a quadratic form determine its algebraic characteristics.

An important characteristic of the forms 12, 13 and 14, when associated with passive networks, is that their values are non-negative no matter what combination of values the variables may have. This property, which is described by the term *positive definite*,* is obviously required in a passive network on physical grounds. We shall later on discuss a simple algebraic way of discovering this property in an arbitrarily given quadratic form.

First, however, we present the analogous derivation on a node basis. The inverse of Eq. 1 which represents equilibrium on a node basis (Eq. 29, Ch. II) reads

$$[Y] \cdot e] = i] \tag{20}$$

Premultiplication by the transpose of $e]$ gives

$$\underline{e} \cdot [Y] \cdot e] = \underline{e} \cdot i] = \underline{e_s} \cdot i_v] + \underline{e_v} \cdot i_s] \tag{21}$$

which is equivalent to Eq. 2, the right-hand side being the power delivered by the sources. In the left-hand side we make use of Eq. 17 of Ch. II and have

$$\underline{e} \cdot [Y] \cdot e] = \underline{v} \cdot [\beta]_t \cdot [Y] \cdot [\beta] \cdot v] \tag{22}$$

With the help of Eq. 27, Ch. II

$$[Y] = [\alpha] \cdot [B] \cdot [\beta]^{-1} = [\beta]_t^{-1} \cdot [B] \cdot [\beta]^{-1} \tag{23}$$

so that Eq. 22 becomes

$$\underline{e} \cdot [Y] \cdot e] = \underline{v} \cdot [B] \cdot v] \tag{24}$$

and Eq. 26 of Ch. II then gives

$$\underline{e} \cdot [Y] \cdot e] = \underline{v} \cdot j] = v_1 j_1 + v_2 j_2 + \cdots + v_b j_b \tag{25}$$

equivalent to the result given by Eq. 6.

More detailed interpretation of the power absorbed by the network branches, the right-hand side of Eq. 25, is had through replacing the matrices in the left-hand side of this equation by their partitioned forms as shown in Eqs. 15 and 28 of Ch. II. This gives

$$P = [e_s \vdots e_v] \cdot \begin{bmatrix} Y_{ll} & \vdots & Y_{ln} \\ \cdots & \cdots & \cdots \\ Y_{nl} & \vdots & Y_{nn} \end{bmatrix} \times \begin{bmatrix} e_s \\ \cdots \\ e_v \end{bmatrix} \tag{26}$$

or

$$P = \underline{e_s} \cdot Y_{ll} \cdot e_s] + \underline{e_v} \cdot Y_{nl} \cdot e_s] + \underline{e_s} \cdot Y_{ln} \cdot e_v] + \underline{e_v} \cdot Y_{nn} \cdot e_v] \tag{27}$$

When only current sources are present as is usually the case on a node basis, this expression for the absorbed power reduces to the last term

* The term "semidefinite" is used if zero as well as positive values are possible.

alone. Making use of Eqs. 110 and 112 of Ch. II, we introduce the node-parameter matrices by the relation (dual to 9 above)

$$[Y_{nn}] = [C]p + [G] + [\Gamma]p^{-1} \tag{28}$$

whereby the expression for absorbed power becomes

$$P = \underline{e_v} \cdot [C] \cdot e_v'] + \underline{e_v} \cdot [G] \cdot e_v] + \underline{\psi_v'} \cdot [\Gamma] \cdot \psi_v] \tag{29}$$

in which

$$\psi_v] = \int e_v] \, dt \tag{30}$$

is the node flux linkage, and again a prime denotes differentiation with respect to time.

We now define the quantities

$$2V = \underline{e_v} \cdot [C] \cdot e_v] = \sum_{s,k=1}^{n} C_{sk} e_s e_k \tag{31}$$

$$2F = \underline{e_v} \cdot [G] \cdot e_v] = \sum_{s,k=1}^{n} G_{sk} e_s e_k \tag{32}$$

$$2T = \underline{\psi_v} \cdot [\Gamma] \cdot \psi_v] = \sum_{s,k=1}^{n} \Gamma_{sk} \psi_s \psi_k \tag{33}$$

The coefficients C_{sk}, G_{sk} and Γ_{sk} are elements in the node parameter matrices $[C]$, $[G]$ and $[\Gamma]$ respectively.

The quantities T, V and F are physically the same energy functions as given by Eqs. 12, 13 and 14 on a loop basis. In Eqs. 31, 32 and 33 they are expressed in terms of node-pair voltages and flux linkages. Algebraically they are quadratic forms defined by their respective parameter matrices; and in a passive system they are positive- or semi-definite.

Analogously to the situation on a loop basis we now find

$$\frac{dV}{dt} = \sum_{s,k=1}^{n} C_{sk} e_s \frac{de_k}{dt} = \underline{e_v} \cdot [C] \cdot e_v'] \tag{34}$$

and

$$\frac{dT}{dt} = \sum_{s,k=1}^{n} \Gamma_{sk} \psi_s \frac{d\psi_k}{dt} = \underline{\psi_v'} \cdot [\Gamma] \cdot \psi_v] \tag{35}$$

so that Eq. 29 is also seen to yield

$$P = 2F + \frac{d}{dt}(T + V) \tag{36}$$

the same as for the loop basis.

2. The Quasi-Orthogonality of Voltage and Current Sources

Consider the following form of equilibrium equation 1 on a loop basis:

$$\begin{aligned} Z_{ll} \cdot i_v] + Z_{ln} \cdot i_s] &= e_s] \\ Z_{nl} \cdot i_v] + Z_{nn} \cdot i_s] &= e_v] \end{aligned} \tag{37}$$

Here we wish to separate the response due to voltage and current sources by writing

$$i_v] = i_v^e] + i_v^i] \tag{38}$$

and

$$e_v] = e_v^e] + e_v^i] \tag{39}$$

where the superscript indicates whether the quantity is due to a voltage or a current source. Correspondingly we can separate Eqs. 37 into the sets

$$\begin{aligned} Z_{ll} \cdot i_v^e] + 0 &= e_s] \\ Z_{nl} \cdot i_v^e] + 0 &= e_v^e] \end{aligned} \tag{40}$$

and

$$\begin{aligned} Z_{ll} \cdot i_v^i] + Z_{ln} \cdot i_s] &= 0 \\ Z_{nl} \cdot i_v^i] + Z_{nn} \cdot i_s] &= e_v^i] \end{aligned} \tag{41}$$

The expression for instantaneous power is similarly written more explicitly in the form

$$\begin{aligned} P &= [e_s \,\vdots\, e_v] \times \begin{bmatrix} i_v \\ \cdots \\ i_s \end{bmatrix} \\ &= \underline{e_s} \cdot i_v^e] + \underline{e_s} \cdot i_v^i] + \underline{i_s} \cdot e_v^e] + \underline{i_s} \cdot e_v^i] \end{aligned} \tag{42}$$

The two middle terms here may be called cross-product terms since one represents power delivered by voltage sources due to a current increment product by current sources, and the other represents power delivered by current sources due to a voltage increment produced by voltage sources. The presence of these cross-product terms is, of course, evidence of the fact that the additive property does not hold in power calculations since the pertinent relations are not linear but instead are quadratic. Superposition does not hold. One cannot calculate power due to the sources taken separately and add these results together, because the sources interfere with each other.

We shall show, however, that under certain circumstances the two middle terms in the expression 42 cancel so that it is true that voltage sources as a group and current sources as a group act independently of each other.

From Eqs. 40 we get

$$e_v^e] = Z_{nl} \cdot i_v^e] = Z_{nl} \cdot Z_{ll}^{-1} \cdot e_s] \tag{43}$$

and from 41

$$i_v^i] = -Z_{ll}^{-1} \cdot Z_{ln} \cdot i_s] \tag{44}$$

so that the two middle terms in Eq. 42 are given by

$$\underline{e_s} \cdot i_v^i] + \underline{i_s} \cdot e_v^e] = -\underline{e_s} \cdot Z_{ll}^{-1} \cdot Z_{ln} \cdot i_s] + \underline{i_s} \cdot Z_{nl} \cdot Z_{ll}^{-1} e_s] \tag{45}$$

Since $[Z]$ is symmetrical (as it is for bilateral networks and consistent cut-set and tie-set matrices) we have

$$[Z_{nl} \cdot Z_{ll}^{-1}]_t = Z_{ll}^{-1} \cdot Z_{ln} \tag{46}$$

and if $e_s]$ and $i_s]$ are identical time functions so that the time operator contained in $[Z]$ yields the same result whether it operates upon $e_s]$ or $i_s]$, then

$$\underline{i_s} \cdot Z_{nl} \cdot Z_{ll}^{-1} \cdot e_s] = \underline{e_s} \cdot Z_{ll}^{-1} \cdot Z_{ln} \cdot i_s] \tag{47}$$

so that Eq. 45 gives

$$\underline{e_s} \cdot i_v^i] + \underline{i_s} \cdot e_v^e] \equiv 0 \tag{48}$$

Under these circumstances the power is given by

$$P = \underline{e_s} \cdot i_v^e] + \underline{i_s} \cdot e_v^i] \tag{49}$$

This relation is, for example, valid in resistive circuits with constant excitation functions (commonly called "dc circuits"). It is true in any bilateral circuits when all sources have the same time dependence, for example, in ordinary *RLC* circuits in which current and voltage sources all are unit impulses or unit step functions occurring simultaneously. It is true in such circuits in the sinusoidal steady state if all sources have the same frequency and are in time phase. Wherever it does apply, the voltage and current sources may be regarded as being *orthogonal* since, like a pair of orthogonal vectors, their interaction is nil.

3. The Transformation of Quadratic Forms

In circuit theory it frequently is expedient or necessary to change from one set of variables to another. It is, therefore, useful to know how the matrix of a given quadratic form is influenced by such a change-of-variable transformation.

Consider the quadratic form

$$F = \sum_{s,k=1}^{n} a_{sk} x_s x_k \tag{50}$$

with the matrix

$$[A] = \begin{bmatrix} a_{11} & \cdots & a_{1n} \\ \cdot & \cdot \quad \cdot \quad \cdot & \cdot \\ a_{n1} & \cdots & a_{nn} \end{bmatrix} \tag{51}$$

In terms of the column matrix

$$x] = \begin{bmatrix} x_1 \\ \cdot \\ \cdot \\ \cdot \\ x_n \end{bmatrix} \tag{52}$$

we may write 50 in the matrix form

$$F = \underline{x} \cdot [A] \cdot x] \tag{53}$$

Now suppose we introduce a new set of variables $y_1 \cdots y_n$ defined by the column matrix

$$y] = \begin{bmatrix} y_1 \\ \cdot \\ \cdot \\ \cdot \\ y_n \end{bmatrix} \tag{54}$$

and related to the x_k's by the linear-transformation equations expressed by the matrix relation

$$[P] \cdot y] = x] \tag{55}$$

in which $[P]$ is regarded as a transformation matrix.

Substitution of 55 and its transpose into 53 gives

$$F = \underline{y} \cdot [P]_t \cdot [A] \cdot [P] \cdot y] \tag{56}$$

If we write this result as

$$F = \underline{y} \cdot [B] \cdot y] \tag{57}$$

or as

$$F = \sum_{s,k=1}^{n} b_{sk} y_s y_k \tag{58}$$

in which the b_{sk} are elements of the matrix $[B]$, then comparison of 56 and 57 shows that

$$[B] = [P]_t \cdot [A] \cdot [P] \tag{59}$$

which is spoken of as a *congruent* transformation of the matrix $[A]$.

For any set of values of the x_k's or corresponding values of the y_k's given by the relation 55 or its inverse, the quadratic form 58 has, of course, the same value as the original one in Eq. 50. The value of a given quadratic form is, therefore, invariant to a congruent transformation of its matrix, and hence, so is its positive definite character.

4. Condition for Positive Definiteness

Suppose the matrix $[A]$ of the quadratic form F in Eq. 50 has the representation

$$[A] = [Q] \cdot [Q]_t \tag{60}$$

If $[Q]$ is nonsingular, we have

$$[Q]^{-1} \cdot [A] \cdot [Q]_t^{-1} = [U] \tag{61}$$

in which $[U]$ is the unit matrix. This result is a special congruent transformation of the matrix $[A]$ yielding a unit matrix.

According to Eqs. 55 and 59 it implies a transformation of variables expressed by

$$y] = [Q]_t \cdot x] \tag{62}$$

and in terms of these new variables the quadratic form F is given by

$$F = y_1^2 + y_2^2 + \cdots + y_n^2 \tag{63}$$

which is called the *canonic* form of F. It is obviously positive for all real values of the variables $y_1 \cdots y_n$; and since real values of y_k correspond to real values of x_k so long as the matrix $[Q]$ in Eq. 62 is real and nonsingular, we can conclude that the quadratic form in Eq. 50 is also positive definite under this same condition.

A necessary and sufficient condition that a quadratic form with the matrix $[A]$ be positive definite is, therefore, that this matrix have the representation 60 in which $[Q]$ is a *real* nonsingular matrix. How do we determine whether this condition is met for a given matrix $[A]$? By attempting the construction of $[Q]$, which can be done in many ways. The following method is simple and straightforward.

Consider the vector set of the matrix $[Q]$ defined by its rows. The operation on the right-hand side of Eq. 60 amounts to forming the Gramian matrix from these row vectors, since the elements a_{sk} of $[A]$ are given in terms of the elements q_{sk} of $[Q]$ by the relation

$$a_{sk} = \sum_{\nu=1}^{n} q_{s\nu} q_{k\nu} = q_s \cdot q_k \tag{64}$$

where q_s and q_k are vectors for the rows s and k, and the right-hand side is their scalar product.

It is clear from this relationship that the orientation of our reference coordinate system relative to the vector set of $[Q]$ is immaterial. Hence we can rotate it so that q_1 coincides with axis 1; q_2 lies in the plane determined by axes 1 and 2; q_3 lies in the subspace defined by axes 1, 2 and 3; and so forth. This orientation amounts to giving $[Q]$ the triangular form

$$[Q] = \begin{bmatrix} q_{11} & 0 & 0 & 0 & \cdots & 0 \\ q_{21} & q_{22} & 0 & 0 & \cdots & 0 \\ q_{31} & q_{32} & q_{33} & 0 & \cdots & 0 \\ \cdot & \cdot & \cdot & \cdot & \cdot & \cdot \\ q_{n1} & q_{n2} & q_{n3} & q_{n4} & \cdots & q_{nn} \end{bmatrix} \tag{65}$$

and the argument just presented shows that nothing is sacrificed by assuming this triangular form for $[Q]$; if we cannot achieve the desired result with this form then it cannot be achieved with any other form either.

The construction of $[Q]$ in 65 from a given matrix $[A]$ is now accomplished by the following relations which are the same as 64, namely:

$$\begin{aligned} a_{11} &= q_{11}^2 \\ a_{12} &= q_{11}q_{21} \\ a_{13} &= q_{11}q_{31} \\ &\cdots\cdots \end{aligned} \tag{66}$$

followed by

$$\begin{aligned} a_{22} &= q_{21}^2 + q_{22}^2 \\ a_{23} &= q_{21}q_{31} + q_{22}q_{32} \\ a_{24} &= q_{21}q_{41} + q_{22}q_{42} \\ &\cdots\cdots\cdots \end{aligned} \tag{67}$$

followed by

$$\begin{aligned} a_{33} &= q_{31}^2 + q_{32}^2 + q_{33}^2 \\ a_{34} &= q_{31}q_{41} + q_{32}q_{42} + q_{33}q_{43} \\ a_{35} &= q_{31}q_{51} + q_{32}q_{52} + q_{33}q_{53} \\ &\cdots\cdots\cdots\cdots \end{aligned} \tag{68}$$

and so forth.

The sequence of relations 66 determine elements in the first column of $[Q]$; those in 67 then determine elements in the second column of $[Q]$; and so forth.

All elements calculated by relations 66 are real if $a_{11} > 0$. All those subsequently calculated by 67 are real if $a_{22} > q_{21}^2$; the ones then calculated by relations 68 are real if $a_{33} > q_{31}^2 + q_{32}^2$; and so forth.

At each stage all quantities are known except one. No solution of simultaneous equations is necessary; the calculations are strictly sequential in nature.

It is clear that the matrix $[Q]$ is not necessarily real. The quadratic form F is positive definite only if $[Q]$ is real. While there are other ways of expressing this condition in terms of properties of the matrix $[A]$, the constructive one given here is computationally easiest to apply in a given numerical problem.

A necessary, though not sufficient, condition for positive definiteness that clearly emerges from the above calculation sequence is that all elements $a_{11}, a_{22}, a_{33}, \ldots$ on the principal diagonal of $[A]$ must be positive. Not only must they be positive but they must be sufficiently large to fulfill the pertinent inequalities given.

As an example, consider

$$[A] = \begin{bmatrix} 2 & 4 & -3 \\ & 3 & 2 \\ & & 5 \end{bmatrix} \tag{69}$$

By 66 we calculate

$$q_{11} = \frac{1}{\sqrt{2}}, \quad q_{21} = \frac{4}{\sqrt{2}}, \quad q_{31} = \frac{-3}{\sqrt{2}} \tag{70}$$

Next, by the first relation of 67 we find

$$q_{22}^2 = 3 - 8 = -5 \tag{71}$$

We can stop here since q_{22} is not real.

The present considerations can also be used to construct a matrix whose quadratic form is sure to be positive definite. Any real nonsingular matrix assumed for $[Q]$ in Eq. 60 yields such an $[A]$-matrix.

5. Energy Functions in the Sinusoidal Steady State

Loop currents and charges are written in the form

$$i_k = \tfrac{1}{2}(I_k e^{j\omega t} + \bar{I}_k e^{-j\omega t}) \tag{72}$$

$$q_k = \frac{1}{2j\omega}(I_k e^{j\omega t} - \bar{I}_k e^{-j\omega t}) \tag{73}$$

where the bar above indicates the conjugate complex value. We find straightforwardly

$$i_s i_k = \tfrac{1}{2} \operatorname{Re}(I_s \bar{I}_k + I_s I_k e^{j2\omega t}) \tag{74}$$

and

$$q_s q_k = \frac{1}{2\omega^2} \operatorname{Re}(I_s \bar{I}_k - I_s I_k e^{j2\omega t}) \tag{75}$$

in which Re, as usual, stands for "real part of."

Substitution into the double sums in Eqs. 12, 13 and 14 gives

$$T = \tfrac{1}{4} \sum_{s,k=1}^{l} L_{sk} I_s \bar{I}_k + \tfrac{1}{4} \operatorname{Re} \left(e^{j2\omega t} \sum_{s,k=1}^{l} L_{sk} I_s I_k \right) \tag{76}$$

$$F = \tfrac{1}{4} \sum_{s,k=1}^{l} R_{sk} I_s \bar{I}_k + \tfrac{1}{4} \operatorname{Re} \left(e^{j2\omega t} \sum_{s,k=1}^{l} R_{sk} I_s I_k \right) \tag{77}$$

and

$$V = \frac{1}{4\omega^2} \sum_{s,k=1}^{l} S_{sk} I_s \bar{I}_k - \frac{1}{4\omega^2} \operatorname{Re} \left(e^{j2\omega t} \sum_{s,k=1}^{l} S_{sk} I_s I_k \right) \tag{78}$$

Observe that the Re operator is not needed for the first terms in these expressions since they are self-conjugate and hence real. Thus, for example, the conjugate of the sum involving $L_{sk} I_s \bar{I}_k$ is one involving $L_{sk} \bar{I}_s I_k$ which is exactly the same because $L_{sk} = L_{ks}$.

The results given by Eqs. 76, 77 and 78 show us what we already know with regard to simple circuits, namely that, in the sinusoidal steady state, each energy function is given by a constant term plus a double-frequency sinusoid. The constant term is evidently the average value, so we can write

$$T_{\text{av}} = \tfrac{1}{4} \sum_{s,k=1}^{l} L_{sk} I_s \bar{I}_k \tag{79}$$

$$F_{\text{av}} = \tfrac{1}{4} \sum_{s,k=1}^{l} R_{sk} I_s \bar{I}_k \tag{80}$$

$$V_{\text{av}} = \frac{1}{4\omega^2} \sum_{s,k=1}^{l} S_{sk} I_s \bar{I}_k \tag{81}$$

It may not be superfluous at this point to remind the reader that by writing the forms 72 and 73 we are tacitly defining the amplitudes I_k as being *peak* values and not root mean squares. That is why we find the factor $\frac{1}{4}$ in the expressions for T_{av} and V_{av} rather than the more familiar factor $\frac{1}{2}$. In Eq. 80 the factor is $\frac{1}{4}$ because F in Eq. 13 is initially defined so that $2F$ and not F is the total rate of energy dissipation. Finally, the factor $1/\omega^2$ appears in Eq. 81 because I/ω is a charge, and electric energy is proportional to elastance times charge squared.

On the node basis we have expressions similar to Eqs. 79, 80, 81 involving node-pair voltage variables and node instead of loop parameters. We leave the details to the reader to work out, and turn our attention rather to a more fundamental point, namely that since the instantaneous energy functions 76, 77 and 78 must not become negative, the amplitude of the double-frequency sinusoid in each of these expressions may never be greater than the value of the constant term. Physically this fact is obvious for any passive network. We shall prove purely by algebraic means that it follows from the positive definite character of the quadratic forms.

Formally we must show that

$$\sum a_{sk} z_s \bar{z}_k \geq |\sum a_{sk} z_s z_k| \tag{82}$$

in which

$$z_k = x_k + jy_k \tag{83}$$

The quantities x_k and y_k are real, the a_{sk} are real coefficients fulfilling the symmetry condition $a_{sk} = a_{ks}$, and the real quadratic form

$$\sum a_{sk} x_s x_k \tag{84}$$

is positive-definite. For simplicity we leave off the limits on the double sums in 82 and 84 but assume that they are summed over both indexes from 1 to n.

Substitution of 83 into 82 gives

$$\sum a_{sk}(x_s + jy_s)(x_k - jy_k) \geq |\sum a_{sk}(x_s + jy_s)(x_k + jy_k)| \tag{85}$$

or

$$\sum a_{sk}[x_s x_k + y_s y_k + j(x_k y_s - x_s y_k)] \geq |\sum a_{sk}[x_s x_k - y_s y_k + j(x_k y_s + x_s y_k)]| \tag{86}$$

Now

$$\sum a_{sk}(x_k y_s - x_s y_k) = 0$$

because of the symmetry condition $a_{sk} = a_{ks}$; and since the remaining form on the left of 86 is positive-definite by hypothesis, the stated inequality will be fulfilled if

$$[\sum a_{sk}(x_s x_k + y_s y_k)]^2 \geq [\sum a_{sk}(x_s x_k - y_s y_k)]^2 + 4[\sum a_{sk} x_s y_k]^2 \tag{87}$$

which is equivalent to

$$(\sum a_{sk} x_s x_k)(\sum a_{sk} y_s y_k) \geq (\sum a_{sk} x_s y_k)^2 \tag{88}$$

If we introduce the column matrices

$$x] = \begin{bmatrix} x_1 \\ \cdot \\ \cdot \\ \cdot \\ x_n \end{bmatrix}; \quad y] = \begin{bmatrix} y_1 \\ \cdot \\ \cdot \\ \cdot \\ y_n \end{bmatrix} \tag{89}$$

and the coefficient matrix

$$[A] = \begin{bmatrix} a_{11} & \cdots & a_{1n} \\ \cdot & \cdots & \cdot \\ a_{n1} & \cdots & a_{nn} \end{bmatrix} \tag{90}$$

then the matrix equivalent of 88 reads

$$(\underline{x} \cdot [A] \cdot x])(\underline{y} \cdot [A] \cdot y]) \geq (\underline{x} \cdot [A] \cdot y])^2 \tag{91}$$

We now find a matrix $[P]$ which transforms $[A]$ congruently to its canonic form; that is to say,

$$[P]_t \cdot [A] \cdot [P] = [U] \tag{92}$$

Since $[A]$ is nonsingular, $[U]$ is a unit matrix.

With the matrix $[P]$ we introduce the change of variable indicated by the relations

$$x] = [P] \cdot x'] \quad \text{and} \quad y] = [P] \cdot y'] \tag{93}$$

Then the inequality in 91 is transformed into

$$(\underline{x}' \cdot [U] \cdot x'])(\underline{y}' \cdot [U] \cdot y']) \geq (\underline{x}' \cdot [U] \cdot y'])^2 \tag{94}$$

or

$$(x_1'^2 + \cdots + x_n'^2)(y_1'^2 + \cdots + y_n'^2) \geq (x_1'y_1' + \cdots + x_n'y_n')^2 \tag{95}$$

If we regard the quantities $x_1' \cdots x_n'$ and $y_1' \cdots y_n'$ as components of vectors x' and y' then the parenthesis expressions on the left are the squared magnitudes of these vectors while the right-hand parenthesis expression is their scalar product. Hence 95 may be written

$$|x'|^2 \cdot |y'|^2 \geq (|x'| \cdot |y'| \cdot \cos\theta)^2 \tag{96}$$

where θ is the angle between the two vectors.

Condition 95, which is known as Schwarz's inequality, is thus obviously true, and hence our proof is completed.

6. Lagrangian Equations

Equilibrium in a linear system can be expressed in terms of the pertinent energy functions which explicitly are functions of the variables chosen to describe the system (like currents and charges or voltages and flux linkages, or in a mechanical system, velocities and displacements) and implicitly are functions of time as the basic independent variable. For an electrical network these equilibrium relations, which are known as Lagrange's equations, must be equivalent to the Kirchhoff laws since these are fundamental to network behavior. The following analysis illustrates this point.

In terms of the relation 12 or 15 for the magnetic stored energy T, we obtain the result

$$\frac{\partial T}{\partial i_k} = \sum_{s=1}^{l} L_{ks} i_s \tag{97}$$

and hence

$$\frac{d}{dt}\frac{\partial T}{\partial i_k} = \sum_{s=1}^{l} L_{ks} \frac{di_s}{dt} \tag{98}$$

while from the forms 13 and 14 we get

$$\frac{\partial F}{\partial i_k} = \sum_{s=1}^{l} R_{ks} i_s \tag{99}$$

and

$$\frac{\partial V}{\partial q_k} = \sum_{s=1}^{l} S_{ks} q_s \tag{100}$$

On a loop basis, to which these relations are pertinent, the power supplied by voltage sources alone is

$$P = e_{s1} i_1 + e_{s2} i_2 + \cdots + e_{sl} i_l \tag{101}$$

so that

$$\frac{\partial P}{\partial i_k} = e_{sk} \tag{102}$$

In view of these results we see that the familiar Kirchhoff voltage equilibrium equations

$$\sum_{s=1}^{l} \left(L_{ks} \frac{di_s}{dt} + R_{ks} i_s + S_{ks} q_s \right) = e_{sk} \qquad \text{for } k = 1, 2, \ldots l \tag{103}$$

can alternately be written

$$\frac{d}{dt} \frac{\partial T}{\partial i_k} + \frac{\partial F}{\partial i_k} + \frac{\partial V}{\partial q_k} = \frac{\partial P}{\partial i_k} \qquad \text{for } k = 1, 2, \ldots l \tag{104}$$

which is the form that Lagrange's equations take for an electrical network characterized on a loop basis.

Analogously on a node basis we get from Eq. 31

$$\frac{\partial V}{\partial e_k} = \sum_{s=1}^{n} C_{ks} e_s \tag{105}$$

and hence

$$\frac{d}{dt} \frac{\partial V}{\partial e_k} = \sum_{s=1}^{n} C_{ks} \frac{de_s}{dt} \tag{106}$$

while from 32 and 33 we have

$$\frac{\partial F}{\partial e_k} = \sum_{s=1}^{n} G_{ks} e_s \tag{107}$$

and

$$\frac{\partial T}{\partial \psi_k} = \sum_{s=1}^{n} \Gamma_{ks} \psi_s \tag{108}$$

The power supplied by current sources in this case is given by

$$P = i_{s1}e_1 + i_{s2}e_2 + \cdots + i_{sn}e_n \tag{109}$$

and so

$$\frac{\partial P}{\partial e_k} = i_{sk} \tag{110}$$

The familiar Kirchhoff current equilibrium equations

$$\sum_{s=1}^{n}\left(C_{ks}\frac{de_s}{dt} + G_{ks}e_s + \Gamma_{ks}\psi_s\right) = i_{sk} \qquad \text{for } k = 1, 2, \ldots n \tag{111}$$

are thus seen also to be expressible as

$$\frac{d}{dt}\frac{\partial V}{\partial e_k} + \frac{\partial F}{\partial e_k} + \frac{\partial T}{\partial \psi_k} = \frac{\partial P}{\partial e_k} \qquad \text{for } k = 1, 2, \ldots n \tag{112}$$

These are Lagrange's equations for an electrical network characterized on a node basis.

In comparing the dual forms for Lagrange's equations as given in 104 and 112 it is significant to note the interchange of roles played by the energy functions T and V. Since these two energy functions are duals, one could have predicted this characteristic of the dual Lagrangian equations.

7. Equilibrium Equations Are the Result of a Minimum Energy Principle

An interesting result of fundamental importance is the fact that in a linear network subjected to the usual voltage and current constraints called sources, the unconstrained currents and voltages at every instant assume values such that the total rate of energy dissipation becomes a minimum, subject to the condition that energy for the whole system is conserved. We will now demonstrate this fact.

First we need to develop some auxiliary relationships having to do with the properties of quadratic forms. Consider the stored magnetic energy function, Eq. 12 or 15,

$$T = \tfrac{1}{2}\sum_{k,s=1}^{l} L_{ks}i_k i_s \tag{113}$$

and form its partial derivative with respect to i_k as in Eq. 97

$$\frac{\partial T}{\partial i_k} = \sum_{s=1}^{l} L_{ks}i_s \tag{114}$$

From these two expressions it becomes clear that we can write the magnetic stored energy function in the form

$$T = \tfrac{1}{2}\sum_{k=1}^{l}\frac{\partial T}{\partial i_k}i_k \tag{115}$$

If we take the total time derivative of 113, as we did in Eq. 16, we have

$$\frac{dT}{dt} = \sum_{k,s=1}^{l} L_{ks} i_k \frac{di_s}{dt} = \sum_{k,s=1}^{l} L_{ks} i_s \frac{di_k}{dt}$$

$$= \sum_{k=1}^{l} \left(\sum_{s=1}^{l} L_{ks} i_s \right) \frac{di_k}{dt} \tag{116}$$

and in view of 114 we obtain

$$\frac{dT}{dt} = \sum_{k=1}^{l} \frac{\partial T}{\partial i_k} \frac{di_k}{dt} \tag{117}$$

On the other hand, if we form the total time derivative of T as given by Eq. 115, noting that the summand is a product of two time functions, we get

$$\frac{dT}{dt} = \frac{1}{2} \sum_{k=1}^{l} \frac{\partial T}{\partial i_k} \frac{di_k}{dt} + \frac{1}{2} \sum_{k=1}^{l} i_k \cdot \frac{d}{dt} \frac{\partial T}{\partial i_k} \tag{118}$$

and with 117 we have

$$\frac{dT}{dt} = \sum_{k=1}^{l} i_k \cdot \frac{d}{dt} \frac{\partial T}{\partial i_k} \tag{119}$$

But from 114 we see that

$$\frac{d}{dt} \frac{\partial T}{\partial i_k} = \sum_{s=1}^{l} L_{ks} \frac{di_s}{dt} \tag{120}$$

which is a function of the current derivatives only (not a function of any current i_k), and hence 119 gives

$$\frac{\partial}{\partial i_k} \frac{dT}{dt} = \frac{d}{dt} \frac{\partial T}{\partial i_k} \tag{121}$$

We can, therefore, write 119 in the alternate form

$$\frac{dT}{dt} = \sum_{k=1}^{l} i_k \cdot \frac{\partial}{\partial i_k} \frac{dT}{dt} \tag{122}$$

In an analogous fashion, starting from Eq. 14, we find

$$\frac{\partial V}{\partial q_k} = \sum_{s=1}^{l} S_{ks} q_s \tag{123}$$

and hence we can write in place of Eq. 14

$$V = \frac{1}{2} \sum_{k=1}^{l} \frac{\partial V}{\partial q_k} q_k \tag{124}$$

Now from Eqs. 17 and 123, noting that $dq_k/dt = i_k$, we have

$$\frac{dV}{dt} = \sum_{k=1}^{l} \frac{\partial V}{\partial q_k} i_k \tag{125}$$

and since $\partial V/\partial q_k$ in 123 is a function of charge only (not a function of any current i_k), this last result yields

$$\frac{\partial}{\partial i_k}\frac{dV}{dt} = \frac{\partial V}{\partial q_k} \tag{126}$$

and therefore we can write 125 in the alternate form

$$\frac{dV}{dt} = \sum_{k=1}^{l} i_k \cdot \frac{\partial}{\partial i_k}\frac{dV}{dt} \tag{127}$$

Finally, from Eq. 13 for the loss function F we have

$$\frac{\partial F}{\partial i_k} = \sum_{s=1}^{l} R_{ks} i_s \tag{128}$$

and hence we get the alternate form (analogous to 115 and 124) which reads

$$F = \tfrac{1}{2}\sum_{k=1}^{l} \frac{\partial F}{\partial i_k} i_k \tag{129}$$

We are now ready to consider the minimization problem.

In this connection we observe first of all that we are here dealing with a so-called "conditioned minimum," not a free minimum. We are not merely interested in finding that set of values for the currents in the expression 14 for which this function becomes a minimum, but we are asked at the same time to consider only such sets of current values for which the conservation of energy (or power) condition 19 or 36 is simultaneously fulfilled.

Lagrange suggested a relatively simple way of dealing with such a problem. We write the condition 19 in the form

$$P - 2F - \frac{d}{dt}(T + V) = 0 \tag{130}$$

We then multiply it by an arbitrary factor λ (called a Lagrangian multiplier) and add this to the function that we wish to minimize, so

$$F + \lambda\left[P - 2F - \frac{d}{dt}(T + V)\right] \tag{131}$$

In view of 130, this resulting function is still equal to F, no matter what the value of λ may turn out to be. It is a function of all the currents $i_1 \cdots i_l$ and of the multiplier λ. To minimize this function (which equals F) with respect to the currents $i_1 \cdots i_l$, we set its partial derivatives with respect to these currents equal to zero and solve the resulting l equations plus Eq. 130 simultaneously for the $l + 1$ unknowns consisting of currents

$i_1 \cdots i_l$ plus the multiplier λ. These values yield the desired conditioned minimum of the function F since fulfillment of the pertinent condition 130 is taken into account in the total minimization process.

The following l equations:

$$\frac{\partial}{\partial i_k}\left\{F + \lambda\left[P - 2F - \frac{d}{dt}(T + V)\right]\right\} = 0 \tag{132}$$

resulting for $k = 1, 2, \ldots l$, plus Eq. 130 are thus to be solved simultaneously for the currents $i_1 \cdots i_l$ and λ. We begin by multiplying the separate equations in the set 132 respectively by $i_1, i_2, \ldots i_l$, and then adding them together to form

$$\sum_{k=1}^{l} \frac{\partial}{\partial i_k}\left\{F + \lambda\left[P - 2F - \frac{d}{dt}(T + V)\right]\right\} i_k = 0 \tag{133}$$

We now make use of relations 122, 127 and 129 together with

$$P = \sum_{k=1}^{l} \frac{\partial P}{\partial i_k} i_k \tag{134}$$

which follows from Eqs. 101 and 102, to convert Eq. 133 into the form

$$2F + \lambda\left[P - 4F - \frac{d}{dt}(T + V)\right] = 0 \tag{135}$$

or

$$2(1 - \lambda)F + \lambda\left[P - 2F - \frac{d}{dt}(T + V)\right] = 0 \tag{136}$$

In view of the condition 130, this immediately gives $\lambda = 1$; and hence the equations 132 become

$$\frac{\partial}{\partial i_k}\left[P - F - \frac{d}{dt}(T + V)\right] = 0 \qquad \text{for } k = 1, 2, \ldots l \tag{137}$$

With the help of Eqs. 121 and 126 this result is readily seen to be equivalent to

$$\frac{d}{dt}\frac{\partial T}{\partial i_k} + \frac{\partial F}{\partial i_k} + \frac{\partial V}{\partial q_k} = \frac{\partial P}{\partial i_k} \qquad \text{for } k = 1, 2, \ldots l \tag{138}$$

which is identical with the Lagrangian equations 104 and hence is also equivalent to the Kirchhoff loop equations 103.

Evidence that the extremum thus determined is a minimum, not a maximum, is provided by the fact that Eq. 128 yields

$$\frac{\partial^2 F}{\partial i_k^2} = R_{kk} \tag{139}$$

which is surely positive; and the required demonstration is thus completed.

Problems

1. Is the following quadratic form positive for all assignable values of the current variables?

$$\begin{aligned} F = 4i_1^2 &+ 2i_1i_2 + i_1i_3 + 3i_1i_4 \\ &+ i_2^2 + 4i_2i_3 + 2i_2i_4 \\ &+ 3i_3^2 + i_3i_4 \\ &+ i_4^2 \end{aligned}$$

2. Tell whether the following matrices characterize quadratic forms that are: (*a*) positive-definite, (*b*) positive-semidefinite, or (*c*) nonpositive.

$$\text{(i)} \begin{bmatrix} 9 & 6 & 3 \\ 6 & 8 & 3 \\ 3 & 3 & 1 \end{bmatrix}; \quad \text{(ii)} \begin{bmatrix} 4 & 6 & 2 & 1 \\ 6 & 10 & 3 & 4 \\ 2 & 3 & 4 & 2 \\ 1 & 4 & 2 & 1 \end{bmatrix}; \quad \text{(iii)} \begin{bmatrix} 16 & 4 & 8 & 4 \\ 4 & 10 & 5 & 7 \\ 8 & 5 & 6 & 2 \\ 4 & 7 & 2 & 9 \end{bmatrix}$$

3. Two coupled inductances l_{11} and l_{22} have a mutual inductance equal to l_{12}. If we define a coupling coefficient $k_{12} = l_{12}/\sqrt{l_{11}l_{22}}$, show that positive stored energy requires the condition $1 - k_{12}^2 \geq 0$ and hence is equivalent to requiring that the coupling coefficient in absolute value not exceed unity.

4. Consider three coupled inductances l_{11}, l_{22}, l_{33} and the mutuals l_{12}, l_{13}, l_{23}. If we define the coupling coefficients $k_{12} = l_{12}/\sqrt{l_{11}l_{22}}$, $k_{13} = l_{13}/\sqrt{l_{11}l_{33}}$, $k_{23} = l_{23}/\sqrt{l_{22}l_{33}}$, show that the positive energy condition requires besides that

$$1 - k_{12}^2 \geq 0, \quad 1 - k_{13}^2 \geq 0, \quad 1 - k_{23}^2 \geq 0$$

that

$$(1 - k_{12}^2) + (1 - k_{13}^2) + (1 - k_{23}^2) \geq 2(1 - k_{12}k_{13}k_{23})$$

and that these conditions are also sufficient.

5. In the sinusoidal steady state, show that the expression for instantaneous power supplied to the network reads

$$P = 2F_{\text{av}} + \frac{1}{2}\,\text{Re}\left[e^{j2\omega t}\sum_{s=1}^{l} E_s I_s\right]$$

in which the quantities E_s are complex amplitudes of the voltage sources. Then show that this instantaneous power is given by the sum of a constant component $P_{\text{av}} = 2F_{\text{av}}$ and a superimposed double-frequency sinusoid with peak amplitude

$$\frac{1}{2}\left|\sum_{s=1}^{l} E_s I_s\right|$$

which is not necessarily smaller than P_{av}.

6. In contrast with the sum in prob. 5, show that

$$\frac{1}{2}\sum_{s=1}^{l} \bar{E}_s I_s = P_{av} + jQ_{av} = \text{complex power}$$

where

$$Q_{av} = 2\omega(V_{av} - T_{av})$$

is the so-called reactive power.

For a single source, show that the peak amplitude of the double-frequency sinusoidal component in the instantaneous power equals the magnitude of the complex power and hence is larger or at least as large as the average power. Under what physical condition is the instantaneous power, therefore, nonnegative?

7. If a_{sk} for $s, k = 1, 2, \ldots n$ are coefficients of a symmetrical matrix $[A]$, show that the function

$$\sum_{k=1}^{n} a_{sk} z_s \bar{z}_k$$

is both real and positive for all *complex* values of the z_k if and only if

$$F = \sum_{s,k=1}^{n} a_{sk} x_s x_k$$

is a positive definite quadratic form in the real variables x_k.

8. Of the following two expressions

(*a*) $$9z_1^2 + 8z_2^2 + 2z_3^2 + 12z_1z_2 + 6z_1z_3 + 6z_2z_3$$

(*b*) $$9\,|z_1|^2 + 8\,|z_2|^2 + 2\,|z_3|^2 + \text{Re}\,[12z_1\bar{z}_2 + 6z_1\bar{z}_3 + 6z_2\bar{z}_3]$$

with $$z_1 = 1 + j; \quad z_2 = 2 - j2; \quad z_3 = 4 + j6$$

which has the larger magnitude, and why?

9. If Z_1 is the input impedance of a passive network show that

$$Z_1 = 4F_{av} + j4\omega(T_{av} - V_{av})$$

where the energy functions are evaluated for 1 ampere at the driving point. On the basis of this result define a resonance condition in the network in terms of stored energies in the electric and magnetic fields.

10. For a network without mutual inductive coupling, characterized on a branch basis in terms of inductances l_k, resistances r_k, and elastances s_k, show that the instantaneous energy functions in the sinusoidal steady state are given by the expressions

$$T = \frac{1}{4}\sum_{k=1}^{\lambda} l_k\,|J_k|^2 + \frac{1}{4}\,\text{Re}\left[e^{j2\omega t}\sum_{k=1}^{\lambda} l_k J_k^2\right]$$

$$F = \frac{1}{4}\sum_{k=\lambda+1}^{\lambda+\rho} r_k\,|J_k|^2 + \frac{1}{4}\,\text{Re}\left[e^{j2\omega t}\sum_{k=\lambda+1}^{\lambda+\rho} r_k J_k^2\right]$$

$$V = \frac{1}{4\omega^2}\sum_{k=\lambda+\rho+1}^{b} s_k\,|J_k|^2 - \frac{1}{4\omega^2}\,\text{Re}\left[e^{j2\omega t}\sum_{k=\lambda+\rho+1}^{b} s_k J_k^2\right]$$

in which J_k denotes the complex current amplitude in the kth branch.

In terms of these results show that in a lossless network, the double-frequency sinusoidal terms have peak amplitudes equal to the respective constant terms, and that the sinusoidal terms in T and in V are in phase opposition so that when T is a maximum $V = 0$ and vice versa. Then show that at a resonance frequency the total stored energy oscillates back and forth between the electric and the magnetic fields. Does this same condition hold for a dissipative network at a pronounced resonance? Why or why not?

Now show that these particular results are still true when mutual inductance is present.

11. In the adjacent circuit

$$e_s(t) = i_s(t) = u_0(t)$$

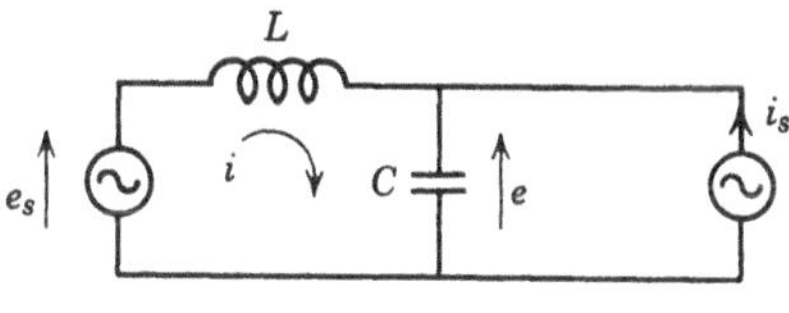

PROB. 11.

Compute the time functions $i(t)$ and $e(t)$ for $t > 0$. From these find an expression for the total energy in the system at any time $t > 0$. Compute separately the energy supplied by the voltage and current sources. Do your results check the expected orthogonality conditions as regards calculation of power due to the simultaneous presence of voltage and current sources? Explain your conclusions.

12. In the circuit shown here:

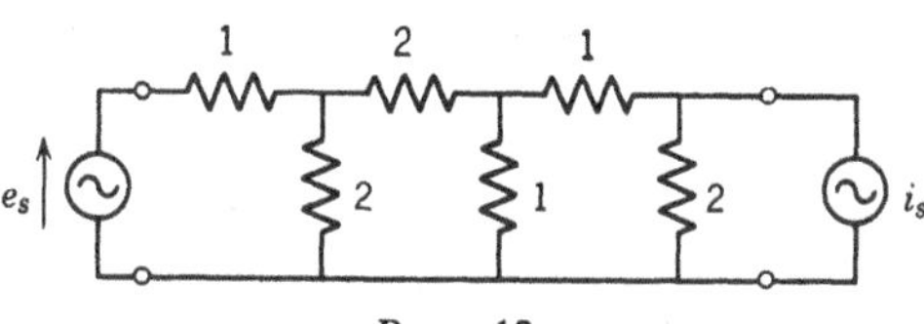

PROB. 12.

element values are in ohms and the sources are constants equal to unity. Calculate:

(*a*) The power supplied by e_s alone.
(*b*) The power supplied by i_s alone.
(*c*) The power increment supplied by e_s over that in (*a*) when i_s is present and the power increment supplied by i_s over that in (*b*) when e_s is present.
(*d*) The net power supplied by both sources simultaneously.

CHAPTER VI

Consideration of Linear Active and Nonbilateral Elements

1. Introductory Remarks

Basic questions regarding topology and the formulation of equilibrium equations when active and nonbilateral elements are embedded in an otherwise passive bilateral linear network have been dealt with in art. 6 of Ch. II. Here we shall concern ourselves with some of the more detailed aspects of the characterization process and with the important question of how we distinguish fundamentally between passive elements and active ones, between nonbilateral elements and bilateral ones; and how we recognize those that are both active and nonbilateral.

It may not be superfluous at this point also to elaborate a bit upon the term "linear" in connection with circuit elements of this sort since practical devices that are active and/or nonbilateral are usually also rather nonlinear. They may, however, be said to be "essentially linear" if operated in the vicinity of a judiciously chosen "operating point" or "quiescent point" with deviations from this point that are sufficiently small to assure a negligible perturbation from linear behavior on the part of the entire network. How such conditions are met is a question that is dealt with in the study of nonlinear circuits. We shall merely assume that they are met; and that the device may, therefore, be treated as being linear and hence characterizable in the same terms as are used for ordinary linear elements. These are the linear volt-ampere relations for the device.

Obviously a vacuum tube or other form of diode rectifier element is excluded completely from our considerations since nonlinearity is its sole reason for existence. The so-called "tunnel" diode, on the other hand, is a two-terminal device presenting (under proper operating conditions) an essentially linear negative resistance. According to a definition to be discussed presently, this device is certainly active; and since its volt-ampere characteristic can be assumed linear for all practical purposes, it is legitimately included among devices that are considered here.

2. Characterization of Active and/or Nonbilateral Devices

For the most part, active and nonbilateral devices are multi-terminal in character, being schematically representable as boxes with three or more protruding terminals. Since most present-day devices of this sort are resistive in character (omitting the effects of parasitic capacitances or inductances which have nothing to do with the basic characterization of the device), we shall limit our discussion to this situation. The pattern of analysis thus established will apply to other cases when they arise.

As pointed out in art. 6, Ch. II the volt-ampere relations for any p terminal-pair device, characterized schematically as a box with $p + 1$ terminals, are given by the relations

$$\begin{aligned} v_1 &= r_{11}j_1 + \cdots + r_{1p}j_p \\ &\cdot \quad \cdot \quad \cdot \quad \cdot \quad \cdot \quad \cdot \\ v_p &= r_{p1}j_1 + \cdots + r_{pp}j_p \end{aligned} \tag{1}$$

or by the inverse set

$$\begin{aligned} j_1 &= g_{11}v_1 + \cdots + g_{1p}v_p \\ &\cdot \quad \cdot \quad \cdot \quad \cdot \quad \cdot \quad \cdot \\ j_p &= g_{p1}v_1 + \cdots + g_{pp}v_p \end{aligned} \tag{2}$$

Topologically the device is regarded as having the structure of a tree with p branches. The quantities v_k and j_k are the tree-branch voltage drops and currents. Since the given characterization of a device of this sort may not be in a form most advantageous for analysis of the total network in which it is to be embedded, it is appropriate to discuss in some detail how we can transform from one characterization to another.

To this end consider the 5-terminal box shown in Fig. 1. The given characterization may be in terms of the potentials of nodes 1, 2, 3, 4 with respect to node 5 as a datum, or it may be in terms of the potentials of nodes 1, . . . 5 with respect to some arbitrary datum—the "floating"-matrix basis discussed in art. 4 of Ch. II—and the desired basis may be pertinent to some tree like the one to which the voltage drops $v_1 \cdots v_4$ are pertinent. On the other hand, the given characterization may be for this or for some starlike tree, while the desired form is the floating-matrix basis, as it might well be if the device is to be embedded in a larger network for which some node other than a terminal of the device is chosen as reference. The method of transforming from one such representation to another is illustrated by the following discussion.

Suppose we begin by assuming the given characterization in terms of node potentials 1, . . . 5 with respect to an arbitrary datum, thus:

$$\begin{aligned} g_{11}e_1 + \cdots + g_{15}e_5 &= i_{s1} \\ \cdot \quad \cdot \quad \cdot \quad \cdot \quad \cdot \quad \cdot \\ g_{51}e_1 + \cdots + g_{55}e_5 &= i_{s5} \end{aligned} \tag{3}$$

Since the currents $i_{s1} \cdots i_{s5}$ sum to zero, each column in the matrix $[g]$ appropriate to this set of equations must likewise sum to zero; the matrix $[g]$ has rank 4 and is, therefore, singular. This fact is, however, not particularly relevant to the following transformation process.

Assume first that we want to obtain from Eqs. 3 a set pertinent to a starlike tree involving four node-to-datum voltages with node 5 as the

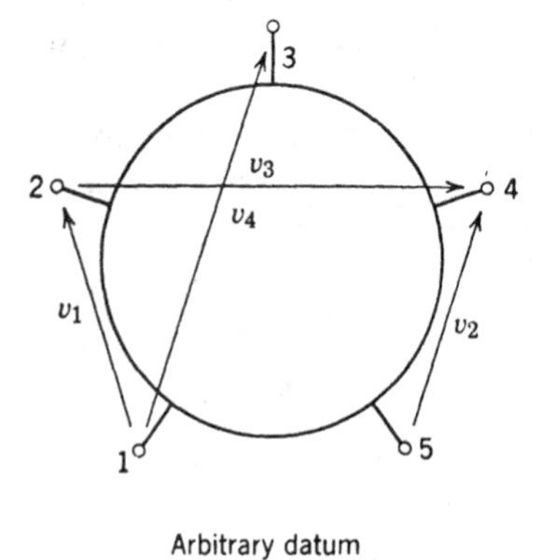

FIG. 1. A 5-terminal box either with one terminal as datum or with an arbitrary datum, in which case it leads to a "floating" matrix.

datum. This transformation is easy. All we need to do is set e_5 equal to zero and then discard the last equation in 3 since i_{s5} now has no physical significance. This abridgment process consists merely of striking out row and column 5 in the matrix $[g]$ to form a new 4-by-4 matrix $[g']$. The resulting equations are simply

$$\begin{array}{c} g_{11}e_1 + \cdots + g_{14}e_4 = i_{s1} \\ \cdot \quad \cdot \quad \cdot \quad \cdot \quad \cdot \quad \cdot \quad \cdot \quad \cdot \quad \cdot \quad \cdot \\ g_{41}e_1 + \cdots + g_{44}e_4 = i_{s4} \end{array} \tag{4}$$

Now suppose we want to transform to the tree to which the voltage drops $v_1 \cdots v_4$ in Fig. 1 are pertinent. If we write the transformation between these quantities and the node potentials in 4 in the matrix form

$$v] = [T] \cdot e] \tag{5}$$

then we see by inspection of Fig. 1 that

$$[T] = \begin{bmatrix} -1 & 1 & 0 & 0 \\ 0 & 0 & 0 & 1 \\ 0 & -1 & 0 & 1 \\ -1 & 0 & 1 & 0 \end{bmatrix} \tag{6}$$

with the inverse

$$[T]^{-1} = \begin{bmatrix} -1 & 1 & -1 & 0 \\ 0 & 1 & -1 & 0 \\ -1 & 1 & -1 & 1 \\ 0 & 1 & 0 & 0 \end{bmatrix} \tag{7}$$

Branch currents (with the same reference arrows as for the voltage drops) are related to the source currents in 4 by the matrix equation

$$i_s] = [T]_t \cdot j] \tag{8}$$

as may readily be verified by inspection of Fig. 1 and the transformation matrix 6. Writing 4 in the matrix form

$$[g'] \cdot e] = i_s] \tag{9}$$

and substituting the inverses of Eqs. 5 and 8 gives

$$[T]_t^{-1} \cdot [g'] \cdot [T]^{-1} \cdot v] = j] \tag{10}$$

which we may write more simply as

$$[\bar{g}] \cdot v] = j] \tag{11}$$

with

$$[\bar{g}] = [T]_t^{-1} \cdot [g'] \cdot [T]^{-1} \tag{12}$$

Thus $[\bar{g}]$ is given by a congruent transformation of the matrix $[g']$ as might have been predicted according to the discussion in art. 3, Ch. V on the basis of power invariance. Equation 11 is the desired result in the form of Eqs. 2.

Now suppose we want to get back from this representation to the floating-matrix form. First we unground the network by raising node 5 off the datum (the reverse process of setting $e_5 = 0$ in Eqs. 3). The matrix of Eq. 5, in which the matrix $e]$ now has five elements and is given by

$$e] = \begin{bmatrix} e_1 \\ e_2 \\ \cdot \\ \cdot \\ \cdot \\ e_5 \end{bmatrix} \tag{13}$$

involves a matrix $[T]$ which is the same as 6 except that a fifth column is added, namely

$$[T] = \begin{bmatrix} -1 & 1 & 0 & 0 & 0 \\ 0 & 0 & 0 & 1 & -1 \\ 0 & -1 & 0 & 1 & 0 \\ -1 & 0 & 1 & 0 & 0 \end{bmatrix} \qquad (14)$$

as we can see by inspection of Fig. 1.

In terms of this revised form of $[T]$, Eqs. 5 and 8 are still appropriate. Premultiplying Eq. 11 by $[T]_t$ and substituting for $v]$ from Eq. 5 therefore yields

$$[T]_t \cdot [\bar{g}] \cdot [T] \cdot e] = i_s] \qquad (15)$$

in which $i_s]$ like $e]$ now is the five-element column matrix

$$i_s] = \begin{bmatrix} i_1 \\ i_2 \\ \cdot \\ \cdot \\ \cdot \\ i_5 \end{bmatrix} \qquad (16)$$

Equation 15 may be written

$$[g] \cdot e] = i_s] \qquad (17)$$

with

$$[g] = [T]_t \cdot [\bar{g}] \cdot [T] \qquad (18)$$

Thus Eqs. 3 are regained.

Observe that since $[T]$ in 14 is nonsquare and hence singular, so is the matrix $[g]$, as a floating matrix should be. The congruent transformation 18 again keeps power-invariant.

Let us consider another simple problem, namely that involving the familiar triode vacuum tube schematically shown in Fig. 2. With respect to node 1 (the cathode) as a datum and the grid and plate voltage drops v_1 and v_2 as indicated, the conductance matrix has the well-known form

$$[\bar{g}] = \begin{bmatrix} 0 & 0 \\ g_m & g_p \end{bmatrix} \qquad (19)$$

in which g_m is the transconductance and g_p the plate conductance of the tube.

Now suppose we want to transform to the floating matrix $[g]$ as we might wish to do when embedding the triode in some passive network structure.

In terms of a node-potential matrix $e]$ involving the potentials e_1, e_2, e_3 of the three nodes in Fig. 2 with respect to an arbitrary datum, the matrix $[T]$ in Eq. 5 is seen to be given by

$$[T] = \begin{bmatrix} -1 & 1 & 0 \\ -1 & 0 & 1 \end{bmatrix} \tag{20}$$

and so the floating matrix $[g]$, according to Eq. 18, becomes

$$[g] = \begin{bmatrix} -1 & -1 \\ 1 & 0 \\ 0 & 1 \end{bmatrix} \times \begin{bmatrix} 0 & 0 \\ g_m & g_p \end{bmatrix} \times \begin{bmatrix} -1 & 1 & 0 \\ -1 & 0 & 1 \end{bmatrix} \tag{21}$$

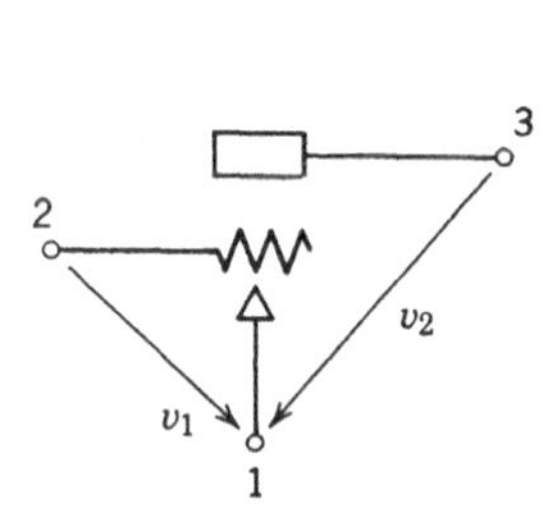

FIG. 2. Schematic of familiar vacuum tube with node 1 (cathode) as datum.

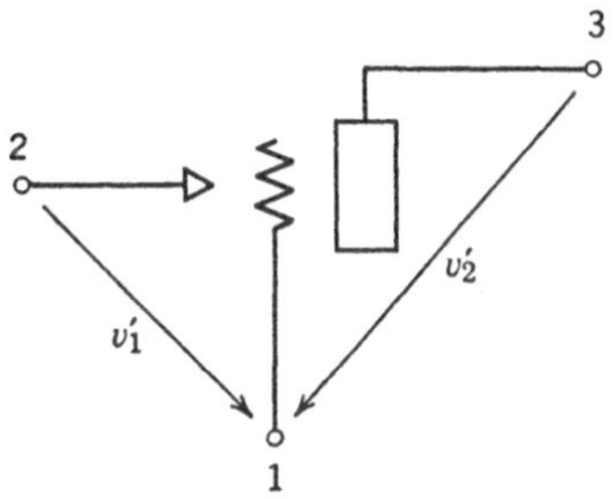

FIG. 3. Alternate orientation of active device.

which evaluates to

$$[g] = \begin{bmatrix} g_m + g_p & -g_m & -g_p \\ 0 & 0 & 0 \\ -(g_m + g_p) & g_m & g_p \end{bmatrix} \tag{22}$$

In the situation shown in Fig. 2, the terminal pairs to which v_1 and v_2 pertain are usually regarded as input and output respectively. Suppose, as is more commonly the case with transistors, we wish to reorient the device so that, let us say, terminal pairs 1-2 and 2-3 become input and output, as is indicated in the redrawn version of the tube shown in Fig. 3.

If we express the voltage drops v_1 and v_2 in Fig. 2 in terms of v_1' and v_2' in Fig. 3 by the matrix equation

$$v] = [T] \cdot v'] \tag{23}$$

we have by inspection

$$[T] = \begin{bmatrix} -1 & 0 \\ -1 & 1 \end{bmatrix} \tag{24}$$

and power invariance requires a corresponding transformation of the branch currents given by

$$j'] = [T]_t \cdot j] \tag{25}$$

In the matrix equation

$$[\bar{g}] \cdot v] = j] \tag{26}$$

pertinent to Fig. 2 with the matrix 19, we premultiply on both sides by $[T]_t$ and substitute for $v]$ from Eq. 23 to get

$$[T]_t \cdot [\bar{g}] \cdot [T] \cdot v'] = j'] \tag{27}$$

which is the desired matrix equation pertinent to the orientation in Fig. 3. The transformed matrix evaluates to

$$\begin{bmatrix} -1 & -1 \\ 0 & 1 \end{bmatrix} \times \begin{bmatrix} 0 & 0 \\ g_m & g_p \end{bmatrix} \times \begin{bmatrix} -1 & 0 \\ -1 & 1 \end{bmatrix} = \begin{bmatrix} g_m + g_p & -g_p \\ -(g_m + g_p) & g_p \end{bmatrix} \tag{28}$$

The effect upon the floating matrix 22 of reorienting the tube is very easily evaluated. We simply interchange rows and columns 1 and 2. The matrix is then appropriate to the orientation in Fig. 3. It is then seen to have the form

$$\begin{bmatrix} 0 & 0 & 0 \\ -g_m & g_m + g_p & -g_p \\ g_m & -(g_m + g_p) & g_p \end{bmatrix} \tag{29}$$

Striking out row and column 1 in this matrix now amounts to grounding node 1 in Fig. 3. The result is matrix 28, as we expect it should be.

3. Definition of Active and/or Nonbilateral Devices

An active network is sometimes defined as one containing sources of energy, while a passive one has only the normal R, L, C elements and no sources. This definition is not only loose and misleading but also useless. If by sources we understand the usual voltage or current constraints, then this definition is completely false. Adding constraints to a passive network does not make it active in the sense that the addition of vacuum tubes or transistors to a network of passive elements alters its potential properties. The latter situation, for example, can yield a network having natural frequencies in the right half of the complex frequency plane or on the j-axis of that plane. The presence of voltage or current constraints produces a response, to be sure, but the effect upon natural frequencies as such is nil.

It is, of course, true that without an energy source in the form of a so-called "power supply," a vacuum tube could not have the effect that it does upon a passive circuit, and so there must be some validity in the statement that active circuits contain sources. However, we are concerned here only with those characteristics of the network that are linear; and the mechanism by which the power supply (usually dc) for a vacuum tube or transistor provides energy for the linear behavior pattern of the associated network is strictly a nonlinear process, as is evidenced by the fact that only nonlinearity can transform energy from one frequency (say zero frequency or dc) to another. While this mechanism of energy transformation is always present in a network containing active devices, it has nothing per se to do with the linear behavior pattern with which we are concerned.

A somewhat more satisfactory basis for the definition of an active device is to say that its association with passive elements allows the resulting network to have natural frequencies with zero real part or positive real part. However, this is hardly a means for defining the active element itself; it is too indirect to be useful.

A simple, clear-cut definition that blends perfectly with methods of linear passive network analysis and extends these methods in a logical manner to include all linear systems is based upon the volt-ampere relations 1 or 2 characterizing the device. If we form the loss function

$$F = v_1 j_1 + v_2 j_2 + \cdots + v_p j_p \tag{30}$$

from Eqs. 1, we get the quadratic form

$$F = \sum_{k,s=1}^{p} r_{ks} j_k j_s \tag{31}$$

and from Eqs. 2 we have

$$F = \sum_{k,s=1}^{p} g_{ks} v_k v_s \tag{32}$$

If either quadratic form is positive-definite, then no set of values for the currents or voltages exists for which the device absorbs negative energy. That is to say, the device will dissipate energy under all conditions imposed by its environment, and hence it must be passive.

On the other hand, if the quadratic form 31 or 32 is not positive definite, then it is surely possible to find an environment (values for the terminal currents or voltages) for which the device absorbs negative energy or delivers positive power to that environment. The device then is unquestionably active.

Since we have simple ways of testing the matrix of a quadratic form for positive definite character, this definition is not only physically meaningful but also is compatible with established methods of linear network analysis.

The volt-ampere characterization in Eqs. 1 or 2 is likewise effective in the definition of nonbilaterality. Thus if the matrix of either of these equations is symmetrical, the device is bilateral; if the matrix is not symmetrical the device is nonbilateral.

Since symmetry of the matrix and positive definiteness of the associated quadratic form are not interrelated, the active character of a device and its possible nonbilaterality have nothing to do with one another. A device can be active without being nonbilateral, or vice versa. However, since practical devices like vacuum tubes or transistors are both active and nonbilateral, one has become used to associating these two properties, at times even to the extent of confusing them. Although, in existing devices they may be inseparable, each can readily be recognized as a property of the matrix associated with pertinent volt-ampere relations like Eqs. 1 or 2.

Regarding the test for positive definiteness, we have some additional discussion to present at this time since that in art. 4 of Ch. V is restricted to quadratic forms having symmetrical matrices. We must extend our considerations to quadratic forms with dissymmetrical matrices since we may want to apply the test for active character to a nonbilateral device.

Any nonsymmetrical matrix may be written as the sum of a symmetrical and a skew-symmetrical matrix. Consider, for example, the resistance matrix of Eqs. 1. We may write

$$\begin{bmatrix} r_{11} & r_{12} & \cdots & r_{1p} \\ r_{21} & r_{22} & \cdots & r_{2p} \\ \cdot & \cdot & \cdot & \cdot \\ r_{p1} & r_{p2} & \cdots & r_{pp} \end{bmatrix} = \begin{bmatrix} a_{11} & a_{12} & \cdots & a_{1p} \\ a_{12} & a_{22} & \cdots & a_{2p} \\ \cdot & \cdot & \cdot & \cdot \\ a_{1p} & a_{2p} & \cdots & a_{pp} \end{bmatrix} + \begin{bmatrix} 0 & b_{12} & \cdots & b_{1p} \\ -b_{12} & 0 & \cdots & b_{2p} \\ \cdot & \cdot & \cdot & \cdot \\ -b_{1p} & -b_{2p} & \cdots & 0 \end{bmatrix} \tag{33}$$

in which

$$\begin{aligned} a_{kk} &= r_{kk} \\ a_{ks} &= \frac{r_{ks} + r_{sk}}{2} \\ b_{ks} &= \frac{r_{ks} - r_{sk}}{2} \end{aligned} \tag{34}$$

If we form the loss function in Eq. 30 with this decomposition of the resistance matrix, it is easy to see that the skew-symmetric component yields nothing, and we have

$$F = \sum_{k,s=1}^{p} a_{ks} j_k j_s \tag{35}$$

The test for positive definiteness is now applied to the symmetrical component in the decomposition 33.

It is obvious physically that the active or passive character of a device is independent of whether its analytical characterization has the form given by Eqs. 1 or 2 which involve inverse matrices. Hence we see that if a given matrix satisfies the test for positive definiteness, then the inverse matrix must also satisfy this test.

For symmetrical matrices this fact is obvious, for if a given matrix has the representation $[A] = [Q] \cdot [Q]_t$ with a real matrix $[Q]$, then its inverse has such a representation. On account of the necessary decomposition 33, a similar result cannot so easily be seen analytically in the case of a dis-symmetrical matrix. It is nevertheless true on physical grounds; and so it is immaterial whether we examine the resistance or the conductance matrix of a device in order to establish its possible active and/or nonbilateral character.

4. Methods of Active Circuit Analysis

A familiar method of analyzing circuits with embedded active devices is based upon replacing the latter by a so-called "equivalent circuit" or "mathematical model" which consists of a combination of ordinary passive elements and a "controlled source," the latter being a voltage or current constraint whose value is proportional to a current or voltage in some branch of the circuit other than that in which this source is located. The choice of nomenclature here is unfortunate, for the basic characteristic of a voltage or current constraint is its complete *independence* of all other conditions in the circuit—the fact that it is dependent upon no other current or voltage value. The name "controlled source" or "controlled constraint" is self-contradictory. Use of the term "source" for "constraints" is, of course, partly responsible for the evolution of this contradictory result.

Actually what is involved here is a mutual parameter and not a source at all. For that matter, we can replace an ordinary resistance by a voltage "source" proportional to the current passing through it. When we have mutual inductive coupling in a circuit, we can replace it by a "controlled source" in one branch proportional to the time derivative of a current in another branch. In a vacuum tube circuit we encounter a mutual resistance or conductance parameter. The replacement of an ordinary parameter by a constraint that is not a constraint hardly seems to be a logical procedure unless there is some advantage to be gained in the resulting analysis.

This practice, which we inherit from our illogical past, is probably largely responsible for the misconception mentioned earlier that active circuits are circuits containing sources. This situation is a good example of how the compounding of misconceptions and illogical procedures can confuse and mislead us.

In order to see more specifically how the method of circuit analysis using the controlled-source model compares with the method presented here, let us consider the vacuum-tube circuit shown in Fig. 4. Suppose we choose the potentials of nodes 1 to 5 with respect to node 0 as variables. Using the matrix 22 as a model, the component contributed to the net $[G]$-matrix by the two tubes is readily seen to be given by

$$\begin{bmatrix} g_{m1}+g_{p1} & -g_{m1} & -g_{p1} & 0 & 0 \\ 0 & 0 & 0 & 0 & 0 \\ -(g_{m1}+g_{p1}) & g_{m1} & g_{p1}+g_{m2}+g_{p2} & -g_{m2} & -g_{p2} \\ 0 & 0 & 0 & 0 & 0 \\ 0 & 0 & -(g_{m2}+g_{p2}) & g_{m2} & g_{p2} \end{bmatrix} \tag{36}$$

in which the significance of the notation is obvious.

We will assume the numerical values (suitably scaled, of course)

$$\begin{aligned} g_{m1} &= 10, \quad g_{p1} = 3 \\ g_{m2} &= 7, \quad g_{p2} = 1 \end{aligned} \tag{37}$$

and let all conductance values in the passive branches be unity. Then we see by inspection that the resultant $[G]$-matrix is given by the sum

$$[G] = \begin{bmatrix} 1 & 0 & 0 & 0 & 0 \\ 0 & 2 & 0 & -1 & 0 \\ 0 & 0 & 2 & 0 & -1 \\ 0 & -1 & 0 & 2 & -1 \\ 0 & 0 & -1 & -1 & 3 \end{bmatrix} + \begin{bmatrix} 13 & -10 & -3 & 0 & 0 \\ 0 & 0 & 0 & 0 & 0 \\ -13 & 10 & 11 & -7 & -1 \\ 0 & 0 & 0 & 0 & 0 \\ 0 & 0 & -8 & 7 & 1 \end{bmatrix} \tag{38}$$

in which the first term is the contribution from passive elements.

The resulting equilibrium equations are

$$\begin{bmatrix} 14 & -10 & -3 & 0 & 0 \\ 0 & 2 & 0 & -1 & 0 \\ -13 & 10 & 13 & -7 & -2 \\ 0 & -1 & 0 & 2 & -1 \\ 0 & 0 & -9 & 6 & 4 \end{bmatrix} \times \begin{bmatrix} e_1 \\ e_2 \\ e_3 \\ e_4 \\ e_5 \end{bmatrix} = \begin{bmatrix} 0 \\ i_s \\ 0 \\ 0 \\ 0 \end{bmatrix} \tag{39}$$

and any desired response can readily be computed.

Attacking the same problem by the controlled-source method, we begin by drawing the schematic shown in Fig. 5, in which each tube in the circuit of Fig. 4 is replaced by its mathematical model* involving the pertinent controlled source. Since the value of each controlled source depends upon an unknown node-pair voltage, it is difficult to get this procedure "off the ground" so to speak.

* A fancy name recently introduced in place of "equivalent circuit."

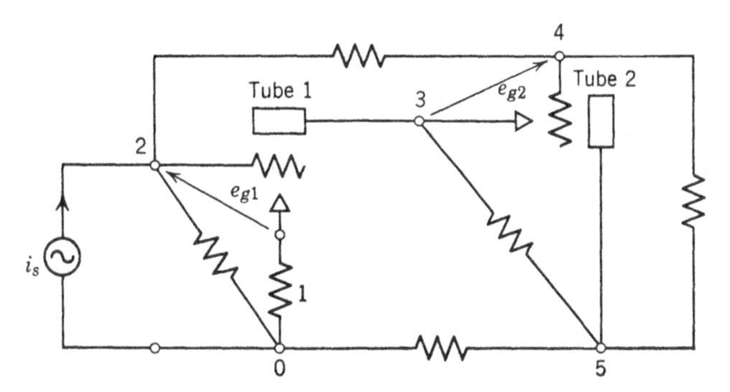

FIG. 4. Example of circuit involving several active elements.

A possible approach here is to choose as variables the node potential differences

$$
\begin{aligned}
e_1' &= e_1 \\
e_2' &= e_2 - e_1 \\
e_3' &= e_3 - e_1 \\
e_4' &= e_4 - e_3 \\
e_5' &= e_5 - e_4
\end{aligned}
\tag{40}
$$

so that two of these appear across the terminal pairs of the controlled sources whose values are proportional to two others. Now we can write node equations fairly straightforwardly and then associate the controlled sources with their respective unknown voltage terms, where they belong in the first place. The process is considerably more roundabout, although it yields the same end result as, of course, it must.

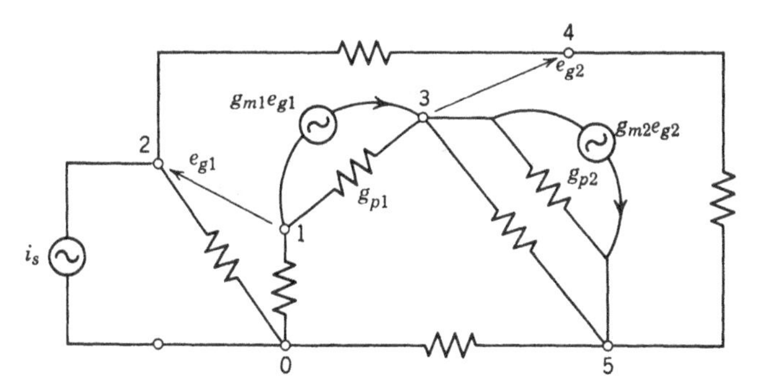

FIG. 5. The circuit of Fig. 4 redrawn so as to accommodate the controlled-source method.

Problems

1. Four-terminal devices having the following matrices on a node-to-datum basis are separately embedded in passive networks. Under suitable environmental conditions, are they capable of delivering energy to their associated passive circuits?

$$(a)\quad \begin{bmatrix} 16 & 8 & 6 \\ 4 & 8.5 & 2 \\ -2 & 5.5 & 2 \end{bmatrix}, \qquad (b)\quad \begin{bmatrix} 10 & 2 & 0 \\ 2 & 3 & 2 \\ 0 & 2 & 1 \end{bmatrix}$$

2. In the following two-terminal-pair network

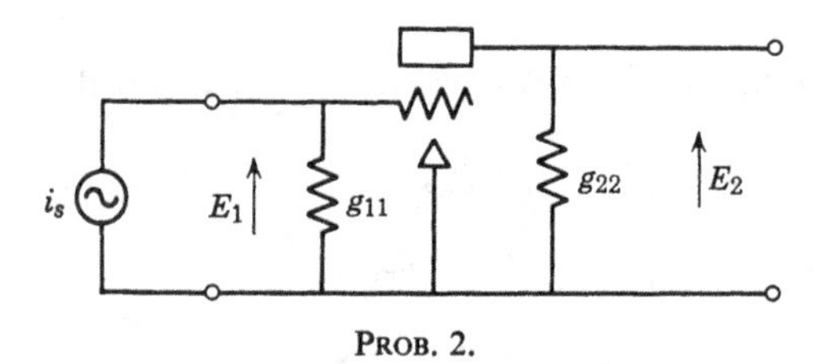

PROB. 2.

the matrix representing the triode alone may be assumed given by Eq. 19 with $g_p = 0$. Find the condition on g_m for which

(*a*) Power absorbed by the passive elements is larger than the power input.
(*b*) Power absorbed by the total network can be negative.
(*c*) Power absorbed in g_{22} is greater than the input power.

3. The adjacent circuit depicts a triod oscillator in rather general terms. At least one of the boxes labeled y_a, y_b, y_c is dissipative.

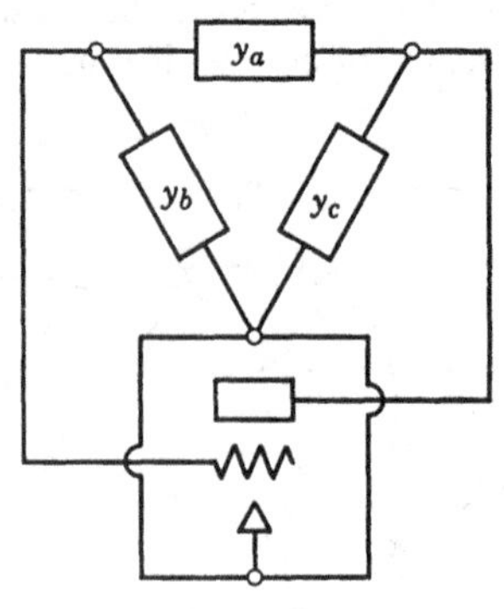

PROB. 3.

Determine several simple circuit combinations for these boxes for which sustained oscillations will result. In each case obtain an expression for the condition under which oscillations can exist and the resulting frequency. The triode may be assumed characterized as in prob. 2.

4.

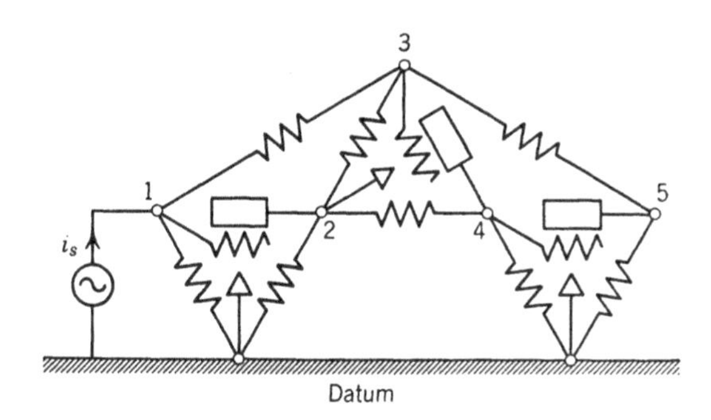

PROB. 4.

For the above circuit, set up node-to-datum equations

(*a*) By the matrix method.
(*b*) By the "controlled source" method.

Each tube may be assumed to have the matrix representation 19 with $g_p = 0$. Choose $g_m = 10$ for all tubes and unity for the eight passive resistances.

5. In the following circuit arrangement:

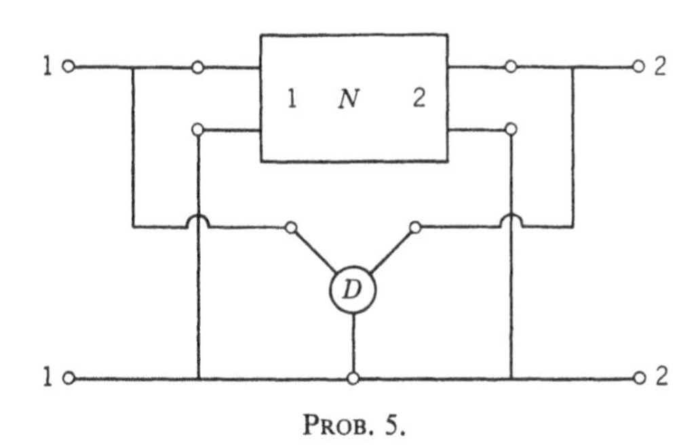

PROB. 5.

N is a passive bilateral network characterized by the short-circuit admittance functions y'_{11}, y'_{22}, y'_{12}, and D is a three-terminal device having an admittance matrix of the form

$$[y] = \begin{bmatrix} 0 & -g \\ g & 0 \end{bmatrix}$$

in which g is positive real.

If z_{11} and z_{21} are open-circuit impedances of the resulting network at terminal pairs 1-1 and 2-2, show that these driving-point and transfer functions can have complex poles even though N contains only R's and C's so that zeros and poles of y'_{11} and y'_{22} as well as poles of y'_{12} are restricted to the negative real axis of the s-plain. Can this resultant network deliver more power to a load than it receives at its input terminals?

6. Repeat prob. 5 if the device has a matrix

$$[y] = \begin{bmatrix} 0 & 0 \\ g & 0 \end{bmatrix}$$

CHAPTER VII

Natural Frequencies and Normal Coordinates

1. The Frozen Distribution at Resonance

Consider a passive bilateral network excited at a single point. Complete solution of the pertinent equilibrium equations yields the voltages and currents in all branches. It is customary to refer to this result as characterizing a current or voltage *distribution function* which describes the way in which, say, the current values vary from branch to branch throughout the network, where the current is largest, where it is smallest, etc.

One of the interesting things that such a distribution function can show is that the current or voltage response is not necessarily largest at or near the source location, or that it becomes steadily smaller for branches farther removed from the source. Indeed, the distribution function can exhibit various maxima and minima; and the largest value of current may even occur in a branch that is farthest from the source.

This distribution pattern, moreover, depends upon the location of the source and, in general, it is completely different for different source locations. To take an extremely simple example, consider a ladder network of resistances excited at one end. Here the current is, of course, largest at the source and drops off steadily as one considers more and more remote branches. If the source is moved to the other end of the ladder, the distribution function is essentially reversed; and if the source is in the center, the distribution function has a maximum there and drops off to either side.

However, an important exception to this rule can occur in any network having finite nonzero natural frequencies. Namely, if the excitation frequency is at or near a natural frequency of the network so that the latter is in resonance, *then the distribution pattern becomes independent of the point at which the excitation is located.* We speak of the distribution as being *frozen.* The whole system is stiff and unyielding. The points of maximum and minimum response remain fixed no matter where the network is driven.

This is an important feature accompanying the resonance phenomenon that has many useful applications; and it is, therefore, needful that we clearly understand how it comes about. The following discussion provides that background.

Suppose we arbitrarily choose a loop basis and begin by writing down the equilibrium equations in matrix form as given by Eq. 108 of Ch. II, which reads

$$([L]p + [R] + [S]p^{-1}) \cdot i_v] = e_s] \tag{1}$$

Let the excitation matrix $e_s]$ contain a single nonzero element which we choose to be a unit impulse located in loop or link No. 1. The particular integral or steady-state part of the solution is then zero and the complementary function or transient part is governed by the homogeneous equation obtained by replacing the right-hand member of Eq. 1 with zero.

As nontrivial solutions to this set of equations we then assume the familiar current expressions

$$i_j = A_j e^{st} \qquad \text{for } j = 1, 2, \ldots l \tag{2}$$

If we use the abbreviation

$$b_{ij} = L_{ij}s + R_{ij} + S_{ij}s^{-1} \tag{3}$$

then substitution of the assumption 2 into the homogeneous form of Eq. 1, and cancellation of the common factor e^{st}, yields the set of algebraic equations

$$\begin{aligned} b_{11}A_1 + b_{12}A_2 + \cdots + b_{1l}A_l &= 0 \\ b_{21}A_1 + b_{22}A_2 + \cdots + b_{2l}A_l &= 0 \\ \cdot \quad \cdot \quad \cdot \quad \cdot \quad \cdot \quad \cdot \quad &\cdot \quad \cdot \\ b_{l1}A_1 + b_{l2}A_2 + \cdots + b_{ll}A_l &= 0 \end{aligned} \tag{4}$$

with the matrix

$$[B] = \begin{bmatrix} b_{11} & \cdots & b_{1l} \\ \cdot & \cdot & \cdot \\ b_{l1} & \cdots & b_{ll} \end{bmatrix} \tag{5}$$

According to the theory of algebraic equations, the homogeneous set 4 possesses nontrivial solutions for the A_j's only if the determinant of $[B]$ is zero. Since the elements 3 of this determinant are functions of the frequency variable s which, so far as the assumptions 2 are concerned, is not yet fixed, we can use this algebraic condition as a means for determining appropriate s-values. We therefore write

$$B = \begin{vmatrix} b_{11} & b_{12} & \cdots & b_{1l} \\ b_{21} & b_{22} & \cdots & b_{2l} \\ \cdot & \cdot & \cdot & \cdot \\ b_{l1} & b_{l2} & \cdots & b_{ll} \end{vmatrix} = 0 \tag{6}$$

Remembering that a determinant of order l is given by a sum of $l!$ terms, each of which is a product of l elements, we see that condition 6 is an algebraic equation of degree $2l$ in s (after multiplying through by s^l so as to eliminate negative powers of s). Roots of this equation, which is variously referred to as the *determinantal* or as the *characteristic* equation, are *natural frequencies* of the network (also called *characteristic* values or modes).

Since Eq. 6 has real coefficients, the roots will either be real or have the form of conjugate complex pairs. In the latter instance each pair defines a frequency (imaginary part) and a damping constant (real part) in a manner that is familiar to us from our experience with simple circuits.

The humble assumption 2 is thus not only justified but becomes expanded into a sum of $2l$ terms of this form, one for each root or natural frequency. Hence the solutions 2 take the form

$$i_j = \sum_{\nu=1}^{2l} A_j^{(\nu)} e^{s_\nu t} \qquad \text{for } j = 1, 2, \ldots l \tag{7}$$

in which the natural frequencies are denoted by $s_1, s_2, \ldots s_{2l}$. Pairs of terms involving conjugate roots have conjugate A_j-values and hence yield a real contribution to the pertinent current i_j.

For the evaluation of the amplitudes $A_j^{(\nu)}$ we now return to Eqs. 4 and recognize that we have all together $2l$ such sets of equations to consider, one set for each s_ν-value. Since the coefficients b_{ij} in these equations are functions of s, we have a different set of coefficients for each value of s.

If all roots of the characteristic equation are complex, which we may as well assume at this point, then we need consider only one set of equations for each pair of conjugate s_ν-values because the corresponding $A_j^{(\nu)}$-values correspondingly occur in conjugate complex pairs.

Now let us consider one such set of the equations 4. It is a homogeneous set, and the pertinent determinant is zero. The matrix $[B]$ has a rank less than l; and unless we are dealing with a highly degenerate situation, the rank is exactly $l - 1$. In that case the equations 4 determine the ratios of the A_j's to one another. Initial conditions in the network (which we will not consider in detail at this time) determine one A_k-value for each set of equations, and the latter then determine all others in terms of this one.

According to the theory of algebraic equations the ratio of A_j-values is given by the corresponding ratio of cofactors in any row of the determinant 6. For example, we may write

$$A_1 : A_2 : \cdots : A_l = B_{11} : B_{12} : \cdots : B_{1l} \tag{8}$$

or

$$A_1 : A_2 : \cdots : A_l = B_{21} : B_{22} : \cdots : B_{2l} \tag{9}$$

and so forth, the notation for cofactors being obvious, as is also the pertinence of these relations to a particular s_ν-value which the notation does not bother to indicate. It is an important property of the matrix $[B]$ that when it has rank $l - 1$, then the ratio of cofactors is the same for each row, so that it doesn't matter which row we choose.

However, the fact that $[B]$ has rank $l - 1$ guarantees only that at least one cofactor is nonzero, and hence there could be rows for which all cofactors are zero. But this situation can happen only if the pertinent network is degenerate, for it means that the natural frequency for which the equations 4 are written does not appear in the transient response of that loop or link to which the row in question pertains. If the network is completely random so that all natural frequencies are excitable from all loops (there is no special element value distribution or topological condition to bring about the isolation of certain natural frequencies from one or more of the chosen loops), then it is not possible for all the cofactors of any row or any column to be zero. We shall tacitly assume that we are dealing with such a nondegenerate case.

With this background regarding transient response and natural frequencies, let us return to the inhomogeneous equation 1 and again assume the excitation to be restricted to a single link, but now let it be located in any link i and be a steady sinusoid expressed in the exponential form

$$e_i = E_i e^{st} \tag{10}$$

in which the frequency variable s is assumed known and we are only interested in the steady state or ultimate response that emerges after any initial transients have died away.

The current response due to the excitation 10 may be written

$$i_j = I_j e^{st} \tag{11}$$

whereupon substitution into the equation set 1 and cancellation of the common factor e^{st} yields

$$\begin{aligned} b_{11}I_1 + b_{12}I_2 + \cdots + b_{1l}I_l &= 0 \\ \cdots\cdots\cdots&\cdots \\ b_{i1}I_1 + b_{i2}I_2 + \cdots + b_{il}I_l &= E_i \\ \cdots\cdots\cdots&\cdots \\ b_{l1}I_1 + b_{l2}I_2 + \cdots + b_{ll}I_l &= 0 \end{aligned} \tag{12}$$

Here the coefficients b_{ij} are again given by the expression 3; however, they are completely known since the complex frequency s is known, being the frequency of the applied source. In contrast with Eqs. 4, the set 12 is inhomogeneous and the pertinent determinant B is not zero.

By Cramer's rule we can write for the resulting current amplitude in loop j

$$I_j = \frac{B_{ij}E_i}{B} \tag{13}$$

and hence the current distribution is expressible by the set of ratios

$$I_1:I_2: \cdots :I_l = B_{i1}:B_{i2}: \cdots :B_{il} \tag{14}$$

Now suppose that the pertinent network has a natural frequency $s_\nu = -\alpha_\nu + j\omega_\nu$ for which $\alpha_\nu \ll \omega_\nu$ so that a pronounced resonance occurs if we choose the frequency s of the sinusoidal source 10 equal to $j\omega_\nu$. This j-axis point in the complex s-plane is very close to the natural frequency s_ν, the separation α_ν being small compared with $|s_\nu|$. Therefore, the determinant B, which is zero for $s = s_\nu$, has a very small value and all current amplitudes I_j given by Eq. 13 are exceptionally large, as is characteristic of a resonant response.

We are not so much interested in the magnitude of the response as we are in the distribution ratios given by Eq. 14. Again since $s \approx s_\nu$, all the cofactors of B have values that are practically the same as they are for $s = s_\nu$; and since (as shown by Eqs. 8 and 9) their ratios are the same for any row i, Eq. 14 shows that the current distribution is independent of the location of the source.

The quantities

$$y_{ji} = \frac{B_{ij}}{B} \tag{15}$$

appearing in Eq. 13 we recognize as the short-circuit driving-point and transfer admittance functions of our network. The zeros of B, or the natural frequencies of the network, are poles of these admittance functions. A *partial fraction expansion* of the rational function 15 (which incidentally is a proper fraction since the denominator is of higher degree than the numerator) places the poles of y_{ji} in evidence and reads

$$y_{ji} = \frac{k_{ji}^{(1)}}{s - s_1} + \frac{k_{ji}^{(2)}}{s - s_2} + \cdots + \frac{k_{ji}^{(2l)}}{s - s_{2l}} \tag{16}$$

Here the coefficients $k_{ji}^{(\nu)}$, called *residues* of the function y_{ji} in its poles, are given by the expression

$$k_{ji}^{(\nu)} = [(s - s_\nu)y_{ji}]_{s=s_\nu} \tag{17}$$

as may readily be seen through multiplying 16 by the factor $s - s_\nu$ and then carrying out the limit process $s \to s_\nu$. All terms on the right-hand side of 16 thus become zero except $k_{ji}^{(\nu)}$, which is given by the expression in Eq. 17. Since B in Eq. 15 contains the factor $s - s_\nu$, it evidently cancels, so that a finite nonzero value results for $k_{ji}^{(\nu)}$.

From what has just been said about the ratios 14 being the same for any row i, we see that the ratios of the residues

$$k_{1i}^{(\nu)}:k_{2i}^{(\nu)}: \cdots :k_{li}^{(\nu)} = B_{i1}:B_{i2}: \cdots :B_{il} \tag{18}$$

have this same property for any natural frequency designated by the superscript ν. The residue matrix

$$[K] = \begin{bmatrix} k_{11} & k_{12} & \cdots & k_{1l} \\ k_{21} & k_{22} & \cdots & k_{2l} \\ \cdot & \cdot & \cdot & \cdot \\ k_{l1} & k_{l2} & \cdots & k_{ll} \end{bmatrix} \tag{19}$$

for any s_ν, therefore, has rank 1 because all rows are proportional.

Because of the reciprocity theorem, $y_{ji} = y_{ij}$ and hence $k_{ji} = k_{ij}$ also, which means that any row of $[K]$ is the same as the corresponding column; $[K]$ is symmetrical. It follows that if we know a single row or column in this matrix, we can at once write down the entire matrix.

When the residue matrix $[K]$ for a set of short-circuit driving-point and transfer admittances has rank 1 for all natural frequencies s_ν, that set of functions is said to be *compact*. We have just shown that the $[y]$-matrix for any nondegenerate network is compact. That is to say, if a network has the property that all natural frequencies are excitable from all of its independent links, then the pertinent $[y]$-matrix is compact. Its residue matrix has rank 1.

By the completely dual procedure, beginning with equilibrium equations on a node basis (the Eqs. 114 of Ch. II) we arrive at precisely the same result with regard to a corresponding set of open-circuit driving-point and transfer impedances having the matrix $[z]$. *It* is compact if the pertinent network has the property that all natural frequencies are excitable from all n independent node pairs.

It might be worth mentioning in this connection that completely general networks of this sort are not commonly found in practice. Ladder networks and other simple structures that are usually preferred because of economy and ease of alignment seldom have all natural frequencies excitable from both input and output terminal pairs. One might almost say that degeneracy is the rule and complete generality is the rare exception. The network having a compact set of impedance or admittance functions occurs only where this feature is essential to achieving a particular characteristic in the desired response.

2. Normal Coordinates

The complete opposite of a network in which all natural frequencies are excitable at all points of entry is one in which the natural frequencies appear isolated in their respective loops or meshes. This network consists of l uncoupled meshes, each placing in evidence one pair of conjugate complex or one real natural frequency or normal mode. Instead of dealing with one l-loop network, we consider l one-loop circuits which together form the total network.

In such a network the characteristic or determinantal equation is factored to begin with; the matrix $[B]$, Eq. 5, is diagonal; and the equation set 4 or 12 consists of l separate equations each involving a single variable. No simultaneous equations need to be solved; and no algebraic equation of degree higher than two need be solved for its roots. Computationally this is the ideal situation. The disturbing feature about it, however, is the fact that this form of network has little practical value since all transfer functions are zero and all driving-point functions can at most involve a polynomial of second degree. Degeneracy has reached its ultimate end and usefulness has fallen to its lowest level, barring only the situation in which we have no network at all.

There is, however, a way in which we can combine the conceptual clarity and mathematical simplicity of this ultimate form of degeneracy with the generality and usefulness of the completely random network. The artifice that accomplishes this end is to say: to each random network there corresponds a fictitious one in which the modes are isolated; and the current variables in this fictitious network are related to the current variables in the actual network by a linear nonsingular transformation. Once this transformation is established we can take advantage of the mathematical simplicity of the fictitious degenerate form in computing any desired response, and then find the response in the actual network by applying the inverse transformation.

Since a linear transformation of the current variables may be interpreted as a change in the reference coordinate system, we can regard this procedure as the transformation to a fictitious coordinate system in which the natural modes of the network are isolated. The coordinates of this system are referred to as the *normal* coordinates of the given network; and the procedure suggested above is spoken of as a *normal coordinate transformation*.

Regarding computational labor, it should be pointed out that the use of a normal coordinate transformation in the process of constructing a desired network response does not produce any simplification, as one might at first be led to believe, since the chore involved in establishing the pertinent transformation offsets any subsequent saving in computational labor that might accrue through its use.

The concept of normal coordinates is not found useful as a labor-saving device; rather, its usefulness lies in the basic understanding of network response that it provides, and in suggesting ways of solving collateral problems like the method of constructing parameter matrices from a prescribed response function in network synthesis (discussed in art. 5 below). We shall now present the detailed procedure involved in a normal coordinate transformation and the conditions under which it is possible.

To this end, consider the equilibrium equations 1. What we wish to accomplish by a coordinate transformation is the simultaneous diagonalization of the matrices $[L]$, $[R]$ and $[S]$. This coordinate transformation will, of course, affect the variables in $i_v]$ as well as the sources in $e_s]$; and it may or may not preserve the invariance of power. Since the practical usefulness of the transformation is enhanced if power relations remain invariant, let us see first of all whether we can accomplish the desired end with this restriction imposed. It turns out, as we shall see, that we must relinquish it only when nonbilateral elements are involved; active as well as passive ones allow the power invariance condition to be included.

If we premultiply both sides of Eq. 1 by the transpose of $i_v]$, we have

$$\underline{i_v} \cdot ([L]p + [R] + [S]p^{-1}) \cdot i_v] = \underline{i_v} \cdot e_s] = P \tag{20}$$

where P is the instantaneous power delivered by the sources as pointed out in art. 1, Ch. V. It is tacitly assumed, of course, that cut-set and tie-set matrices are consistent; otherwise the right-hand side of Eq. 20 does not represent power supplied by the sources. We also remind ourselves that this consistency results in symmetrical parameter matrices.

Now if we introduce the transformation of current variables given by

$$i_v] = [T] \cdot i_v'] \tag{21}$$

and note that its transpose reads

$$\underline{i_v} = \underline{i_v'} \cdot [T]_t \tag{22}$$

substitution into Eq. 20 yields

$$\underline{i_v'} \cdot [T]_t \cdot ([L]p + [R] + [S]p^{-1}) \cdot [T] \cdot i_v'] = P \tag{23}$$

Power invariance is thus seen to require a congruent transformation of the parameter matrices (as is also shown in art. 3, Ch. V). In the new coordinate system we have the matrices

$$[L'] = [T]_t \cdot [L] \cdot [T] \tag{24}$$

$$[R'] = [T]_t \cdot [R] \cdot [T] \tag{25}$$

$$[S'] = [T]_t \cdot [S] \cdot [T] \tag{26}$$

If the substitution 21 is to achieve a normal coordinate transformation then the resultant matrices 24, 25 and 26 should be diagonal. In order to see how this result can be achieved, we will break the transformation down into simpler steps. First we consider the matrix $[L]$ alone and find a congruent transformation with a matrix $[T_1]$ that carries it into the canonic form

$$[T_1]_t \cdot [L] \cdot [T_1] = [U_l] \tag{27}$$

Here $[U_l]$ is a unit matrix of order l.

As pointed out previously (art. 4, Ch. V), this result is possible only if $[L]$ is nonsingular and characterizes a positive definite quadratic form. The latter condition is fulfilled if all inductance elements are passive. The nonsingular character, on the other hand, is not necessarily fulfilled, even if all current variables are dynamically independent.* If not $[L]$, but another one of the three matrices $[L]$, $[R]$ or $[S]$ is nonsingular, we can use it instead, but for the present argument we may as well assume that $[L]$ is this one.

The same congruent transformation with the matrix $[T_1]$ carried out on $[R]$ and $[S]$ leaves these matrices symmetrical but converts them into forms which we may denote by $[R_1]$ and $[S_1]$.

We now make use of the fact that any real symmetrical matrix (singular or nonsingular) may be transformed congruently to the diagonal form involving its latent roots by means of a pertinent orthogonal modal matrix.† For the matrix $[S_1]$ suppose this transformation reads

$$[T_2]_t \cdot [S_1] \cdot [T_2] = [D] \tag{28}$$

in which

$$[D] = \begin{bmatrix} d_1 & 0 & 0 & \cdots & 0 \\ 0 & d_2 & 0 & \cdots & 0 \\ \cdot & \cdot & \cdot & \cdot & \cdot \\ 0 & 0 & 0 & \cdots & d_l \end{bmatrix} \tag{29}$$

and $d_1, d_2, \ldots d_l$ are the latent roots of $[S_1]$.

This same congruent transformation with the matrix $[T_2]$ applied to $[U_l]$ in Eq. 27 leaves it unchanged because the transpose of the orthogonal

* In this connection it is significant to point out that diagonalization may be possible if the sum and difference of two matrices are positive and nonsingular (see "Synthesis of Multiterminal Two-Element-Kind Networks" by F. S. Boxall, *Technical Report No. 95*, Nov. 1, 1955, Electronics Research Laboratory, Stanford University, Stanford, California, p. 14 *et seq*.

† See *The Mathematics of Circuit Analysis* (by E. A. Guillemin, M.I.T. Press and John Wiley and Sons, 1949 (hereafter referred to as *MCA*), pp. 111–122.

matrix $[T_2]$ is also its inverse. We now have achieved the simultaneous diagonalization of two of the three matrices in Eq. 1 or Eq. 20.

We cannot continue this process in any way to diagonalize also the third matrix.* In the present argument this matrix is $[R]$, but it might be any other depending upon which one of the three is nonsingular to start with and which one we choose next to diagonalize by the orthogonal transformation 28.

The only possibility for achieving a normal coordinate transformation of the equilibrium Eqs. 1 by this procedure is for the matrices $[R_1]$ and $[S_1]$ to have the same modal matrix $[T_2]$, so that we have in addition to 28 the result

$$[T_2]_t \cdot [R_1] \cdot [T_2] = [Q] \tag{30}$$

with

$$[Q] = \begin{bmatrix} q_1 & 0 & 0 & \cdots & 0 \\ 0 & q_2 & 0 & \cdots & 0 \\ \cdot & \cdot & \cdot & \cdot & \cdot \\ 0 & 0 & 0 & \cdots & q_l \end{bmatrix} \tag{31}$$

The latent roots $q_1 \cdots q_l$ of $[R_1]$ appearing here are completely unrelated to the latent roots $d_1 \cdots d_l$ of $[S_1]$. In other words, the two matrices $[R_1]$ and $[S_1]$ may have altogether different latent roots although they possess the same orthogonal modal matrix.

A special case occurs if $q_1 = q_2 = \cdots = q_l = q$, for then $[R_1] = q[U_l]$ or

$$[R] = q[L] \tag{32}$$

which states that the resistance and inductance parameter matrices are proportional. The condition leading to 30 and 31 is considerably more general than this proportionality condition, although it also is rather specialized.

In the completely general situation on a loop or node basis, the normal coordinate transformation cannot be carried out in this manner with a real transformation matrix regardless of whether power invariance is included, since it is not possible with any real transformation to convert all three of the matrices in the power expression 23 to diagonal forms, or, what amounts to the same thing, to reduce the three quadratic forms in this expression to sums of squares.

When the matrices $[R_1]$ and $[S_1]$ do have the same modal matrix $[T_2]$ so that the results expressed by Eqs. 28, 29, 30 and 31 are possible, then

* In this connection see, however, the semidiagonalization procedure discussed in art. 3 below.

the three transformed matrices 24, 25 and 26 are diagonal. The resultant transformation matrix is given by

$$[T] = [T_1] \cdot [T_2] \tag{33}$$

and the resultant parameter matrices are

$$[L'] = [U_l] \tag{34}$$

$$[R'] = [Q] \tag{35}$$

$$[S'] = [D] \tag{36}$$

The first of these is the unit matrix of order l, the other two are the diagonal matrices 29 and 31.

The equilibrium equations in normal coordinates are expressed by

$$([U_l]p + [Q] + [D]p^{-1}) \cdot i_v'] = e_s'] \tag{37}$$

in which the excitation matrix is seen to be

$$e_s'] = [T]_t \cdot e_s] \tag{38}$$

the inverse of which, namely

$$e_s] = [T]_t^{-1} \cdot e_s'] \tag{39}$$

is the companion relation to Eq. 21 expressing the transformation of variables.

The characteristic or determinantal Eq. 6 now has the special form

$$\begin{bmatrix} s + q_1 + d_1 s^{-1} & 0 & \cdots & 0 \\ 0 & s + q_2 + d_2 s^{-1} & \cdots & 0 \\ \cdot & \cdot & \cdot & \cdot \\ 0 & 0 & \cdots & s + q_l + d_l s^{-1} \end{bmatrix} = 0 \tag{40}$$

from which it is clear that the natural frequencies are given by the separate quadratic equations

$$s + q_j + d_j s^{-1} = 0 \qquad \text{for } j = 1, 2, \ldots l \tag{41}$$

Although the class of networks permitting this normal coordinate transformation is rather restricted, we see that their natural frequencies are not. With the constants q_j and d_j in the factor 41 completely arbitrary (except that they be real, and also positive if the network is passive), we see that the possible s_ν-values which Eq. 40 can yield are as varied as those of *any* network containing the same kind of elements.

By the completely dual procedure, beginning with node equations 114 of Ch. II, we arrive at an analogous result which is precisely the same in character. No real or power-invariant transformation for the simultaneous diagonalization of pertinent parameter matrices is in general possible here either, not even in the case of passive bilateral networks.

We can, however, achieve a power-invariant normal coordinate transformation for a completely general linear network with any loss function and with active as well as passive elements, by starting from equilibrium equations expressed on a mixed basis, as presented in art. 5, Ch. II. The matrix 133 given there, which we repeat here for convenience:

$$\left\{\begin{bmatrix} L & \vdots & 0 \\ \cdots & \cdots & \cdots \\ 0 & \vdots & C \end{bmatrix} p + \begin{bmatrix} R & \vdots & \gamma_{ln} \\ \cdots & \cdots & \cdots \\ -\gamma_{nl} & \vdots & G \end{bmatrix}\right\} \times \begin{bmatrix} i_v \\ \cdots \\ e_v \end{bmatrix} = \begin{bmatrix} e_s \\ \cdots \\ i_s \end{bmatrix} \tag{42}$$

is our starting point.

The reason for choosing the mixed basis for expressing equilibrium is clear since the form of this matrix equation is essentially that of a system characterized by two energy functions. We now do not need to diagonalize more than two parameter matrices simultaneously since the inductive and capacitive elements are contained in one matrix while loss elements and the "connection submatrix" γ_{ln} are contained in the other.

As we have just seen, however, the simultaneous diagonalization of these two matrices by a congruent transformation requires both to be symmetrical besides stipulating that one of them be positive definite (negative definite would also do but we can skip this) and nonsingular. Even if we assume, as we shall to start with, that no nonbilateral elements are present, then the matrix containing loss elements, in partitioned form, is skew-symmetrical, and so these conditions are not met.

We should, however, be aware of the fact that it is illogical to expect the desired coordinate transformation to be real. Since the latent roots involved are natural frequencies of our network, and these are surely complex in any general *RLC* situation, we expect that the pertinent congruent transformation will involve a matrix with complex elements.

In anticipation of these things it is, therefore, not inconsistent to do a little multiplying by the operator $j = \sqrt{-1}$ and put Eq. 42 into the modified but equivalent form

$$\left\{\begin{bmatrix} L & \vdots & 0 \\ \cdots & \cdots & \cdots \\ 0 & \vdots & C \end{bmatrix} p + \begin{bmatrix} R & \vdots & j\gamma_{ln} \\ \cdots & \cdots & \cdots \\ j\gamma_{nl} & \vdots & G \end{bmatrix}\right\} \times \begin{bmatrix} ji_v \\ \cdots \\ e_v \end{bmatrix} = \begin{bmatrix} je_s \\ \cdots \\ i_s \end{bmatrix} \tag{43}$$

Regarding the required nonsingular character of the first matrix, we will assume that if any linear constraint relations exist among the loop currents or node-pair voltages, these have been used to eliminate dynamically superfluous variables so that all are now independent. Discussion in art. 4 below will show that the nonsingular character of this matrix is thus always achievable.

If we are considering a passive system or one in which active (but bilateral) elements are resistive in basic character, then the positive definiteness of the quadratic form for the stored energy is assured, as is the symmetry of the matrix containing R and G. Hence the problem of carrying out a normal coordinate transformation (congruently so as to keep power invariant) is now routine.

As in the preceding discussion, we can carry it out in two steps by first transforming the matrix with L and C to its canonic form, and then subjecting the resulting loss matrix to an orthogonal transformation with its modal matrix, thus converting it to the diagonal form while leaving the canonic matrix unaltered.

These two steps may be combined into a single congruent transformation with a matrix $[T]$ of order b characterizing the coordinate transformation indicated by the relations

$$\begin{bmatrix} ji_v \\ \cdots \\ e_v \end{bmatrix} = [T] \cdot \begin{bmatrix} ji_v' \\ \cdots \\ e_v' \end{bmatrix} \quad \text{and} \quad \begin{bmatrix} je_s \\ \cdots \\ i_s \end{bmatrix} = [T]_t^{-1} \cdot \begin{bmatrix} je_s' \\ \cdots \\ i_s' \end{bmatrix} \tag{44}$$

Equilibrium equation 43 is thus converted into

$$\{[U_b]p + [D]\} \times \begin{bmatrix} ji_v' \\ \cdots \\ e_v' \end{bmatrix} = \begin{bmatrix} je_s' \\ \cdots \\ i_s' \end{bmatrix} \tag{45}$$

where $[U_b]$ is a unit matrix of order $b = l + n$. More specifically

$$[T]_t \cdot \begin{bmatrix} L & \vdots & 0 \\ \cdots & \cdot & \cdots \\ 0 & \vdots & C \end{bmatrix} \cdot [T] = [U_b] \tag{46}$$

and

$$[T]_t \cdot \begin{bmatrix} R & \vdots & j\gamma_{ln} \\ \cdots & \cdot & \cdots \\ j\gamma_{nl} & \vdots & G \end{bmatrix} \cdot [T] = [D] \tag{47}$$

with

$$[D] = \begin{bmatrix} d_1 & 0 & 0 & \cdots & 0 \\ 0 & d_2 & 0 & \cdots & 0 \\ \cdot & \cdot & \cdot & \cdot & \cdot \\ 0 & 0 & 0 & \cdots & d_b \end{bmatrix} \tag{48}$$

In contrast with Eq. 29, $[D]$ here is of order b and its diagonal elements $d_1 \cdots d_b$ are complex since they are the natural frequencies of our network. The transformation matrix $[T]$ now has complex elements also, as is anticipated since the loss matrix in Eq. 43 is no longer real.

Equation 45 expresses equilibrium in the normal coordinates to which the primed variables refer, and Eqs. 44 are the pertinent transformation relations for the variables and sources.

Regarding energy relations we observe that premultiplication on both sides of Eq. 43 by

$$[-j\,\underline{i_v} \,\vdots\, \underline{e_v}] \tag{49}$$

the transposed conjugate of the column matrix for variables in 43, gives

$$\underline{i_v} \cdot [Lp] \cdot i_v] + \underline{e_v} \cdot [Cp] \cdot e_v] + \underline{i_v} \cdot [R] \cdot i_v] + \underline{e_v} \cdot [G] \cdot e_v] = \underline{i_v} \cdot e_s] + \underline{e_v} \cdot i_s] \tag{50}$$

because

$$-\underline{e_v} \cdot \gamma_{nl} \cdot i_v] + \underline{i_v} \cdot \gamma_{ln} \cdot e_v] = 0 \tag{51}$$

as is evident from the fact that these terms are each other's transpose, but are matrices of order 1. It is also evident, of course, from the fact that the skew-symmetrical component of the loss matrix contributes nothing to the energy or power, as shown in art. 3 of the preceding chapter.

The right-hand side of Eq. 50 represents power supplied by the sources; the first two terms on the left are time rate of energy storage and the remaining two terms are rate of energy dissipation in the lossy elements.

Analogous energy relations are obtained for the transformed Eq. 45 through multiplication on both sides by the transposed conjugate of the column matrix for the primed variables. We thus obtain a conservation-of-energy expression for the network in terms of its normal coordinate representation.

If diagonalization of the matrices in Eqs. 42 or 43 is our only objective and we are not interested in or are willing to dispense with power invariance, then we are not limited to a congruent transformation but can pre- and post-multiply by any nonsingular matrices that accomplish the desired end. In this situation the positive definiteness of neither of the two matrices nor their symmetry is required, and so we can include equilibrium equations for any linear systems that are nonbilateral as well as active if we wish.

For the present, however, suppose that active or nonbilateral elements are resistive in character so that the matrix with L and C in 43 is still symmetrical and pertinent to a positive definite quadratic form. Assume further that any linear constraint relations among the variables have been used to eliminate dynamically superfluous ones so that this quadratic form (as shown in art. 4 below) is nonsingular. Then we can straightforwardly construct a matrix $[P]$ which congruently transforms the first matrix in 43 to its canonic form and the second one to some other dissymmetrical matrix $[B]$, thus

$$[P]_t \times \begin{bmatrix} L & \vdots & 0 \\ \cdot\;\cdot & \cdot & \cdot\;\cdot \\ 0 & \vdots & C \end{bmatrix} \times [P] = [U_b] = \text{unit matrix of order } b \tag{52}$$

and

$$[P]_t \times \begin{bmatrix} R & \vdots & j\gamma_{ln} \\ \cdot\;\cdot & \cdot & \cdot\;\cdot \\ j\gamma_{nl} & \vdots & G \end{bmatrix} \times [P] = [B] = \text{dissymmetrical matrix of order } b \tag{53}$$

A collinear transformation* of $[B]$ with an appropriate matrix $[Q]$ now yields a diagonal matrix $[D]$ (whose elements are the latent roots of $[B]$) and leaves the canonic matrix 52 unchanged. This collinear transformation of $[B]$ reads

$$[Q]^{-1} \cdot [B] \cdot [Q] = [D] = \begin{bmatrix} d_1 & 0 & 0 & \cdots & 0 \\ 0 & d_2 & 0 & \cdots & 0 \\ \cdot & \cdot & \cdot & \cdots & \cdot \\ 0 & 0 & 0 & \cdots & d_b \end{bmatrix} \tag{54}$$

in which the diagonal elements, as in Eq. 48, are the complex natural frequencies of our linear system, and the matrix $[Q]$ is in general complex.

The variables in Eq. 43 are transformed as indicated in the relations

$$\begin{bmatrix} ji_v \\ \cdots \\ e_v \end{bmatrix} = [P] \cdot [Q] \cdot \begin{bmatrix} ji_v' \\ \cdots \\ e_v' \end{bmatrix} \quad \text{and} \quad \begin{bmatrix} je_s \\ \cdots \\ i_s \end{bmatrix} = [P]_t^{-1} \cdot [Q] \cdot \begin{bmatrix} je_s' \\ \cdots \\ i_s' \end{bmatrix} \tag{55}$$

* See *MCA*, pp. 111–115.

although these are not of any particular interest except to show that power is now not invariant. We have

$$[-j\underline{i_v} \vdots \underline{e_v}] \cdot \begin{bmatrix} je_s \\ \cdots \\ i_s \end{bmatrix} = [-j\underline{i'_v} \vdots \underline{e'_v}] \cdot [Q]_t \cdot [Q] \cdot \begin{bmatrix} je'_s \\ \cdots \\ i'_s \end{bmatrix} \tag{56}$$

which fails to yield power invariance because $[Q]$ is not orthogonal as it is when $[B]$ is symmetrical.

Use of Eqs. 52, 53, 54 and 55 in 43 again gives

$$\{[U_b]p + [D]\} \cdot \begin{bmatrix} ji'_v \\ \cdots \\ e'_v \end{bmatrix} = \begin{bmatrix} je'_s \\ \cdots \\ i'_s \end{bmatrix} \tag{57}$$

which, like Eq. 45, represents the equilibrium of our network in normal coordinate form, however, without power invariance but for a more general class of network elements.

3. A Useful Geometrical Interpretation for Normal Coordinates and Network Response

The intimate relationship between normal coordinates and transient response is rather obvious since natural frequencies which characterize the transient behavior of a network are placed in evidence by these coordinates. The fact that normal coordinates also play a dominant role in sinusoidal steady-state response is not so evident and, therefore, merits special consideration, particularly since the pertinent interpretation of the normal coordinates themselves adds significantly to a sound understanding of their function in the mechanism of network response.

For a passive bilateral network let us consider again the homogeneous algebraic equations 4 involving amplitudes A_j of the transient currents in Eq. 2 and coefficients given by the expression 3. As pointed out in art. 1, existence of a nontrivial and nondegenerate solution requires the pertinent matrix $[B]$, Eq. 5, to have rank $l - 1$. If we regard the rows of $[B]$ as defining a set of vectors $b_1, b_2, \ldots b_l$ in l-dimensional space, then this vector set must occupy an $(l - 1)$-dimensional subspace. The desired solution, represented by a vector A with the components, $A_1, A_2, \ldots A_l$, is simultaneously orthogonal to all vectors $b_1, b_2, \ldots b_l$; that is to say, A is orthogonal to the $(l - 1)$-dimensional subspace occupied by the vector set of $[B]$. Equations 4, therefore, determine only a direction in the l-dimensional space, namely the orientation of the vector A.

Since we have a distinct set of equations 4 for each natural frequency s_ν, we find in this way as many directions in the l-dimensional space as there are natural frequencies, recognizing that each natural frequency is defined by a conjugate pair of s_ν-values. In a general situation in which l such pairs are involved, the homogeneous equations 4 determine all together l directions in the l-dimensional vector space. These are defined by the unit vectors

$$t_\nu = \frac{A^{(\nu)}}{|A^{(\nu)}|} \qquad \text{for } \nu = 1, 2, \ldots l \tag{58}$$

where consecutive numbering here refers to one of each pair of conjugate complex s_ν-values. Components of these unit vectors are complex numbers as are components of the vectors $b_1 \cdots b_l$. The absolute-value sign in 58 pertains to space coordinates only. We are dealing with a complex vector space.

A set of coordinate axes having the directions of these unit vectors *are the normal coordinates of the given network*. Since these directions are in general not mutually orthogonal, the normal coordinates form an *affine* system, although the set of reference coordinates is Cartesian.

Components

$$t_{j\nu} = \frac{A_j^{(\nu)}}{|A^{(\nu)}|} \tag{59}$$

of the unit vectors 58, which are their projections upon the reference axes, are recognized as being the direction cosines of the normal coordinates in the reference system. The matrix

$$[T] = \begin{bmatrix} t_{11} & \cdots & t_{1l} \\ \cdot & \cdot & \cdot \\ t_{l1} & \cdots & t_{ll} \end{bmatrix} \tag{60}$$

is, therefore, seen to characterize the transformation 21 expressing current variables i_j in the Cartesian system in terms of current variables i_j' in the affine system.

Introducing this transformation in the homogeneous form of equilibrium equation 1 yields the transformed set of equations

$$([L]p + [R] + [S]p^{-1}) \cdot [T] \cdot i_\nu'] = 0 \tag{61}$$

If, for the currents in the normal coordinates, we assume

$$i_\nu' = A_\nu' e^{st} \tag{62}$$

then we get for the amplitudes $A_1' \cdots A_l'$ a set of homogeneous algebraic equations (like Eqs. 4) with the matrix

$$[B'] = [B] \cdot [T] \tag{63}$$

For the elements of this matrix we have

$$b'_{i\nu} = \sum_{j=1}^{l} b_{ij}t_{j\nu} = L'_{i\nu}s + R'_{i\nu} + S'_{i\nu}s^{-1} \tag{64}$$

By substituting from Eq. 3, this yields

$$b'_{i\nu} = s\sum_{j=1}^{l} L_{ij}t_{j\nu} + \sum_{j=1}^{l} R_{ij}t_{j\nu} + s^{-1}\sum_{j=1}^{l} S_{ij}t_{j\nu} \tag{65}$$

The vectors t_ν in Eq. 58 defined by the columns of $[T]$ (the transposed vector set of $[T]$) are mutually orthogonal to all vectors defined by the rows of $[B]$ for $s = s_\nu$. More specifically, for $s = s_1$ the first column in $[T]$ is orthogonal to all rows in $[B]$; for $s = s_2$ the second column in $[T]$ is orthogonal to all rows in $[B]$; and so forth. Therefore, all elements in the νth column of $[B']$ (the elements 65 for $i = 1, 2, \ldots l$) contain the factor $s - s_\nu$, and so we can write

$$[B'] = [B_1] \cdot \begin{bmatrix} s - s_1 & & & 0 \\ & \cdot & & \\ & & \cdot & \\ & & & \cdot \\ 0 & & & s - s_l \end{bmatrix} \tag{66}$$

where an abbreviated method is used for writing a diagonal matrix involving the factors $s - s_1 \cdots s - s_l$, and the elements in $[B_1]$ are recognized to be linear functions of s (polynomials of degree 1).

Since the columns in $[B']$, Eq. 63, are linear combinations of the columns in $[B]$, and since $[B]$ is symmetrical, the columns in $[\bar{T}]$ (a bar indicates the conjugate value) or the rows in $[\bar{T}]_t$ are mutually orthogonal to the columns in $[B_1]$ for $s = \bar{s}_\nu$. That is to say, the first row in $[\bar{T}]_t$ (the transposed conjugate of $[T]$) is orthogonal to all columns in $[B]$ and hence also to all columns in $[B_1]$ for $s = \bar{s}_1$; the second row in $[\bar{T}]_t$ is orthogonal to all columns in $[B_1]$ for $s = \bar{s}_2$; and so forth. Hence, in the product $[\bar{T}]_t \cdot [B_1]$, all elements of the νth row contain the factor $s - \bar{s}_\nu$, and so we can write

$$[\bar{T}]_t \cdot [B_1] = \begin{bmatrix} s - \bar{s}_1 & & & 0 \\ & \cdot & & \\ & & \cdot & \\ & & & \cdot \\ 0 & & & s - \bar{s}_l \end{bmatrix} \cdot [H] \tag{67}$$

in which the elements of $[H]$ are constants.

Equations 63, 66 and 67 now enable us to write

$$[\bar{T}]_t \cdot [B] \cdot [T] = \begin{bmatrix} s-\bar{s}_1 & & 0 \\ & \ddots & \\ 0 & & s-\bar{s}_l \end{bmatrix} \times [H] \times \begin{bmatrix} s-s_1 & & 0 \\ & \ddots & \\ 0 & & s-s_l \end{bmatrix} \tag{68}$$

Since the matrix $[B]$ is symmetrical and real for real values of s, the transposed conjugate of the left-hand side of Eq. 68 is equal to itself; therefore the right-hand side must have the same property, and hence $[H]$ must be Hermitian.

From this result we get

$$[B]^{-1} = [T] \cdot \begin{bmatrix} \frac{1}{s-s_1} & & 0 \\ & \ddots & \\ 0 & & \frac{1}{s-s_l} \end{bmatrix} \cdot [H]^{-1} \cdot \begin{bmatrix} \frac{1}{s-\bar{s}_1} & & 0 \\ & \ddots & \\ 0 & & \frac{1}{s-\bar{s}_l} \end{bmatrix} \cdot [\bar{T}]_t \tag{69}$$

The elements of this matrix are the short-circuit driving-point and transfer admittance functions y_{ji}. Considering their partial fraction expansion, as shown in Eq. 16, we can evaluate the residue matrix $[K]$, Eq. 19, appropriate to $s = s_\nu$, by inspection of the right-hand side of Eq. 69.

Thus, for $s \to s_\nu$, the νth diagonal term in the first diagonal matrix swamps all others, and we get

$$[K]_{s=s_\nu} = \begin{bmatrix} 0 & \cdot & t_{1\nu} & \cdot & 0 \\ 0 & \cdot & t_{2\nu} & \cdot & 0 \\ \cdot & \cdot & \cdot & \cdot & \cdot \\ \cdot & \cdot & \cdot & \cdot & \cdot \\ \cdot & \cdot & \cdot & \cdot & \cdot \\ 0 & \cdot & t_{l\nu} & \cdot & 0 \end{bmatrix} \cdot [H]^{-1} \cdot \begin{bmatrix} \frac{1}{s_\nu-\bar{s}_1} & & 0 \\ & \ddots & \\ 0 & & \frac{1}{s_\nu-\bar{s}_l} \end{bmatrix} \cdot [\bar{T}]_t \tag{70}$$

while at the conjugate pole where $s \to \bar{s}_\nu$, we have

$$[K]_{s=\bar{s}_\nu} = [T] \cdot \begin{bmatrix} \dfrac{1}{\bar{s}_\nu - s_1} & & 0 \\ & \ddots & \\ 0 & & \dfrac{1}{\bar{s}_\nu - s_l} \end{bmatrix} \cdot [H]^{-1} \cdot \begin{bmatrix} 0 & 0 & \cdots & 0 \\ \cdot & \cdot & \cdot & \cdot \\ \bar{t}_{1\nu} & \bar{t}_{2\nu} & \cdots & \bar{t}_{l\nu} \\ \cdot & \cdot & \cdot & \cdot \\ 0 & 0 & \cdots & 0 \end{bmatrix} \tag{71}$$

which is the transpose conjugate and hence the conjugate of the expression 70, as it should be.

We also observe from expression 70 that the residue matrix has rank 1, since only the νth column of the matrix $[T]$ is involved. This same circumstance shows that the columns in $[K]$ are multiples of the νth column in $[T]$, as the ratios expressed by Eq. 18 involving cofactors of the determinant B indicate, inasmuch as the elements 59 in the νth column of $[T]$ fulfill these same ratios according to Eqs. 8 and 9.

The result expressed by Eq. 68, although not a normal coordinate transformation in the usual sense, certainly accomplishes the same end regarding isolation of the natural frequencies since these, in the form of conjugate complex pairs, are assigned to the l axes whose direction cosines are the elements in respective columns of the matrix $[T]$. The transformation 68 fails, of course, in the simultaneous diagonalization of the matrices $[L]$, $[R]$ and $[S]$, although it accomplishes a sort of semi-diagonalization of unique form. The diagonalization becomes complete if $[H]$ turns out to be a diagonal matrix, which will be found to agree with the condition under which the diagonal form for the determinant B given by Eq. 40 can be achieved.*

The chief reason for presenting this geometrical interpretation of the normal coordinate transformation is that we can now visualize a number of things about network behavior with greater ease and clarity. For example, in a nondegenerate situation in which all natural frequencies are excitable from all loops or links, the set of unit vectors 58, which we can crudely visualize as a bunch of arrows emanating from the origin of our Cartesian reference coordinates, have arbitrary orientations so that each has projections (components) on *all* of the reference axes. An excitation in any one reference axis produces oscillations in all normal coordinates, and through these (by the mechanism of projection) its effect is transmitted to all the other reference coordinates.

* In any two-element-kind situation it may be pointed out that the present approach achieves a normal coordinate transformation without requiring either parameter matrix separately to be nonsingular.

The mechanism of network response thus becomes visualizable in simple geometrical terms. Degeneracies occur if one or more of the unit vectors 58 coincide with some of the reference coordinates.

Suppose that the unit vector t_1 coincides with reference axis 1. Since the reference axes are orthogonal, t_1 is now orthogonal to all reference axes except axis 1, and hence its particular natural frequency is not excitable except from axis 1; if excited there, it cannot be transmitted to any of the other reference axes. Recognizing that the term "reference axis" is the geometrical counterpart of "loop" or "link," we readily see the physical implication of this statement.

Each unit vector 58 has various projections upon the reference axes; some large, some small, some even zero. These projections portray a current distribution function for the network—the function described in art. 1 above. The set of vectors 58 give us l such distribution functions, one for each complex conjugate pair of natural frequencies.

Any sinusoidal steady-state response can also be visualized in terms of the excitation of normal coordinates, which act as a coupling mechanism between different loops of the network. The phenomenon of resonance occurs when the applied frequency is close to a natural frequency, thus exciting an unusually large amplitude in one of the normal coordinates which then transmits this oscillation to all other loops in amounts proportional to the projections of this normal coordinate upon the corresponding reference axes.

In general, when the applied frequency is arbitrary, all normal coordinates are moderately excited; none predominates over the others. The extent to which each normal coordinate is excited or the distribution of amplitudes among these coordinates depends upon the point of excitation; and so one expects that the response among the other reference coordinates also depends upon the point of excitation. The distribution function changes as the location of the sources is varied.

For a resonance condition, however, one normal coordinate is so highly excited that we may neglect all the others by comparison. Thus, with essentially only one normal coordinate excited, the distribution of response among the reference coordinates is invariant to the source location. The distribution is frozen, as pointed out in art. 1. Now we have a simple geometrical picture showing how this result comes about.

We need one more piece of information to make this picture complete, namely the extent to which each normal coordinate becomes excited for a sinusoidal source of given frequency and location. That this distribution of excited amplitudes among the normal coordinates is not merely dependent upon the relative geometrical orientations of reference and normal coordinate axes is strikingly evident in the resonance situation where one normal coordinate is singled out to the almost complete exclusion of the others, notwithstanding that, by geometrical projectivity alone, this condition cannot come to pass.

Evidently this distribution function depends also upon the source *frequency* relative to the various natural frequencies. Precisely this, is the distribution that is placed in evidence by the partial fraction expansion 16 of the appropriate steady-state response function, which we repeat for convenience:

$$y_{ji} = \frac{k_{ji}^{(1)}}{s - s_1} + \frac{k_{ji}^{(2)}}{s - s_2} + \cdots + \frac{k_{ji}^{(2l)}}{s - s_{2l}} \tag{72}$$

Here we can assume that the first l terms pertain to one of each pair of conjugate natural frequencies and the remaining l terms to the other of each pair. Then the result made evident by Eq. 70 enables us to write

$$t_{j\nu} = \frac{k_{ji}^{(\nu)}}{\sqrt{(k_{1i}^{(\nu)})^2 + (k_{2i}^{(\nu)})^2 + \cdots + (k_{li}^{(\nu)})^2}} = \frac{k_{ji}^{(\nu)}}{|k_i^{(\nu)}|} \tag{73}$$

for any fixed value of the index i. Except for a normalization factor, we thus see that the residues $k_{ji}^{(\nu)}$ of the steady-state response functions y_{ji} are the direction cosines of the normal coordinates!

Terms in the expression 72 yield the amplitudes with which various normal coordinates are excited. In this connection it must be remembered that with every excitation frequency $s = j\omega$ in the sinusoidal steady state there is associated the conjugate frequency $s = -j\omega$, so that we must consider both in order to obtain the resultant excitation of any normal coordinate. We need compute this resultant, however, only for the first l terms in 72 since the corresponding results for the remaining l terms merely yield a conjugate value which can be written down at once if desired.

As a result of these interpretations the partial fraction expansion of steady-state response functions takes on a new and broader significance, not only for the evaluation of transient response (as will be amply discussed in later chapters) but also for computation of sinusoidal steady-state response, which likewise is regarded as effected through the excitation of normal coordinates. In this regard the resonance phenomenon in particular receives an interpretation of striking geometrical simplicity and clarity.*

* For additional detailed discussion along these lines see: E. A. Guillemin, *Communication Networks*, Vol. I, John Wiley and Sons, 1931, Ch. VII entitled, "The Vector Interpretation of the Transient Solution," pp. 248–283. Also the paper: "Making Normal Coordinates Coincide with the Meshes of an Electrical Network" by E. A. Guillemin, *IRE Proceedings*, Vol. 15, Nov. 1927, p. 935.

From a function theoretical point of view these facts are not surprising since the partial fraction expansion is a characterization in terms of the life-giving elements (the poles) of a function while normal coordinates afford a characterization in terms of the life-giving elements (the natural frequencies) of networks.

4. Prediction of the Number of Natural Frequencies

It is useful to be able to determine directly from the topology and element-kind distribution in a network, how many natural frequencies or normal modes it possesses. That is to say, we would like to be able to predict the degree of the characteristic equation without going through all the steps needed to derive it. As brought out by the discussion in art. 1 above and implied there by the expression 7 for the impulse response, this number (or degree) equals $2l$ in a completely nondegenerate situation in which all l of the loop currents are dynamically independent.

Since the randomly encountered network almost always embodies some degree of degeneracy, the extent of which is not obvious by inspection, it is clear that the number of topologically independent currents (which is l) is not directly useful in a determination of the degree of the characteristic equation. The number of *dynamically* independent currents, which is at most equal to l but frequently is less than this number, is the determining factor in this situation.

The expression 7 for the impulse response shows that the number of arbitrary constants of integration (the amplitudes $A_j^{(\nu)}$ for $\nu = 1, 2, \ldots 2l$) also equals the degree of the characteristic equation or the number of natural frequencies $s_1, s_2, \ldots s_{2l}$. As pointed out in art. 1 by Eqs. 8 and 9, the homogeneous equations 4 merely determine the ratios of the $A_j^{(\nu)}$ for $j = 1, 2, \ldots 2l$ (from one loop current to another); the final determination of all of these amplitudes requires that one of them for each ν-value be fixed by some physical condition apart from the algebraic manipulations.

A unique solution, therefore, requires that there be just exactly ν such physical conditions available—no more and no less. These are the *initial conditions* which describe the state of the network at the instant at which the exciting impulse strikes. This number of independent initial conditions, therefore, also equals the degree of the characteristic equation which we wish to determine.

Physically the state of a network at any given instant is described or fixed by the currents in the inductances and charges in the capacitances that exist at that instant. Offhand one would say, therefore, that the number of initial conditions equals the number of inductances plus the number of capacitances in the network, for these are the numbers of currents and charges that are involved. There are, however, certain topological arrangements of inductances or capacitances for which the currents or charges are not independent. For example, if a set of capacitance branches forms a closed loop, the voltages in these branches must add to zero and the charges must have values that meet this condition, which amounts to a constraint relation among these charges or among the corresponding currents.

More specifically, let us suppose that branches 1, 3, 4 and 7 in a given network form a closed loop of capacitances. Denoting these by c_1, c_3, c_4, c_7 and the charges correspondingly, we have

$$\frac{q_1}{c_1} + \frac{q_3}{c_3} + \frac{q_4}{c_4} + \frac{q_7}{c_7} = 0 \tag{74}$$

or in terms of the pertinent branch currents which are the derivatives of these charges, we get

$$\frac{j_1}{c_1} + \frac{j_3}{c_3} + \frac{j_4}{c_4} + \frac{j_7}{c_7} = 0 \tag{75}$$

Since the branch currents are expressible in terms of loop currents (by the columns of a tie-set matrix), this condition turns out to be a linear constraint relation among the loop-current variables.

Similarly, if branches 2, 5, 7 and 9 form a closed loop of inductances l_2, l_5, l_7, l_9 we have

$$l_2\frac{dj_2}{dt} + l_5\frac{dj_5}{dt} + l_7\frac{dj_7}{dt} + l_9\frac{dj_9}{dt} = 0 \tag{76}$$

or, after integration,

$$l_2j_2 + l_5j_5 + l_7j_7 + l_9j_9 = \text{constant} \tag{77}$$

which again yields a linear constraint among the loop currents. When the inductances are mutually coupled, the appropriate relation involves more terms, but the constraint relation exists in essentially the same form.

We refer to the topological conditions yielding these constraints as *capacitance tie sets* and *inductance tie sets*. *Resistance tie sets* analogously result in constraint relations among the loop-current variables.

The duals of these situations similarly yield constraint relations among the node-pair voltage variables. Thus we may have capacitive branches 1, 3, 6, 8 and 9 forming a cut set. Since the sum of currents in these branches must be zero, we get

$$c_1\frac{dv_1}{dt} + c_3\frac{dv_3}{dt} + c_6\frac{dv_6}{dt} + c_8\frac{dv_8}{dt} + c_9\frac{dv_9}{dt} = 0 \tag{78}$$

where the $v_1, v_3 \ldots$ are the pertinent voltage drops. After integration, this condition reads

$$c_1v_1 + c_3v_3 + c_6v_6 + c_8v_8 + c_9v_9 = \text{constant} \tag{79}$$

Expressing these branch voltages in terms of node-pair voltages (by the columns of a cut-set matrix) converts this relation into a linear constraint among node-pair voltage variables.

We can thus recognize the existence of such constraint relations in terms of topological situations that we refer to as *inductance cut sets, capacitance cut sets* and *resistance cut sets.*

At this point in our reasoning we are inclined to state that the number of dynamically independent variables on any basis (loop, node or mixed) is equal to the total number of variables in terms of which the equilibrium equations for that basis are established, diminished by the number of independent constraints among these respective variables. On a loop basis this equals the number l diminished by the number of independent inductance, capacitance and resistance tie sets; on a node basis it is the number n diminished by the number of independent inductance, capacitance and resistance cut sets; and on a mixed basis it is the number $l + n$ diminished by the total number of constraints among both loop currents and node-pair voltages.

We observe that the number of dynamically independent variables obtained by this method of determination is not in general the same on one basis as it is on another. On a mixed basis, for instance, this method yields as many dynamically independent variables for a given network as are involved in the loop and node bases combined (which is actually not too surprising since the mixed basis is, after all, a combination of the loop and node methods).

In this regard one should bear in mind first of all that the mixed basis is the only one for which the situation is clear-cut regarding dynamic variables, and that the loop and node methods obscure the true state of affairs. On a mixed basis, each separate equilibrium equation contains only constant terms and those involving first-order derivatives, while each equation on a loop or node basis may additionally contain an integral term representing respectively a loop charge or a node-pair flux linkage, or it may contain only the integral term. So far as dynamic variables are concerned, those currents or voltages for which both derivatives and integrals occur should receive a weight of 2 and those for which only a derivative or only an integral occurs should have a weight of 1.

On this basis we will find the same number of dynamically independent variables regardless of how we express the equilibrium of our network. Indeed, we *must* arrive at this conclusion, for the number of dynamically independent variables equals the so-called *degrees of freedom* of the system which in turn equals the number of current and charge values needed to fix the state of the network at any instant.

This number, we now recognize, is equal to the total number of inductances and capacitances (or energy storage elements) diminished by the total number of independent inductance and capacitance tie sets *and* cut sets. We can alternately say that it is equal to the total number of variables on a mixed basis (loop currents plus node-pair voltages) diminished by the total number of independent constraints among these variables (equal to the number of independent inductance, capacitance *and* resistance tie sets and cut sets).

The latter specification is fraught with possible redundancy inasmuch as we may, in some situations, find ourselves adding to and subtracting from the total number with which the previous statement begins, a number of independent resistance tie sets and cut sets, this being the amount by which the total number of variables on a mixed basis exceeds the number of inductance and capacitances. For example, consider the extreme case of a pure resistance network. Here all l voltage-law equations and all n current-law equations are linear constraints; the system has zero degrees of freedom in a dynamic sense. We can arrive at this result either by saying that the total number of inductances and capacitances equals zero, the total number of inductance and capacitance tie sets and cut sets equals zero, and so the difference equals zero, or that the total number of variables equals $l + n$, the total number of resistance tie sets and cut sets equals $l + n$, and so the difference equals zero.

It is evidently more sensible to arrive at the number of degrees of freedom (or independent initial conditions, or the degree of the characteristic equation, or the number of natural frequencies) by subtracting from the total number of storage elements the number of independent inductance and capacitance tie sets and cut sets. Our next objective is to present a simple and unambiguous method of determining these tie sets and cut sets.

Tie sets are combinations of branches forming closed paths. If such a closed path can be formed from branches involving only inductances, then this loop still exists if all capacitance and resistance branches are removed or open-circuited. On the other hand, a closed path cannot consist exclusively of inductances unless it exists when all capacitances and resistances are removed. The removal or open-circuiting of all R and C branches is, therefore, a necessary and sufficient means to obtain a subgraph which places the inductance tie sets in evidence.

We have but to draw this subgraph and determine the number of its independent loops or links by the well-known topological method in order to establish unambiguously the number of independent inductance tie sets. We may refer to this subgraph as an *open-circuit L-graph*.

We can similarly construct an *open-circuit C-graph* by removing from the graph for the total network all branches corresponding to inductances and resistances. The number of independent capacitance tie sets equals the number of loops in this subgraph as determined by the pertinent topology. Finally, we can form an *open-circuit R-graph* by removing all inductances and capacitances, and determine by the same well-established procedure the number of independent resistance tie sets contained in the given network.

A completely dual procedure applies to the determination of inductance, capacitance and resistance cut sets. Thus a cut set is found by picking up some of the nodes and pulling them away from the remaining ones, noting the branches that become stretched in the process, or by passing a "cutting surface" through the total network graph and noting the branches that are "cut" by this surface (those that are cut an even number of times are not cut at all and those that are cut an odd number of times are cut only once).

If a cut set can consist of branches involving only inductances, then it still exists if all capacitance and resistance branches are short-circuited by joining the nodes at their end points and then removing these elements altogether. On the other hand, a cut set cannot consist exclusively of inductances unless it exists when all capacitance and resistance branches are short-circuited. The short-circuiting (and removal) of all R and C branches is, therefore, a necessary and sufficient means to obtain a subgraph which places the inductance cut sets in evidence.

We have but to draw this subgraph and determine its topologically independent node pairs in order to establish unambiguously the number of independent inductance cut sets. The subgraph involved here is referred to as a *short-circuit L-graph*.

By similarly constructing a *short-circuit C-graph* (short-circuiting and removing all R's and L's) or a *short-circuit R-graph* (short-circuiting and removing all L's and C's) we can with the same degree of ease and preciseness determine the number of independent capacitance and resistance cut sets contained in the given network.

The importance of the adjective "independent" should be carefully noted in these statements. The topological procedures just described for determining inductance, capacitance and resistance tie sets and cut sets automatically yields the number of independent ones, while groping for these by inspection of the given network would not only leave doubt as to whether some may have been overlooked, but would fail to establish clearly the independence of those that had been found.

The procedure for establishing the number of degrees of freedom, or independent initial conditions, or the degree of the characteristic equation, or the number of natural frequencies of a given network may now be summarized formally in the following manner (preferably having in mind the mixed basis for the equilibrium equations):

(*a*) Draw separately an open-circuit L-graph and an open-circuit C-graph. By the conventional topological procedure find the number of independent loops l_L in the L-graph and the number of independent loops l_C in the C-graph. The total number of inductance and capacitance tie sets is then

$$l_s = l_L + l_C \tag{80}$$

(*b*) Draw separately a short-circuit L-graph and a short-circuit C-graph. By the conventional topological procedure find the number of independent node pairs n_L in the L-graph and the number of independent node pairs n_C in the C-graph. The total number of inductance and capacitance cut sets is then

$$n_s = n_L + n_C \tag{81}$$

(*c*) Let the number of inductive branches be denoted by b_L, the number of capacitive branches by b_C. Then

$$m = b_L + b_C - l_s - n_s \tag{82}$$

equals the degree of the characteristic equation, or the number of natural frequencies counting conjugate complex frequencies as though they were two modes, which is sensible because they are represented in the complex s-plane by two distinct points.

The number m given by Eq. 82 also equals the total number of dynamically independent variables on the mixed or any other basis, since this number represents the degrees of freedom of the system.

For passive bilateral networks characterized by two kinds of elements (the so-called *LC*-, *RC*- or *RL*-networks) this result, and the topological methods for establishing it, have been known for a long time,* the motivation for these ideas having been provided by normal coordinate reasoning which prompts the conclusion that loop and node methods must ultimately lead to the same number of independent equilibrium equations notwithstanding that $l \neq n$, since a given system having m degrees of freedom cannot be characterized by various numbers of independent variables depending on which one of several algebraic procedures is chosen to obtain these equations.

More recently, interest in these things has revived. The topological

* See the footnote on p. 165 of *SPN*.

method for determining the number of natural frequencies has been extended to *RLC* networks in essentially the manner presented here.* Significant in this connection is the demonstration† that a normal coordinate transformation is always possible for a linear system with arbitrary loss as discussed in the preceding articles, for it is this result that provides a sound basis for the contention that the degrees of freedom or the number of dynamically independent variables is a fundamental invariant in such a system.

It is of some significance to emphasize the rather obvious fact that natural frequencies involved in the foregoing discussions are the finite, nonzero ones. Any zero modes, if present, can be recognized by inspection and are of minor practical significance since the ever-present incidental dissipation in actual circuits prevents them from being realized.

Also we must observe that the network excitation must be either a voltage source in series with a branch or a current source across a node pair. Otherwise the topology is disturbed by insertion of the source and the results obtained will be in error unless this effect is properly accounted for.

Finally, it is not superfluous to mention that degeneracies can occur on account of inherent topological and/or element-value symmetries, causing modes to coincide. These situations are not amenable to the foregoing method of mode determination.

In connection with two-element-kind networks, one can give a simple topological proof of the fact that the number of dynamically independent variables is the same on a loop or node basis. We will give the detailed argument with regard to an *LC*-network; the analogous proofs for *RL*- and *RC*-networks are left as an exercise for the reader.

Consider formation of the short-circuit *L*-graph by successively short-circuiting the capacitances in the given *LC*-network. Each time a capacitance branch is eliminated, the number of nodes is reduced by one. We tentatively conclude, therefore, that the number of independent nodes or node pairs in the short-circuit *L*-graph is $n - b_C$ since the number of independent nodes in the given network is n and the number of capacitance branches is b_C. However, if k of these capacitive branches form a closed loop, then the number of nodes on this loop (which is also k) is reduced to 1 after only $k - 1$ of the capacitances are short-circuited. Shorting the last capacitance does not further reduce the number of nodes.

* See "The Degrees of Freedom in *RLC* Networks" by A. Bers, *IRE Trans. Circuit Theory*, Vol. CT-6, March 1959, No. 1, p. 91.

† In this connection see also, "The Normal Coordinate Transformation of a Linear System with an Arbitrary Loss Function," *M.I.T. Res. Lab. Electron., Quarterly Progress Report for Jan. 15, 1960*, p. 236, or *J. Math. Phys.*, Vol. 39, No. 2, July 1960.

If the given network contains l_C independent capacitance loops (or tie sets), then it becomes clear that the number of independent nodes in the short-circuit L-graph are given by

$$n_L = n - b_C + l_C \tag{83}$$

Analogously we find that the number of independent nodes in the short-circuit C-graph is given by

$$n_C = n - b_L + l_L \tag{84}$$

Equation 81 then yields for the total number of inductance and capacitance cut sets

$$n_s = n_L + n_C = 2n - (b_L + b_C) + (l_L + l_C) \tag{85}$$

Use of Eq. 80 and the fact that $b_L + b_C = b = l + n$ in this LC-network, we find the significant result

$$n - n_s = l - l_s \tag{86}$$

which implies the statement made above.

Prediction of the number of natural frequencies can also be done in an alternate way that is more direct than the method of reasoning given thus far. This alternate way makes use of the representation of the determinant of a network as a sum of all the link sets, the proof of which is given in art. 2, Ch. III.

The impedances of separate branches are given by $l_k s$, r_k or $1/c_k s$ according to whether they are inductive, resistive, or capacitive respectively, s being the complex frequency variable. Each term in the complete expression for the determinant is a product of l branch impedances since l is the number of branches in a link set. Thus each term is a constant times some power of s; and it is clear, therefore, that the determinant is a polynomial in s. This polynomial equated to zero is the determinantal or characteristic equation, as also pointed out in art. 1 above.

The link set involving the largest number of inductive branches and the smallest number of capacitive branches yields the highest power of s, while the term involving the largest number of capacitive and the smallest number of inductive branches yields the lowest power of s. The difference is the degree of the characteristic equation.

If one link set consists entirely of inductances and another consists entirely of capacitances, then it is obvious that the degree is $2l$, the highest that it can be. In this case one term has the degree l and another has the degree $-l$.

An upper bound for the term of highest degree is evidently b_L, the number of inductive branches, and a lower bound for the term of lowest degree is $-b_C$, the number of capacitive branches. The upper bound can be realized only if: (*a*) there is at least one link set that contains no capacitive branches; (*b*) the corresponding tree contains no inductive branches; and (*c*) $b_L \leq l$. The reason for (*a*) is obvious, since capacitive branches introduce negative powers of s. Stipulation (*b*) merely insures that none of the available inductive branches are wasted by being in the tree instead of among the links where they can help boost the degree of the pertinent term; and (*c*) is self-evident since l is an upper bound that cannot be exceeded no matter how many inductive branches there are. Actually, if (*a*) and (*b*) are fulfilled, then (*c*) is automatically fulfilled because all of the inductive branches then are among the links and these cannot exceed themselves in number. We can content ourselves with conditions (*a*) and (*b*).

Similarly we arrive at the conclusion that the lower bound of the term of lowest degree can be realized only if: (*a*) there is at least one link set that contains no inductive branches; and (*b*) the corresponding tree contains no capacitive branches.

Regarding the upper bound b_L, we can test for condition (*a*) by drawing the open-circuit C-graph. If it has no closed paths then this condition is fulfilled; on the other hand if it has l_C loops or links, then obviously no link set can exist that does not include at least this number of capacitive branches. The upper bound b_L is then decreased by l_C.

Regarding a test for condition (*b*), we draw the short-circuit L-graph and see whether it has any independent nodes or tree branches. If it does not, then clearly a tree exists in the total graph which includes no inductive branches since short-circuiting all C-branches and all R-branches causes all nodes to coincide (the necessary and sufficient condition that the short-circuit L-graph shall have zero independent nodes). If the short-circuit L-graph has n_L independent nodes then the upper bound b_L is decreased by n_L.

We thus see that, in general, the highest degree among all link sets is $b_L - n_L - l_C$. In an exactly analogous manner we arrive at the conclusion that the lowest degree among all link sets is $-(b_C - n_C - l_L)$. Hence the degree of the characteristic equation becomes

$$\begin{aligned} m &= b_L - n_L - l_C + b_C - n_C - l_L \\ &= b_L + b_C - (n_L + n_C) - (l_L + l_C) \end{aligned} \tag{87}$$

which agrees with the result obtained previously.*

* This derivation was given by Harry Lee in a seminar held by the Networks Group at M.I.T. in March 1958. Since then several publications have appeared giving essentially the same results.

5. Equivalent Networks

In the practical use of networks we are interested in realizing a box with p terminal pairs having a prescribed open-circuit impedance or short-circuit admittance matrix of order p, the situations of most common occurrence being those for which $p = 1$ or 2. The number of independent loops l or of independent node pairs n of the network within the box is almost always larger than p, usually quite a bit larger, especially when $p = 1$ or 2.

In such cases it is possible to find numerous network geometries and element-value distributions all of which yield the same $[z]$- or $[y]$-matrix at the accessible terminal pairs. These are referred to as *equivalent networks.*

Different members of such networks are characterized by different parameter matrices on a loop or node basis notwithstanding that their terminal-pair matrices are alike. Of course, if p were equal to l or to n (depending on the mode of characterization) then the terminal-pair matrix would be a linear combination of the parameter matrices (like $[z] = [L]s + [R] + [S]s^{-1}$); no variation in the parameter matrices would be possible and no equivalent network could exist. It is the fact that the order of the parameter matrices can be larger than the order of the terminal-pair matrix that allows different parameter matrices to yield the same terminal-pair matrix and hence make possible the realization of equivalent networks.

In the equivalent network problem we are concerned with the construction of a variety of networks all of which have a stated terminal-pair behavior. In most cases no specifications are made regarding numbers of loops or node pairs for the network within the box so long as these do not become excessively large. Hence equivalent networks always exist. The problem is to find effective ways for their determination.

A logical way of dealing with this problem is to devise methods of transforming parameter matrices so as to keep the terminal-pair matrix for the implied network invariant. An appropriately chosen linear transformation of the loop-current or node-pair voltage variables offers such a possible approach.

Consider again the forced equilibrium equations represented by the matrix of Eq. 1, and introduce a linear transformation of variables given by

$$i_v] = [A] \cdot i_v'] \tag{88}$$

Assume that pliers-type entries in links $1 \cdots p$ are the accessible terminal pairs and choose a structure for the matrix $[A]$ that has the partitioned form

$$[A] = \begin{bmatrix} u_p & \vdots & 0 \\ \cdots & \cdots & \cdots \\ a_{qp} & \vdots & a_{qq} \end{bmatrix} \tag{89}$$

The submatrix u_p is a unit matrix of order p; a_{qp} and a_{qq} represent the last q rows of $[A]$ partitioned into groups of p and q columns, with $p + q = l$, the order of $[A]$. Correspondingly, only the first p elements in the excitation matrix $e_s]$ are nonzero.

The special form chosen for $[A]$ stipulates that $i_j \equiv i_j'$ for $j = 1, 2, \ldots p$, which keeps the volt-ampere relations at the accessible terminal pairs invariant and hence insures that the pertinent matrices $[y]$ or $[z]$ likewise remain invariant to the transformation 88. The submatrices a_{qp} and a_{qq} are unrestricted except by realizability conditions for the resulting parameter matrices.

These are found by introducing the transformation 88 into the quadratic forms, given by Eqs. 12, 13 and 14 in Ch. V, representing energy functions for the pertinent network. As shown in art. 3 of that chapter, the resulting parameter matrices are obtained from the following congruent transformations of the given ones:

$$[L'] = [A]_t \cdot [L] \cdot [A] \tag{90}$$

$$[R'] = [A]_t \cdot [R] \cdot [A] \tag{91}$$

$$[S'] = [A]_t \cdot [S] \cdot [A] \tag{92}$$

An entirely analogous procedure is applicable to parameter matrices appropriate to a node basis. The equivalent networks are found by applying to the resultant parameter matrices the realization methods discussed in Ch. IV.

Difficulty in the use of this method is experienced in choosing the arbitrary part of the transformation matrix $[A]$, Eq. 89, in such a way that the realizability conditions discussed in Ch. IV are fulfilled. These are rather restrictive because the methods given there seek to realize the desired network without ideal transformers and without mutual inductive coupling, which are practically desirable restrictions.

If we do not impose these restrictions then the only condition, so far as passive bilateral networks are concerned, is that the quadratic forms having the matrices 90, 91 and 92 be positive definite. If we start from a physical network then the given matrices $[L]$, $[R]$ and $[S]$ obviously satisfy this condition, and so do the resultant matrices since (as pointed out in art. 3, Ch. V) this property of a quadratic form is invariant to a linear transformation of its variables with any real nonsingular matrix.

In consideration of the realizability conditions discussed in Ch. IV, it is found helpful to break the parameter matrix transformation down into a cascade of simple ones by constructing the matrix $[A]$ as a product of elementary transformation matrices.* This decomposition is indicated by writing

$$[A] = [T_1] \times [T_2] \times [T_3] \times \cdots \tag{93}$$

The total transformation carried out, for example, on the matrix $[L]$, takes the form

$$[L_1] = [T_1]_t \cdot [L] \cdot [T_1] \tag{94}$$

$$[L_2] = [T_2]_t \cdot [L_1] \cdot [T_2] \tag{95}$$

$$[L_3] = [T_3]_t \cdot [L_2] \cdot [T_3] \tag{96}$$

and so forth.

Since the elementary transformations correspond to: interchange of rows and columns, multiplication of rows and columns by factors, and addition of row to row and column to column, we can carry out these operations directly upon the given parameter matrices, thus avoiding matrix multiplications altogether. At the same time we are in a position to observe more closely the effect of each contemplated operation so that the problem of staying within a framework of realizability conditions is more easily dealt with.

The elementary operations thus contemplated or carried out must, of course, correspond to a transformation matrix meeting the form requirements displayed in Eq. 89; otherwise the intended invariance of the pertinent matrix $[z]$ or $[y]$ will not be met. That is to say, the manipulation of rows and columns must leave the first p rows of the transformation matrix unchanged from those of a unit matrix.

A useful elementary transformation results from the multiplication of a row and corresponding column by a factor. Much can be done by this technique alone in the way of bringing about a more favorable distribution of element values. Physically this process corresponds to the insertion of ideal transformers within the network, and for this reason we speak of it as "internal impedance leveling" since it amounts to changing the impedance levels of internal meshes.

A few simple examples will illustrate these statements. In Fig. 1, circuit (*a*) is an inverted "ell" in one of two possible orientations; circuit (*c*) shows the opposite orientation of this "ell"; and circuit (*b*) is an intermediate form. These three circuits are related by an impedance-level

* See *MCA*, pp. 54–58.

transformation as indicated in the following matrix manipulations:

$$n \rightarrow \begin{bmatrix} L_1 & L_1 \\ L_1 & (L_1 + L_2) \end{bmatrix} \rightarrow \begin{bmatrix} n^2L_1 & nL_1 \\ nL_1 & (L_1 + L_2) \end{bmatrix} \rightarrow \begin{bmatrix} n^2L_1 & nL_1 \\ nL_1 & nL_1 \end{bmatrix} \qquad (97)$$
$$\uparrow$$
$$n$$

On the left is the inductance matrix of circuit (*a*) for a loop basis choosing loop currents as indicated. After multiplication of row and column 1 by the factor n, it is converted to the form shown in the center.

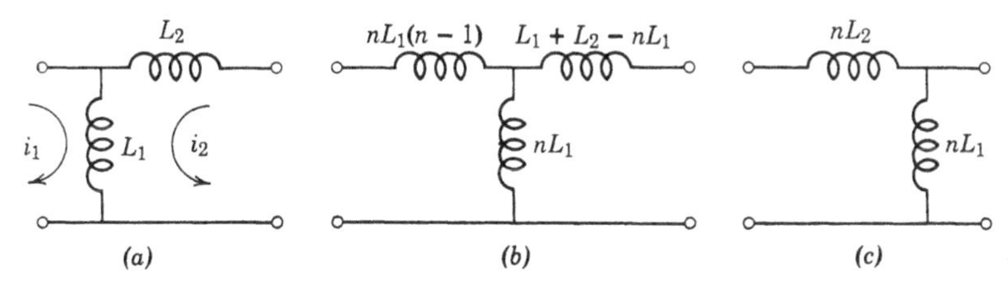

FIG. 1. Illustrating an impedance-level transformation according to the operations indicated in 97.

By inspection, this matrix is seen to yield circuit (*b*) with the indicated element values. If negative values are to be avoided, the factor n is restricted to lie within the range

$$1 \le n \le 1 + \frac{L_2}{L_1} \qquad (98)$$

If we choose n equal to the upper limit of this range, the right-hand matrix of 97 becomes appropriate and the pertinent circuit is that shown in Fig. 1, part (*c*) which is referred to as the completely transformed version of circuit (*a*).

Since n is larger than unity, this transformation raises the impedance level at the input terminals. The factor by which the input impedance becomes multiplied is evidently n^2.

A similar situation involving capacitive circuits is shown in Fig. 2. Corresponding matrix manipulations on a node basis read

$$n \rightarrow \begin{bmatrix} C_1 & -C_1 \\ -C_1 & C_1 + C_2 \end{bmatrix} \rightarrow \begin{bmatrix} n^2C_1 & -nC_1 \\ -nC_1 & C_1 + C_2 \end{bmatrix} \rightarrow \begin{bmatrix} n^2C_1 & -nC_1 \\ -nC_1 & nC_1 \end{bmatrix} \qquad (99)$$
$$\uparrow$$
$$n$$

Here the matrix on the left characterizes circuit (*a*) for node potentials with respect to the base line as a datum. The central matrix corresponds in the same manner to circuit (*b*) which involves no negative elements if

$$1 \leq n \leq 1 + \frac{C_2}{C_1} \tag{100}$$

The completely transformed version of circuit (*a*), resulting for the maximum allowable value of n, is circuit (*c*) pertinent to the right-hand matrix in 99.

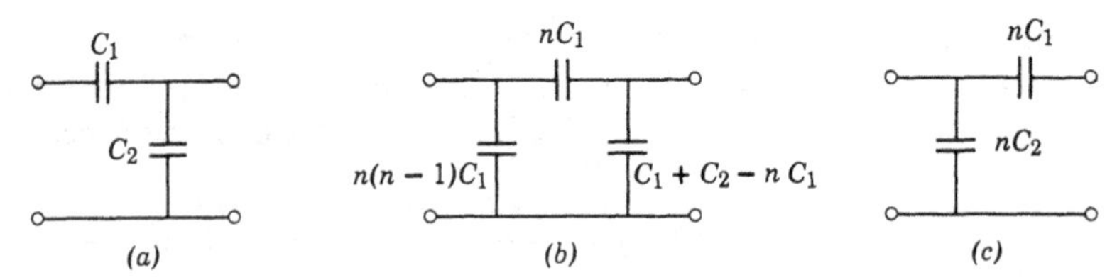

FIG. 2. Illustrating an admittance-level transformation according to the operations indicated in 99.

Since the node basis implies admittances, and n is again larger than unity, the impedance level at the input in this case is lowered, the factor by which the impedance there is multiplied being $1/n^2$.

It is interesting to note that the symmetrical π-circuit in Fig. 3 results for the geometric mean value of n, namely

$$n = \sqrt{1 + \frac{C_2}{C_1}} \tag{101}$$

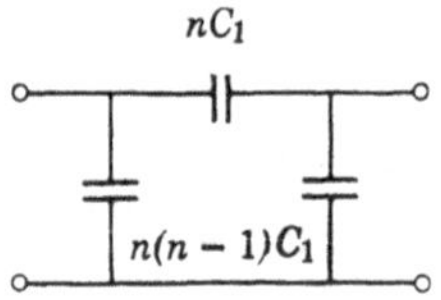

FIG. 3. Symmetrical circuit resulting for geometric mean of extreme values in 100, namely the value 101.

An interesting application of these results is to the bandpass-filter network shown in part (*a*) of Fig. 4 in which the element values on a 1-ohm impedance level are given by

$$\begin{aligned} L &= \frac{w}{\omega_0^2}, \quad C = \frac{1}{w} \\ L_1 &= \frac{2}{w}, \quad C_1 = \frac{w}{2\omega_0^2} \end{aligned} \tag{102}$$

where ω_0 denotes the midband frequency and w the bandwidth, both in radians per second. Since

$$\frac{L_1}{L} = \frac{C}{C_1} = 2\left(\frac{\omega_0}{w}\right)^2 \tag{103}$$

this circuit becomes impossible to construct if the Q of the filter, namely ω_0/w, is large. For $\omega_0/w = 100$, for example, the ratios 103 are 20,000. If the impedance level is chosen so that one of the inductances is reasonable then the other is impossibly large or impossibly small.

By splitting C_1 in the series branch into two equal parts, and decreasing the impedance level of the center mesh by a factor $(\omega_0/w)^2$, the capacitive "ell" circuits on either side are transformed into symmetrical pi's like the one in Fig. 3, and circuit (*b*) in Fig. 4 results. The values of L and C are the same as before, while $C_0 = 1/\omega_0$ farads. The ratio $C_0/C = w/\omega_0$ is still somewhat small, but this can easily be accommodated practically; otherwise the element values are now very uniform. An impossible situation is thus converted into a very reasonable one.

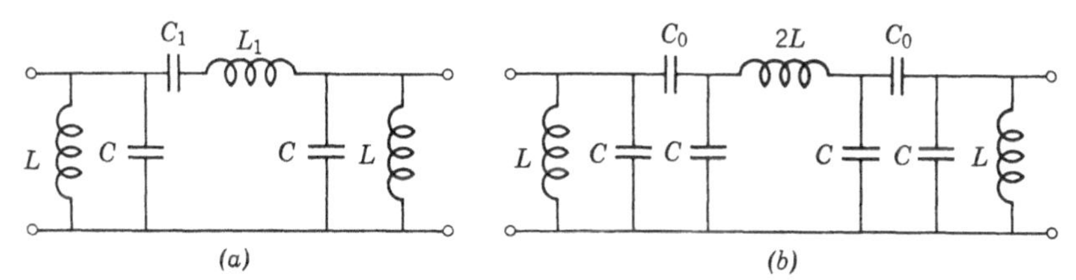

FIG. 4. Application of internal impedance-level transformations to a bandpass filter.

A somewhat different series of transformations carried out upon this same filter network is illustrated in Fig. 5. Circuit (*a*) at the top is the tee-circuit (also called "mid series" section) corresponding to the pi-circuit in part (*a*) Fig. 4. These two circuits are simply duals of each other, and since dual networks have identical transfer functions (apart from a constant multiplier) they can be used interchangeably.

Circuit (*b*) in Fig. 5 is obtained by two operations upon circuit (*a*). The inductive ell within the dotted circle is converted to its opposite form by dropping the impedance level at the input; and the check-marked capacitance in series with the voltage source E_{s1} is converted into an equivalent current source in parallel with this capacitance. Considering the impedance-level change as well as the source transformation, we have $I_{s1} = K_1 s E_{s1}$ where K_1 is an appropriate constant.

From (*b*) to (*c*) we interchange the shunt L and C at the input and make another source transformation, this time involving the arrow-marked inductance. This gives $E_{s2} = K_2 s^2 E_{s1}$.

The ell of capacitances within the dotted circle in circuit (*c*) is now converted to its opposite form; L and C in the resulting series branch on the left are then interchanged, and another ell-circuit conversion carried out. These operations yield circuit (*d*) which has a higher impedance level at the input than circuit (*c*).

Finally, L and C in the first series branch in (*d*) are interchanged and another source conversion is carried out involving the check-marked capacitance. This step yields $I_{s2} = K_3 s^3 E_{s1}$.

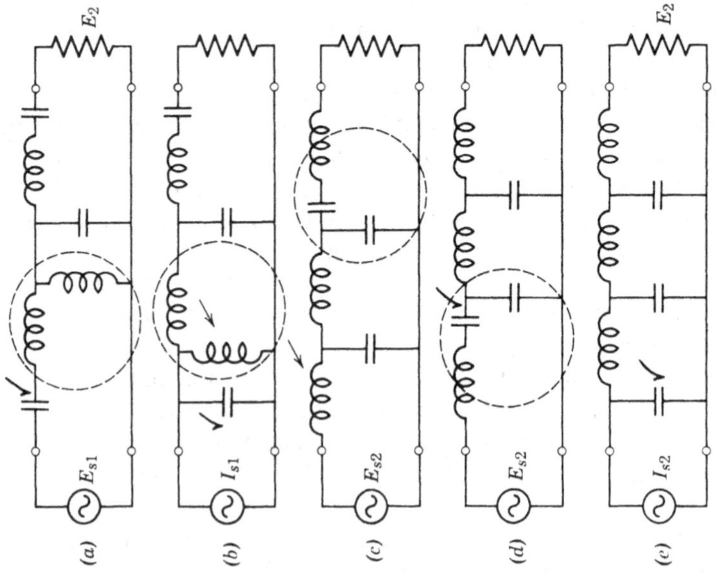

FIG. 5. Series of source and impedance-level transformations carried out on a bandpass filter. These yield all transmission zeros at infinity.

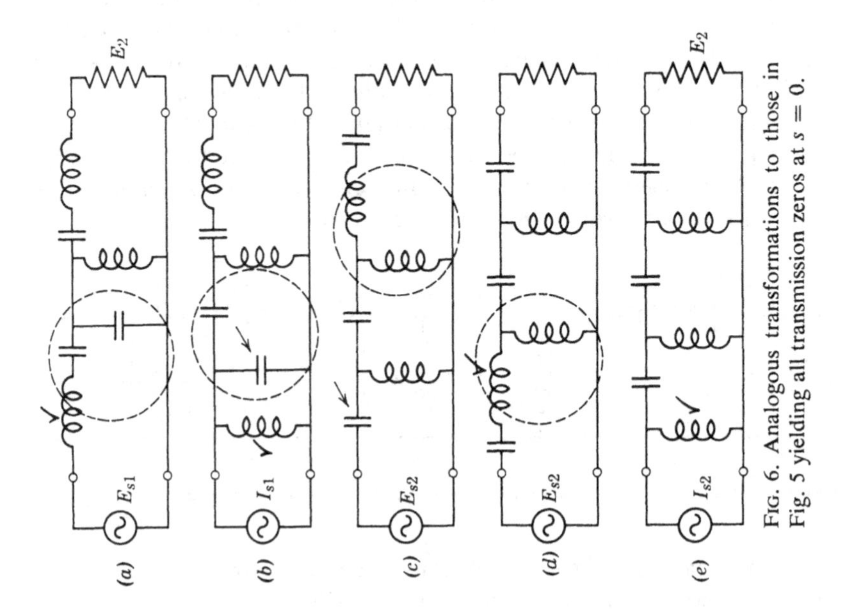

FIG. 6. Analogous transformations to those in Fig. 5 yielding all transmission zeros at $s = 0$.

In the final circuit of Fig. 5, circuit (*e*), the transfer function E_2/I_{s2} is E_2/E_{s1} of circuit (*a*) multiplied by $1/K_3s^3$. In a narrow-band filter this modification has little effect in the vicinity of the passband which is usually the range of principal interest. Whereas circuit (*a*) has its six transmission zeros (i.e. zeros of its transfer function) equally divided between $s = 0$ and $s = \infty$, circuit (*e*) has then all located at $s = \infty$.

An analogous series of impedance-level and source transformations may be carried out so as to yield a resultant bandpass characteristic which is the original one multiplied by s^3/K_3, so that all transmission zeros fall

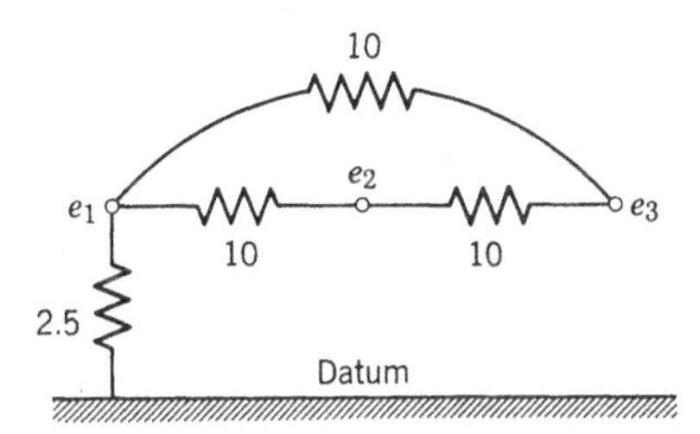

FIG. 7. Example illustrating how number of nodes can be changed by a transformation on the node basis.

at $s = 0$. These transformations are illustrated by the successive sketches in Fig. 6. They follow the same pattern as do the transformations in Fig. 5, and hence require no additional comment. Here again, the bandpass characteristic of the filter is essentially unaltered in the vicinity of the passband in any narrow-band application.

The resulting circuits in Fig. 5, part (*e*) and Fig. 6, part (*e*) are interesting in that one would not guess that these could exhibit a bandpass behavior only slightly different from that of the circuit in part (*a*) of either figure. One of these looks like a lowpass and the other like a highpass filter circuit. In dealing with equivalent or semi-equivalent networks "looks" can be deceiving.

It might be thought that this linear transformation method allows us to obtain only that class of equivalent networks having the original number of topologically independent loops or node pairs since these are invariants in the transformation pertinent to either basis. Upon closer study we see, however, that such a conclusion is false. First of all, if we use a loop basis, l to be sure remains invariant but n does not; and on a node basis the reverse is true. Hence by switching alternately from a loop to a node basis and back again, we can vary both l and n as much as we like.

A second possibility is illustrated by the simple example in Fig. 7 which shows a single resistance of 2.5 ohms augmented into a 3-node network involving three additional 10-ohm resistances but presenting at its input terminals (datum to node 1) still the same 2.5 ohms. On the indicated node-to-datum basis, the conductance matrix reads

$$[G] = \begin{bmatrix} 0.6 & -0.1 & -0.1 \\ -0.1 & 0.2 & -0.1 \\ -0.1 & -0.1 & 0.2 \end{bmatrix} \tag{104}$$

If we multiply row and column 1 by 1.8 and row and column 3 by 2, this matrix is converted into

$$[G'] = \begin{bmatrix} 0.6 & -0.18 & -0.20 \\ -0.18 & 0.648 & -0.36 \\ -0.20 & -0.36 & 0.80 \end{bmatrix} \tag{105}$$

yielding the network of Fig. 8 in which element values again are given in

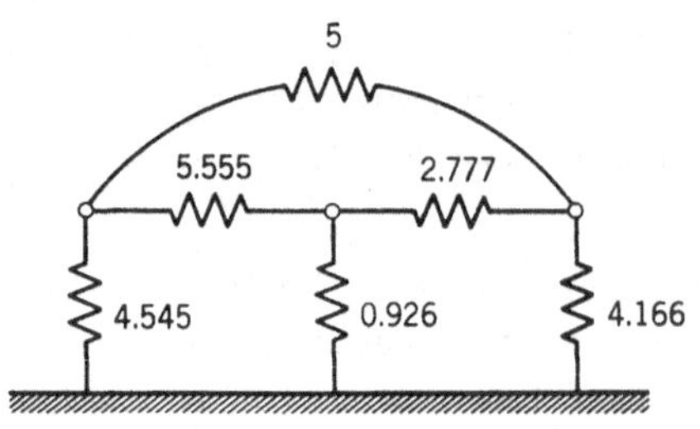

FIG. 8. Transformed network in Fig. 7 according to matrix transformations indicated in Eqs. 104 and 105.

ohms. Although it is not immediately evident, the input resistance is still 2.5 ohms as our linear transformation method predicts it should be.

6. Construction of Parameter Matrices from Given Response Functions

In the synthesis of networks, the terminal-pair matrix $[z]$ or $[y]$ is specified and its realization is sought. A possible approach is to devise a way of constructing parameter matrices appropriate to the given terminal-pair matrix and then use the realization techniques discussed in Ch. IV. The collateral problem of finding equivalent networks representing alternative physical forms for the solution thus becomes an integral part of the procedure based upon that approach, since the flexibility inherent in the process of parameter-matrix construction involves essentially the same algebraic manipulations as occur in the linear-transformation method just described.

As might be expected, the normal coordinate transformation provides the necessary link between a prescribed terminal-pair matrix and the implied parameter matrices, for on the one hand it places in evidence the natural frequencies which are poles of the pertinent impedances or admittances, while on the other it is related through a linear transformation to the parameter matrices of the network.

In the equilibrium equations 43 appropriate to a passive bilateral network we make the familiar substitutions

$$i_v = I_v e^{st}, \quad e_v = E_v e^{st}, \quad i_s = I_s e^{st}, \quad e_s = E_s e^{st} \tag{106}$$

and introduce the notation

$$[U] = \begin{bmatrix} jE_s \\ \cdots \\ I_s \end{bmatrix}, \quad [V] = \begin{bmatrix} jI_v \\ \cdots \\ E_v \end{bmatrix} \tag{107}$$

and

$$[W] = \begin{bmatrix} Ls + R & \vdots & j\gamma_{ln} \\ \cdots & \cdots & \cdots \\ j\gamma_{nl} & \vdots & Cs + G \end{bmatrix} \tag{108}$$

These equations are then abbreviated in the form

$$[W] \cdot [V] = [U] \tag{109}$$

and Eqs. 46, 47 and 48 yield

$$[T]_t \cdot [W] \cdot [T] = \begin{bmatrix} d_1 + s & 0 & \cdots & 0 \\ 0 & d_2 + s & \cdots & 0 \\ \cdots & \cdots & \cdots & \cdots \\ 0 & 0 & \cdots & d_b + s \end{bmatrix} = [W_d] \tag{110}$$

Solution of Eq. 109 involves the inverse matrix

$$[W]^{-1} = [T] \cdot [W_d]^{-1} \cdot [T]_t = [X] \tag{111}$$

If we partition this inverse matrix as indicated by writing

$$[X] = \begin{bmatrix} x_{ll} & \vdots & jX_{ln} \\ \cdots & \cdots & \cdots \\ -jX_{nl} & \vdots & X_{nn} \end{bmatrix} \tag{112}$$

then the inverse of 109 with 107 substituted gives

$$\begin{aligned} X_{ll}E_s] + X_{ln}I_s] &= I_v] \\ X_{nl}E_s] + X_{nn}I_s] &= E_v] \end{aligned} \tag{113}$$

Observe that elements in the submatrix X_{ll} are short-circuit driving-point and transfer admittances, those in X_{nn} are open-circuit driving-point and transfer impedances, and those in X_{ln} and X_{nl} are dimensionless transfer functions.

Equations 110 and 111 show that a typical element x_{ji} of the matrix $[X]$ is given by the expression

$$x_{ji} = \sum_{\nu=1}^{b} \frac{k_{ji}^{(\nu)}}{d_\nu + s} \tag{114}$$

in which

$$k_{ji}^{(\nu)} = t_{j\nu} t_{i\nu} \tag{115}$$

the $t_{j\nu}$ being elements of the transformation matrix $[T]$.

Equation 114 we recognize as a partial fraction expansion of the impedance (or admittance or dimensionless-transfer) function x_{ji}. The residues $k_{ji}^{(\nu)}$ are related to elements in the jth and ith rows of transformation matrix $[T]$ by the simple expression 114. This situation enables us to construct the matrix $[T]$ to suit the requirements of a stated x_{ji} function. For a driving-point function $j = i$ the residues fix only a single row of $[T]$; for a transfer function, only the products of respective elements in two rows are fixed. The remainder of the matrix $[T]$ is then free to choose so as to yield parameter matrices fulfilling realizability conditions.

According to Eqs. 108 and 110 we find

$$\begin{bmatrix} L & \vdots & 0 \\ \cdots & \cdots & \cdots \\ 0 & \vdots & C \end{bmatrix} = [T]_t^{-1} \cdot [T^{-1}] \tag{116}$$

and

$$\begin{bmatrix} R & \vdots & j\gamma_{ln} \\ \cdots & \cdots & \cdots \\ j\gamma_{nl} & \vdots & G \end{bmatrix} = [T]_t^{-1} \cdot \begin{bmatrix} d_1 & 0 & \cdots & 0 \\ 0 & d_2 & \cdots & 0 \\ \cdot & \cdot & \cdot & \cdot \\ 0 & 0 & \cdots & d_b \end{bmatrix} \cdot [T]^{-1} \tag{117}$$

The fact that these relations involve the inverse of $[T]$, while the relation 115 for the residues involves $[T]$ itself, is not a particularly desirable situation. When only a single driving-point or transfer function is prescribed, the problem is still not too difficult. More generally, some cut-and-try may be unavoidable.

Since the synthesis procedure discussed in art. 9, Ch. IV is appropriate here, we recognize that $l = n$ and that γ_{ln} is a unit matrix of order n.

Although a number of free choices are available in the construction of the parameter matrices L, R, C, G we can introduce additional freedom by adding surplus terms in the partial-fraction expansion 113 involving zero residues. This step evidently increases the order of the transformation matrix $[T]$ (and hence increases also the orders of the parameter matrices) and introduces some zero elements in $[T]$, a process that is analogous to the introduction of surplus factors (or augmentation) in the conventional realization techniques.

When dealing with two-element-kind networks, the more familiar node or loop basis is adequate. Consider a passive bilateral RC network characterized on the node basis in the usual manner in terms of a capacitance matrix $[C]$ and a conductance matrix $[G]$, both of order n, the number of tree branches in a pertinent graph. The node admittance matrix is then given by

$$[Y] = [C]s + [G] \tag{118}$$

and the equilibrium equations read

$$[Y] \cdot E] = I_s] \tag{119}$$

in which $E]$ is a column matrix involving the n node-pair voltage variables and $I_s]$ is a column matrix with elements equal to the current sources bridged across node pairs (tree branches).

A normal coordinate transformation here is always straightforwardly possible.* We can think of accomplishing it by first transforming $[C]$ (which we assume to be nonsingular) congruently to its canonic form which is the unit matrix of order n because the pertinent quadratic form is positive definite. The same transformation carried out upon $[G]$ leaves it symmetrical and nondiagonal. We then transform this resultant $[G]$-matrix congruently with its orthogonal modal matrix and obtain the diagonal form involving its latent roots which are the natural frequencies of our passive bilateral network. The effect of this transformation upon the canonic form of $[C]$ is, of course, nil.

If we combine the two transformations into a single one with the real nonsingular matrix $[T]$, then we can summarize the total transformation by writing

$$[T]_t \cdot [Y] \cdot [T] = [Y_d] = \begin{bmatrix} g_1 + s & 0 & \cdots & 0 \\ 0 & g_2 + s & \cdots & 0 \\ \cdot & \cdot & \cdot & \cdot \\ 0 & 0 & \cdots & g_n + s \end{bmatrix} \tag{120}$$

The diagonal matrix $[Y_d]$ characterizes the network in its normal coordinates, and $g_1, g_2, \ldots g_n$ are the negatives of the natural frequencies

* The method discussed in art. 3 is, for example, directly applicable here and yields at once the appropriate transformation matrix $[T]$ whether $[C]$ or $[G]$ is assumed to be nonsingular or not.

of the network. Since they are the latent roots of a real symmetrical matrix, their required positive realness is assured.

The response of our *RC* network is characterized by the inverse of the matrix $[Y]$ in Eq. 119. According to Eq. 120, this inverse is given by

$$[Y]^{-1} = [Z] = [T] \cdot [Y_d]^{-1} \cdot [T]_t \tag{121}$$

where

$$[Y_d]^{-1} = [Z_d] = \begin{bmatrix} \dfrac{1}{g_1 + s} & 0 & \cdots & 0 \\ 0 & \dfrac{1}{g_2 + s} & \cdots & 0 \\ \cdot & \cdot & \cdot & \cdot \\ 0 & 0 & \cdots & \dfrac{1}{g_n + s} \end{bmatrix} \tag{122}$$

In terms of the elements $t_{j\nu}$ of the matrix $[T]$ we see that a typical element in the open-circuit impedance matrix $[Z]$ is given by the expression

$$z_{ji} = \sum_{\nu=1}^{n} \frac{k_{ji}^{(\nu)}}{g_\nu + s} \tag{123}$$

with

$$k_{ji}^{(\nu)} = t_{j\nu} t_{i\nu} \tag{124}$$

These results are formally like those in Eqs. 114 and 115; however, here the elements $t_{j\nu}$ are real whereas in the general *RLC* case they are complex.

The relevant parameter matrices are found from Eqs. 118 and 120 to be given by

$$[C] = [T]_t^{-1} \cdot [T]^{-1} \tag{125}$$

and

$$[G] = [T]_t^{-1} \cdot \begin{bmatrix} g_1 & 0 & \cdots & 0 \\ 0 & g_2 & \cdots & 0 \\ \cdot & \cdot & \cdot & \cdot \\ 0 & 0 & \cdots & g_n \end{bmatrix} \cdot [T]^{-1} \tag{126}$$

in which $g_1 \cdots g_n$ are determined from the poles of the specified impedance function 123.

In order to simplify the relationship between elements in $[T]$ and those in $[T]^{-1}$, this matrix can be assumed to have a triangular or any other special form. The only conditions that we need to consider are: (1) obtaining the specified values for the residues 124; and (2) fulfilling realizability criteria for the parameter matrices 125 and 126 according to the realization methods discussed in art. 3, Ch. IV.

7. Networks with Embedded Active Devices

Let us now suppose that the RC network just considered has embedded in it one or more resistive elements that are active and/or nonbilateral. This circumstance alters things only in that the matrix $[G]$ is no longer symmetrical and perhaps that its associated quadratic form is not necessarily positive definite. The synthesis procedure in this case involves only a minor variation from that just given.

We begin again with the relations expressed by Eqs. 118 and 119, and subject $[Y]$ to a congruent transformation with an appropriate matrix $[A]$ that carries $[C]$ into its canonic form and transforms $[G]$ into some other nonsymmetric matrix, say $[\hat{G}]$. If $g_1, g_2, \ldots g_n$ are the latent roots of $[\hat{G}]$, and $[B]$ denotes the pertinent modal matrix* fulfilling the relation

$$[B]^{-1} \cdot [\hat{G}] \cdot [B] = \begin{bmatrix} g_1 & 0 & \cdots & 0 \\ 0 & g_2 & \cdots & 0 \\ \cdot & \cdot & \cdot & \cdot \\ 0 & 0 & \cdots & g_n \end{bmatrix} \tag{127}$$

then the diagonalization of $[Y]$, analogous to that in Eq. 120, is now given by

$$[P]_t \cdot [Y] \cdot [Q] = [Y_d] = \begin{bmatrix} g_1 + s & 0 & \cdots & 0 \\ 0 & g_2 + s & \cdots & 0 \\ \cdot & \cdot & \cdot & \cdot \\ 0 & 0 & \cdots & g_n + s \end{bmatrix} \tag{128}$$

in which

$$[P] = [A] \cdot [B]_t^{-1} \quad \text{and} \quad [Q] = [A] \cdot [B] \tag{129}$$

Equation 121 is now replaced by

$$[Y]^{-1} = [Z] = [Q] \cdot [Y_d]^{-1} \cdot [P]_t \tag{130}$$

Equations 122 and 123 hold as before, and instead of 124 we now have for the residues

$$k_{ji}^{(\nu)} = q_{i\nu} p_{i\nu} \tag{131}$$

in which $q_{j\nu}$ and $p_{i\nu}$ denote elements of $[Q]$ and $[P]$ respectively.

Since the latent roots of a real, nonsymmetrical matrix are in general complex, we see that the z_{ji} in Eq. 123 now may have complex poles; and we also see that this result is due to the nonbilaterality of our RC network since this property is responsible for the nonsymmetry of $[G]$ or $[\hat{G}]$.

Comparing Eqs. 124 and 131, we also observe that, whereas the residues of passive bilateral driving-point impedances must be positive real

* See *MCA*, pp. 111–115.

(since the squares of real elements in $[T]$ are involved), such impedances may have negative real or complex residues when active nonbilateral elements are embedded in the RC network because the elements of $[P]$ and $[Q]$, even when these are real, can independently have either algebraic sign.

Regarding parameter matrices, we now have

$$[C] = [A]_t^{-1} \cdot [A]^{-1} = [P]_t^{-1} \cdot [Q]^{-1} \tag{132}$$

and from 128 we get

$$[G] = [P]_t^{-1} \cdot \begin{bmatrix} g_1 & 0 & \cdots & 0 \\ 0 & g_2 & \cdots & 0 \\ \cdot & \cdot & \cdot & \cdot \\ 0 & 0 & \cdots & g_n \end{bmatrix} \cdot [Q]^{-1} \tag{133}$$

Since this conductance matrix represents not only the passive resistive branches but also the embedded active nonbilateral devices (vacuum tubes or transistors) it should be regarded as a sum

$$[G] = [G_1] + [G_2] \tag{134}$$

in which $[G_1]$ is symmetrical and realizable by an $(n + 1)$-node network with passive elements while $[G_2]$ has the structure and element values pertinent to one or more physical vacuum tubes or transistors connected to the same node pattern. That is to say, $[G]$ has a decomposition like that given by Eq. 38 in Ch. VI for the example illustrated there in Fig. 4. This decomposition of $[G]$ into additive parts, since it obviously is not unique, injects additional flexibility into the construction of matrices $[P]$ and $[Q]$ or $[A]$ and $[B]$.

In this connection it should be observed that while the latent roots of a real dissymmetrical matrix like $[\hat{G}]$, Eq. 126, are in general complex, they need not be, and neither need the matrix $[B]$ be complex. When it is complex, then $[P]$ and $[Q]$ in 129 are, of course, also complex; but one can approach the synthesis problem by constructing either real or complex matrices $[P]$ and $[Q]$. In this regard we have two alternate possibilities to consider:

(*a*) We can construct real matrices $[P]$ and $[Q]$, forming a dissymmetrical matrix $[\hat{G}]$ with real latent roots. However, any assumed driving-point or transfer impedance having the partial-fraction expansion 123 with residues given by 131 may have zeros in the complex plane. Hence a voltage transfer ratio like $E_2/E_1 = z_{12}/z_{11}$ can have zeros and poles in the complex plane. The real latent roots $g_1 \cdots g_n$ of $[\hat{G}]$, which are poles of z_{11} and z_{12}, have no direct influence upon the transfer ratio E_2/E_1 and may be chosen in any way that facilitates realizability of the pertinent parameter matrices.

(*b*) We can assume complex poles for a desired transfer impedance z_{12} and prescribe these as the latent roots of $[\hat{G}]$. The construction of $[P]$ and $[Q]$ then involves complex elements since those of the matrix $[B]$ are complex, although $[A]$ remains real; and, of course, $[G]$, according to Eq. 133, must be real. Assuming a triangular form for $[P]$ and $[Q]$ and hence for $[P]^{-1}$ and $[Q]^{-1}$ again facilitates the construction of these matrices although their complex character adds to the computational labor, particularly where some cut-and-try is involved. Such difficulties may, however, be resolved with the help of appropriate computing aids.

When active and nonbilateral elements are embedded in a general *RLC* network, then we can still obtain similar results by making use of the diagonalization procedure discussed in the final paragraphs of art. 2 appropriate to such a situation. Equilibrium is expressed on a mixed basis by Eq. 43, and the desired diagonalization is accomplished by Eqs. 52, 53 and 54 involving the transformation of variables and sources given in Eq. 55.

For the substitutions 106 and the notation given in Eqs. 107 and 108 we again have our equilibrium equations in the form of Eq. 109. In place of 110 and 111, however, we now find

$$[Q]^{-1}\cdot[P]_t\cdot[W]\cdot[P]\cdot[Q] = \begin{bmatrix} d_1+s & 0 & \cdots & 0 \\ 0 & d_2+s & \cdots & 0 \\ \cdot & \cdot & \cdot & \cdot \\ 0 & 0 & \cdots & d_b+s \end{bmatrix} = [W_d] \tag{135}$$

and

$$[W]^{-1} = [P]\cdot[Q]\cdot[W_d]^{-1}\cdot[Q]^{-1}\cdot[P]_t = [X] \tag{136}$$

With the matrix $[X]$ partitioned as indicated in Eq. 112, we again have solutions in the form of Eqs. 113; and the elements of $[X]$ are given by the partial-fraction expansion 114. However, in place of Eq. 115, the residues are now given by the expression

$$k_{ji}^{(\nu)} = h_{j\nu}o_{j\nu} \tag{137}$$

in which $h_{j\nu}$ and $o_{j\nu}$ are elements of the matrices

$$[H] = [P]\cdot[Q] \quad \text{and} \quad [O] = [P]\cdot[Q]_t^{-1} \tag{138}$$

An approach to the synthesis of active and/or nonbilateral networks through construction of parameter matrices from given rational driving-point and transfer functions may thus be formed to follow essentially the same pattern as for passive bilateral networks. In place of relations 115 and 116 we now have

$$\begin{bmatrix} L & \vdots & 0 \\ \cdot & \cdot & \cdot \\ 0 & \vdots & C \end{bmatrix} = [P]_t^{-1} \cdot [P]^{-1} \tag{139}$$

and

$$\begin{bmatrix} R & \vdots & j\gamma_{ln} \\ \cdot & \cdot & \cdot \\ j\gamma_{nl} & \vdots & G \end{bmatrix} = [P]_t^{-1} \cdot [Q] \cdot \begin{bmatrix} d_1 & 0 & \cdots & 0 \\ 0 & d_2 & \cdots & 0 \\ \cdot & \cdot & \cdot & \cdot \\ 0 & 0 & \cdots & d_b \end{bmatrix} \cdot [Q]^{-1} \cdot [P]^{-1} \tag{140}$$

Here again we consider only $l = n$ and γ_{ln} equal to a unit matrix of order n.

8. Transformation Theory Applied to Active and Nonbilateral Networks

The quasi-normal coordinate transformation of equilibrium equations on a loop basis discussed in art. 3 for a passive bilateral network can readily be extended to active and nonbilateral networks. The essential difference in the latter case is that the matrix $[B]$, Eq. 5, is not symmetrical.

If we define a transformation matrix $[T^*]$ that bears the same relation to $[B]_t$ that $[T]$ bears to $[B]$, then essentially the same detailed procedure as that given in art. 3, yields the result

$$[\bar{T}^*]_t \cdot [B] \cdot [T] = \begin{bmatrix} s - \bar{s}_1 & & 0 \\ & \ddots & \\ 0 & & s - \bar{s}_l \end{bmatrix} \cdot [H] \cdot \begin{bmatrix} s - s_1 & & 0 \\ & \ddots & \\ 0 & & s - s_l \end{bmatrix} \tag{141}$$

which differs from the one stated by Eq. 68 in that the first matrix on the left is not the transposed conjugate of $[T]$, and the matrix $[H]$, although constant, is neither real nor symmetrical.

This result, which incidentally can also be gotten on a node basis, is, of course, not a true normal coordinate transformation. In terms of equilibrium equations on a mixed basis, on the other hand, the transformation of variables and sources expressed by Eqs. 55 yielding the complete diagonalization of parameter matrices as shown by Eq. 57 is for all practical purposes a normal coordinate transformation achievable for general active and nonbilateral networks. In terms of it we can make an interesting statement about such networks.

Since Eqs. 45 and 57 expressing equilibrium in the normal coordinates are identical in form, notwithstanding the fact that 45 is derived from a passive bilateral network and 57 from a more general network containing active and nonbilateral devices, it is clear that, for any given stable network containing active and/or nonbilateral elements, there exists at least one passive bilateral one in which the variables and sources are related to those of the given network by a nonsingular linear transformation.

More specifically, we can identify the primed variables in Eqs. 44 with those in Eqs. 55 since these pertain respectively to the normal coordinate Eqs. 45 and 57 which are identical in form. The unprimed variables in Eqs. 44, on the other hand, pertain to a passive bilateral (abbreviated PB) network while the unprimed variables in Eqs. 55 pertain to a general nonpassive and/or nonbilateral (abbreviated non-PB) network. Indicating this distinction by appropriate subscripts, a comparison of Eqs. 44 and 55 enables us to write

$$\begin{bmatrix} ji_v \\ \cdots \\ e_v \end{bmatrix}_{\text{PB}} = [T]\cdot[Q]^{-1}\cdot[P]^{-1}\cdot\begin{bmatrix} ji_v \\ \cdots \\ e_v \end{bmatrix}_{\text{non-PB}} \tag{142}$$

and

$$\begin{bmatrix} je_s \\ \cdots \\ i_s \end{bmatrix}_{\text{PB}} = [T]_t^{-1}\cdot[Q]^{-1}\cdot[P]_t\cdot\begin{bmatrix} je_s \\ \cdots \\ i_s \end{bmatrix}_{\text{non-PB}} \tag{143}$$

This result is perhaps not too interesting from a computational point of view inasmuch as the establishment of pertinent transformation matrices (the triple products in Eqs. 142 and 143) involves considerable work. Nevertheless, once these two matrices are known, one can analyze all aspects of the given network containing active and nonbilateral devices by considering an equivalent problem involving the passive bilateral reference network. Its behavior and that of the given network are simply connected by separate linear transformations of the independent and dependent variables.

That is to say, the sources in the passive bilateral reference network are not the same as those in the given network, but they are uniquely related to one another. The same statement holds regarding response in the two networks. Of significance is the fact that the sources in the passive bilateral reference network do not in any way depend upon current or voltage variables in either network. In other words, no "controlled sources" are involved or need be considered in this approach to the study of general linear systems.

It is, however, important to observe that this attitude is possible only if the given network is stable in spite of the embedded active devices, for the natural frequencies of the two networks must obviously be the same, and hence no passive reference network can exist unless all natural frequencies lie in the left half of the complex frequency plane. Such networks are referred to as "incidentally active" or also as "potentially active" networks. The quadratic form for the overall loss function is positive- or semi-definite. Even though active elements are present, their passive element environment is such that the resulting network is stable.

Off hand one might be somewhat puzzled as to just how such incidentally active networks differ from ordinary passive ones inasmuch as their natural frequencies are no different. Of what good is it to embed the active elements if their active character is swamped by the passive environment? What use can be made of these more general linear networks?

There are essentially two ways in which incidentally active networks serve a useful function: (*a*) they can be used to realize driving-point impedances not realizable with passive bilateral networks; and (*b*) they can realize completely general forms of driving-point and transfer functions even though the associated passive elements are only *RC* or *RL*.

Besides being required to have all zeros and poles restricted to the left half-plane the driving-point impedance of a passive bilateral network must additionally have a real part that is non-negative on the j-axis of the complex frequency plane. The necessity of this requirement is obvious from the fact that a passive network must absorb positive power at all frequencies in the sinusoidal steady state, and this would not be so if the j-axis real part of any driving-point impedance could assume a negative value.

It can be shown that such a function $Z(s)$ has a positive real part for all points in the right half of the s-plane. In fact, this property plus the rather obvious requirement that $Z(s)$ be real for real values of s, turns out to be sufficient to completely characterize the driving-point impedance of a passive bilateral network.* A rational function fulfilling these requirements is called "positive-real" (abbreviated p.r.). Driving-point impedances of passive bilateral networks must be p.r. functions.

When active devices are embedded, however, then the j-axis real part is no longer required to be consistently non-negative although poles and zeros are still restricted to the left-half plane if the network is both open-circuit and short-circuit stable. There are practical situations where such a driving-point impedance can serve a useful purpose. The incidentally active network can realize it.

* For a detailed discussion of this item see *SPN*, Ch. I. A brief consideration is also given in the following Ch. VIII.

There also are practical situations where we want to realize the transfer function of a passive bilateral *RLC* network but wish to avoid inductances altogether (usually because of their size and weight and sensitivity to inductive interference). The incidentally active network can accomplish this task also, as a comparison of expression 124 for a passive *RC* network with 131 for such a network with embedded active and nonbilateral devices shows. Thus, as pointed out in the previous article, the residues of impedances for incidentally active *RC* networks can have arbitrary complex values.

In this connection a collateral comment is appropriate to the effect that one must not jump to the conclusion that any network is active or even incidentally active if it contains some negative elements R, L or C. If we recall the problem of realizing a network from its specified parameter matrices as discussed in Ch. IV, and are simultaneously mindful of the linear-transformation method for solving the equivalent network problem as presented in art. 5 above, we become aware of the possibility that a given passive bilateral network with all positive elements may very easily be transformed into an equivalent one involving some negative elements.

For example, for the transformation 97 of networks in Fig. 1, consider a value of n that lies outside the range 98. Circuit (*b*) in Fig. 1 acquires a negative inductance in one of its series branches. A similar comment applies to the circuits in Fig. 2 and the range for n given by 100; or suppose we multiply the rows and columns No. 2 and 3 of matrix 104 by factors larger than 2. The resulting resistive network in the topological form of Fig. 8 then displays some negative resistances, notwithstanding that the input resistance is still the "good old" passive and positive 2.5 ohms.

This state of affairs creates the following problem. Suppose we are confronted with a circuit diagram containing only R's, L's and C's but with some negative values. If we do not know that it is derivable by a nonsingular transformation from a network with all positive elements, how can we tell whether indeed it is so derivable?

The straightforward answer is to say: Let us write down the quadratic form for the energy function that involves the negative element (or elements) and apply the test for positive definiteness. If the form is positive- or semi-definite, then we can conclude that an equivalent network with all positive elements (and possibly some ideal transformers) exists. If not, then the given network contains active elements (like tunnel diodes, for instance).

This procedure can run into a snag, however, unless one is on the look-out for certain degeneracies and knows how to deal with them. The following example illustrates these things.

Consider the circuit in Fig. 9 in which element values are in henrys and farads. Since two negative capacitances are involved, there is some question as to whether this circuit is passive in the sense that all driving-point impedances at pliers- or soldering-iron-type entries are p.r.

Suppose we try to investigate this situation by writing the quadratic form for stored electric energy to see whether it is positive definite. All we need to do is write down the pertinent parameter matrix. On a loop basis, choosing the meshes as indicated we have

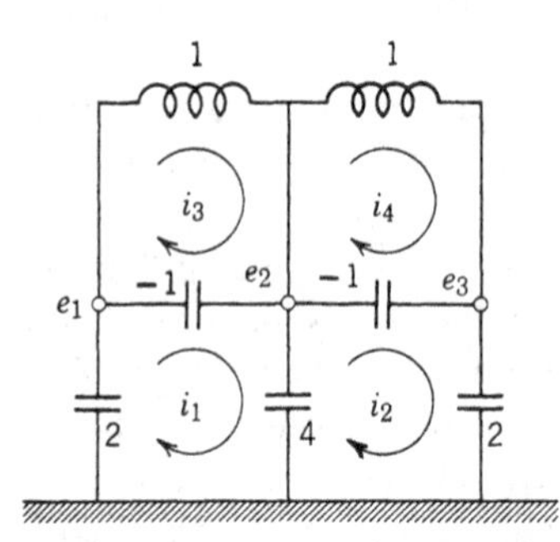

FIG. 9. Circuit containing negative elements which is nevertheless passive.

$$[S] = \begin{bmatrix} -\frac{1}{4} & -\frac{1}{4} & 1 & 0 \\ -\frac{1}{4} & -\frac{1}{4} & 0 & 1 \\ 1 & 0 & -1 & 0 \\ 0 & 1 & 0 & -1 \end{bmatrix} \tag{144}$$

Immediately we are puzzled by the fact that all principal diagonal elements are negative. Normally, this fact tells us at once that the quadratic form is not positive definite. However, let us write down the capacitance matrix on a node basis for the indicated node-to-datum set of node-pair voltages. We have by inspection

$$[C] = \begin{bmatrix} 1 & 1 & 0 \\ 1 & 2 & 1 \\ 0 & 1 & 1 \end{bmatrix} \tag{145}$$

To test this matrix by the method of art. 4, Ch. V we consider the representation $[C] = [A] \cdot [A]_t$ and assume a triangular form for $[A]$. By inspection we get

$$[A] = \begin{bmatrix} 1 & 0 & 0 \\ 1 & 1 & 0 \\ 0 & 1 & 0 \end{bmatrix} \tag{146}$$

Since this matrix is real we can conclude that the quadratic form having the matrix $[C]$ in 145 can assume no negative values. The matrix $[A]$, however, has a zero determinant and hence so does $[C]$. The quadratic form is, therefore, semidefinite; it can become zero for certain nonzero values of the variables.

This latter property is best seen by transforming the quadratic form with the matrix 145 to its canonic form. By the well-known process of operating upon rows we see that premultiplication of [C] by the matrix

$$[T] = \begin{bmatrix} 1 & 0 & 0 \\ -1 & 1 & 0 \\ 1 & -1 & 1 \end{bmatrix} \tag{147}$$

produces all zeros below the principal diagonal. The congruent transformation

$$[T] \cdot [C] \cdot [T]_t = \begin{bmatrix} 1 & 0 & 0 \\ 0 & 1 & 0 \\ 0 & 0 & 0 \end{bmatrix} = [C'] \tag{148}$$

yields the desired canonic form, and we see that one of the three node potentials is dynamically superfluous since two node potentials suffice to characterize the quadratic form.

The change of variable

$$e] = [T]_t e'] \tag{149}$$

accompanies the congruent transformation 148, and so we have

$$\begin{aligned} e_1 &= e_1' - e_2' + e_3' \\ e_2 &= \phantom{e_1' - {}} e_2' - e_3' \\ e_3 &= \phantom{e_1' - e_2' - {}} e_3' \end{aligned} \tag{150}$$

The canonic form given by

$$2V = \underline{e'} \cdot [C'] \cdot e'] \tag{151}$$

evidently becomes zero for $e_1' = e_2' = 0$, which the equations 150 show is equivalent to

$$e_1 = e_3, \quad e_2 = -e_3 \tag{152}$$

For the original matrix 145 we have

$$2V = e_1^2 + 2e_2^2 + e_3^2 + 2e_1e_2 + 2e_2e_3 \tag{153}$$

Substituting the values 152 into 153 we do indeed obtain zero for the energy function V.

The canonic form, obtained by substituting 149 into 151 or 150 into 153, is found to be

$$2V = (e_1')^2 + (e_2')^2 \tag{154}$$

It is evident that the canonic form "tells the story."

Although in this example the form 153 for the given parameter matrix 145 already reveals its semidefinite character, the elimination of dynamically superfluous variables through transformation to the pertinent canonic form is always best.

We can incidentally recognize the existence of a linear constraint among the variables e_1, e_2, e_3 by noting that the short-circuit C-graph for the circuit of Fig. 9 has one independent node pair.

Similarly, we observe that the open-circuit C-graph has two independent loops, which indicates that two linear constraint relations exist among the loop currents i_1, i_2, i_3, i_4 which can be used to eliminate two of these variables. We readily find that these constraint relations, expressed in terms of the mesh charges, read

$$q_3 = q_4 = \tfrac{1}{4}(q_1 + q_2) \tag{155}$$

In terms of the matrix 144 we get

$$2V = -\tfrac{1}{4}q_1^2 - \tfrac{1}{4}q_2^2 - q_3^2 - q_4^2 - \tfrac{1}{2}q_1q_2 + 2q_1q_3 + 2q_2q_4 \tag{156}$$

Substituting the constraint relations 155 makes this

$$2V = \tfrac{1}{8}q_1^2 + \tfrac{1}{8}q_2^2 + \tfrac{1}{4}q_1q_2 \tag{157}$$

implying an elastance matrix for which we can write

$$8[S] = \begin{bmatrix} 1 & 1 \\ 1 & 1 \end{bmatrix} = [A] \cdot [A]_t \tag{158}$$

with

$$[A] = \begin{bmatrix} 1 & 0 \\ 1 & 0 \end{bmatrix} \tag{159}$$

The semidefinite character of the quadratic form is again evident.

Now we are also in a position to answer the question as to whether a driving-point impedance for a particular entry is p.r. Suppose we cut into the first shunt-capacitance branch on the left of the circuit in Fig. 9. By so doing we interfere with one of the equations yielding a constraint relation among the mesh charges. Instead of the two relations 155 we have only the one involving q_4. Using this one alone, in Eq. 156 we get

$$2V = -\tfrac{5}{16}q_1^2 + \tfrac{3}{16}q_2^2 - q_3^2 - \tfrac{1}{8}q_1q_2 + 2q_1q_3 \tag{160}$$

which evidently is not positive definite. Hence an impedance at this cut is *not* p.r.

The same conclusion evidently holds for a cut in any other capacitive branch. We can, however, cut into the inductive branches or solder leads to any node pair and find a p.r. impedance function. We can now make a rather general statement in this regard, as follows.

On a node basis, form the pertinent quadratic form F, T or V, according to which involves negative elements, and determine the independent constraint relations among the node-pair variables. Through their use, find the abridged quadratic form in terms of dynamically independent variables. If this form (or forms, if more than one involves negative elements) is positive- or semi-definite, then all soldering-iron-type impedances are p.r. for which the implied current source does not participate in the cut set of a constraint relation. If the implied current source does participate in the cut set of a constraint relation (as, for example, it does for a driving point between node 1 and datum in the circuit of Fig. 9) then the pertinent soldering-iron-type impedance is p.r. only if the abridged quadratic form obtained by ignoring that constraint relation is still positive- or semi-definite (as it is in the example of Fig. 9).

If one or more of the pertinent abridged quadratic forms using *all* constraint relations fails to be positive- or semi-definite, then all soldering-iron-type impedances fail to be p.r. since the non-negativeness of all associated energy functions is a *necessary* condition for the p.r. character of any driving-point impedance of a given network.

A completely dual statement pertinent to a loop basis applies to pliers-type impedances; and, incidentally, since the p.r. character of a driving-point impedance also insures that the corresponding admittance is p.r., it follows that the network under consideration is both open-circuit and short-circuit stable for the points of entry considered.

Problems

1. Show in detail that the transformation given in art. 3 for the general RLC case does accomplish a normal-coordinate transformation according to the conventional interpretation of that process.

2. In the LC case, show that the result given by Eq. 68 assumes the expected conventional form for this two-element-kind situation.

3. In the following network, element values are given in henrys, farads and ohms:

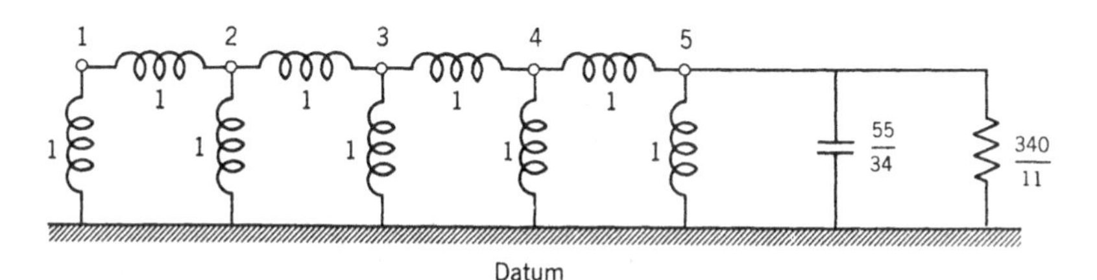

PROB. 3.

Assume a sinusoidal current source having unit amplitude and a frequency of 1 radian per second applied first between datum and node 1, then between datum and node 2, etc. Each time calculate the voltage amplitudes at all nodes and thus show that the distribution remains the same for this resonance frequency. Now consider a frequency of 2 radians per second and calculate the voltage distribution for the source at node 1 and again at node 5. Note that now the distributions are not the same.

4. For the adjacent circuit:

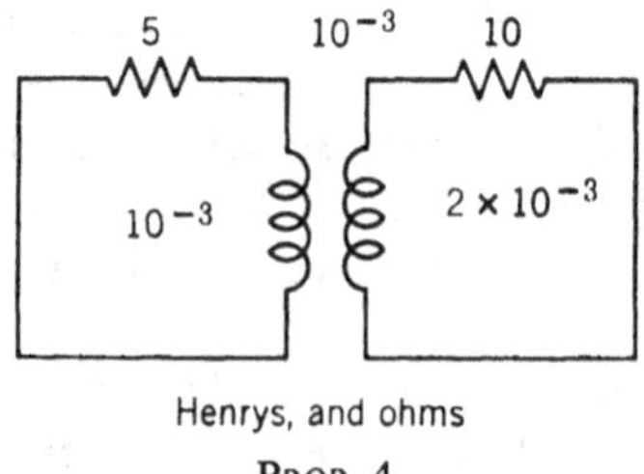

Henrys, and ohms

PROB. 4.

(*a*) Find and plot the normal coordinates.

(*b*) Find the response for a unit step voltage applied in mesh 1 and check the resulting amplitude distribution geometrically by forming projections of the normal coordinates determined in part (*a*).

(*c*) Calculate the admittance functions y_{11}, y_{22}, y_{12}; expand these in partial fractions, and from the residus thus formed determine elements of the transformation matrix $[T]$. Check this matrix with that found in part (*a*).

5. For the following circuit and element values:

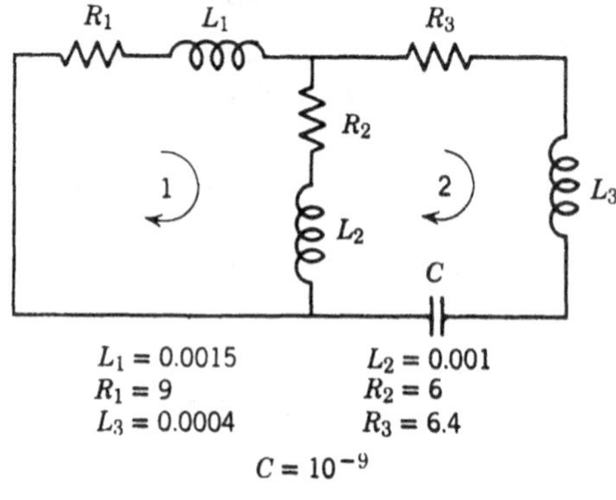

$C = 10^{-9}$

PROB. 5.

determine and plot the normal coordinates. What can you say about the distribution functions for this network excited by a unit step voltage in mesh 1? In mesh 2?

6. For a given network, the elements in the first row of the transformation matrix $[T]$, Eq. 60, are given by

$$3 \quad 2 \quad 2 \quad 3 \quad 1$$

The corresponding natural frequencies are

$$s = -1, \quad -2, \quad -3, \quad -4, \quad -5$$

Within a constant multiplier, write down the partial fraction expansion for the impedance function z_{11}.

7. The third row of the residue matrix K, Eq. 19, for a certain network reads

$$1 \quad 2 \quad 3 \quad 2 \quad 1$$

Construct the complete matrix $[K]$ for the pertinent natural frequency.

8. The first row of the residue matrix $[K]$ for a certain network is given by

$$\begin{array}{cccc l} 4 & 3 & 2 & 1 & \text{for } s = s_1 \\ 3 & 2 & 1 & 2 & \text{for } s = s_2 \\ 4 & 2 & 1 & 1 & \text{for } s = s_3 \\ 2 & 1 & 2 & 1 & \text{for } s = s_4 \end{array}$$

Construct the transformation matrix $[T]$, Eq. 60.

9. A certain 3-node RC network has the following open-circuit impedance functions in partial-fraction expansion form:

$$z_{11} = \frac{1}{s+1} + \frac{2}{s+2} + \frac{3}{s+3}$$

$$z_{12} = \frac{1}{s+1} - \frac{1}{s+2} + \frac{2}{s+3}$$

$$z_{13} = \frac{2}{s+1} - \frac{1}{s+2} + \frac{1}{s+3}$$

Find the pertinent parameter matrices $[C]$ and $[G]$.

10. Suppose parameter matrices $[C]$ and $[G]$ are specified and we construct a matrix $[T]^{-1}$ (perhaps in triangular form) to satisfy Eq. 125. From its inverse $[T]$ we then calculate the natural frequencies by Eq. 126. Is this result acceptable? Why or why not?

11. In a certain RC network the natural frequencies are

$$s_1 = -1, \quad s_2 = -2, \quad s_3 = -3$$

The first of these can be excited from any node pair; the second from node pairs 2 and 3 only; and the third can be excited only from node pair 3. Residues of all the z_{ik} are $+1$. Find the parameter matrices $[C]$ and $[G]$.

12. Given the following 4-node RC network in which all element values are unity:

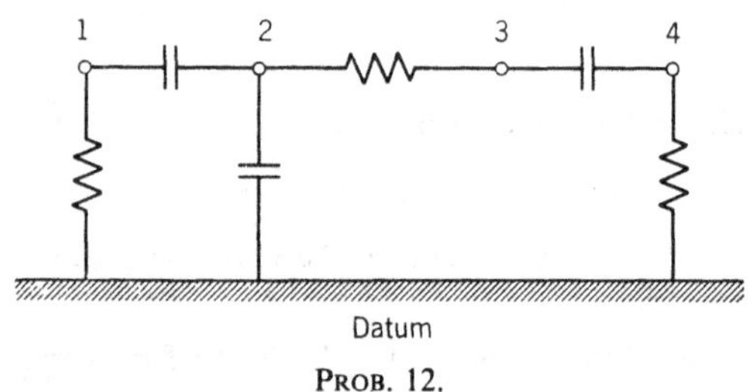

PROB. 12.

Construct the transformation matrix $[T]$, Eq. 60, according to the method discussed in art. 3.

CHAPTER VIII

The Impedance or Admittance Concept

1. Basic Definition of the Admittance Function; Forced Behavior

Our starting point in the present discussion is the differential equation expressing a specific excitation-response relationship for a given linear network. Let us assume that the excitation is a voltage and the response a current, and thus arbitrarily choose the loop basis to fix the pertinent relationship. The dual situation obtained on a node basis, or one in which both excitation and response are currents or voltages, we regard as a trivial one which the reader can supply when the need arises.

Since the network is linear and the parameters are constant, the pertinent differential equation is likewise linear with constant coefficients. Its form reads

$$\left(a_m \frac{d^m}{dt^m} + a_{m-1} \frac{d^{m-1}}{dt^{m-1}} + \cdots + a_1 \frac{d}{dt} + a_0\right) i_j(t)$$
$$= \left(b_k \frac{d^k}{dt^k} + b_{k-1} \frac{d^{k-1}}{dt^{k-1}} + \cdots + b_1 \frac{d}{dt} + b_0\right) e_i(t) \quad (1)$$

in which m and k are fixed integers and the coefficients $a_0 \cdots a_m$ and $b_0 \cdots b_k$ are functions of the network elements alone.

The equilibrium equations from which this specific equation is derived are expressed in matrix form by Eq. 1 of Ch. VII. The voltage excitation is assumed to be located in link i and the current response in loop or link j. If $i = j$ we have a driving-point situation, and if $i \neq j$, a transfer relation is involved. In that event Eq. 1 is valid only so long as $e(t)$ is the excitation and $i(t)$ the response, while if $e(t)$ and $i(t)$ pertain to the same terminal pair, their roles regarding excitation and response may be interchanged (as shown in art. 4 of Ch. III).

The excitation, which is assumed to be dead for $t < 0$, is given by the exponential function

$$e_i(t) = E_i e^{st} \qquad \text{for } t \geq 0 \quad (2)$$

and the network is at rest until the occurrence of this event. The complete expression for the ensuing response $i_j(t)$, as in the simple examples with which we are familiar, consists of the sum of a particular integral (forced response) and a complementary function (natural behavior) which satisfies the corresponding homogeneous equation (Eq. 1 for $e = 0$). For the moment we are interested only in the particular integral which is given by

$$i_j(t) = I_j e^{st} \qquad \text{for } t \geq 0 \tag{3}$$

If these relations pertinent to the particular integral are substituted into the matrix equation 1 of Ch. VII, the discussion given there shows that the complex amplitudes E_i and I_j are related by the equation

$$BI_j = B_{ij}E_i \tag{4}$$

in which B is the network determinant with elements

$$b_{ij} = L_{ij}s + R_{ij} + S_{ij}s^{-1} \tag{5}$$

and B_{ij} is a cofactor corresponding to deletion of row i and column j in this determinant.

On the other hand, if we substitute 2 and 3 into Eq. 1 we get, after cancellation of the common exponential factor,

$$(a_m s^m + a_{m-1}s^{m-1} + \cdots + a_1 s + a_0)I_j$$
$$= (b_k s^k + b_{k-1}s^{k-1} + \cdots + b_1 s + b_0)E_i \tag{6}$$

Since the determinant B and its cofactors B_{ij} are polynomials in the complex frequency variable s, a comparison of Eqs. 4 and 6 makes evident the identifications

$$B = a_m s^m + a_{m-1}s^{m-1} + \cdots + a_1 s + a_0 \tag{7}$$

and

$$B_{ij} = b_k s^k + b_{k-1}s^{k-1} + \cdots + b_1 s + b_0 \tag{8}$$

The pertinent admittance function (Eq. 15, Ch. VII) is defined as the ratio

$$y_{ji}(s) = \frac{B_{ij}}{B} = \frac{I_j}{E_i} = \frac{b_k s^k + b_{k-1}s^{k-1} + \cdots + b_1 s + b_0}{a_m s^m + a_{m-1}s^{m-1} + \cdots + a_1 s + a_0} \tag{9}$$

Its basic function is to characterize the particular integral or forced response for an exponential excitation given by Eq. 2. It is a rational function (quotient of finite polynomials) of the complex frequency s determined by the excitation.

A significant observation is the fact that the admittance function 9 determines the differential equation 1 since the same set of coefficients is involved in both of these relations. Thus one might determine the admittance function 9 by the method discussed in art. 1 of Ch. VII or by any other expedient, and from this result write down the pertinent differential equation. Since the latter expresses equilibrium of the physical network and completely characterizes the transient as well as the steady-state behavior (if it exists), we recognize at the outset that the admittance function has a much deeper significance than merely that of determining the steady-state part (particular integral) of the response to a sinusoidal excitation. The following discussion elaborates this thought.

If the polynomials in 9 are written in the factored forms

$$B_{ij} = b_k(s - s_a)(s - s_b) \cdots (s - s_k) \tag{10}$$

$$B = a_m(s - s_1)(s - s_2) \cdots (s - s_m) \tag{11}$$

The admittance function 9 is given by the expression

$$y_{ji}(s) = \frac{H(s - s_a)(s - s_b) \cdots (s - s_k)}{(s - s_1)(s - s_2) \cdots (s - s_m)} \tag{12}$$

which places its zeros $s_a, s_b, \ldots s_k$, and poles $s_1, s_2, \ldots s_m$ in evidence. The constant multiplier

$$H = \frac{b_k}{a_m} \tag{13}$$

is alternately referred to as the *level factor* of the admittance function. Apart from this level factor, the admittance function is uniquely determined by its zeros and poles; and precisely these same data determine the differential equation 1 except for a relatively unimportant constant multiplier.

When a driving-point function is involved, the admittance (resp. impedance) completely characterizes the network for that point of access. When two terminal pairs are involved and the network is bilateral, three admittance functions (two driving-point and one transfer) are needed to characterize the network, but again these three functions completely determine the network behavior, transient as well as steady-state, at these points of access.

For a p terminal-pair bilateral network, $p(p - 1)/2$ admittance functions $y_{ji}(s)$, the elements of a symmetrical short-circuit admittance matrix Y of order p, uniquely determine the behavior of that network under all excitation conditions and for all time following a rest condition; and the principle of duality enables us at once to make a similar statement involving a symmetrical open-circuit impedance matrix $[Z]$ with elements $z_{ji}(s)$.

When bilaterality is not fulfilled, these matrices involve p^2 distinct elements and so it takes a larger number of admittance or impedance functions to characterize the network at p points of access, but the same degree of generality regarding the completeness of characterization still holds.

Since the determinant B in Eq. 7 is the common denominator of all driving-point and transfer admittance functions pertinent to a given network we see that these all have the same poles. That is to say, all pliers-type driving point admittances for any linear network have the same poles, and these are the same as the poles of any pliers-type transfer admittance (ratio of current in one branch to voltage inserted in another). The zeros, which are determined by the cofactor 8, are in general different for each function.

By duality we can state that the open-circuit driving-point and transfer impedances for any linear network have the same poles. In other words, every soldering-iron-type driving-point impedance for a given network has the same poles, and these are also the poles of all soldering-iron-type transfer impedances (ratio of voltage across one node pair to current applied across another).

These statements assume, of course, that the network is nondegenerate so that all natural frequencies are excitable from all creatable driving points, since the poles of either the short-circuit admittances or the open-circuit impedances are these natural frequencies. This item is now discussed in further detail.

2. Natural Behavior and the Complete Response

The excitation is alternately referred to as the forcing function; and the particular integral, which emulates it, is correspondingly referred to as the *forced response*. The total response, on the other hand, is influenced among other things by characteristics or tendencies which are peculiar to the network itself and hence express its own natural or force-free behavior. These are found by asking for a nontrivial solution to the differential equation 1 when the excitation $e_1(t)$ is identically zero.

For such a nontrivial solution we tentatively assume

$$i_j(t) = A_j e^{st} \tag{14}$$

in which neither A nor s are known, but substitution into Eq. 1 for $e_i \equiv 0$ gives

$$(a_m s^m + a_{m-1} s^{m-1} + \cdots + a_1 s + a_0) A_j e^{st} = 0 \tag{15}$$

The requirement that our solution be nontrivial demands $A_j \neq 0$, and so we have the condition

$$a_m s^m + a_{m-1} s^{m-1} + \cdots + a_1 s + a_1 = 0 \tag{16}$$

In other words, the nontriviality requirements yields the characteristic equation 16 determining s-values for which function 14 satisfies the homogeneous equation 1 (for $e_i \equiv 0$). These s-values are the characteristic values or natural frequencies $s_1, s_2, \ldots s_m$ placed in evidence by the factored form 11 of the polynomial in this equation

Whereas the tentative assumption made in Eq. 14 modestly considers a single exponential term, we thus see that the force-free behavior is actually given by a function of the form

$$i_j(t) = \sum_{\nu=1}^{m} A_j^{(\nu)} e^{s_\nu t} \tag{17}$$

as is also pointed out in art. 1, Ch. VII (and indicated there by Eq. 7).

The complete solution for this loop current is now obtained by forming the sum of 3 (the particular integral) and 17 (the complementary function). Suppose we indicate this distinction between $i_j(t)$ in 3 and in 17 by adding a suffix p (for particular integral) to the former and a suffix 0 to the latter which is the force-free or natural response. Then we have for the net response

$$\begin{aligned} i_j(t) &= i_{j0}(t) + j_{jp}(t) \\ &= A_j^{(1)} e^{s_1 t} + A_j^{(2)} e^{s_2 t} + \cdots + A_j^{(m)} e^{s_m t} + I_j e^{st} \end{aligned} \tag{18}$$

in which I_j is found from Eq. 9 and the amplitudes $A_j^{(1)} \cdots A_j^{(m)}$ of the transient terms are determined from an equal number of initial conditions (as pointed out in art. 4, Ch. VII). This determination is discussed in detail in Ch. X.

If we are considering a driving-point situation for which $e(t)$ and $i(t)$ in Eq. 1 pertain to the same terminal pair, then we may alternately regard the current as excitation and the voltage as response without invalidating this equation. The excitation, which we again assume to be dead for $t < 0$, is now given by Eq. 3, and the particular integral for the response by Eq. 2 with the index i replaced by j. The complex amplitudes I_j and E_j are related by the admittance function 9 which is designated a driving-point function $y_{jj}(s)$ but is otherwise unchanged in functional form.

The complementary function or force-free behavior is now a nontrivial solution to Eq. 1 for the stipulation $i_j(t) \equiv 0$. Here we make the tentative assumption

$$e_j(t) = B_j e^{st} \tag{19}$$

analogous in form to Eq. 14 pertinent to voltage excitation. Again neither B_j nor s are known at this point, but substitution into Eq. 1 for $i_j \equiv 0$ gives

$$(b_k s^k + b_{k-1} s^{k-1} + \cdots + b_1 s + b_0) B_j e^{st} = 0 \tag{20}$$

The nontriviality requirement $B_j \neq 0$ now yields the characteristic equation

$$b_k s^k + b_{k-1} s^{k-1} + \cdots + b_1 s + b_0 = 0 \tag{21}$$

Its roots are the values $s_a, s_b, \ldots s_k$ that are placed in evidence by the factored form 10 of the polynomial in this equation.

We thus see that, in place of the modest assumption 19, the force-free behavior is given by the function

$$e_j(t) = \sum_{\nu=a,\,b,\,\ldots}^{k} B_j^{(\nu)} e^{s_\nu t} \tag{22}$$

Zeros of the polynomial 10, in contrast with those of the polynomial 11, are now the characteristic values or natural frequencies.

The complete response is formed by the sum of particular integral 2 and complementary function 22. If we distinguish these by adding suffixes p and 0 as we did in the previous solution for current, we can write for the net response

$$\begin{aligned} e_j(t) &= e_{j0}(t) + e_{jp}(t) \\ &= B_j^{(a)} e^{s_a t} + B_j^{(b)} e^{s_b t} + \cdots + B_j^{(k)} e^{s_k t} + E_j e^{st} \end{aligned} \tag{23}$$

E_j is found from Eq. 9 (in which the suffix i is replaced by j) and the amplitudes $B_j^{(a)} \cdots B_j^{(k)}$ of the transient terms are now the integration constants which give solution 23 the required flexibility to meet arbitrarily specified initial conditions.

In this driving-point situation it is significant to observe that the natural frequencies for a voltage excitation are poles of the admittance function 9 while, for a current excitation, they are zeros of this function. More specifically, the response 18 for an applied voltage involves the characteristic equation 16 whose roots are the zeros $s_1, s_2, \ldots s_m$ of the denominator polynomial in 9; and the response function 23 for an applied current contains the natural frequencies $s_a, s_b, \ldots s_k$ which are roots of the characteristic equation 21 involving the numerator polynomial in 9. Since the force-free behavior or transient terms in 18 result for $e_j \equiv 0$ whereas in 23 they correspond to $i_j \equiv 0$, we see that poles of the driving-point admittance 9 are natural frequencies of the network with its input terminals short-circuited and that zeros of this function, which are poles of the driving-point impedance, are natural frequencies of the network with its input terminals open-circuited.

Stated consistently in terms of an impedance function this result recognizes that zeros and poles of this function may be identified with natural frequencies of the network resulting respectively under conditions of short-circuit and open-circuit constraint at the driving-point terminals. As will be shown presently, this physical interpretation for the zeros and poles of an impedance enables one in simple cases to write down the appropriate expression for this driving-point function by inspection.

3. Impedances in Terms of Energy Functions; p.r. Character

In order to study the properties of an impedance function in a rather general manner, it is effective to express it in terms of energy functions associated with the pertinent network. In this regard the forms given by Eqs. 12, 13 and 14 in Ch. V are not appropriate because they involve the instantaneous currents instead of their complex amplitudes; and those in Eqs. 79, 80 and 81 for the sinusoidal steady state are only moderately useful because they imply $s = j\omega$ whereas we do not wish to restrict the complex frequency variable to the j-axis. For this reason we prefer to introduce a somewhat modified set of energy functions more appropriate to our immediate objective, and to this end we begin with the equilibrium equations on a loop basis as given by matrix 108 of Ch. II, as follows:

$$([L]p + [R] + [S]p^{-1}) \cdot i_v] = e_s] \tag{24}$$

which we can rewrite in the more explicit algebraic form

$$\sum_{j=1}^{l} (L_{ij}p + R_{ij} + S_{ij}p^{-1}) \cdot i_j = e_i; \qquad i = 1, 2, \ldots l \tag{25}$$

For the assumptions

$$e_i = E_i e^{st} \quad \text{and} \quad i_j = I_j e^{st} \tag{26}$$

These differential equations yield the following algebraic equations in terms of complex voltage and current amplitudes:

$$\sum_{j=1}^{l} (L_{ij}s + R_{ij} + S_{ij}s^{-1})I_j = E_i; \qquad i = 1, 2, \ldots l \tag{27}$$

If we multiply on both sides by $\bar{I}_i$ (the conjugate of I_i) and sum over the index i (which means multiply the respective equations for $i = 1$, $i = 2$, etc. by $\bar{I}_1$, $\bar{I}_2$, etc. and then add all resulting equations together) we can write the result as

$$sT_0 + F_0 + \frac{V_0}{s} = \sum_{i=1}^{l} E_i\bar{I}_i \tag{28}$$

where we have introduced the functions (quadratic forms)

$$T_0 = \sum_{i,j=1}^{l} L_{ij}\bar{I}_iI_j = \sum_{i,j=1}^{l} L_{ij}I_i\bar{I}_j \tag{29}$$

$$F_0 = \sum_{i,j=1}^{l} R_{ij}\bar{I}_iI_j = \sum_{i,j=1}^{l} R_{ij}I_i\bar{I}_j \tag{30}$$

$$V_0 = \sum_{i,j=1}^{l} S_{ij}\bar{I}_iI_j = \sum_{i,j=1}^{l} S_{ij}I_i\bar{I}_j \tag{31}$$

We are restricting the discussion in this article to passive bilateral networks and are assuming that conditions leading to symmetrical parameter matrices are fulfilled. Hence equivalence of the alternate forms in each of these expressions is obvious since $L_{ij} = L_{ji}$; i and j are merely summation indexes and hence may be interchanged, because the summation extends independently over each of these two indexes.

T_0 is the only one of these three functions that has the dimensions of energy; F_0 is power, and V_0 is time-rate-of-change of power. For the sake of simplicity, we shall nevertheless refer to them as energy functions. For $s = j\omega$, they are related to the average energy functions 79, 80 and 81 in Ch. V as follows:

$$T_0 \to 4T_{av}, \quad F_0 \to 4F_{av}, \quad V_0 \to 4\omega^2 V_{av} \tag{32}$$

For any complex value of the frequency variable s, they have no physical meaning. Mathematically they are clearly defined for any complex s since they are explicit functions of the current amplitudes I_j which in turn are explicitly determined for any s-value by Eqs. 27.

Assume now that the network has p points of access so that only p of the voltage amplitudes E_i in Eqs. 27 are nonzero. The numbering of loops is so chosen that the first p pertain to the accessible terminal pairs. Making use of the notation in Eq. 5, the matrix $[B]$ of Eqs. 27 is partitioned as follows:

$$[B] = \begin{bmatrix} b_{pp} & \vdots & b_{pq} \\ \cdots & \cdots & \cdots \\ b_{qp} & \vdots & b_{qq} \end{bmatrix} \tag{33}$$

where $p + q = l$. The appropriately abridged form of Eqs. 27 may then be written

$$\sum_{j=1}^{p} z_{ij} I_j = E_i; \qquad i = 1, 2, \ldots p \tag{34}$$

in which coefficients z_{ij} are elements of the open-circuit impedance matrix

$$[Z] = [b_{pp} - b_{pq} \cdot b_{qq}^{-1} \cdot b_{qp}] \tag{35}$$

as shown in art. 4, Ch. I.

Substituting Eq. 34 into 28, noting that $E_i \equiv 0$ for $i > p$, now yields

$$sT_0 + F_0 + \frac{V_0}{s} = \sum_{i,j=1}^{p} z_{ij} \bar{I}_i I_j = \sum_{i,j=1}^{p} z_{ij} I_i \bar{I}_j \tag{36}$$

where equivalence of the two forms on the right again follows from the symmetry condition $z_{ij} = z_{ji}$.

A basic property of impedance functions may now be deduced from this result if we observe first that the quadratic forms T_0, F_0, V_0, given by Eqs. 29, 30 and 31, have positive real values for all complex values of the amplitudes I_i and hence for all values of the complex frequency variable s. This fact is seen if we write

$$I_i = x_i + jy_i, \quad \bar{I}_i = x_i - jy_i \tag{37}$$

Then we have, for example,

$$T_0 = \sum_{i,j=1}^{l} L_{ij}(x_i x_j + y_i y_j) + j \sum_{i,j=1}^{l} L_{ij}(x_i y_j - x_j y_i) \tag{38}$$

The second of these two sums is zero because $L_{ij} = L_{ji}$. Hence it follows that T_0 is real. The first term is the sum of two quadratic forms with the same matrix $[L]$, and these have positive values for all real values of the x_i and y_i because the network is assumed to be passive. Hence T_0 can have only positive real values; and the same is obviously true also for the functions F_0 and V_0.

Substituting 37 into the double sum in 36, we find in an analogous manner that

$$\sum_{i,j=1}^{p} z_{ij} I_i \bar{I}_j = \sum_{i,j=1}^{p} z_{ij}(x_i x_j + y_i y_j) \tag{39}$$

while for any complex $s = \sigma + j\omega$ we have

$$sT_0 + F_0 + \frac{V_0}{s} = \sigma T_0 + F_0 + \frac{\sigma V_0}{\sigma^2 + \omega^2} + j\left(\omega T_0 - \frac{\omega V_0}{\sigma^2 + \omega^2}\right) \tag{40}$$

Equation 36 states that 39 equals 40. From the positive realness of T_0, F_0, V_0 we see that 40 is real for real values of s and can have only positive real values in the right half of the s-plane. Therefore 39 is real for real values of s and can have only positive real values in the right half of the s-plane for all real values of the x's or y's. However, this statement is neither stronger nor weaker than to say that the function

$$Z(s) = \sum_{i,j=1}^{p} z_{ij} x_i x_j \tag{41}$$

is real for real values of s and can have only positive real values for right half-plane s-values and for all real values of the x_i's. The fact that the x's and y's in 39 are independent makes this conclusion obvious.

We can summarize this result by the statement:

$Z(s)$ in 41 is real for real values of s

and

$$\text{Re}\,[Z(s)] \geq 0 \qquad \textit{for } \text{Re}\,[s] \geq 0 \tag{42}$$

A function having these properties is called a *positive real function* (abbreviated *p.r. function*). All the detailed properties of driving-point and transfer impedances can be derived from this basic result. Some of the more important of these properties are considered in the next three articles of this chapter.

4. Loci of Constant Real and Imaginary Parts near a Pole or Zero of Finite Order

Preliminary to studying some of the detailed properties of impedances resulting from the basic p.r. property stated in 42, it is necessary to recognize the behavior of a rational function in the immediate vicinity of one of its poles or zeros. This we shall now discuss by evaluating loci of constant real and imaginary parts in the vicinity of a pole or zero of finite order.

A pole of order n at the point $s = s_\nu$ is given by the function

$$z(s) = \frac{k}{(s - s_\nu)^n} \tag{43}$$

In order to make a study in the vicinity of s_ν simpler, we introduce the substitution

$$s - s_\nu = \rho e^{j\theta} \tag{44}$$

The polar plot of ρ versus θ as θ varies from zero to 2π is a locus enclosing the point $s = s_\nu$ if ρ remains nonzero; it passes through the point $s = s_\nu$ if and when ρ passes through zero.

Recognizing that the coefficient k in 43 is in general complex, we write

$$k = me^{j\phi} \tag{45}$$

and then obtain 43 in the form

$$z(s) = r + jx = \frac{m}{\rho^n} e^{-j(n\theta - \phi)} \tag{46}$$

yielding

$$r = \frac{m}{\rho^n} \cos (n\theta - \phi); \quad x = -\frac{m}{\rho^n} \sin (n\theta - \phi) \tag{47}$$

Corresponding loci, according to these expressions, as they appear in the s-plane, are shown in Figs. 1 and 2 for $n = 1$ and $n = 2$ respectively. The origin in each figure corresponds to the point $s = s_\nu$ at which the pole is located. The angle ϕ, as Eq. 45 shows, is the angle of the complex coefficient k; we see that it determines the angular orientation of each family of loci.

For a simple pole, the loci are a set of tangent circles which are small for large values of r or x, and large for small values of these real and imaginary parts. The values $r = 0$ and $x = 0$ correspond to a straight line (degenerate circle of infinite radius) separating the region for which r or x is positive from that in which these functions are negative. Each of these regions, therefore, is a half-plane. If we think of the complex k coefficient (for $n = 1$ this is the residue of the impedance function in the pertinent pole) as a vector emanating from the point $s = s_\nu$, as indicated in these sketches, then, for the real part, the region of positive values is the half-plane toward which k points. The loci for constant imaginary part, it is interesting to note, are the real-part loci rotated through 90° in the clockwise direction.

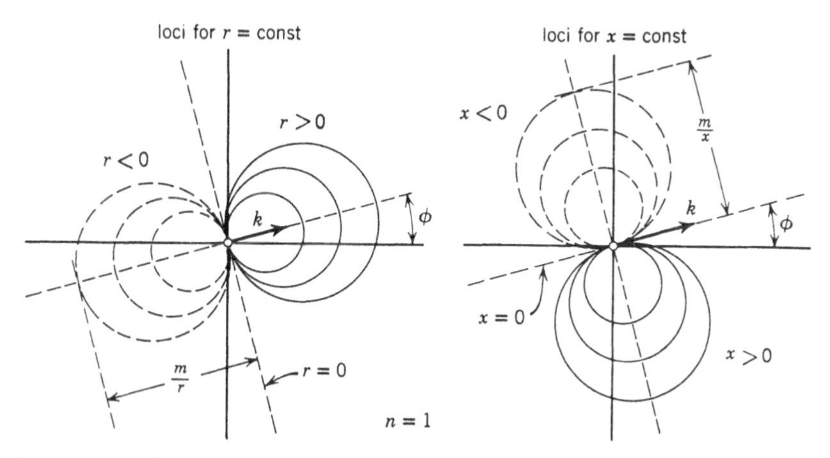

FIG. 1. Loci of constant real and imaginary parts of an impedance in the vicinity of a simple pole.

For a pole of order 2, the loci are the lemniscates shown in Fig. 2. Here the pattern is twice as repetitive; the regions of positive and negative values lie in angular segments 45° apart. The straight lines (degenerate loci for $r = 0$ and $x = 0$) are also twice as plentiful.

The manner in which these patterns change as the pole order becomes larger is more or less clear and need not be further elaborated.

Returning to the plot for the constant real part in the vicinity of a simple pole (as shown in Fig. 1), it is interesting to liken these loci to paths of constant altitude (niveau lines) in a mountainous terrain. As we imagine ourselves walking along the line $r = 0$ from the bottom toward the top of this figure, the terrain is constantly sloping upward from left to right. At first, when we are far from the point $s = s_\nu$, this upward slope is very gentle, but it becomes steeper as we approach the point $s = s_\nu$. That is to say, we find ourselves traveling along a hillside at constant altitude, and as we get nearer to the point $s = s_\nu$ this hillside becomes steeper and steeper.

At the point s_ν it is infinitely steep; the hillside becomes a vertical wall on our right, while on our left lies a chasm of infinite depth. Once past

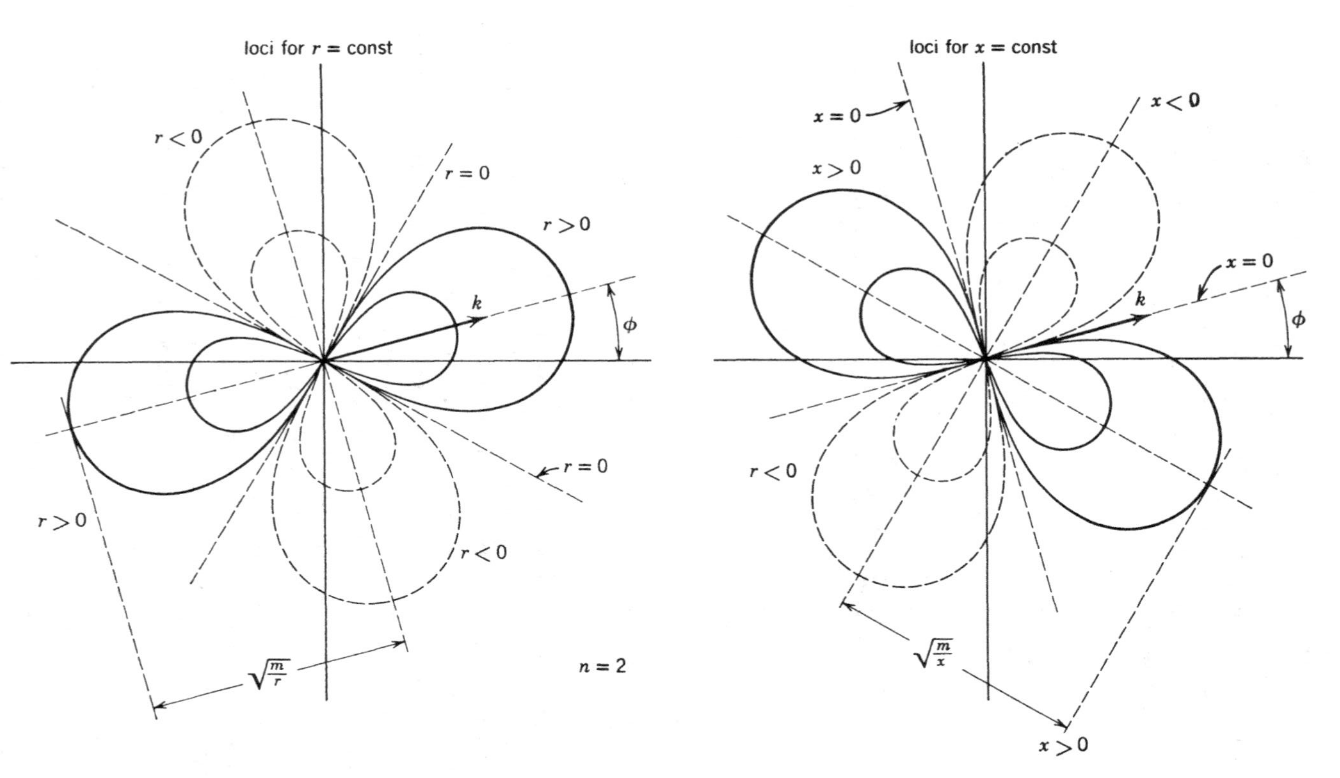

FIG. 2. Loci of constant real and imaginary parts of an impedance in the vicinity of a second-order pole.

this precarious point the terrain becomes gradually less steep again until we find ourselves traveling at constant altitude through a gently sloping meadow as we did at the beginning of our journey.

If, instead of walking along the niveau line $r = 0$, we choose a straight-line path parallel to this one, at a fixed distance α to the right of it, we will be making an ascent as we approach the point $s = s_\nu$, and a descent beyond it, the highest point on this itinerary occurring where we are

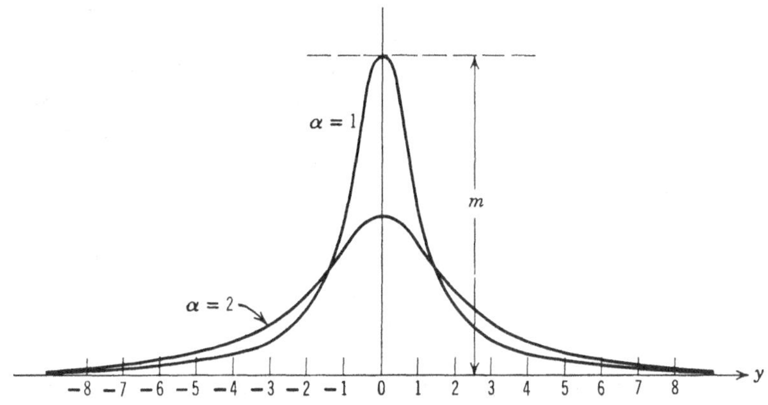

FIG. 3. Real-part variation along a line perpendicular to the residue k of a simple pole for two distances, $\alpha = 1$ and $\alpha = 2$, within the positive region.

directly opposite the pole position. A plot of altitude (value of r) versus distance along this path is shown in Fig. 3 for two values of α. The distance coordinate is denoted by y and is measured in units of α with an origin directly opposite the pole position. For large distances from the pole position these curves merge, as one expects that they should.

This situation corresponds to considering the real part of expression 43 along the j-axis of the s-plane for a positive-real value of the residue $k = m$ and a simple left half-plane pole at $s_\nu = -\alpha + j\omega_\nu$. Letting $\omega = \omega_\nu = y$, we then have for the real part of $z(s)$

$$r = \frac{m\alpha}{\alpha^2 + y^2} = \frac{m}{\alpha} \cdot \frac{1}{1 + (y/\alpha)^2} \tag{48}$$

which yields the plots in Fig. 3.

It is significant to note that the net area under the curve r versus $y = \omega - \omega_\nu$ is a constant independent of the value α. We have for this area

$$\int_{-\infty}^{\infty} \frac{m\alpha}{\alpha^2 + y^2}\,dy = m \int_{-\infty}^{\infty} \frac{d(y/\alpha)}{1 + (y/\alpha)^2}$$

$$= m \tan^{-1}\left(\frac{y}{\alpha}\right)\Big]_{-\infty}^{\infty} = m\pi \tag{49}$$

As the pole is allowed to move closer and closer to the j-axis, α becomes smaller and smaller; the curves in Fig. 3 if plotted versus $y = \omega - \omega_\nu$ become taller and slimmer but always enclose the same area given by the value 49. It is clear, therefore, that the function $r(\omega)$ approaches an impulse of value $m\pi$ located at the pole position which, in the limit, is on the j-axis. That is to say, when an impedance function has a simple pole on the j-axis with positive-real residue m, then the j-axis real part of that impedance contains an impulse of value $m\pi$ located at the pole position.

Returning once more to the loci for $r = \text{const}$ in Fig. 1, this result means that the locus labeled $r = 0$ is a zero niveau line everywhere except at the point $s = s_\nu$ where the situation is indeterminate since, as described above, the hillside becomes infinitely steep and hence degenerates into a vertical wall. This indeterminacy or degeneracy is evaluated by the process of moving along a line parallel to $r = 0$ and noting how the terrain changes with distance α from the niveau line $r = 0$, as depicted in Fig. 3. Important is the fact that we move along the niveau line $r = 0$ normal to the residue k at the point $s = s_\nu$ in order to encounter an impulse of value $m\pi$ in the function r at this point. For a simple j-axis pole and a positive-real residue, this niveau line is the j-axis.

For a simple pole anywhere in the s-plane and a complex residue with arbitrary angle, the niveau line $r = 0$ still passes through the pole at right angles to the residue vector k, even though the pertinent impedance function has other poles also; and r as a function of distance along the niveau line (which is no longer necessarily a straight line) contains an impulse of value $m\pi$.

Of minor collateral interest is the fact that the families of curves for $r = \text{const}$ and $x = \text{const}$ in Figs. 1 and 2 are mutually orthogonal. That is to say, if, for any value of n, the family for $x = \text{const}$ is superimposed upon the family for $r = \text{const}$, these two sets of curves intersect at right angles everywhere. This property, which is well known in the theory of conformal mapping,* follows directly from the Cauchy-Riemann equations which the real and imaginary parts of any complex function possessing a unique derivative must fulfill.

If the point $s = s_\nu$ is a zero instead of a pole, then in its immediate vicinity the impedance behaves like the function

$$z(s) = a(s - s_\nu)^n \tag{50}$$

where n is the order of the zero and the coefficient a has some complex value. Again making use of the substitution 44, and letting

$$a = he^{j\phi} \tag{51}$$

we have

$$z(s) = r + jx = h\rho^n e^{j(n\theta+\phi)} \tag{52}$$

* See, for example, *MCA*, pp. 258–259.

yielding
$$r = h\rho^n \cos(n\theta + \phi); \quad x = h\rho^n \sin(n\theta + \phi) \tag{53}$$

The loci for r = const and x = const again are represented by orthogonal families of curves corresponding to positive or negative values of r and x within angular segments. These are shown in Fig. 4 for $n = 1$ and $n = 2$ and an assumed angle $\phi = 0$. The shaded regions correspond to positive values for r. Those for positive x are advanced by 90° and 45°

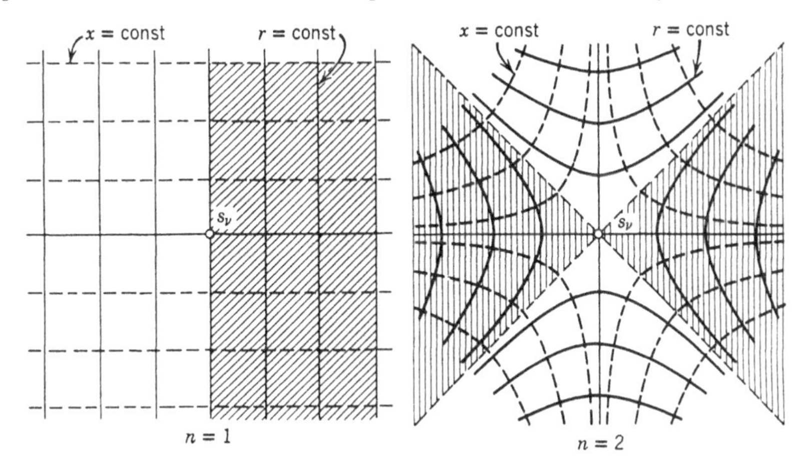

FIG. 4. Loci of constant real and imaginary parts of an impedance in the vicinity of a simple zero ($n = 1$) and in the vicinity of a double-order zero ($n = 2$), also called a saddle point.

respectively. Again the loci for x = const are those for r = const rotated through an angle $\pi/2n$, this time in the counterclockwise direction.

If we regard these loci as niveau lines in a mountainous terrain, the plot for $n = 2$ has the character of a "pass," that is to say, a point where a valley crosses a ridge. We may also describe it by saying that it has the form of a saddle. Thus, if we move along the vertical line through s_ν, we find ourselves in a valley, climbing up to a ridge at s_ν, and then descending into the valley beyond. Along a horizontal line through s_ν, we move along a ridge, descend to the point s_ν which is the mountain pass, and then climb again beyond this point.

The point s_ν is spoken of as a "saddle point;" or, using a hydrodynamic analogy in which one set of loci is regarded as representing the flow lines of a liquid and the orthogonal set as loci of constant velocity, one speaks of s_ν as a "point of stagnation" for obvious physical reasons, namely, the liquid stagnates at the point $s = s_\nu$.

The so-called *order* of the saddle point is the integer $n = 1$ from which it is clear that n must be at least 2 in order that the loci may be given this interpretation. The simple zero for $n = 1$ in Fig. 4 obviously is not a saddle point. The sketch for $n = 2$ in this figure depicts a saddle point of order 1; similar sketches for saddle points of higher order have a correspondingly higher order of angular repetition.

It might be added that while a zero of order 2 or more is a saddle point, the latter is not necessarily a zero of the function whose real and imaginary parts are thus portrayed, because we can add any constant to the real part without affecting the picture (this merely fixes the altitude of the mountain pass) while such an additive constant spoils an otherwise perfect zero.

5. Properties of Driving-Point and Transfer Impedances

Making use of the p.r. property stated in 42, we will now derive some more detailed characteristics of impedances. To begin with, it is useful to observe that if a given function is p.r. then its reciprocal must be p.r. also, and vice versa. Thus if $Z(s) = R + jX$ is p.r., then the corresponding admittance given by

$$Y(s) = \frac{1}{R + jX} = \frac{R - jX}{R^2 + X^2} \tag{54}$$

obviously fulfills the conditions stated in 42. It is real for real s, and its real part is positive wherever R is positive.

From the fact, brought out in the previous article, that real and imaginary parts change their algebraic signs $2n$ times as we traverse around a pole or zero of order n, it follows at once that a p.r. function can have neither poles nor zeros in the right half of the s-plane. Since the driving-point impedance of a passive network is a p.r. function, this result makes good physical sense because the zeros and poles (as shown in art. 2 above) are short-circuit and open-circuit natural frequencies. If their real parts were positive, then the natural or force-free behavior would grow exponentially, which is manifestly impossible in a passive system.

From the real-part loci in Figs. 1 and 2 we can also see that poles may be tolerated on the j-axis, but only if they are simple and if the residues there are real and positive, for terms representing such poles in the partial-fraction expansion of an impedance function will evidently not violate conditions 42.

Polynomials in numerator and denominator of a driving-point impedance are thus restricted by the condition that they may have no right half-plane zeros. These are referred to as *Hurwitz* polynomials. Moreover, any j-axis zeros that they may have must be simple.

This condition, although necessary, is evidently not sufficient since it alone does not guarantee that residues in j-axis poles are positive real, or that the real part of the impedance is non-negative on or to the right of the j-axis. Investigation of the latter condition is aided by the following considerations.

If, in the relation 36, we consider a single driving-point so that $p = 1$, then we get for the pertinent driving-point impedance

$$z_{11}(s) = sT_0 + F_0 + \frac{V_0}{s} \tag{55}$$

in which the quadratic forms (given by Eqs. 29, 30, 31) are tacitly understood to be evaluated for unit current at the driving point. Since the current distribution throughout the network at any frequency is independent of the value of the input current, these forms are simply proportional to the square of the input current which we can normalize at 1 ampere if we wish.

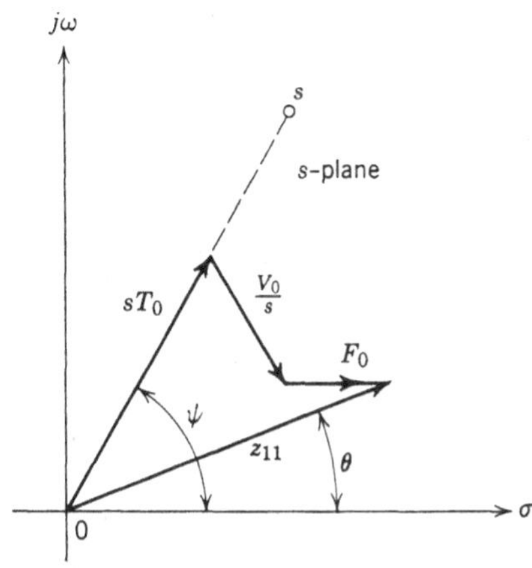

FIG. 5. Illustrating the polar form of the positive-real property.

The positive realness of T_0, F_0 and V_0 for all values of the complex frequency variable s enables us to recognize an interesting property of $z_{11}(s)$ by inspection, as shown in the s-plane sketch of Fig. 5. Here s is any complex frequency in the right half-plane. The vector sT_0 has the same direction as a line from the origin to the point s; the vector V_0/s has the conjugate angle; and F_0 has a zero angle. The vector sum of these three equals $z_{11}(s)$ which clearly has an angle that is smaller in absolute value than the angle of s, even when the latter is negative. Moreover, these angles are equal only where they are identically equal as in the rather trivial cases for which F_0 and V_0 or F_0 and T_0 are absent (purely inductive or purely capacitive networks) or if s is any point on the positive-real axis in which case both angles are zero.

This result, which we may regard as the polar form of the p.r. condition stated in 42, immediately shows that if the real part of $z_{11}(s)$ is non-negative at some j-axis point $s = j\omega$, then it is surely non-negative at the point $s = j\omega + \alpha$ which lies α units directly to the right of the j-axis point. Thus the real part of $z_{11}(s)$ is positive wherever the angle θ in absolute value is $\pi/2$ or less. By hypothesis this is so at a point $s = j\omega$. In moving to the point $s = j\omega + \alpha$, the angle ψ of s decreases in absolute value, and the angle θ must, therefore, also decrease in absolute value by at least as much. Truth of the statement is thus evident.

A tacit assumption is involved here, namely that $z_{11}(s)$ must remain bounded. So far as the right half-plane is concerned, this condition is already considered; if there are any j-axis poles, their contribution to $z_{11}(s)$ can be separated as in a partial-fraction expansion. The above reasoning is then straightforwardly applied to the remainder of $z_{11}(s)$; and as for the terms representing j-axis poles, their simplicity plus the positive realness of pertinent residues satisfies the p.r. condition as already pointed out.

It thus becomes clear that if a rational function, real for real s, is analytic in the right half-plane (has no poles there), has j-axis poles that are simple and yield positive-real residues, and has a j-axis real part that is non-negative exclusive of the immediate vicinity of poles, then it must be p.r. and can be regarded as the driving-point impedance of a physical passive network.

When the network contains active elements, then its driving-point impedances need not be p.r. although their poles are still restricted to the left half-plane if the network is open-circuit stable, and the zeros of any such impedance are similarly restricted if a short-circuit constraint at the pertinent terminal pair leaves the network stable.* The j-axis real part may become negative, and j-axis poles need not be simple or residues real and positive.

For p.r. impedances the simplicity requirement for j-axis poles (and zeros since these are poles of the reciprocal function which must be p.r. also) applied to the points $s = 0$ and $s = \infty$ yields the result that the highest and lowest powers of numerator and denominator polynomials cannot differ by more than unity. This is a condition that is obvious by inspection for any given rational function.

Since Hurwitz polynomials are formed from products of factors $(s + a)$ and $(s^2 + as + b)$ in which all coefficients are real, positive and nonzero, it is obvious that another condition (necessary but not sufficient) on the polynomials is that all coefficients be real, positive and nonzero.†

* Stability here implies that the force-free response, when excited, does not become unbounded.

† An exception to the "nonzero" part of this statement occurs if the polynomial is marginally Hurwitz in the sense that all zeros lie upon the j-axis. In this case it is a product of factors $s^2 + b$ so that all odd-power terms are zero (missing). If additionally a zero at $s = 0$ is present, then s is also a factor and all even-power terms are missing.

This condition can also be seen by inspection for any given rational function. We begin to see that driving-point impedances for passive networks are a rather restricted kind of rational function.

Transfer impedances, on the other hand, are not quite so narrowly defined, as may readily be seen from the expression 41 which, for $p = 2$, reads

$$Z(s) = z_{11}x_1^2 + 2z_{12}x_1x_2 + z_{22}x_2^2 \tag{56}$$

Since this function must be p.r. we see that z_{12}, like z_{11} or z_{22}, can have no right half-plane poles, a result that agrees with conclusions reached in the previous chapter. However, the zeros of z_{12} are not so restricted. They may lie anywhere in the complex s-plane and have any finite multiplicity, so long as the total expression 56 remains p.r. for all real x-values.

Although j-axis poles in z_{12} must be simple and their residues real in order that these conditions may be fulfilled for $Z(s)$, the residues of z_{12} in such poles need not be positive. Thus a residue of $Z(s)$ is given in terms of residues of z_{11}, z_{22}, z_{12} by the expression

$$k = k_{11}x_1^2 + 2k_{12}x_1x_2 + k_{22}x_2^2 \tag{57}$$

which must be a positive-definite quadratic form in the variables x_1, x_2. Hence, in addition to $k_{11} \geq 0$ and $k_{22} \geq 0$, the determinant of the residue matrix

$$\begin{bmatrix} k_{11} & k_{12} \\ k_{21} & k_{22} \end{bmatrix} \tag{58}$$

must be positive. This requirement yields (for simple j-axis poles) the so-called *residue condition*

$$k_{11}k_{22} - k_{12}^2 \geq 0 \tag{59}$$

showing that residues in z_{12} may be negative.

When both driving-point impedances z_{11} and z_{12} are fixed, the requirements upon z_{12} may still be rather restrictive in spite of these more lenient conditions. On the other hand, if these driving-point impedances can be freely chosen (except that they be p.r., of course) then the only restrictions upon z_{12} are: (*a*) its poles must lie in the left half-plane or on the j-axis; (*b*) j-axis poles must be simple and have real residues.

These conditions allow the degree of numerator polynomial in z_{12} to be anything from zero to one more than the degree of the denominator polynomial. Any higher degree would give z_{12} a multiple-order pole at $s = \infty$, which is a j-axis point. A zero in z_{12} at $s = \infty$, on the other hand, can have any multiplicity. For example, if the numerator in z_{12} is a constant, the order of this zero equals the degree of the denominator.

Since a transfer impedance z_{12} is not required to be a p.r. function, its real part need not be positive in the right half of the s-plane or on the j-axis.

If the network has embedded active devices, the restrictions upon z_{12} are the same as for passive networks so long as stability is required, as is usually the case. If the active device is nonbilateral, then z_{12} and z_{21} are not necessarily the same, but this fact has nothing to do with restrictions upon either of these functions. Stable active networks offer no wider possibilities for transfer impedances (or other transfer functions) than do passive bilateral networks except that a larger constant multiplier (yielding signal amplification) is possible when active devices are involved.

As pointed out in Ch. IV, however, an important practical advantage of networks with embedded active devices is the fact that perfectly general-transfer-function behavior can be had even though only two-element-kind networks (like *RC* networks, for example) are employed.

6. Two-Element-Kind Driving-Point Impedances; the Hurwitz Test

When networks are composed of only passive bilateral elements, and these are restricted to two kinds like L and C, R and C, or R and L, then the associated impedance functions assume more specialized forms that we will briefly describe since they are of practical importance in the design of filters and other networks widely used in communication circuits.

For *LC* networks which are lossless, the driving-point impedance is called a *reactance function*. Force-free behavior on open circuit or short circuit is undamped; hence all zeros and poles are simple and lie on the j-axis; residues are real and positive. The discussion of Fig. 1 shows that for $s = j\omega$ the real part of a reactance function is zero except for impulses at the pole positions.

Using the mountainous-terrain analogy, the j-axis behavior of the imaginary part may also be seen by inspection of the pertinent loci in Fig. 1. The j-axis here corresponds to the dotted line at right angles to the locus for $x = 0$ (normal to k). As we proceed along this line from the bottom to the top of the figure, the terrain rises continuously, reaching infinite height at the pole. Here there is an abrupt drop to minus infinity followed by a continuous rise toward elevation again.

Any reactance function consists simply of a linear superposition of single-pole functions like the one depicted in Fig. 1, and hence the j-axis imaginary part is seen to be continuously increasing everywhere in the positive direction along this axis except at pole positions where the function is discontinuous and drops from $+\infty$ to $-\infty$. It is clear, therefore, that zeros and poles must alternate along the j-axis.

A reactance function may be described by saying that it has simple zeros and poles alternating along the j-axis with symmetry about the origin.

This symmetry, which results because polynomials with real coefficients must have zeros in conjugate pairs, requires that the origin be either a zero or a pole (never a finite nonzero value) and that the same be true of the point at $s = \infty$. Alternately, a reactance function may be said to have only simple poles symmetrically located about $s = 0$, with positive real residues, since the simplicity of zeros and their alternation with the poles follows from this requirement.

These properties of a reactance function may also be seen to follow directly from Eqs. 40 and 55 which, for $F_0 \equiv 0$, yield the following expression for the real part of $z_{11}(s)$:

$$r_{11}(\sigma, \omega) = \sigma T_0 + \frac{\sigma V_0}{\sigma^2 + \omega^2} \tag{60}$$

This function is positive in the right half-plane, negative in the left half-plane, and zero on the j-axis. The real-part loci in Figs. 1 and 2 tell us by inspection that this real-part property can be had only if the pertinent impedance function has exclusively simple poles on the j-axis with positive real residues.

It is clear from this description of the reactance function that it just barely meets the p.r. requirement with nothing to spare. It is a *marginally* p.r. function. Anything less in the way of its real-part behavior would be unacceptable.

The fact that the real part is identically zero on the j-axis (except for a finite number of impulses whose existence can be recognized only through an appropriate-limit process) requires $z_{11}(s)$ to be an odd rational function; that is to say, it must be the ratio of two polynomials of which one is odd (contains only odd powers of s) and the other is even (contains only even powers of s). If a polynomial $P(s)$ is separated into its even and odd parts by writing

$$P(s) = m(s) + n(s) \tag{61}$$

in which $m(s)$ and $n(s)$ are respectively even and odd, then a reactance function has the form

$$z_{11}(s) = \frac{m(s)}{n(s)} \tag{62}$$

The zeros of $m(s)$ and $n(s)$ must, of course, be simple and they must separate each other on the j-axis.

It is interesting and useful to recognize that if $z_{11}(s)$ in Eq. 62 is a reactance function then $P(s)$ in Eq. 61 is a Hurwitz polynomial, and vice versa. We can show this relationship in the following way. Consider the rational function

$$Z(s) = \frac{m(s)}{P(s)} = \frac{1}{1 + (n/m)} \tag{63}$$

which may be regarded physically as the impedance of a network formed by placing 1 ohm in parallel with the LC circuit realizing $z_{11}(s)$. The j-axis real part of this impedance is given by

$$\text{Re}\,[Z(j\omega)] = \left[\frac{m^2}{m^2 - n^2}\right]_{s=j\omega} \tag{64}$$

as may readily be seen, noting that, for $s = j\omega$, $m - n$ is the conjugate of $m + n$.

The denominator in expression 64 is the square of an absolute value and hence positive real; the numerator likewise is positive real since $m(j\omega)$ is real and its square is positive. Hence the j-axis real part of the impedance $Z(s)$ in Eq. 63 is non-negative for *any* polynomial $P(s)$, whether Hurwitz or not. However, $Z(s)$ is p.r. only if $P(s)$ *is* Hurwitz. We can say that the Hurwitz character of $P(s)$ is a *necessary and sufficient* condition for $Z(s)$ to be p.r., and conversely that the p.r. character of $Z(s)$ is a necessary and sufficient condition for $P(s)$ to be Hurwitz.

Now $Z(s)$ in Eq. 63 is not p.r. unless its reciprocal $1 + (n/m)$ is p.r. This rational function has poles where $m = 0$; and since m is even, its zeros in general are distributed in the s-plane with symmetry about both real and imaginary axes (called *quadrantal* symmetry). A special form of quadrantal symmetry places all of the zeros on the j-axis, and obviously this form alone is admissible if $1 + (n/m)$ is to be p.r. Moreover, all these j-axis zeros must be simple so that the poles of $1 + (n/m)$ will be simple; and the residues of this function in these poles, which are the residues of n/m, must be real and positive. This is just the condition for which n/m or m/n is a reactance (or a susceptance) function. Hence the above statement, which is commonly used as a basis for testing a given polynomial for Hurwitz character, is proved.

Consider next the impedance of an RC network for which Eq. 55 takes the form

$$z_{11}(s) = F_0 + \frac{V_0}{s} \tag{65}$$

The imaginary part of this complex function is given by

$$x(\sigma, \omega) = \frac{-\omega V_0}{\sigma^2 + \omega^2} \tag{66}$$

which is negative everywhere in the upper half of the s-plane, positive in the lower half of this plane, and zero on the real axis. This one-signed character in either half-plane makes it immediately clear, from loci like those in Figs. 1, 2 and 4, that zeros and poles are restricted to lie upon the real axis; and, of course, this must be the negative real axis since $z_{11}(s)$

must meet p.r. conditions. Moreover, the detailed character of the plots in Figs. 1 and 2 shows that only simple poles are permitted and that the residues must be real and positive.

Along the real axis this $z_{11}(s)$-function (like any p.r. function) is real; and if we want to recognize its behavior along this axis we have but to look at the real-part loci in Fig. 1 rotated so that k is real and positive. As we proceed along the real axis in the positive direction (left to right) $r(\sigma, 0) = z_{11}(\sigma)$ is continuously decreasing, reaching $-\infty$ at the pole position where it jumps to $+\infty$ and then decreases again. This behavior is like that of a reactance function along the j-axis except that the slope is negative instead of positive. We see, therefore, that simple zeros must alternate with simple poles on the negative real axis.

Since an RL admittance is the dual of an RC impedance, it follows at once that all that has been said about an RC impedance, applies unaltered to an RL admittance; and, therefore, we see that an RL impedance also has all simple zeros and poles restricted to the negative real axis. What then is the difference between an RL and an RC impedance?

If we go back to expression 55 we have for the RL impedance

$$z_{11}(s) = F_0 + sT_0 \tag{67}$$

and for its imaginary part

$$x(\sigma, \omega) = \omega T_0 \tag{68}$$

which is positive in the upper half-plane, negative in the lower one, and zero on the real axis. A glance at Fig. 1 tells us that now all residues must be real and *negative*.

Since the origin of the s-plane is situated upon both real and imaginary axes, and j-axis poles must in any p.r. function have positive residues, we see that an RC impedance can have a pole at $s = 0$, but an RL impedance cannot. Whereas the smallest critical frequency (pole or zero) in an RC impedance is always a pole, in an RL impedance it is always a zero because the value of this smallest critical frequency may be *arbitrarily* small. This distinction between RL and RC impedances is also appreciated from simple-minded physical reasoning since the impedance of a capacitance at zero frequency is infinite while that of an inductance is zero.

7. Construction of Impedance Functions by Inspection

In this article we wish to point out that the interpretation of zeros and poles of a driving-point impedance as being natural frequencies for open- and short-circuit constraints respectively enables one in simple cases to write down an appropriate analytic expression for the impedance by inspection of a given network schematic. For transfer impedances the zeros have a different interpretation, although that for the poles is the same since the

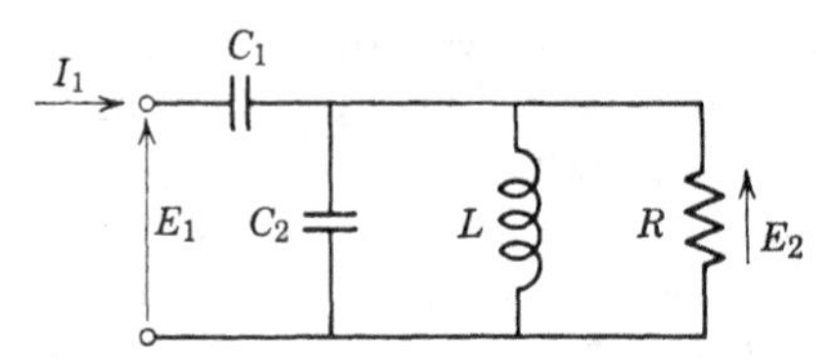

FIG. 6. Example illustrating the construction of a driving-point impedance by inspection.

poles themselves are the same. The zeros of a transfer impedance (also called transmission zeros) are frequencies at which the output is isolated from the input. In simple circuits, these frequencies can also be recognized by inspection, and we can then determine the appropriate function with much less algebraic manipulation than is involved in conventional procedures. These ideas are elaborated in the following paragraphs.

Consider the simple circuit in Fig. 6 and suppose that we wish to obtain the driving-point and transfer impedances $z_{11}(s) = E_1/I_1$ and $z_{12}(s) = E_2/I_1$. We will denote zeros with odd-integer subscripts and poles with even integers. On open circuit this network has the natural frequencies

$$s_{2,4} = -\alpha_2 \pm j\omega_2 \tag{69}$$

with

$$\alpha_2 = \frac{1}{2RC_2}, \quad \omega_2 = \sqrt{\omega_{20}^2 - \alpha_2^2}, \quad \omega_{20}^2 = \frac{1}{LC_2} \tag{70}$$

while on short circuit the capacitance C_1 is in parallel with C_2 and we have the frequencies

$$s_{1,3} = -\alpha_1 \pm j\omega_1 \tag{71}$$

with

$$\alpha_1 = \frac{1}{2R(C_1 + C_2)}, \quad \omega_1 = \sqrt{\omega_{10}^2 - \alpha_1^2}, \quad \omega_{10}^2 = \frac{1}{L(C_1 + C_2)} \tag{72}$$

These results are easily seen by inspection from our familiarity with parallel RLC circuits. The frequencies 69 are poles of $z_{11}(s)$ and those of 71 are zeros.

Since the points $s = 0$ and $s = \infty$ may also be critical frequencies, it is useful to sketch the degenerate forms of the circuit pertinent to the

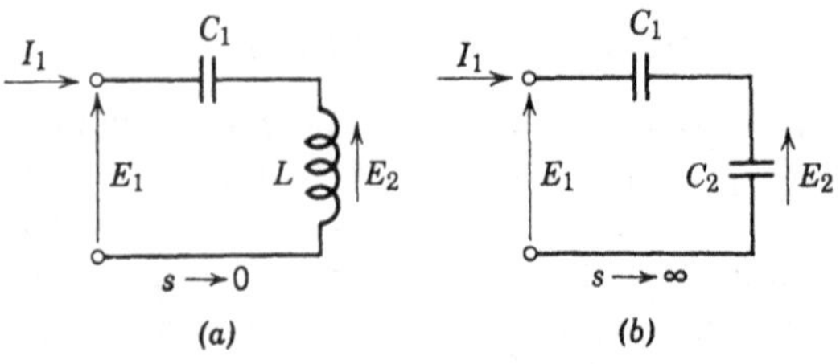

FIG. 7. Degenerate forms of the circuit of Fig. 6 for $s = 0$ and for $s = \infty$.

vicinities of these points as is done in parts (*a*) and (*b*) of Fig. 7. Here part (*a*) is appropriate to very small and part (*b*) to very large frequencies. We see by inspection that

$$[z_{11}(s)]_{s\to 0} \to \frac{1}{C_1 s} \tag{73}$$

and that

$$[z_{11}(s)]_{s\to\infty} \to \frac{C_1 + C_2}{C_1 C_2 s} \tag{74}$$

Equation 73 shows that the impedance has a pole at $s = 0$, and 74 shows that it has a zero at $s = \infty$. The pole at $s = 0$ introduces a factor s in the denominator of the expression for $z_{11}(s)$, while the zero at infinity means that the numerator polynomial is one degree lower than the denominator polynomial.

These results enable us to write

$$z_{11}(s) = \frac{H(s - s_1)(s - s_3)}{s(s - s_2)(s - s_4)} \tag{75}$$

Together with Eqs. 69 through 72 this form yields

$$[z_{11}(s)]_{s\to 0} \to \frac{Hs_1s_3}{ss_2s_4} = \frac{H\omega_{10}^2}{s\omega_{20}^2} = \frac{HC_2}{(C_1 + C_2)s} \tag{76}$$

and

$$[z_{11}(s)]_{s\to\infty} \to \frac{H}{s} \tag{77}$$

Comparison of Eq. 73 with 76 or of 74 with 77 gives the value of the constant multiplier

$$H = \frac{C_1 + C_2}{C_1 C_2} \tag{78}$$

The constant multiplier can always be evaluated by noting the asymptotic behavior of the impedance at either $s = 0$ or $s = \infty$. By considering both, we obtain a useful check on the result.

Regarding the transfer impedance $z_{12}(s)$ we observe first that it also has the poles given by Eqs. 69 and 70 because its poles are the same open-circuit natural frequencies. Its zeros are frequencies for which $E_2 = 0$ in spite of a nonzero I_1. By inspection of Fig. 6 we see that this condition occurs only at $s = 0$ and at $s = \infty$. More specifically we see from the degenerate circuits of Fig. 7 that

$$[z_{12}(s)]_{s\to 0} \to Ls \tag{79}$$

and

$$[z_{12}(s)]_{s\to\infty} \to \frac{1}{C_2 s} \tag{80}$$

The numerator of this impedance, which must be one degree lower than the denominator, contains only the factor s. The transfer impedance is, therefore, seen to have the form

$$z_{12}(s) = \frac{Hs}{(s - s_2)(s - s_4)} \tag{81}$$

from which we get

$$[z_{12}(s)]_{s\to 0} \to \frac{Hs}{s_2 s_4} = HLC_2 s \tag{82}$$

and

$$[z_{12}(s)]_{s\to\infty} \to \frac{H}{s} \tag{83}$$

Comparison of 79 with 82 or of 80 with 83 evaluates and checks the constant multiplier

$$H = \frac{1}{C_2} \tag{84}$$

Let us consider next the problem of finding the transfer impedance $z_{12}(s) = E_2/I_1$ for the circuit of Fig. 8 which is a somewhat generalized

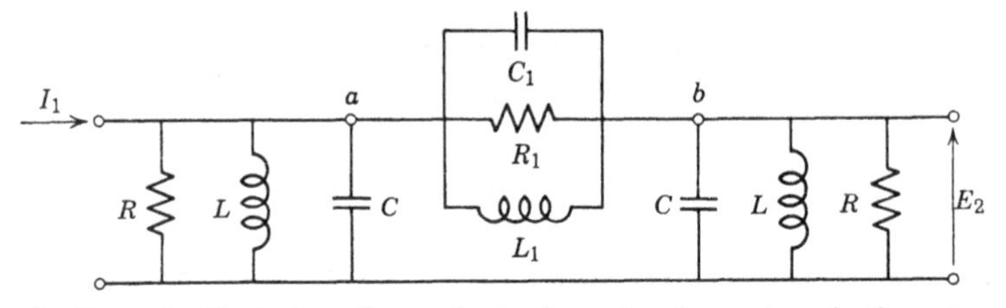

FIG. 8. Example illustrating the method of constructing a transfer impedance by inspection.

version of the familiar double-tuned circuit. Poles of the input impedance, which are also poles of $z_{12}(s)$, evidently occur where the parallel RLC combination at either end is an infinite impedance. These frequencies are

$$s_{2,4} = -\alpha_2 \pm j\omega_2 \tag{85}$$

with

$$\alpha_2 = \frac{1}{2RC}, \quad \omega_2 = \sqrt{\omega_{20}^2 - \alpha_2^2}, \quad \omega_{20}^2 = \frac{1}{LC} \tag{86}$$

A second pair of open-circuit natural frequencies is recognized by noting that the two identical RLC circuits in series are in parallel with the $R_1L_1C_1$ circuit across terminals a–b, thus yielding a resultant parallel combination of $R_0L_0C_0$ with

$$R_0 = \frac{2RR_1}{2R + R_1}, \quad L_0 = \frac{2LL_1}{2L + L_1}, \quad C_0 = \frac{C}{2} + C_1 \tag{87}$$

The second pair of poles in the function $z_{12}(s)$, therefore, are

$$s_{6,8} = -\alpha_6 \pm j\omega_6 \tag{88}$$

with

$$\alpha_6 = \frac{1}{2R_0C_0}, \quad \omega_6 = \sqrt{\omega_{60}^2 - \alpha_6^2}, \quad \omega_{60}^2 = \frac{1}{L_0C_0} \tag{89}$$

As for zeros of the transfer impedance, we see that one pair obviously corresponds to the natural frequencies of the parallel $R_1L_1C_1$ combination, for when it is an infinite impedance, E_2 is zero. Hence we have

$$s_{1,3} = -\alpha_1 \pm j\omega_1 \tag{90}$$

with

$$\alpha_1 = \frac{1}{2R_1C_1}, \quad \omega_1 = \sqrt{\omega_{10}^2 - \alpha_1^2}, \quad \omega_{10}^2 = \frac{1}{L_1C_1} \tag{91}$$

To determine the remaining zeros of the transfer impedance we consider the degenerate forms of this circuit appropriate to $s \to 0$ and $s \to \infty$ as

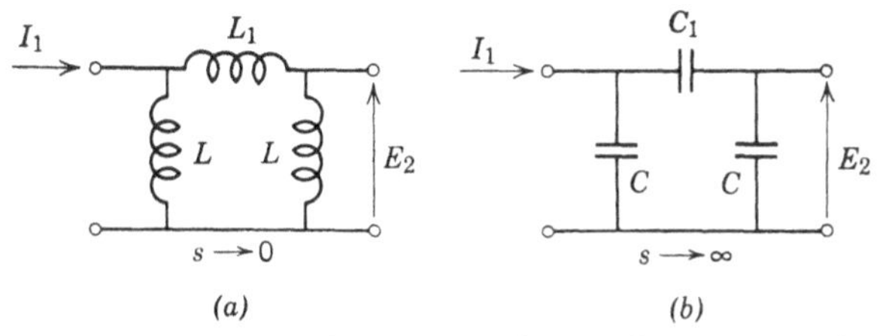

FIG. 9. Degenerate forms of the circuit of Fig. 8 for $s = 0$ and for $s = \infty$.

shown respectively in parts (*a*) and (*b*) of Fig. 9. By inspection we see that

$$[z_{12}(s)]_{s\to 0} \to \frac{L^2s}{2L + L_1} \tag{92}$$

and

$$[z_{12}(s)]_{s\to\infty} \to \frac{C_1}{(2C_1 + C)Cs} \tag{93}$$

Hence there is one transmission zero at $s = 0$ and one at $s = \infty$. Our transfer impedance, therefore, has the form

$$z_{12}(s) = \frac{Hs(s - s_1)(s - s_3)}{(s - s_2)(s - s_4)(s - s_6)(s - s_8)} \tag{94}$$

in which the finite nonzero critical frequencies are given in Eqs. 85, 86; 88, 89; 90, 91.

From the asymptotic relations

$$[z_{12}(s)]_{s\to 0} \to \frac{Hss_1s_3}{s_2s_4s_6s_8} = \frac{Hs\omega_{10}^2}{\omega_{20}^2\omega_{60}^2} = \frac{HsL^2C(2C_1 + C)}{C_1(2L + L_1)} \tag{95}$$

and

$$[z_{12}(s)]_{s\to\infty} \to \frac{H}{s} \tag{96}$$

we get through comparison with 92 and 93

$$H = \frac{C_1}{C(2C_1 + C)} \tag{97}$$

Without writing a single circuit equation we have thus evaluated the transfer impedance for this pair of identically tuned RLC circuits coupled

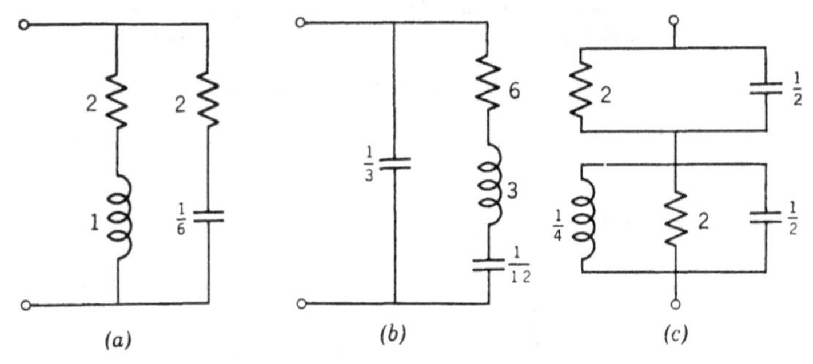

FIG. 10. Examples illustrating the method of constructing driving-point impedances by inspection.

in a very general manner. If $R_1 = \infty$ and $C_1 = 0$, the circuits are inductively coupled; if $R_1 = \infty$ and $L_1 = \infty$, they are capacitively coupled; and if $L_1 = \infty$ and $C_1 = 0$, the coupling is resistive. The usual situation of interest is one for which the coupling is slight; $C_1 \ll C$ for the capacitive case and $L_1 \gg L$ for the inductive. In terms of the pertinent pole-zero constellation in the s-plane and its dependence upon circuit parameters as given by Eqs. 85 through 93, one may with little effort study a large variety of special cases.

Let us now consider the numerical examples shown in Fig. 10, for which analytic expressions for the driving-point impedances are desired. Element values are in henrys, farads and ohms.

For circuit (a) the poles or open-circuit natural frequencies are those of a series RLC circuit with $R = 4$, $L = 1$, $C = \frac{1}{6}$, yielding the quadratic factor $(s^2 + 4s + 6)$. On short circuit we have an RL and an RC circuit with $R/L = 2$ and $1/RC = 3$ respectively. Hence we have the zero factors $(s + 2)$ and $(s + 3)$. For $s = \infty$ the impedance equals 2 and for

$s = 0$ its value is also 2. These results we see by inspection since at $s = \infty$ the inductance is an open circuit and the capacitance is a short circuit, while at $s = 0$ the reverse is true. Circuit (*a*), therefore, has the impedance

$$Z_a(s) = \frac{2(s + 2)(s + 3)}{(s^2 + 4s + 6)} \tag{98}$$

In circuit (*b*) we have a series *RLC* loop for both open- and short-circuit constraints, the two situations differing only in the value of the capacitance involved. Thus the poles are given by the factor $s^2 + 2s + 5$ and the zeros by $s^2 + 2s + 4$. The shunt capacitance produces a zero at $s = \infty$ while at $s = 0$ we see that the impedance has a pole. More specifically we have

$$[Z_b(s)]_{s \to 0} \to \frac{12}{5s} \tag{99}$$

and

$$[Z_b(s)]_{s \to \infty} \to \frac{3}{s} \tag{100}$$

It thus becomes clear that this impedance is given by the expression

$$Z_b(s) = \frac{3(s^2 + 2s + 4)}{s(s^2 + 2s + 5)} \tag{101}$$

Lastly we consider circuit (*c*) in Fig. 10. The finite nonzero poles are evidently the natural frequencies of the parallel *RC* and *RLC* circuits considered separately. These introduce the denominator factors $s + 1$ and $s^2 + s + 8$. On short circuit this network degenerates into a single parallel *RLC* combination with $R = 1$, $L = \frac{1}{4}$, $C = 1$, giving the quadratic factor $s^2 + s + 4$. The asymptotic behavior shows by inspection that

$$[Z_c(s)]_{s \to 0} \to 2 \tag{102}$$

and

$$[Z_c(s)]_{s \to \infty} \to \frac{4}{s} \tag{103}$$

It is clear from these results that

$$Z_c(s) = \frac{4(s^2 + s + 4)}{(s + 1)(s^2 + s + 8)} \tag{104}$$

When circuits are more elaborate it is no longer possible to write down the natural frequencies by inspection. After all, these are in general given by the roots of an algebraic equation of higher degree (the characteristic or determinantal equation). On the other hand, the method discussed in art. 4 of Ch. VII enables one in any case to determine the number of zero factors $(s - s_1)$, $(s - s_3)$, . . . as well as the number of pole factors

$(s - s_2), (s - s_4), \ldots$ involved in the numerator and denominator respectively; and the asymptotic behavior can always be recognized by inspection, thus determining the constant multiplier and any factor s demanded by the required behavior at $s = 0$ or at $s = \infty$. Apart from the actual numerical values of the finite nonzero poles and zeros, an expression for the driving-point impedance of any given network can thus be written down by an inspection procedure.

Problems

1. Given the impedance function

$$Z(s) = \frac{P(s)}{Q(s)} = \frac{a_0 + a_1 s + a_2 s^2 + \cdots + a_m s^m}{b_0 + b_1 s + b_2 s^2 + \cdots + b_n s^n}; \qquad a_k > 0, \quad b_k > 0$$

(*a*) Write down the differential equation expressing equilibrium of the pertinent system.
(*b*) State limitations on the integers m and n if (i) $Z(s)$ is a driving-point function, (ii) it is a transfer function.
(*c*) May either the terminal voltage or current be excitation or response?

2. The equation of equilibrium of a certain network relative to stated points of access reads

$$\frac{d^5 i}{dt^5} + 2\frac{d^4 i}{dt^4} + 3\frac{d^3 i}{dt^3} + 2\frac{d^2 i}{dt^2} + \frac{di}{dt} + 6i = \frac{d^3 e}{dt^3} + 2\frac{d^2 e}{dt^2} + 4\frac{de}{dt} + 3e$$

(*a*) Do e and i refer to the same terminal pair? Why or why not?
(*b*) Which variable is the excitation and which the response?
(*c*) Suppose the third derivative term for i has a negative coefficient. Is this acceptable? Might one of the e-terms be negative? Explain.

3. If for a passive bilateral network we write

$$z_{ij}(j\omega) = r_{ij}(\omega) + jx_{ij}(\omega)$$

Show that the quadratic form in the real variables x_k

$$R(\omega) = \sum_{i,j=1}^{p} r_{ij}(\omega) x_i x_j$$

must be positive or semidefinite. For two terminal pairs ($p = 2$) this result yields the j-axis real-part condition

$$r_{11} r_{22} - r_{12}^2 \geq 0$$

from which it is clear that a transfer impedance need not have a non-negative j-axis real part.

4. For RC or RL networks, show that a residue condition like the one expressed by Eq. 59 must be fulfilled at all poles on the negative real axis.

5. From the real-part loci in Fig. 1, show that the j-axis real part of an RC impedance must be a continuously decreasing function of ω, and for an RL impedance it must be continuously increasing. Thus observe that it is sufficient for an RC impedance to show that it is non-negative for $s = \infty$, and for an RL impedance that it is non-negative for $s = 0$ in order to assure the non-negative j-axis real-part condition for such driving-point functions. Discuss in detail.

6. Show that two terminal-pair RC realizability conditions are fulfilled if impedances z_{11} and z_{22} have only simple poles with positive real residues on the negative real axis and z_{12} fulfills a residue condition in the form of Eq. 59. In

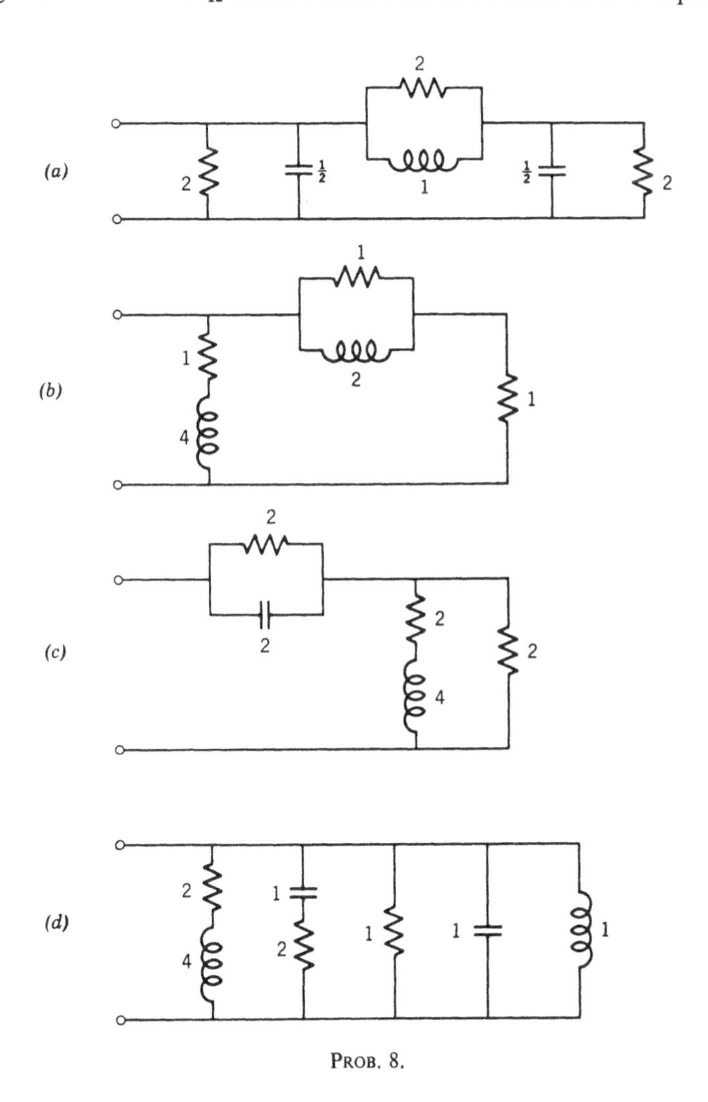

PROB. 8.

particular, show that the j-axis real-part condition stated in prob. 3 is in this case automatically fulfilled and hence need not separately be investigated.

7. Which of the following are realizable passive bilateral driving-point impedances? Why or why not?

$$(a)\ Z(s) = \frac{s+2}{(s+1)(s^2+2s+2)} \qquad (b)\ Z(s) = \left(\frac{s+1}{s+2}\right)^4$$

$$(c)\ Z(s) = \frac{(s+1)(s+3)}{(s+2)(s+4)} \qquad (d)\ Z(s) = \frac{(s+1)(s+3)^2}{(s+2)(s+4)}$$

$$(e)\ Z(s) = \frac{(s^2+1)(s^2+3)}{(s^2+2)(s^2+4)} \qquad (f)\ Z(s) = \frac{s(s^2+1)(s^2+3)}{(s^2+2)(s^2+4)}$$

$$(g)\ Z(s) = \frac{(s^2+1)(s^2+3)}{s(s^2+2)(s^2+4)}$$

8. For the networks illustrated on page 247 write down the driving-point impedances by inspection. Element values are in henrys, farads, ohms.

CHAPTER IX

Response in the Frequency Domain

Knowing that all lumped network response functions—impedances, admittances, or dimensionless ratios—are rational functions of the complex frequency variable s, is of inestimable value in many ways. Existing mathematical knowledge regarding both the properties of these functions and the graphical or analytic methods for portraying them effectively becomes immediately available for the study of network problems. More importantly, the entire problem of network characterization is placed into one compact package labeled "rational functions." Even the so-called "distributed-parameter" networks like transmission lines or wave guides are essentially included here since their describing functions, some of which are meromorphic or transcendental, can be approximated in the finite s-plane by rational functions so that the behavior patterns which they imply are merely limiting forms of those obtained from rational functions. In this chapter we shall discuss some analytic and graphical methods that are especially effective in portraying the most pertinent properties of these functions.

1. s-Plane Geometry; Residue and Resonance Evaluation

One of the most useful techniques for the characterization of rational functions is the s-plane representation in terms of so-called frequency factors which have the appearance of vectors drawn in this plane. As an example, the function

$$Z(s) = \frac{H(s - s_1)(s - s_3)}{(s - s_2)(s - s_4)(s - s_6)} \tag{1}$$

is characterized in this manner in part (a) of Fig. 1 for a j-axis value of the variable s. The frequency factors $s - s_\nu$ are represented by arrows emanating from the various critical frequencies $s_1, s_2, \ldots$, all terminating upon the variable point s. Zeros here are again denoted by odd-integer subscripts and poles by even integers.

From a plot of this sort one can perceive the general behavior pattern of the function $Z(s)$ regarding both magnitude and angle variations as the point s is allowed to move along some path (for example, the j-axis) in the s-plane. More specifically, the magnitude of $Z(s)$ is influenced only by variations in the lengths of these arrows, while their angles determine the angle of $Z(s)$ which is given simply by the sum of angles of zero factors minus the sum of angles of pole factors. The whole problem of behavior determination is reduced to a problem in geometry which, in many simple cases, can be solved by inspection.

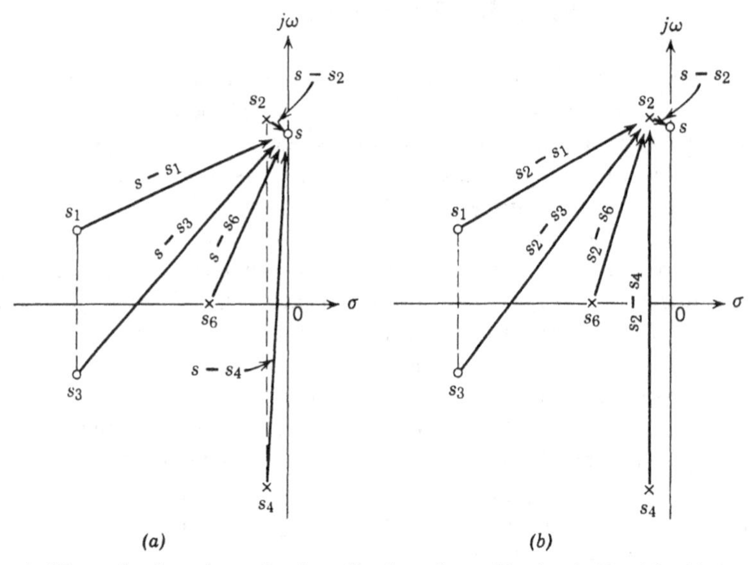

FIG. 1. Illustrating how the evaluation of an impedance like that in Eq. 1 is aided by its geometrical representation in the s-plane.

Another advantage which this geometrical representation has is that it makes the formulation of approximations appropriate to any given situation rather obvious. For example, if we are interested in the behavior of $Z(s)$, Eq. 1, in the vicinity of resonance occurring near the pole s_2, we can see from the geometry in part (a) of Fig. 1 that the behavior of $Z(s)$ is influenced primarily by variation of the factor $s - s_2$, since the remaining factors are more or less constant in length and in angle throughout the frequency range of interest which is confined essentially to the interval in which the angle of $s - s_2$ varies between $-\pi/4$ and $+\pi/4$, these limits being the so-called half-power points.

Considering all factors in $Z(s)$ except $s - s_2$ to be constant (equal to some appropriate average value) for the study of this resonance problem is immediately suggested by the s-plane geometry. We can even make an estimate of the percentage error involved in this approximation; all we need for this purpose is a ruler and a protractor for measuring angles.

If the impedance function $Z(s)$ were given by some messy expression in terms of the R's, $j\omega L$'s and $j\omega C$'s of the pertinent network branches, as it is when formulated according to a rather common approach to circuit analysis, the possibility of an approximate analysis would be completely obscured by algebraic unwieldiness. Moreover, we wouldn't know how to begin formulating an approximate analysis, let alone estimating its accuracy.

Choosing appropriate constant values for all factors except $s - s_2$ in formulating the resonance behavior of $Z(s)$ is done by letting the tips of the corresponding arrows fall upon some fixed point in the resonance vicinity. In Fig. 1, part (*b*) this point is the pole s_2 which appears to be a rather logical choice. The fixed part of the impedance function $Z(s)$ in Eq. 1 is thus given by

$$k = \left[\frac{H(s - s_1)(s - s_3)}{(s - s_4)(s - s_6)}\right]_{s=s_2} \tag{2}$$

and the impedance for the resonance vicinity is represented by the approximate expression

$$Z(s) \approx \frac{k}{s - s_2} \tag{3}$$

At this point it is interesting to observe that the constant k in Eq. 2, which can alternately be expressed by the relation

$$k = [(s - s_2) \cdot Z(s)]_{s=s_2} \tag{4}$$

is the *residue* of the complex function $Z(s)$ in the simple pole at $s = s_2$, as Eq. 17 in Ch. VII shows. The complete partial-fraction expansion of the impedance 1 reads

$$Z(s) = \frac{k^{(2)}}{s - s_2} + \frac{k^{(4)}}{s - s_4} + \frac{k^{(6)}}{s - s_6} \tag{5}$$

which yields the approximate relation 3 for $s \approx s_2$ where the first term dominates. k in Eq. 2 or 4 is the same as $k^{(2)}$ in Eq. 5.

The geometrical picture in Fig. 1, part (*b*), applies, therefore, to the evaluation of residues in the partial-fraction expansion of a rational function. By means of this picture, residue evaluation becomes a geometrical problem. For rough work a ruler and protractor yield quick answers and save tedious computations besides lending a circumspection to the work that helps prevent errors and provides a means for avoiding absurd results due to calculation errors.

2. Minimum-Reactive and Minimum-Phase Functions; Real-Part Sufficiency

Certain classes of driving-point and transfer functions are distinguished according to their distribution of zeros and poles in the complex frequency plane. For example, driving-point impedances and admittances (p.r. functions) are divided into two classes: Those that do *not* have j-axis poles are called *minimum-reactive* or *minimum-susceptive* impedances or admittances respectively; and those having one or more j-axis poles are *non*-minimum-reactive or *non*-minimum-susceptive driving-point functions. Except for impulses at j-axis poles, the j-axis real parts of these two classes of functions for otherwise fixed pole-zero patterns are evidently identical. The non-minimum functions have j-axis imaginary (reactive) parts equal to those of the minimum functions plus additive terms due to j-axis poles. The terminology used here is, therefore, self-explanatory.

A given p.r. impedance may have j-axis poles but no j-axis zeros. In that event it is non-minimum-reactive, but the corresponding admittance *is* minimum-susceptive. Similarly, an admittance may be non-minimum-susceptive while the corresponding impedance is minimum-reactive. If a p.r. function has both zeros and poles on the j-axis then neither the impedance nor the admittance is minimum-reactive or minimum-susceptive.

These distinctions among p.r. functions are important in connection with the rather interesting possibility of constructing such a function when only its j-axis real part is known. To show how this is done, we point out first that any function $Z(s)$ is readily separable into its even and odd parts. If we write this separation in the form

$$Z(s) = \text{Ev}\,(Z) + \text{Odd}\,(Z) \tag{6}$$

then by definition

$$Z(-s) = \text{Ev}\,(Z) - \text{Odd}\,(Z) \tag{7}$$

and hence

$$\text{Ev}\,(Z) = \tfrac{1}{2}[Z(s) + Z(-s)] \tag{8}$$

$$\text{Odd}\,(Z) = \tfrac{1}{2}[Z(s) - Z(-s)] \tag{9}$$

On the j-axis (i.e., for $s = j\omega$) these two parts are respectively real and imaginary. Therefore we see that

$$\text{Re}\,[Z(j\omega)] = \tfrac{1}{2}[Z(s) + Z(-s)]_{s=j\omega} \tag{10}$$

This j-axis real part of $Z(s)$ is an even rational function of ω which may be written in the form

$$\text{Re}\,[Z(j\omega)] = \frac{A_0 + A_1\omega^2 + \cdots + A_n\omega^{2n}}{B_0 + B_1\omega^2 + \cdots + B_n\omega^{2n}} = \frac{A(\omega^2)}{B(\omega^2)} \tag{11}$$

and since $s = j\omega$ implies $s^2 = -\omega^2$, the last two equations allow us to write

$$\frac{A(-s^2)}{B(-s^2)} = \tfrac{1}{2}[Z(s) + Z(-s)] = f(s) \tag{12}$$

Since $Z(s)$ can at most have a simple pole at $s = \infty$, and this pole (if present) must be contained in Odd (Z), the degree of the polynomial $A(-s^2)$ cannot exceed that of $B(-s^2)$. The asymptotic value of $f(s)$ for $s \to \infty$ is either zero or a positive constant.

From this result we recognize that the j-axis real part determines the sum of $Z(s)$ and $Z(-s)$. This sum can readily be separated into its component parts if $Z(s)$ is minimum-reactive. In this case $Z(s)$ has poles only in the left half-plane and $Z(-s)$ has poles only in the right half-plane since they are images of the poles of $Z(s)$ about the point $s = 0$.

If we expand the function $f(s)$ into partial fractions it thus becomes clear that the terms for left half-plane poles belong to $Z(s)$ and those for right half-plane poles to $Z(-s)$, while the asymptotic value of $f(s)$ for $s \to \infty$, if present, is an additive constant assignable to both $Z(s)$ and $Z(-s)$.

Observe that the algebraic function 11 for the j-axis real part derived from the expression in Eq. 10 cannot contain impulses, even if $Z(s)$ has j-axis poles. Therefore, if we start from a non-minimum-reactive impedance, form the function $f(s)$ in Eq. 12, and, from its partial fraction expansion construct a function $Z(s)$ by the procedure just described, this one will be minimum-reactive. It will be the original $Z(s)$ function minus its j-axis poles. Presence of such poles can be recognized only if the pertinent impulses are supplied with a given real-part function 11.

Thus we may say that if a j-axis real part is specified, *inclusive* of any j-axis impulses that it may have, then the pertinent impedance function $Z(s)$ can be constructed whether it is minimum-reactive or not. The above algebraic procedure yields the minimum-reactive portion of $Z(s)$, and from the given impulses and their locations we can write down terms for the j-axis poles since the residues there are related to values of the impulses as shown in Eq. 49 of Ch. VIII.

When completely specified, we recognize therefore that the j-axis real part uniquely determines the total impedance function $Z(s)$. This result, which may seem a little odd at first, loses its mysterious aspect if we recognize from Eq. 12 that what we call the real "part" is actually not part of the pertinent function at all but rather it is the whole function with a certain redundancy in the form of a repetition of all poles in the right half-plane. We form the desired function by removing this redundancy. We do not create a "whole" from a "part."

Observe that uniqueness of the method depends upon the fact that $Z(s)$ has no poles in the right half-plane, wherefore $Z(-s)$ has none in the left half-plane. Thus a clear-cut separation in the poles of $f(s)$ can be made. There is no ambiguity as to which poles belong to $Z(s)$ and which to $Z(-s)$.

It is, however, not necessary that zeros of the pertinent impedance (or admittance) function be restricted to the left half-plane, and hence the same procedure can be applied to transfer functions (impedances, admittances, or dimensionless ratios).

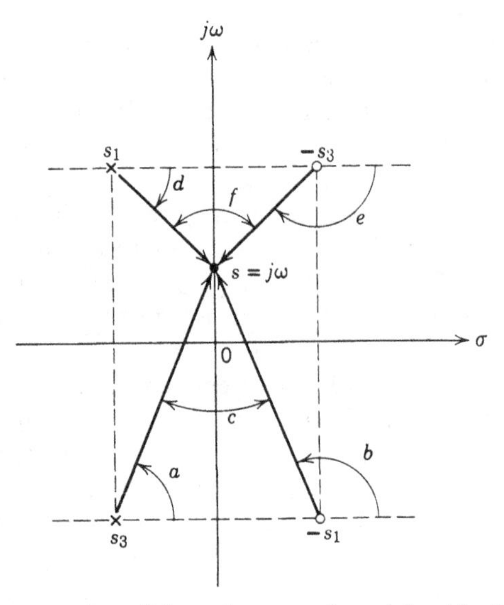

FIG. 2. s-plane representation of the pole-zero pattern defined by 13 which is a basic "all-pass" component.

In connection with transfer functions a distinction is made between those that do and those that do not have right half-plane zeros, since this circumstance has a significant influence upon the resulting phase shift associated with a given j-axis magnitude function. Suppose, for example, that s_1 and s_3 are a pair of conjugate left half-plane zeros in the transfer impedance $z_{12}(s)$ so that its numerator contains the factors $s - s_1$ and $s - s_3$. If these zeros are reflected into the right half-plane, the pertinent factors become $s + s_1$ and $s + s_3$ and the effect upon $z_{12}(s)$ is to be multiplied by the ratio

$$\frac{(s + s_1)(s + s_3)}{(s - s_1)(s - s_3)} \tag{13}$$

which has unit magnitude for all $s = j\omega$.

Hence the j-axis magnitude of $z_{12}(s)$ is unaffected by this shift of zeros from the left to the right half-plane, but the lag angle as a function of ω is significantly increased, as reference to the s-plane sketch in Fig. 2

shows. Here the angles of the factors $s + s_1$ and $s + s_3$ are denoted by b and e respectively, while those of $s - s_1$ and $s - s_3$ are d and a. The angle subtended at the j-axis point s by the upper and lower half-plane factors are indicated by the letters f and c. According to a familiar theorem in geometry, we can write that

$$e = d + f \quad \text{and} \quad b = a + c \tag{14}$$

Therefore, the net lag angle contributed by the upper half-plane factors is

$$-d + e = f \tag{15}$$

and that contributed by the lower half-plane factors is

$$a - b = -c \tag{16}$$

Hence the total lag angle introduced by the ratio 13 is

$$\beta = f - c \tag{17}$$

By inspection we see that this angle is zero for $s = 0$ and increases to the value 2π as s moves up the j-axis toward the point $s = j\infty$. That is to say, by reflecting the original left half-plane zeros from the left to the right half-plane, the j-axis magnitude of the transfer impedance $z_{12}(s)$ remains unchanged but its phase-angle variation is increased by 2π radians. For this reason, transfer functions are designated as *minimum-phase* or *non-minimum-phase* according to whether all zeros are in the left half-plane or some lie in the right half-plane respectively.

3. Potential-Theory Analogies; Logarithmic and Dipole Potentials

In a network-design problem, the desired behavior for a driving-point or transfer function is specified, and a method for construction of an appropriate rational function meeting these specifications as well as realizability conditions is needed. This is the first step in the solution to any network-synthesis problem. The second step is concerned with realization of the rational function by an appropriate network schematic involving ideal elements. Construction of a physical network constitutes the third and final step.

Although the second step can be carried out exactly, the first and third usually involve approximations and compromises. Step one in particular is spoken of as dealing with an approximation problem; here the desired characteristics, which may be given analytically or graphically, are approximated by a rational function which must additionally meet pertinent realizability conditions.

Various techniques are available in this connection, including digital and analog computing devices. The latter are based upon recognition of certain parallelisms between forms in which rational functions can be expressed and those characterizing other physical phenomena. A particularly effective analogy of this sort makes use of properties of the electric or magnetic scalar potential function in a two-dimensional space which is identified with the complex s-plane. We shall discuss briefly some of the features of this interpretation which is helpful primarily because it provides a basis for physical visualization of the problem and thus by intuition often leads to a possible method of solution.

A two-dimensional potential problem with which electrical engineers have some degree of familiarity is that associated with the study of transmission lines where longitudinal uniformity of the physical structure is assumed and interest centers largely in the electric and magnetic field configuration in the cross-sectional plane. The scalar electric-potential function in this plane is two-dimensional and hence logarithmic; that is to say, the potential due to a single filamental charge (longitudinally uniform linear distribution) varies as the logarithm of the reciprocal distance from this filament. At a distance r from the filament, the potential equals $K \ln (1/r)$ where K is a constant proportional to the charge per unit length carried by the filament.

As pointed out in art. 4 of Ch. III, the transmission properties of networks are often expressed in terms of the logarithm of a pertinent transfer impedance or its inverse. Specifically, if we write

$$\gamma = \ln \frac{1}{z_{12}} = \alpha + j\beta \tag{18}$$

then

$$\alpha = \ln \left| \frac{1}{z_{12}} \right| = \ln \frac{\text{magnitude of input}}{\text{magnitude of output}} \tag{19}$$

and

$$\beta = -\measuredangle z_{12} = \text{negative angle of } z_{12} \tag{20}$$

Here α is positive when the output is smaller in absolute value than the input, which is the normal state of affairs in a passive network; and β is positive when the output lags the input, which is also normal. α is called the attenuation function; its units are *nepers*. β is the phase lag or phase function and its units are radians.

If we have a transfer impedance of the form

$$z_{12}(s) = \frac{(s - s_1)(s - s_3)(s - s_5) \cdots}{(s - s_2)(s - s_4)(s - s_6) \cdots} \tag{21}$$

where we have dropped the constant multiplier because it is of no consequence in the present discussion, then the attenuation function has the representation

$$\alpha = \ln \frac{1}{|s - s_1|} + \ln \frac{1}{|s - s_3|} + \ln \frac{1}{|s - s_5|} + \cdots$$
$$- \ln \frac{1}{|s - s_2|} - \ln \frac{1}{|s - s_4|} - \ln \frac{1}{|s - s_6|} - \cdots \quad (22)$$

Here the quantities $|s - s_1|$, $|s - s_2|$, etc. are distances in the complex frequency plane; they are the distances from some common variable point s to the zeros or poles. Being mindful of the two-dimensional potential mentioned above, we can regard α as that potential function if we assume filaments carrying positive charges to be located at the zeros and filaments carrying negative charges to be located at the pole positions. All charges are alike in absolute value and a multiplicative constant (or scale factor) remains arbitrary.

Instead of thinking in terms of the electric potential due to filament charges we can as well consider the scalar magnetic potential due to filamental currents since its functional dependence upon coordinates in the cross-sectional plane is the same. Electric potential is easier to measure, however, and hence physical devices built to simulate the situation given in Eq. 22 are commonly based upon the electrical analog and take the form of an electrolytic tank or a semiconducting sheet.*

If the s-plane is represented by a semiconducting sheet, one can simulate the terms in Eq. 22 by touching the sheet with pinpointed probes at the points s_ν, feeding a fixed amount of current into each probe at the zeros $s_1, s_3, \ldots$ and withdrawing the same amount at each pole $s_2, s_4, \ldots$. If the number of finite zeros is less than the number of poles, then a number of zeros, equal to the difference, are at $s = \infty$. One must in this instance feed this difference current into the sheet at some remote point or set of points.

The potential measured at any point in the semiconducting sheet (with respect to a convenient datum—the remote point will do) by means of a roving probe (located at the variable point s) is proportional to the function α. Within a constant multiplier, one may thus determine the attenuation as a function of s for any given pole-zero distribution by physical means.

* See W. W. Hansen and O. C. Lundstrom, "Experimental Determination of Impedance Functions by the Use of an Electrolytic Tank," *Proc. IRE*, Vol. 33, Aug. 1945, pp. 528–534. Also A. R. Boothroyd, E. C. Cherry and R. Makar, "An Electrolytic Tank for the Measurement of Steady-State Response, Transient Response, and Allied Properties of Networks," *Proc. Inst. Elec. Engrs. (London)*, Pt. I, Vol. 96, May 1949, pp. 163–177. Also R. E. Scott, "Network Synthesis by Use of Potential Analogs," *Proc. IRE*, Vol. 40, Aug. 1952, pp. 970–973. Many additional references will be found there.

If the real and imaginary parts of the rational function itself rather than its logarithm are to be represented, we make use of the analogy between the potential of an electric or magnetic dipole and the real or imaginary parts of a simple pole of a rational function. An electric dipole consists of a pair of charges $+q$ and $-q$ separated by a relatively small distance d. The so-called *moment* of the dipole is a vector pointing from the negative to the positive charge (the *axis* of the dipole) having a magnitude $m = qd$.

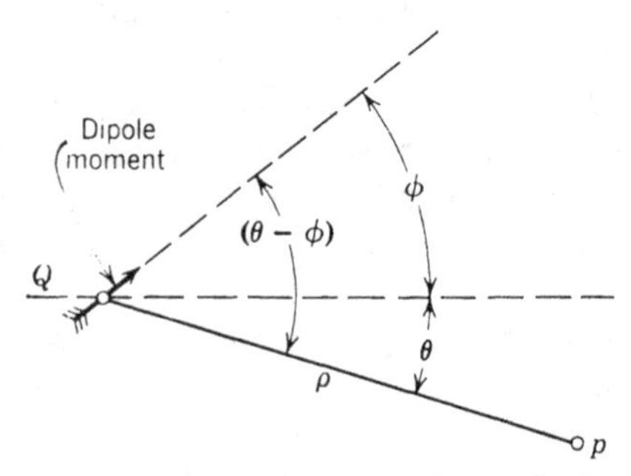

FIG. 3. Relative to the interpretation of a simple pole as a dipole moment according to potential theory.

Theoretically one should consider the limit resulting from $q \to \infty$ and $d \to 0$ such that m remains constant at a desired value, but actually we may think of q and d as being finite and nonzero just so long as the distance to the observation point is large compared with d.

Figure 3 shows a dipole, with arbitrary orientation of its moment, located at some point Q. The observation point is P, and its distance from the dipole is indicated by ρ. In appropriate units, the potential at P is given by the expression

$$\text{dipole potential} = \frac{m \cos(\theta - \phi)}{\rho} \tag{23}$$

Comparison with Eq. 4 of Ch. VIII shows that the real part contributed by one of the simple poles of a rational function is given by a relation having precisely this same form. Regarding the complex value of the residue k as a vector, we see from Eq. 45, Ch. VIII that it may be identified with the vector moment of the dipole. The imaginary part in Eq. 47 of Ch. VIII is obtained as the potential of the dipole rotated through 90° in the clockwise direction.

If a given rational function has a number of simple poles in the finite s-plane with known positions and complex residues, we can replace each by an electric dipole, with angular orientation equal to the pertinent residue angle and moment equal to the residue magnitude, whereupon the resultant potential at some variable point s may be interpreted as the real part of the rational function at that point; and the corresponding imaginary part is found by simply rotating all dipoles through 90° in the clockwise direction.

The semiconducting sheet and pinpointed probes may also be used here. A pair of closely spaced probes, of which one feeds and the other withdraws current, simulates a dipole and represents each pole of the rational function. Intensity of the current is proportional to the magnitude of the residue, and angular orientation of the dipole equals the angle of the residue. In this arrangement the rational function is represented by its poles and residues; in the other situation, based upon the analogy between terms in Eq. 22 and the logarithmic potentials of single filaments, the rational function is represented by its poles and zeros.

In the pole-zero analog, the roving probe picks up a potential proportional to the logarithm of the magnitude of the function or its reciprocal according to whether the pole probes or the zero probes feed current into the sheet. The latter circumstance is of little importance since it affects only the algebraic sign of the potential.

Phase angle, or the imaginary part of the logarithm of the rational function, is not quite as easily determined in the pole-zero analog. A basis for its determination is given by the Cauchy-Riemann equation

$$\frac{d\beta}{d\omega} = \frac{d\alpha}{d\sigma} \tag{24}$$

from which the phase angle at some j-axis point is given by

$$\beta = \int_0^\omega \frac{d\alpha}{d\sigma}\, d\omega \tag{25}$$

Here $d\alpha/d\sigma$ represents the potential gradient in a direction normal to the j-axis. For any j-axis point it is proportional to the current density normal to this axis at that point. The phase β for any $s = j\omega$, according to Eq. 25, is thus seen to be proportional to the total current crossing the j-axis between $s = 0$ and the pertinent point $s = j\omega$, which is not readily measurable.

We are, however, not so much interested in analog devices as we are in the physical interpretations which their theoretical basis provides. For example, it is physically immediately obvious that a zero or pole pattern symmetrical about the j-axis produces no net current across that axis because the correspondingly symmetrically located probes feed equal currents into the sheet. Hence it is at once clear that such a pattern produces no phase shift and affects only the loss function. If in such a pattern all probes in one half-plane are removed, the loss function is unaffected except for a factor 1/2, but a phase shift now takes place.

Analogously, if we have poles on one side of the j-axis and zeros in symmetrically located positions, the j-axis is an equipotential locus and hence the loss function is obviously a constant. We do, however, have current crossing the j-axis, and so this pattern produces phase shift only. If we now remove all the probes on one side of the j-axis, the current crossing it is merely halved; its distribution along the axis is not changed. Hence such an "all-pass" pole-zero pattern clearly yields the same phase function, except for a factor 1/2 when all of the critical frequencies in one

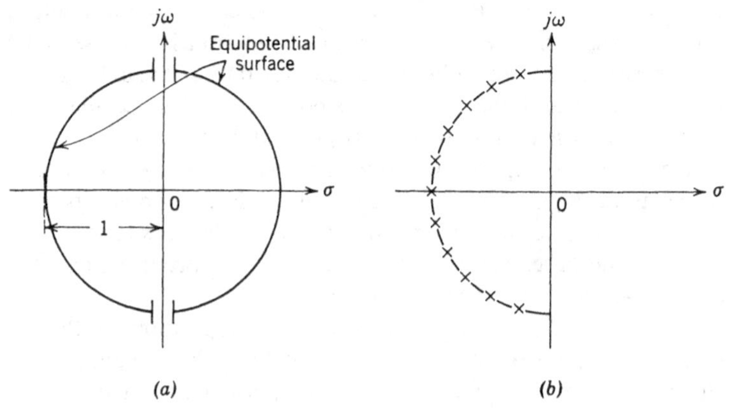

FIG. 4. Quantization of the Faraday "ice pail" yields the pole pattern of a lowpass filter.

half-plane are removed. However, the j-axis does not remain an equipotential locus, and hence the associated loss function does not remain constant.

We can, of course, obtain these same results without the aid of the physical picture lent by these potential-theory interpretations, but they are so much more meaningful, and evident by inspection, when seen in this light. We can furthermore make use of certain well-known experimental evidence in potential theory to establish useful principles and points of view. Faraday's famous ice-pail experiment, for example, gives us a basic principle that is useful in solving the approximation problem for filter networks. According to this experiment the potential inside a conducting vessel is constant; or we can say that the potential inside a closed (or substantially closed) equipotential surface is constant, since the conducting vessel merely serves as a convenient mechanism for providing the almost-closed equipotential surface.

With reference to part (a) of Fig. 4, the cylindrical surface axially normal to the s-plane is assumed to provide such an equipotential locus. The cylinder is supposed to carry some total charge Q per unit of axial length, which distributes uniformly over the surface and produces a potential outside that may be computed as though Q were a filamental distribution normal to the s-plane at $s = 0$. If we assume Q to be negative and regard the potential along $s = j\omega$ as a loss function α, we have

$$\alpha = -K \ln \frac{1}{\omega} = K \ln \omega \qquad \text{for } \omega > 1 \tag{26}$$

where K is an appropriate constant multiplier. For $|\omega| < 1$, Faraday's ice-pail experiment tells us that the potential remains constant at the value it has for $\omega = 1$ where the variable point enters the cylinder. A plot of

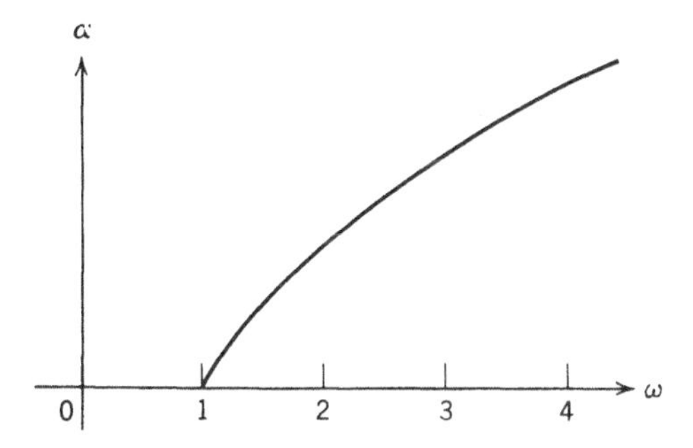

FIG. 5. Lowpass attenuation characteristic resulting from the pole pattern in Fig. 4.

this function is shown in Fig. 5, in which the scale factor for α is proportional to the charge Q.

If we now imagine the cylinder replaced by parallel wires uniformly spaced circumferentially, each carrying the same fraction of the total charge Q, the potential function will be only slightly changed in form if the spacing between wires is small. The total charge Q, instead of being uniformly distributed circumferentially, is divided into a finite number of equal lumps concentrated at equally spaced points. This process is spoken of as "quantization," the "quanta" being the equal portions of charge carried by the single filaments or wires.

Networkwise this process of quantization is equivalent to the transition from a distributed-parameter network characterized by transcendental impedance functions to a lumped-parameter network characterized by rational impedance functions. The poles of the latter may now be identified, according to the potential analog previously discussed, with the filaments carrying lumped charges. If we do this, and subsequently discard the right half-plane poles (which, as pointed out above, merely changes the potential or loss function α by a factor $\frac{1}{2}$), we get the picture in part (*b*), Fig. 4, which still has essentially the loss function shown in Fig. 5.

One should carefully observe in this connection that the uniform spacing of poles circumferentially is a result of the circular form of the original cylindrical surface in Fig. 4, part (*a*). Had this cylindrical surface been assumed in any other form, the total charge Q would not have distributed uniformly in the circumferential dimension. Since each term in Eq. 22 must of necessity have the *same* multiplier, the analogy between poles (respectively, zeros) and charge-carrying filaments implies that all filaments carry *equal* charges. Hence, the only way in which an unequal circumferential charge distribution can be accommodated in the quantization process is to make the circumferential pole density (reciprocal spacing) proportional to the circumferential charge density.

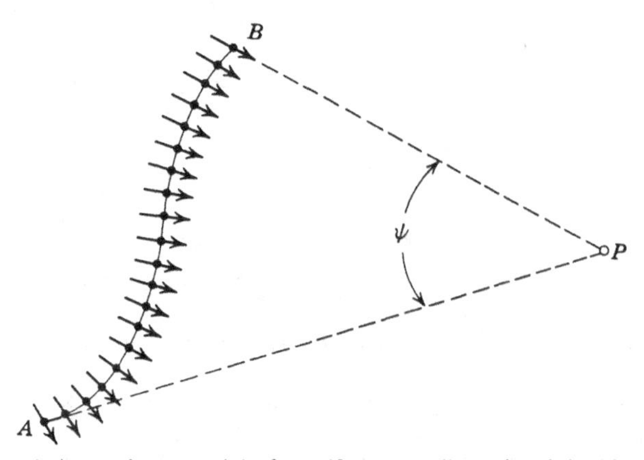

FIG. 6. Relative to the potential of a uniform two-dimensional double stratum.

Thus we see that our "ice pail" in Fig. 4, part (*a*) may have any shape other than that of a circular cylinder provided we determine, for that shape, the corresponding charge-density distribution, and quantize accordingly. Variation of the potential function outside the ice pail will, of course, depend upon its shape, although inside it will always have the same constant behavior.

Returning to the dipole-potential analog sketched in Fig. 3, it is interesting to consider the situation shown in Fig. 6 in which many identical dipoles are uniformly spaced along, and normally oriented to, some curve or path with the end points A, B. Since the configuration is uniform in the dimension normal to the paper, it amounts to a surface distribution of dipoles and is equivalent to a membrane with constant charge density of equal magnitude but opposite sign on its two sides. This distribution is called a *double stratum*. The potential produced by it at some point P is proportional to the angle ψ subtended by the end points A, B, independently of the shape of path connecting them. The truth of this simple result is easily shown.

With reference to the sketch in Fig. 7, dl is an increment in the curved path connecting A and B. If m, as in Eq. 23, is the moment of a single dipole and the density of dipoles per unit length of path is denoted by D, then the dipole moment per increment dl is $mD\,dl$, and its direction is normal to this path increment. The angle between this normal direction and the line PQ joining dl with the point P is $\theta - \varphi$ as in Fig. 3.

From the geometry in Fig. 7 we see that $dl \cos(\theta - \varphi)$ is the projection of the path increment upon a line perpendicular to PQ and hence the

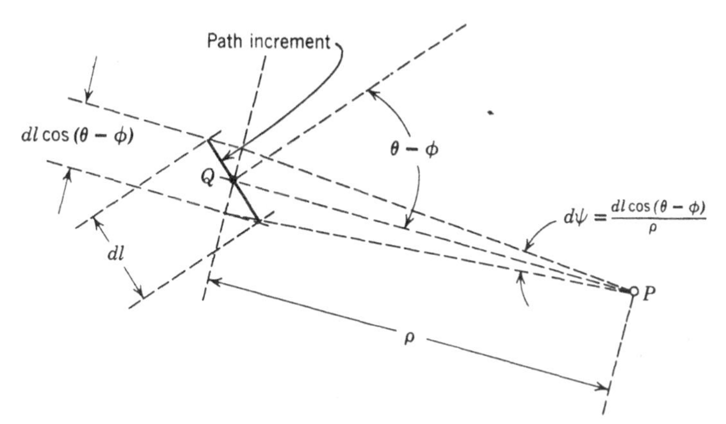

FIG. 7. Contribution of an element of the double stratum in Fig. 6 to the net potential at point P.

angular increment which dl subtends at the point P is

$$d\psi = \frac{dl \cos(\theta - \varphi)}{\rho} \tag{27}$$

Therefore the potential at P contributed by the path increment dl, according to formula 23, is given by

$$\frac{mD\,dl \cos(\theta - \varphi)}{\rho} = mD\,d\psi \tag{28}$$

Integration over the path in Fig. 6 from A to B obviously yields $mD\psi$ for the net potential at P; and it is clear that the shape of the path has no influence so long as the charge distribution on the double stratum is uniform.

If the plane of Fig. 6 is identified with the complex s-plane, the electric dipoles are replaced by simple poles of a rational function, and their moments by the residues of this function in these poles. The net real part at some variable point s (equivalent to the point P) then is proportional to the subtended angle ψ alone. Since a distribution of discrete poles is an approximation to a uniform distribution, this result is somewhat crude in most practical applications, but it provides a fast estimate of a result that otherwise would require a long and tedious calculation.

If we want the corresponding imaginary part, we have but to rotate all residues through 90° in the clockwise direction. In the analogous potential problem this operation produces a *dipole chain*, and since adjacent positive and negative charges in this chain cancel, we have left only a single plus charge at A in Fig. 6 and a single minus charge at B. Calculation of the resultant potential at P for this situation is simple, and hence so is calculation of the imaginary part of our rational function having the assumed pole distribution.

4. Conformal Transformations and Some Useful Pole-Zero Configurations

It is important to recognize that the use of potential theory does not make the approximation problem easier to solve unless it yields a situation for which the solution is recognizable by inspection. Otherwise we merely find ourselves trading one computationally awkward problem for another equally awkward one.

Primarily, potential theory provides a useful mechanism for intuitive reasoning and mental visualization. It also provides a number of specific solutions in terms of certain classical boundary value problems for which solutions are available in the rather extensive literature on potential theory and conformal mapping. An example of this sort, appropriate to our present discussion, is given by the well-known conformal transformation

$$w = s \pm \sqrt{s^2 + 1} \quad \text{or} \quad s = \tfrac{1}{2}\left(w - \frac{1}{w}\right) \tag{29}$$

which maps the s-plane doubly upon the w-plane, once upon the interior of the unit circle of that plane and again upon the exterior of this unit circle. The unit circle itself becomes the doubly traversed interval of the j-axis corresponding to $-1 < \omega < 1$.

Concentric circles and radial lines in the w-plane become ellipses and hyperbolas respectively in the s-plane. Letting

$$w = re^{j\beta} \tag{30}$$

the family of confocal and orthogonal ellipses and hyperbolas in the s-plane result for constant values of r and β respectively, as shown in Fig. 8. The s-plane is a two-sheeted Riemann surface; the points $s = \pm j$ are branch points, and the j-axis interval connecting them may be interpreted as a branch cut.

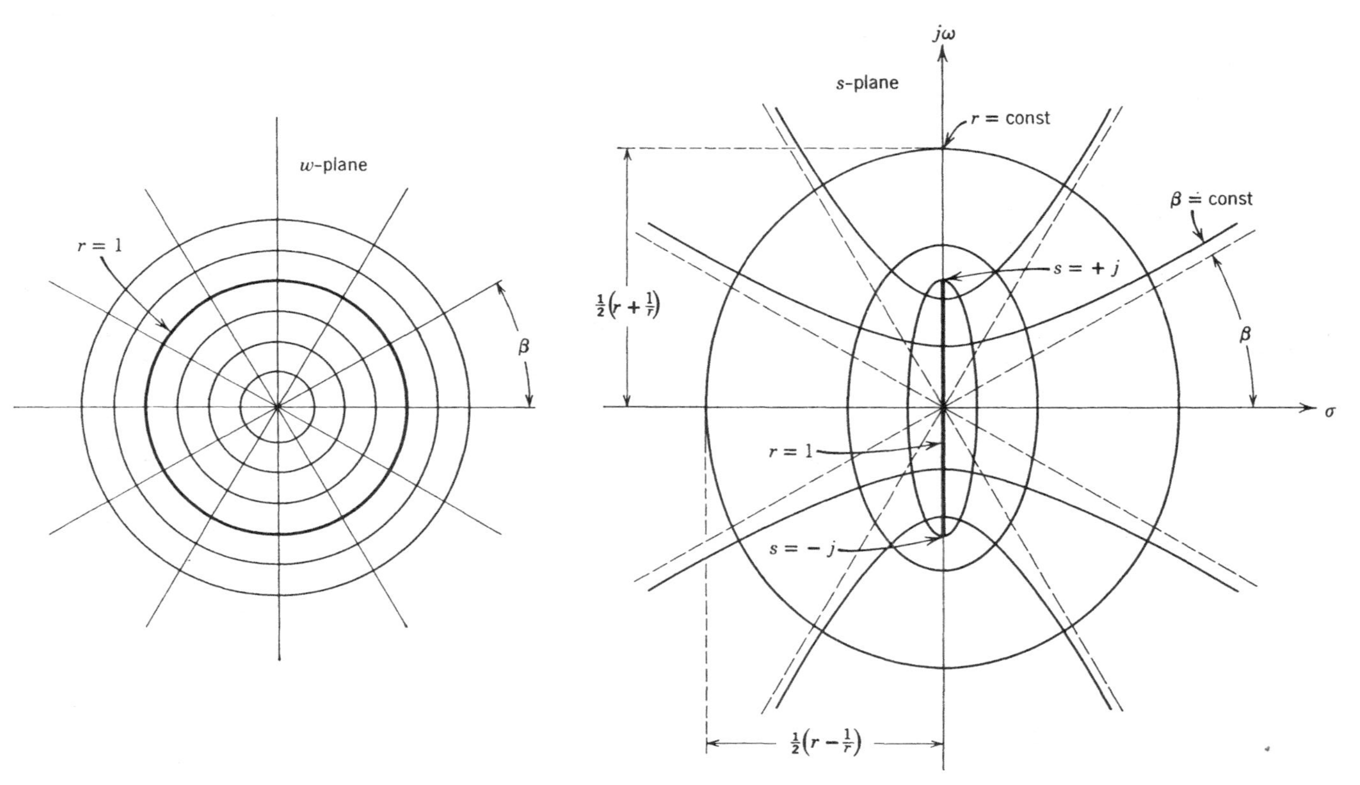

FIG. 8. Conformal representation of the transformation given by Eq. 29.

If we regard the circular cylinder designated by $r = 1$ in the w-plane as an equipotential surface, then the corresponding surface in the s-plane becomes a sheet whose normal intersection with this plane coincides with the branch cut. A net charge Q on the circular cylinder of the w-plane distributes uniformly in a circumferential direction, as is also indicated in this field map by drawing the flux lines $\beta =$ const with uniform spacing. On the corresponding equipotential surface in the s-plane, the total charge Q distributes nonuniformly, the charge density for any ω-value being proportional to the density of flux lines $\beta =$ const at that point.

Variation of the potential along the j-axis of the s-plane may be obtained by noting first that, for a negative charge on the cylinder of the w-plane, we have

$$\alpha = -K \ln \frac{1}{r} = K \ln r \tag{31}$$

or

$$r = e^{\alpha/k} \tag{32}$$

Equations 29 and 30 thus yield

$$s = \sinh \left(\frac{\alpha}{k} + j\beta\right) \tag{33}$$

giving

$$\sigma + j\omega = \sinh \left(\frac{\alpha}{k}\right) \cos \beta + j \cosh \left(\frac{\alpha}{k}\right) \sin \beta \tag{34}$$

whence

$$\sigma = \sinh \left(\frac{\alpha}{k}\right) \cos \beta \tag{35}$$

$$\omega = \cosh \left(\frac{\alpha}{k}\right) \sin \beta \tag{36}$$

For $\sigma = 0$ ($s = j\omega$) we have

either

$$\alpha = 0, \quad \beta = \sin^{-1}\omega \qquad \text{for } -1 < \omega < 1 \tag{37}$$

or

$$\beta = \pm\frac{\pi}{2}, \quad \alpha = K \cosh^{-1} \omega \qquad \text{for } |\omega| > 1 \tag{38}$$

These two relations are pertinent respectively to the passband and stop-band. The characteristics which they yield are shown in Fig. 9.

They are unattainable in a lumped network since they would require a continuous pole distribution having the density variation β in the interval $-1 < \omega < 1$ of the j-axis. A lumped network impedance cannot have a singularity of this sort; it can have only poles, and these must not lie upon the j-axis if the impedance is to remain bounded for $s = j\omega$.

We can, of course, approximate these characteristics in a lumped network by placing poles along some other equipotential contour, that is to say, upon an elliptic contour in the s-plane. Since the latter is doubly mapped upon the w-plane, one ellipse in the s-plane corresponds to *two* circles in the w-plane, one larger and one smaller than the unit circle. A certain number of poles on an elliptic contour of the s-plane yields twice as many poles in the w-plane, since each s-plane pole corresponds to two poles on the same radial line in the w-plane, one inside and one outside the unit circle, their radii having reciprocal values.

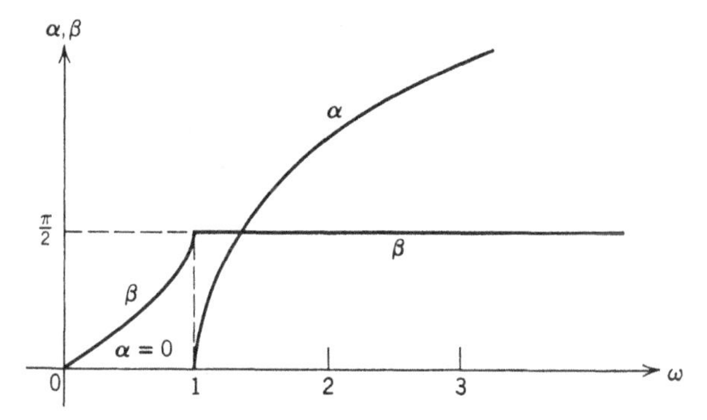

FIG. 9. Attenuation and phase lag as functions of ω corresponding to the conformal plot in Fig. 8.

Since circumferential charge distribution on a circle in the w-plane is uniform, quantization there is simple. Suppose we consider placing $2n$ poles on an elliptic contour in the s-plane, n in the left and n at symmetrically located points in the right half-plane. This distribution corresponds to $2n$ poles on a circle of radius $r > 1$ in the w-plane and $2n$ additional poles on a circle of radius $1/r$. It is convenient to define the radii of these circles by letting $r = e^{\pm\delta}$ and choosing some value for δ.

On the circle $r = 1$, corresponding in the s-plane to the passband region $-1 < \omega < 1$, we no longer have a constant potential. It is not difficult to see, however, that the potential, and hence the loss function, has equiripple character in this region, since the poles are uniformly spaced on the two circles of the w-plane. Their corresponding locations in the s-plane can be calculated from Eqs. 32, 35 and 36 by letting $\beta = (\pi/2) + (\nu\pi/2n)$ for ν equal to odd-integer values. Left half-plane poles at the points $s_\nu = \sigma_\nu + j\omega_\nu$ are thus determined by the relations

$$\left.\begin{aligned}\sigma_\nu &= \sinh\delta\cos\frac{(n+\nu)\pi}{2n}\\ \omega_\nu &= \cosh\delta\sin\frac{(n+\nu)\pi}{2n}\end{aligned}\right\}\quad \text{for } \nu = 1, 3, \ldots 2n-1 \tag{39}$$

The parameter δ controls the ripple amplitude in the passband interval $-1 < \omega < 1$ in a manner that can be determined as follows. The substitution $r = e^{\delta}$ together with Eq. 30 amounts to a further change of variable

$$w = e^{\gamma} \qquad \text{with } \gamma = \delta + j\beta \tag{40}$$

which transforms the circles and radial lines of the w-plane in Fig. 8 into a rectangular grid in the γ-plane in such a way that the entire w-plane is mapped upon a horizontal strip 2π units wide in the γ-plane, the region within the unit circle of the w-plane being mapped upon the left half-plane portion of this strip and the region outside of the unit circle upon the right half-plane portion.

The $2n$ poles in the s-plane, which are doubly mapped upon the w-plane, become a double infinity of poles in the γ-plane, namely an infinite string of uniformly spaced poles on each of two vertical lines spaced $+\delta$ and $-\delta$ units from the imaginary axis of that plane, as is indicated in the sketch in Fig. 10. The pertinent γ-values are

$$\gamma_{\nu} = \pm\delta + j\frac{\pi}{2}\left(1 + \frac{\nu}{n}\right) \tag{41}$$

for all positive and negative *odd*-integer values of ν.

When n is even, no poles lie upon the real axis, but when n is odd, two of them do, namely the two in Eq. 41 for $\nu = -n$. For this reason, two positions for the real axis are shown in Fig. 10.

Suppose we denote by $H(s)$ a polynomial having the $2n$ zeros on an elliptic locus in the s-plane. These are the n left half-plane zeros defined by Eqs. 39 and their images in the right half-plane. Thus, if a polynomial having only the left half-plane zeros is denoted by $P(s)$, we have $H(s) = P(s)P(-s)$. $H(j\omega)$ then equals the squared j-axis magnitude of $P(s)$. Since the pole pattern in Fig. 10 follows from that in the s-plane by a change of the *independent* variable, a polynomial $H(\gamma)$ having a double infinity of zeros for all the poles in the γ-plane, nevertheless has all the *same* values as $H(s)$ except that they occur in different points of the plane; and furthermore, since the j-axis of the γ-plane corresponds to the j-axis of the s-plane, it follows that we can just as well study the j-axis behavior of $H(\gamma)$ as the j-axis behavior of $H(s)$ in order to discover the maxima and minima in its equiripple character.

$H(\gamma)$ is, strictly speaking, not a polynomial any longer inasmuch as the number of its zeros is not finite. We shall see presently that it is a transcendental function. The zero factors of $H(\gamma)$, according to Eq. 41, are given by

$$\gamma - \delta - j\frac{\pi}{2}\left(1 + \frac{\nu}{n}\right) \to 1 - \frac{n(\gamma - \delta)}{j\dfrac{\pi}{2}(n + \nu)} \tag{42}$$

and

$$\gamma + \delta - j\frac{\pi}{2}\left(1 + \frac{\nu}{n}\right) \rightarrow 1 - \frac{n(\gamma + \delta)}{j\frac{\pi}{2}(n + \nu)} \tag{43}$$

where the arrow means that the two expressions are zero at the same points which, in the formation of a polynomial, is all that matters since a multiplying constant remains arbitrary anyway.

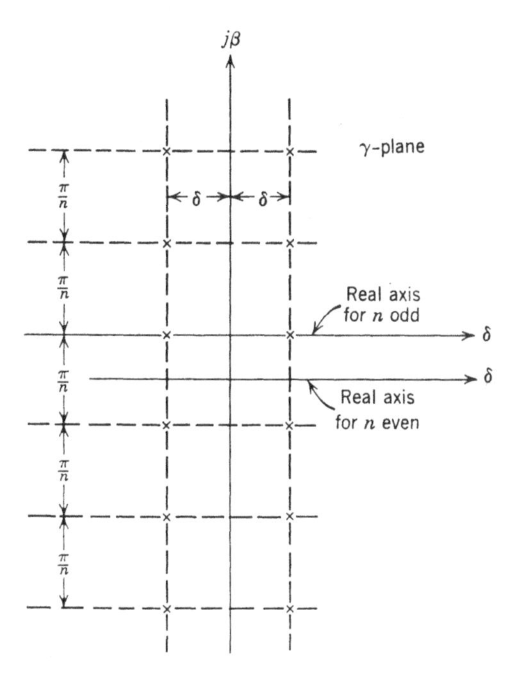

FIG. 10. Quantization in the γ-plane corresponding to uniform pole distribution upon concentric circles in the ω-plane of Fig. 8.

For n even, these factors yield all the zeros for all positive and negative odd-integer values of ν. For n odd, however, Fig. 10 shows that we have additionally to consider the factors* $n(\delta - \gamma)$ and $n(\delta + \gamma)$.

Apart from a constant multiplier the polynomial $H(\gamma)$ is formed from an infinite product of these factors. Fortunately the infinite products

* These factors are written in this form because their product $(\delta + \gamma)(\delta - \gamma) = \delta^2 - \gamma^2$ for $\gamma = j\beta$ reads $\delta^2 + \beta^2$ which is *positive* as it must be since $H(j\beta)$ is the square of an absolute value and hence must be positive. In 42 and 43 it doesn't matter whether we write the factors $\gamma - \delta$ or $\delta - \gamma$ since they occur in pairs of conjugate values anyway.

involved here are recognizable as representations for the entire transcendental functions cosh and sinh. Specifically we have for *n even*

$$\prod_{\pm\nu \text{ odd}}^{\infty} \left[1 - \frac{n(\gamma - \delta)}{j\frac{\pi}{2}(n + \nu)}\right] = \cosh n(\gamma - \delta) \tag{44}$$

$$\prod_{\pm\nu \text{ odd}}^{\infty} \left[1 - \frac{n(\gamma + \delta)}{j\frac{\pi}{2}(n + \nu)}\right] = \cosh n(\gamma + \delta) \tag{45}$$

and for *n odd*

$$n(\delta - \gamma) \prod_{\pm\nu \text{ odd}}^{\infty} \left[1 - \frac{n(\gamma - \delta)}{j\frac{\pi}{2}(n + \nu)}\right] = -\sinh n(\gamma - \delta) \tag{46}$$

$$n(\delta + \gamma) \prod_{\pm\nu \text{ odd}}^{\infty} \left[1 - \frac{n(\gamma + \delta)}{j\frac{\pi}{2}(n + \nu)}\right] = \sinh n(\gamma + \delta) \tag{47}$$

Within a constant multiplier we can, therefore, write for n even

$$H(\gamma) = \cosh n(\gamma + \delta) \cosh n(\gamma - \delta) = \sinh^2 n\delta + \cosh^2 n\gamma \tag{48}$$

and for *n odd*

$$H(\gamma) = -\sinh n(\gamma + \delta) \sinh n(\gamma - \delta) = \sinh^2 n\delta - \sinh^2 n\gamma \tag{49}$$

Since the transformations 29 and 40 yield a relation between s and γ that reads

$$s = \sinh \gamma \tag{50}$$

we see that (apart from an appropriate multiplier)

$$H(s) = \sinh^2 n\delta + \cosh^2 (n \sinh^{-1} s) \qquad \text{for } n \text{ even} \tag{51}$$

and

$$H(s) = \sinh^2 n\delta - \sinh^2 (n \sinh^{-1} s) \qquad \text{for } n \text{ odd} \tag{52}$$

For $s = j\omega$ we now have

$$H(j\omega) = \sinh^2 n\delta + \cos^2 (n \sin^{-1} \omega) \qquad \text{for } n \text{ even} \tag{53}$$

$$H(j\omega) = \sinh^2 n\delta + \sin^2 (n \sin^{-1} \omega) \qquad \text{for } n \text{ odd} \tag{54}$$

The second term in each of these expressions represents the square of a finite polynomial (for finite n) that oscillates between $+1$ and -1 within the range $-1 < \omega < 1$. A detailed study shows that, for n even or odd, this polynomial (which is a *Tschebyscheff polynomial* of the first kind) can alternately be represented by the single expression

$$T_n(\omega) = \cos(n \cos^{-1} \omega) \tag{55}$$

and hence we can write more simply

$$H(j\omega) = \sinh^2 n\delta + T_n^2(\omega) \tag{56}$$

for n even or odd.

The ripple amplitude in the passband is now easily established. We have

$$H_{max} = 1 + \sinh^2 n\delta; \quad H_{min} = \sinh^2 n\delta \tag{57}$$

Since only the ratio of these two expressions is of interest, the constant multiplier which was left dangling in the above manipulations is of no consequence. If we remind ourselves that H represents the *squared* absolute value of the polynomial $P(s)$ for $s = j\omega$, we see that a desired ripple ratio is given by

$$\begin{aligned} A = \sqrt{H_{min}/H_{max}} &= \frac{\sinh n\delta}{\sqrt{1 + \sinh^2 n\delta}} \\ &= \frac{\sinh n\delta}{\cosh n\delta} = \tanh n\delta \end{aligned} \tag{58}$$

which can also be written

$$A = \frac{1 - e^{-2n\delta}}{1 + e^{-2n\delta}} \quad \text{or} \quad e^{-2n\delta} = \frac{1 - A}{1 + A} \tag{59}$$

from which we can compute a ripple from a given δ-value or compute a δ-value that will yield a specified ripple.

A far simpler situation is presented by the uniformly spaced poles on a circular contour as shown in Fig. 4, part (*b*). For n left half-plane poles in $P(s)$, the polynomial $H(s) = P(s)P(-s)$ has $2n$ zeros equally spaced on the unit circle with symmetry about the j-axis and no poles on this axis. About the positive j-axis as an angular reference, these are the $2n$th roots of -1. Hence we have

$$H(s) = 1 + (js)^{2n} \tag{60}$$

since $H(s) = 0$ gives

$$(js)^{2n} = -1 = e^{j\nu\pi} \tag{61}$$

or for ν odd

$$js_\nu = e^{j\nu\pi/2n}; \quad s_\nu = e^{j(\pi/2)(\nu/n-1)} \tag{62}$$

From Eq. 60 we have for $s = j\omega$

$$H(j\omega) = 1 + \omega^{2n} \tag{63}$$

and for the response function of our network, which is given by the reciprocal of the polynomial $P(s)$, we have the j-axis magnitude

$$\frac{1}{|P(j\omega)|} = \frac{1}{\sqrt{1 + \omega^{2n}}} \tag{64}$$

Plots of this so-called *Butterworth* function are shown for several values of the integer n in Fig. 11. An interesting property is that all curves pass through the value 0.707 (the half-power value) for $\omega = 1$ (the cutoff frequency) regardless of the integer n. For increasing values of n the

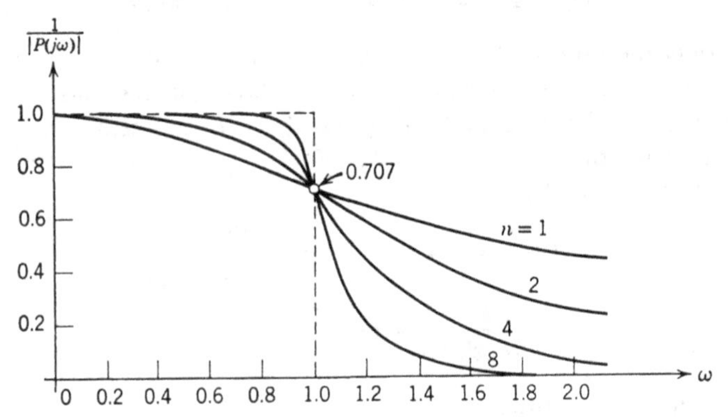

FIG. 11. Butterworth response functions for various orders (numbers of discrete poles in a given distribution).

function approximates more and more closely to the dotted rectangle which may be regarded as the amplitude response characteristic of an ideal lowpass filter.

The integer n is spoken of as the *order* of a given Butterworth function. In this connection it is interesting to observe that the Butterworth function of order 1 has the form of a simple resonance curve in the vicinity of the pole frequency. Although the present discussion is confined to the lowpass class, appropriate pole patterns for the bandpass and other filter classes are readily obtainable by means of suitable transformations of the independent frequency variable. These are discussed in the next article.

5. Frequency Transformations

It has already been seen in our more elementary discussions that if, in a normalized lowpass characteristic (one for which the cutoff is normalized at 1 radian per second, as in the examples of the preceding article), we replace ω by $1/\omega$, the frequency scale is inverted about the cutoff point $\omega = 1$, and the function becomes an appropriate approximant to an ideal highpass filter characteristic. In the same simple manner, it is possible to transform a lowpass approximant to any desired filter class.

Since the process operates only upon the independent variable, it has no effect upon the tolerance with which the pertinent function approximates a given ideal characteristic. Thus a certain ripple amplitude in the lowpass approximant remains unchanged when that function is transformed to its bandpass counterpart by means of a suitable change in its independent variable.

This change of variable can, moreover, be so chosen that its effect upon the given lowpass approximant may be interpreted directly in terms of the modifications it produces in the structure and element values of the lowpass network realization (the so-called *lowpass prototype* network). In this way one can avoid altogether the need for forming the pertinent bandpass or other more complicated approximant and applying to it a realization procedure which correspondingly becomes computationally more tedious to carry out.

Suppose for the purposes of this discussion we denote the complex frequency variable for the lowpass function by the symbol $\lambda = \bar{\sigma} + j\bar{\omega}$, and that for the transformed function by $s = \sigma + j\omega$. What we are looking for is a relation between λ and s that converts a given lowpass $Z(\lambda)$ into an approximant $Z(s)$ appropriate to whatever filter class we may be interested in. The pertinent relation, as we shall see, is simply

$$\lambda = z(s) \tag{65}$$

in which $z(s)$ is a reactance function (recall art. 6 in Ch. VIII). It is not difficult to appreciate why this simple function fulfills all of the requirements mentioned above.

Two points in the λ-plane, namely $\lambda = 0$ and $\lambda = \infty$, are of particular interest. The point $\lambda = 0$ is where the passband of the lowpass characteristic is centered, and the vicinity of $\lambda = \infty$ is where the lowpass approximating function is described by its asymptotic behavior. This is where the stopband is centered. The function $\lambda = z(s)$ essentially is required to make these two λ-values fall at certain specified points on the j-axis of the s-plane in some alternating manner since in any filter class the passband and stopbands alternate. A reactance function has precisely this required behavior.

Thus the so-called lowpass-highpass transformation $\lambda = 1/s$ interchanges the origin and infinity. It makes $\lambda = 0$ fall at $s = \infty$, and $\lambda = \infty$ at $s = 0$. The passband of the approximating function becomes centered at $s = \infty$ and its stopband at $s = 0$; a lowpass characteristic is thereby changed to highpass.

If we want the lowpass characteristic changed to bandpass, then we want $z(s)$ to have a zero at a finite nonzero s-value and poles (points where $\lambda = \infty$) at $s = 0$ and $s = \infty$. The reactance function appropriate to a series LC combination has precisely this character. Specifically the function

$$\lambda = \frac{s^2 + \omega_0^2}{s} \tag{66}$$

transforms the point $\lambda = 0$ into the two j-axis points $s = \pm j\omega_0$, and the point $\lambda = \infty$ into the points $s = 0$ and $s = \infty$. Hence the function 66 becomes an appropriate lowpass-bandpass transformation.

A lowpass-bandstop transformation is evidently given by the reciprocal of 66, namely

$$\lambda = \frac{s}{s^2 + \omega_0^2} \tag{67}$$

which transforms $\lambda = 0$ into $s = 0$ and $s = \infty$, and the point $\lambda = \infty$ into $s = \pm j\omega_0$. A lowpass to low-and-bandpass transformation is effected by the reactance function

$$\lambda = \frac{s(s^2 + \omega_0^2)}{(s^2 + \omega_\infty^2)} = \frac{s(\omega_0^2 - \omega_\infty^2)}{(s^2 + \omega_\infty^2)} + s \tag{68}$$

for which the point $\lambda = 0$ is carried over into the three points $s = 0$, $s = \pm j\omega_0$; and the point $\lambda = \infty$ is likewise triplicated since it corresponds to $s = \infty$ and $s = \pm j\omega_\infty$. Transformations which are capable of carrying a lowpass approximant over into any desired filter class can thus be written down without difficulty.

All of these transformations additionally have the property that the network modifications which they imply are obvious by inspection if we are familiar with the physical realizations of the reactance functions involved. In this connection it is rather obvious that functions 66, 67 and 68 are realized respectively by the LC combinations in sketches (*a*), (*b*) and (*c*) of Fig. 12 where element values are in henrys and farads.

Thus the lowpass-bandpass transformation 66 changes an inductance having the reactance $L\lambda$ into a series LC combination determined by a reactance function equal to 66 with its impedance level multiplied by L; and it changes a capacitance having the susceptance $C\lambda$ into a parallel LC combination determined by a susceptance function 66 with its admittance level multiplied by C. That is to say, each inductance in the lowpass prototype network having a value of L henrys has placed in series with it a capacitance of $1/L\omega_0^2$ farads; and a capacitance in the lowpass prototype network having a value of C farads has placed in parallel with it an inductance of $1/C\omega_0^2$ henrys.

The lowpass-bandstop transformation 67 may similarly be interpreted directly in terms of modifications in the lowpass prototype network. Since this function is the reciprocal of 66, duality shows that the interpretations are essentially the same as for the lowpass-bandpass situation with the roles of L and C interchanged. Specifically, a lowpass L becomes a capacitance of $1/L$ farads in parallel with an inductance of L/ω_0^2 henrys; and a lowpass C becomes an inductance of $1/C$ henrys in series with a capacitance of C/ω_0^2 farads.

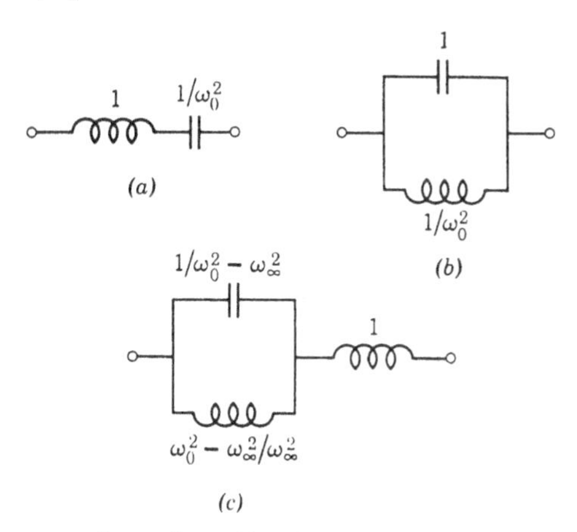

FIG. 12. Frequency transformations 66, 67, 68 realized physically by network configurations (*a*), (*b*), (*c*) respectively as shown here.

More elaborate transformations like 68 and others nevertheless are readily interpreted networkwise since the pertinent functions need merely be regarded as reactances or susceptances that have their impedance or admittance levels respectively multiplied by the lowpass prototype L and C values.

It remains to see what some of the more detailed properties of these transformations are, and how a particular situation calling for specific cutoff frequencies is accommodated. If we rewrite the lowpass-bandpass transformation 66 in terms of j-axis values, it reads

$$\bar{\omega} = \frac{\omega^2 - \omega_0^2}{\omega} \tag{69}$$

a plot of which is shown in Fig. 13.

The cutoff frequency of the lowpass function is denoted by $\bar{\omega}_c$, and the cutoff frequencies of the corresponding bandpass by ω_1 and ω_2. These are given in terms of $\bar{\omega}_c$ and ω_0 by the roots of the following equation:

$$\omega^2 - \bar{\omega}_c\omega - \omega_0^2 = 0 \tag{70}$$

which is obtained by setting $\bar{\omega} = \bar{\omega}_c$ in Eq. 69. Reference to Fig. 13 shows that the roots of Eq. 70 are $-\omega_1$ and ω_2. By the theory of equations, the sum of these roots equals the coefficient of the linear term with its sign changed, and their product equals the constant term. Hence we have

$$\omega_2 - \omega_1 = w = \bar{\omega}_c \tag{71}$$

and

$$\omega_1\omega_2 = \omega_0^2 \tag{72}$$

where we have denoted the bandwidth of the bandpass by w.

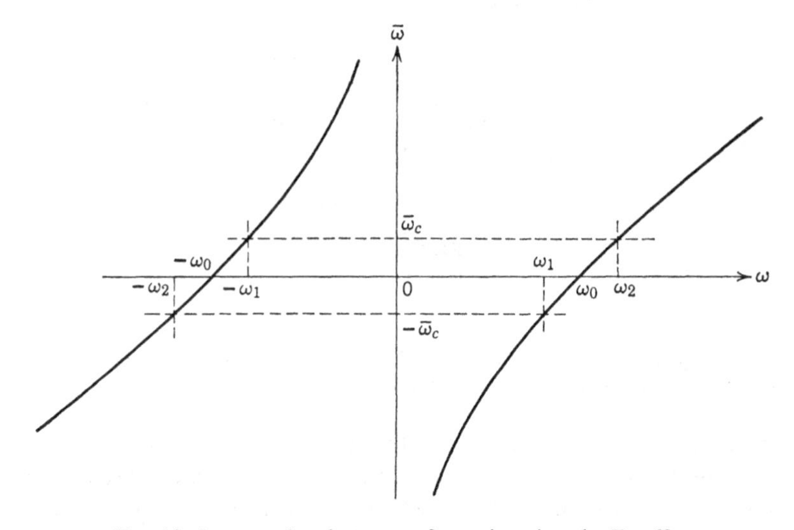

FIG. 13. Lowpass-bandpass transformation given by Eq. 69.

Equation 71 shows that the transformation 66 or 69 has the property of keeping the bandwidth of the filter invariant; and Eq. 72 shows that the arithmetic symmetry of the lowpass filter characteristic (which is an even function of $\bar{\omega}$) is transformed into a bandpass characteristic having geometrical symmetry about the center frequency ω_0, since any two frequencies whose geometric mean value is ω_0 correspond to a pair of $\bar{\omega}$ values like $\pm\bar{\omega}_c$.

In applying the transformation to a specific example, the procedure is first to frequency-scale the lowpass prototype network to have a cutoff $\bar{\omega}_c$ equal to the bandwidth of the desired bandpass; then, picking the prescribed values of ω_1 and ω_2 consistent with this bandwidth 71, compute ω_0^2 from Eq. 72. This fixes the transformation.

In the more elaborate lowpass to low-and-bandpass transformation 68, which has the j-axis form

$$\bar{\omega} = \frac{\omega(\omega^2 - \omega_0^2)}{\omega^2 - \omega_\infty^2} \tag{73}$$

we may denote the transformed lowpass cutoff frequency by ω_1, and those of the bandpass by $\omega_2 < \omega_3$. A pertinent sketch like the one in Fig. 14 shows that the quantities ω_1, ω_3, and $-\omega_2$ are roots of the equation

$$\omega^3 - \bar{\omega}_c\omega^2 - \omega_0^2\omega + \bar{\omega}_c\omega_\infty^2 = 0 \tag{74}$$

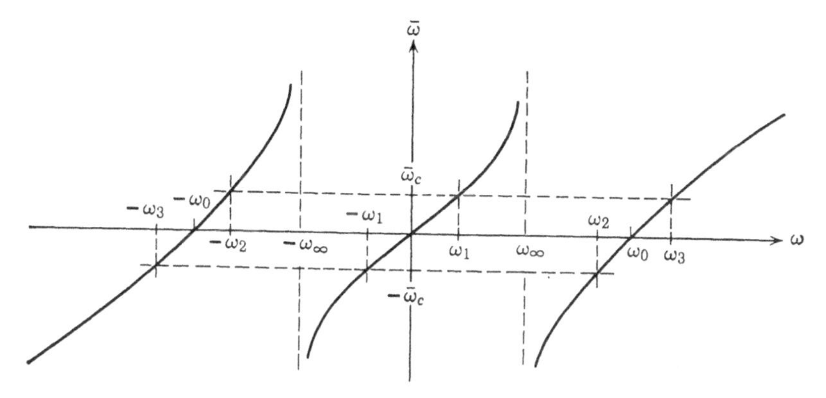

FIG. 14. Low-and-bandpass transformation given by Eq. 73.

By the theory of equations we can now write

$$\omega_3 - \omega_2 + \omega_1 = \bar{\omega}_c \tag{75}$$

$$\omega_1\omega_2 - \omega_1\omega_3 + \omega_2\omega_3 = \omega_0^2 \tag{76}$$

$$\omega_1\omega_2\omega_3 = \bar{\omega}_c\omega_\infty^2 \tag{77}$$

The first of these again shows the invariance of total bandwidth (a property which is generally true for transformations of this type*). The procedure for determining a specific transformation is to frequency-scale the lowpass prototype network to a value of $\bar{\omega}_c$ equal to the *combined* bandwidth of the desired transformed filter characteristic; then picking values for the cutoff frequencies ω_1, ω_2, ω_3 consistent with this bandwidth 75, we obtain the appropriate values for the parameters ω_0^2 and ω_∞^2 from Eqs. 76 and 77. The desired transformation 68 is thus established.

* This result was originally presented by R. Feldtkeller in an article entitled "Zur Frequenzabhangigkeit der Reaktanzen," *Elek. Nachr.-Tech.*, Vol. 13, No. 12, Dec. 1936, pp. 401–404. The above presentation follows a method of proof given by Hans Salinger in a "Note on the Frequency Behavior of Reactances," *Proc. IRE*, Vol. 26, No. 1, Jan. 1938, pp. 107–111.

Problems

1. A polynomial $p(s)$ has the zeros shown in the adjacent s-plane sketch.

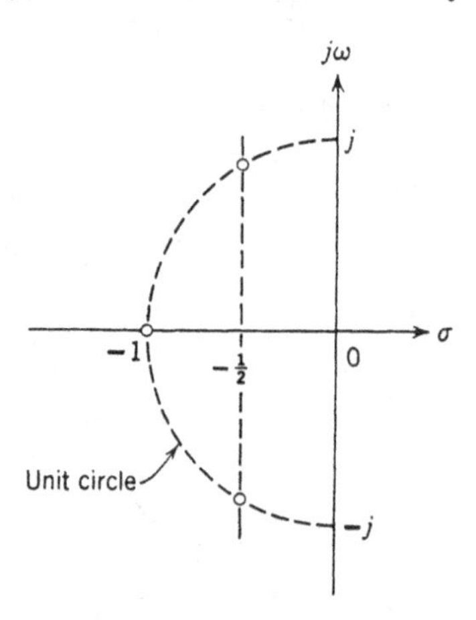

PROB. 1.

(*a*) Write it as a product of factors with real coefficients.
(*b*) Write the expression for $|p(j\omega)|^2$ as a function of ω.

2. Repeat prob. 1 for a polynomial $p(s)$ having zero factors $s - s_\nu$ with

$$s_\nu = -\sigma_\nu + j\omega_\nu = je^{j(\nu\pi/2n)}; \qquad \nu = 1, 3, \ldots 2n - 1 \quad \text{and} \quad n = 4, 5.$$

3. A polynomial with real coefficients has left half-plane zeros at points which are images about the unit circle. Show:

(*a*) That its coefficient values have symmetry about a vertical center line as in

$$p(s) = a + bs + cs^2 + ds^3 + cs^4 + bs^5 + as^6$$

(*b*) That it may alternately be written as a polynomial in the variable

$$\lambda = s + \frac{1}{s}$$

(*c*) that the polynomial defined in prob. 2 belongs to this class.

4. Consider the polynomial defined in prob. 2 in the form given in prob. 3, part (*a*), and show that $a = 1$ while

$$b = \sum_{\nu=1}^{2n-1} \sin \frac{\nu\pi}{2n} = \frac{1}{\sin \pi/2n}$$

5. Consider a transfer impedance

$$z_{12}(s) = \frac{1}{p(s)}$$

with $p(s)$ as defined in prob. 2. Show that:

(*a*) The total phase lag of this function from $\omega = 0$ to $\omega = \infty$ is $n\pi/2$.

(*b*) The phase shift from $\omega = 0$ to $\omega = 1$ is exactly half this value.

(*c*) The phase *slope* at $\omega = 0$ is given by

$$\left(\frac{d\beta}{d\omega}\right)_{\omega=0} = \sum_{\nu=1}^{2n-1} \sigma_\nu = \frac{1}{\sin \pi/2n}$$

while at $\omega = 1$ it is given by

$$\left(\frac{d\beta}{d\omega}\right)_{\omega=1} = \frac{1}{2} \sum_{\nu=1}^{2n-1} \frac{1}{\sigma_\nu}$$

6. The adjacent sketch shows the pole-zero plot of a given $Z(s)$. The constant multiplier is unity.

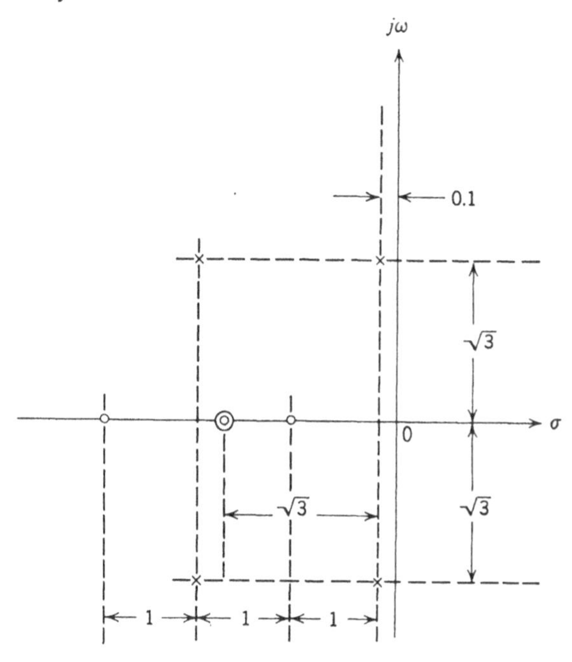

PROB. 6.

(*a*) Write down the partial fraction expansion.

(*b*) Determine an expression for $Z(s)$ appropriate for the resonance vicinity.

(*c*) Could this $Z(s)$ be the driving-point or transfer impedance of a passive network? Why or why not?

(*d*) By interchanging the double zero with a simple one, could it become a passive network impedance, and if so, which zeros would you interchange?

7. Expand the following impedance in partial fractions, combining conjugate complex terms:

$$Z(s) = \frac{(s+3)^3}{(s+1)(s+2)(s^2+s+1)}$$

8. Given the even-part function

$$\operatorname{Ev} Z(s) = \frac{s^2(s^2-1)}{2(s^2+4)}$$

construct $Z(s)$.

9. A lowpass transfer function for a filter network is to be constructed to the following specifications: Passband variation not to exceed 1 decibel. Attenuation at 1.5 times cutoff to be at least 40 decibels.

How many poles will be required for (*a*) a Butterworth function; (*b*) a Tschebyscheff function?

10. The following circuit realizes a third-order Butterworth transfer impedance for a normalized lowpass prototype filter. Find the corresponding circuit, indicating all element values, for a bandpass filter having a bandwidth of 10,000 cycles per second and a center frequency of 1 megacycle.

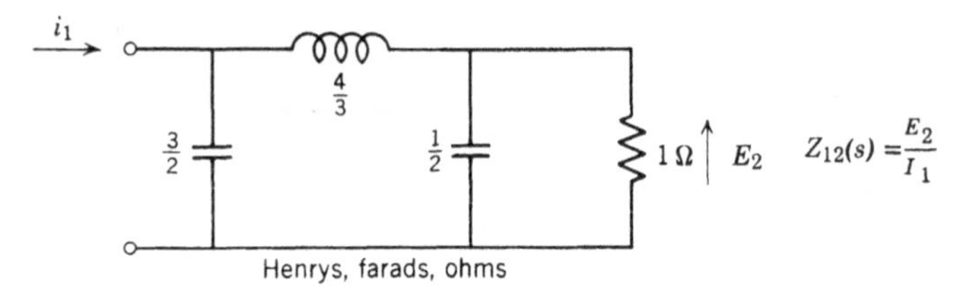

PROB. 10.

11. What kind of networks will realize the following impedances?

(*a*) $$Z(s) = \frac{1}{s+1} + \frac{2}{s+2} + \frac{3}{s+3}$$

(*b*) $$Z(s) = \frac{-1}{s+1} - \frac{2}{s+2} - \frac{3}{s+3} + 3$$

Find the network realizing impedance (*a*). Reform the expression (*b*) so that no minus signs appear and the network becomes recognizable by inspection. What condition fixes the constant term in (*b*)?

CHAPTER X

Response in the Time Domain

1. The Transform Concept; Time-Domain Response in Closed Form

Our starting point in this chapter is the same as in the last one, namely the differential equation expressing equilibrium in terms of a current response $i(t)$ in some branch of the network and a voltage excitation $e(t)$ at the same or at some other point. For convenience we repeat this equation here in the form

$$\left(a_n \frac{d^n}{dt^n} + a_{n-1} \frac{d^{n-1}}{dt^{n-1}} + \cdots + a_1 \frac{d}{dt} + a_0\right) i(t)$$
$$= \left(b_m \frac{d^m}{dt^m} + b_{m-1} \frac{d^{m-1}}{dt^{m-1}} + \cdots + b_1 \frac{d}{dt} + b_0\right) e(t) \quad (1)$$

In the present discussion we wish to construct the *complete* solution for an excitation defined by

$$\begin{aligned} e(t) &= 0 && \text{for } t < 0 \\ e(t) &= Ee^{s_0 t} && \text{for } t \geq 0 \end{aligned} \quad (2)$$

assuming the network to be at rest before $t = 0$.

We will accomplish this task in two steps; in the first, we formulate a solution to the simpler equation

$$\left(a_n \frac{d^n}{dt^n} + a_{n-1} \frac{d^{n-1}}{dt^{n-1}} + \cdots + a_1 \frac{d}{dt} + a_0\right) i(t) = e(t) \quad (3)$$

In the second, we use this result to obtain the desired solution to Eq. 1.

As pointed out previously, the solution to Eq. 3 consists essentially of two parts: the particular integral, and the complementary function. The particular integral or forced response we will write in the form

$$i_p(t) = B_0 e^{s_0 t} \quad (4)$$

where it is tacitly understood that this function applies to $t \geq 0$ only, since for $t < 0$ the response is assumed to be dead.

The complementary function or force-free response is a nontrivial solution to the homogeneous equation

$$\left(a_n \frac{d^n}{dt^n} + a_{n-1} \frac{d^{n-1}}{dt^{n-1}} + \cdots + a_1 \frac{d}{dt} + a_0\right) i(t) = 0 \tag{5}$$

It is tentatively assumed to be given by

$$i_0(t) = Be^{st} \tag{6}$$

whereupon substitution into Eq. 5 yields

$$(a_n s^n + a_{n-1} s^{n-1} + \cdots + a_1 s + a_0) Be^{st} = 0 \tag{7}$$

and a nontrivial result is seen to demand

$$a_n s^n + a_{n-1} s^{n-1} + \cdots + a_1 s + a_0 = 0 \tag{8}$$

Denoting the roots of this equation (the natural frequencies) by $s_1, s_2, \ldots s_n$, we recognize that the tentative assumption 6 needs to be revised to read

$$i_0(t) = B_1 e^{s_1 t} + B_2 e^{s_2 t} + \cdots + B_n e^{s_n t} \tag{9}$$

and the total solution to Eq. 3 for $t > 0$ becomes

$$i(t) = i_p(t) + i_0(t) = B_0 e^{s_0 t} + B_1 e^{s_1 t} + \cdots + B_n e^{s_n t} \tag{10}$$

We must next evaluate the amplitudes B_0, B_1, B_2, $\ldots B_n$. For this purpose we return to the differential Eq. 3, and consider the conditions that must be fulfilled at the initial instant $t = 0$ when the excitation is suddenly applied. Before this instant the right-hand member of Eq. 3, according to the definition of $e(t)$ given in Eq. 2, is zero; immediately after $t = 0$ the excitation voltage equals E (which is not necessarily a real amplitude). At the instant $t = 0$, therefore, the excitation is discontinuous; that is to say, it suddenly jumps from the value zero to the value E.

Since the equilibrium Eq. 3 must hold at $t = 0$ as well as at any other instant, it is necessary for the function on the left-hand side of this equation to be discontinuous at $t = 0$ also. Its value must jump from zero to E at this instant.

Off hand, we might suppose that any one or all of the $n + 1$ terms comprising the function on the left of Eq. 3 can participate in this jump. However, a little thought reveals that only the highest derivative term can participate, since a discontinuity in any lower derivative would require the higher ones to contain singularity functions like an impulse or doublet, etc. Since $e(t)$ by definition does not contain such functions, and since there can be no cancellation among the left-hand terms of Eq. 3 because all coefficients $a_0 \cdots a_n$ are positive (these are coefficients of a Hurwitz polynomial, see art. 5, Ch. VIII), we must conclude that only the nth derivative term on the left of Eq. 3 is discontinuous at $t = 0$; all lower derivative terms remain zero *through* the initial instant. Hence we have

$$i(t) = \frac{di}{dt} = \frac{d^2 i}{dt^2} = \cdots = \frac{d^{n-1} i}{dt^{n-1}} = 0; \qquad \frac{d^n i}{dt^n} = E \qquad \text{for } t = 0 \quad (11)$$

where by $t = 0$ we now mean the instant immediately *after* the excitation is applied.

These $n + 1$ so-called initial conditions suffice to determine the $n + 1$ amplitudes in Eq. 10, for they yield the following set of simultaneous algebraic equations:

$$\begin{aligned} s_0^0 B_0 + s_1^0 B_1 + s_2^0 B_2 + \cdots + s_n^0 B_n &= 0 \\ s_0 B_0 + s_1 B_1 + s_2 B_2 + \cdots + s_n B_n &= 0 \\ s_0^2 B_0 + s_1^2 B_1 + s_2^2 B_2 + \cdots + s_n^2 B_n &= 0 \\ \cdots\cdots\cdots\cdots\cdots\cdots \\ s_0^n B_0 + s_1^n B_1 + s_2^n B_n + \cdots + s_n^n B_n &= E \end{aligned} \quad (12)$$

Observe that right-hand members are zero in all but the last of these equations. The coefficients s_0^0, s_1^0, etc., which have the value unity (because any quantity raised to the zero power equals one) are inserted merely to give the complete set of coefficients a homogeneous appearance.

The very special form of determinant appearing in these equations (called a *Vandermonde determinant*) enables us to express the solutions in closed form with relatively little effort.*

We begin by defining a polynomial

$$(s - s_0)(s - s_1) \cdots (s - s_{\nu-1})(s - s_{\nu+1}) \cdots (s - s_n) \quad (13)$$

with zeros equal to the n roots of the following equation:

$$s^n + H_{n-1}^{(\nu)} s^{n-1} + H_{n-2}^{(\nu)} s^{n-2} + \cdots + H_1^{(\nu)} s + H_0^{(\nu)} = 0 \quad (14)$$

in which the coefficients $H_k^{(\nu)}$ are the well-known *symmetric functions* of these roots as expressed by the relations

$$\begin{aligned} H_0^{(\nu)} &= (-1)^n s_0 s_1 \cdots s_{\nu-1} s_{\nu+1} \cdots s_n \\ H_1^{(\nu)} &= (-1)^{n-1} \times (\text{sum of } n s_k\text{'s taken } n-1 \text{ at a time}) \\ H_2^{(\nu)} &= (-1)^{n-2} \times (\text{sum of } n s_k\text{'s taken } n-2 \text{ at a time}) \\ &\cdots\cdots\cdots\cdots \\ H_{n-1}^{(\nu)} &= (-1) \times (s_0 + s_1 + \cdots + s_{\nu-1} + s_{\nu+1} + \cdots + s_n) \end{aligned} \quad (15)$$

* The method of solution given here is due to Dr. M. V. Cerillo of the M.I.T. Research Laboratory of Electronics.

Observe that Eq. 13 defines $n + 1$ polynomials since the missing factor $s - s_\nu$ may involve any one of the quantities $s_0, s_1, \ldots s_n$. Correspondingly Eq. 15 defines $n + 1$ sets of symmetric functions.

According to Cramer's rule, amplitude B_ν in the equation set 12 is given by

$$B_\nu = \frac{\overbrace{\begin{vmatrix} s_0^0 & s_1^0 & \cdot & 0 & \cdot & s_n^0 \\ s_0^1 & s_1^1 & \cdot & 0 & \cdot & s_n^1 \\ \cdot & \cdot & \cdot & \cdot & \cdot & \cdot \\ \cdot & \cdot & \cdot & \cdot & \cdot & \cdot \\ \cdot & \cdot & \cdot & \cdot & \cdot & \cdot \\ s_0^n & s_1^n & \cdot & E & \cdot & s_n^n \end{vmatrix}}^{(\nu+1)\text{th column}}}{\begin{vmatrix} s_0^0 & s_1^0 & s_2^0 & \cdots & s_n^0 \\ s_0^1 & s_1^1 & s_2^1 & \cdots & s_n^1 \\ \cdot & \cdot & \cdot & \cdot & \cdot \\ s_0^n & s_1^n & s_2^n & \cdots & s_n^n \end{vmatrix}} \tag{16}$$

To the $H_0^{(\nu)}$-multiplied first row in each of these determinants we now add the $H_1^{(\nu)}$-multiplied second row, the $H_2^{(\nu)}$-multiplied third row, . . . the $H_{n-1}^{(\nu)}$-multiplied nth row (which is the next-to-last row), and finally the last row as it stands. The remaining rows in each determinant we leave as they are.

Since the symmetric functions satisfy Eq. 14, we see that all elements except the $(\nu + 1)$th in the revised first row of each determinant vanish. Therefore B_ν equals simply the ratio of these two nonvanishing elements, since their corresponding cofactors are the same in numerator and denominator and hence cancel. We thus have

$$B_\nu = \frac{E}{H_0^{(\nu)}s_\nu^0 + H_1^{(\nu)}s_\nu^1 + H_2^{(\nu)}s_\nu^2 + \cdots + H_{n-1}^{(\nu)}s_\nu^{n-1} + s_\nu^n} \tag{17}$$

which is equivalent to

$$B_\nu = \frac{E}{(s_\nu - s_0)(s_\nu - s_1)\cdots(s_\nu - s_{\nu-1})(s_\nu - s_{\nu+1})\cdots(s_\nu - s_n)} \tag{18}$$

If we denote the polynomial in Eqs. 7 or 8 by $P(s)$, we can also write this expression in the compact form

$$B_\nu = \left[\frac{(s - s_\nu)E}{(s - s_0)P(s)}\right]_{s=s_\nu} \qquad \text{for } \nu = 0, 1, 2, \ldots n \tag{19}$$

Together with Eq. 10, this result represents the complete solution to differential equation 3 for the excitation function specified in Eq. 2 and initial rest conditions.

A solution to the differential equation 1 for the same conditions can now readily be formulated, since the right-hand side of this equation may be regarded as a sum of excitation functions which are constant-multiplied derivatives of the excitation function in Eq. 3. We have but to add the corresponding constant-multiplied derivatives of solution 10 together to form the desired solution to Eq. 1.

So long as the integer m is less than n, the highest derivative of $i(t)$ in the solution 10 encountered in this process is still continuous at $t = 0$. If $m = n$, the highest derivative of $i(t)$ contains a step function; and for $m = n + 1$, it contains an impulse. According to the discussion in art. 5 of Ch. VIII, m cannot be larger than $n + 1$. In any event, the successive derivatives of the solution to Eq. 3 can give us no trouble, and so the contemplated formulation of a solution to Eq. 1 is justified.

As we shall see, the process of constructing a solution can always be handled in such a way that a value of m equal to or larger than n need never be considered. We shall, therefore, proceed on this assumption.

If we focus our attention upon a particular term in Eq. 10 involving B_ν and note that each successive differentiation multiplies this amplitude by s_ν, we see that the process just described has the effect of multiplying B_ν by

$$b_0 + b_1 s_\nu + b_2 s_\nu^2 + \cdots + b_m s_\nu^m = Q(s_\nu) \tag{20}$$

in which we have introduced the polynomial

$$Q(s) = b_0 + b_1 s + b_2 s^2 + \cdots + b_m s^m \tag{21}$$

If we introduce the amplitude factors

$$A_\nu = B_\nu \cdot Q(s_\nu) \tag{22}$$

and recognize that the rational function

$$Y(s) = \frac{Q(s)}{P(s)} \tag{23}$$

according to Eq. 9 of Ch. VIII or Eq. 15 of Ch. VII is simply the pertinent admittance function, we find from Eq. 19 the following expression for the amplitudes 22:

$$A_\nu = \left[(s - s_\nu) \cdot Y(s) \cdot \frac{E}{(s - s_0)}\right]_{s=s_\nu} \qquad \text{for } \nu = 0, 1, 2, \ldots n \tag{24}$$

and the desired solution to Eq. 1 reads

$$i(t) = A_0 e^{s_0 t} + A_1 e^{s_1 t} + A_2 e^{s_2 t} + \cdots + A_n e^{s_n t} \tag{25}$$

Our first objective in this chapter is thus accomplished.

From this result it is a simple matter to obtain the response of the network to a unit step or impulse excitation. Thus the function $e(t)$ defined by Eqs. 2 becomes a unit step if we let $E = 1$ and $s_0 = 0$. Formula 24 then reads

$$A_\nu = \left[(s - s_\nu) \cdot \frac{Y(s)}{s}\right]_{s=s_\nu} \qquad \text{for } \nu = 1, 2, \ldots n \tag{26}$$

while 25 is more appropriately written

$$i(t) = Y(0) + A_1 e^{s_1 t} + A_2 e^{s_2 t} + \cdots + A_n e^{s_n t} \tag{27}$$

and, as usual, this function represents the response only for $t \geq 0$; for $t < 0$, $i(t) = 0$. The first term, which is constant, represents the steady-state response, since the remaining exponential terms become zero for large values of t.

The unit-impulse response is the time derivative of 27. If $m < n$ (as we are tentatively assuming it to be) then this function is continuous at $t = 0$. However, the derivative is not necessarily continuous at this point unless $m < (n - 1)$; for $m = (n - 1)$, the impulse response is discontinuous at $t = 0$. Nevertheless, since function 27 itself is continuous at this point, the process of differentiation applies only to $t > 0$ and hence the first term, being constant, contributes nothing to the result while the amplitudes A_ν in the remaining terms become multiplied by s_ν.

We can accommodate this result by writing the unit impulse response in the form

$$i(t) = k_1 e^{s_1 t} + k_2 e^{s_2 t} + \cdots + k_n e^{s_n t} \tag{28}$$

in which the amplitudes are given by the formula

$$k_\nu = [(s - s_\nu) \cdot Y(s)]_{s=s_\nu} \tag{29}$$

These we recognize as being the residues of $Y(s)$ appearing in its partial-fraction expansion

$$Y(s) = \frac{k_1}{s - s_1} + \frac{k_2}{s - s_2} + \cdots + \frac{k_n}{s - s_n} \tag{30}$$

which is applicable only if $m < n$ so that $Y(s) \to 0$ for $s \to \infty$.

When this condition is not fulfilled, that is to say, when $m = n$ or $n + 1$, then $Y(s)$ is an improper fraction. In that case we divide denominator into numerator by long division until we have a remainder that is one degree less than the degree of the denominator. Algebraically this process is expressed by writing

$$Y(s) = \alpha_1 s + \alpha_0 + Y'(s) \tag{31}$$

in which $Y'(s)$ now fulfills the condition $m = n - 1$ and the additive linear polynomial in 31 is called the *rational entire remainder* function.

Physically it represents a parallel RC combination for which the response can be written down by inspection and added to function 28. Thus it is clear that the assumption $m < n$ made above involves essentially no restriction in the generality of our discussion, and we shall therefore continue to make it.

It is, incidentally, clear that values for the residues obtained from formula 29 are the same whether we use $Y(s)$ or $Y'(s)$ since in this evaluation the additive terms in 31 yield nothing. In evaluating residues in a situation where $m \geq n$ it is entirely optional whether we do this in terms of the given $Y(s)$ or carry out the long-division process first and then find residues of the proper fraction $Y'(s)$. Formula 29 yields the same result either way.

In terms of these results we are now in a position to introduce a useful interpretation. Comparing 28 with 30 we recognize that to each term in 30 of the form $k_\nu/(s - s_\nu)$ there corresponds a term $k_\nu e^{s_\nu t}$ in 28. The former is a function of frequency, the latter a function of time. We can say that $k_\nu/(s - s_\nu)$ is the *frequency-domain representation* of $k_\nu e^{s_\nu t}$ or that the latter is the *time-domain representation* of $k_\nu/(s - s_\nu)$.

This attitude implies that actually only one function is involved; the expression $k_\nu e^{s_\nu t}$ represents this function in a domain in which time is the independent variable while the form $k_\nu/(s - s_\nu)$ represents it in a domain in which frequency is the independent variable. The situation is in many respects not unlike that resulting from a transformation of the independent variable, such as introducing a frequency transformation or a conformal transformation as discussed in the previous chapter. The detailed aspects are, of course, more subtle than this (as will be elaborated in the following chapters), but in a certain sense we can nevertheless speak of $k_\nu/(s - s_\nu)$ as the *transform* of $k_\nu e^{s_\nu t}$ if we like, and regard the latter as the *inverse transform* of the frequency function $k_\nu/(s - s_\nu)$.

$Y(s)$ is the frequency-domain representation of the network while the impulse response 28 is its time-domain representation. Just as $Y(s)$ describes the network in the frequency domain, so the impulse response 28 describes it in the time domain.

Analogously we may say that the factor $E/(s - s_0)$ in the bracket expression 24 represents the voltage excitation function in the frequency domain while $Ee^{s_0 t}$ as in Eq. 2 describes it in the time domain because these two functions have the same forms as do respective terms in the expressions 28 and 30.

It is of collateral interest to call attention to the fact that the admittance

$$y_\nu = \frac{k_\nu}{s - s_\nu} \tag{32}$$

corresponding to a term in the partial-fraction expansion 30, obviously stems from an equilibrium equation that reads

$$\frac{di}{dt} = s_\nu i = k_\nu e(t) \tag{33}$$

which is the equilibrium relation of a series RL circuit with $L = 1/k_\nu$ and $R = -s_\nu/k_\nu$. Recalling the unit impulse response of that simple circuit, we find the solution of 33 for $e(t)$ equal to a unit impulse to be

$$i(t) = k_\nu e^{s_\nu t} \qquad \text{for } t > 0 \tag{34}$$

We thus have an independent derivation of the results stated by Eqs. 28, 29 and 30.

So far we have overlooked the fact that the voltage excitation function defined by Eq. 2 is complex and hence is not representative of a physical situation. We can, of course, meet this objection by taking the real part of solution 25, corresponding to taking the real part of the function in Eq. 2. It is, however, computationally easier to revise our excitation function to have the form

$$\begin{aligned} e(t) &= 0 && \text{for } t < 0 \\ e(t) &= \tfrac{1}{2}(Ee^{s_0 t} + \bar{E}e^{\bar{s}_0 t}) && \text{for } t \geq 0 \end{aligned} \tag{35}$$

where the bar over a quantity indicates its conjugate value.

Since the network is linear, we add solutions for the separate terms in 35 to form a resultant solution having the form

$$i(t) = A_0 e^{s_0 t} + \bar{A}_0 e^{\bar{s}_0 t} + A_1 e^{s_1 t} + A_2 e^{s_2 t} + \cdots + A_n e^{s_n t} \tag{36}$$

in which amplitudes are evaluated by the expression

$$A_\nu = [(s - s_\nu) \cdot Y(s) \cdot E(s)]_{s=s_\nu} \qquad \text{for } \nu = 0, 1, 2, \ldots n \tag{37}$$

with the transform of source function 35 given by

$$E(s) = \frac{1}{2}\left[\frac{E}{s - s_0} + \frac{\bar{E}}{s - \bar{s}_0}\right] \tag{38}$$

Solution 36 for the real excitation 35 is a real function of the time. The first two terms representing the particular integral are a conjugate pair; and the remaining terms forming the complementary function are real for real s_ν-values and conjugate complex for pairs of conjugate s_ν-values (which is not so in the solution 25 because the corresponding $E(s)$ term in formula 24 as compared with formula 37 is not real for real s).

Observe that the transform of the source given by $E(s)$ in Eq. 38, like $Y(s)$, is a rational function (quotient of polynomials) in the frequency variable s. Its numerator and denominator polynomials are respectively of first and second degree. Like the polynomials in $Y(s)$, these have real coefficients also. $E(s)$ is real for real values of s. Moreover, if $e(t)$ in Eq. 35 is stable, s_0 has a negative real part. Hence $E(s)$, like $Y(s)$, is analytic in the right half of the s-plane.

Because of these facts we can do all sorts of interesting things regarding the interpretation of solution 36 and formula 37. For example, we can interchange the identities of $E(s)$ and $Y(s)$. We can suppose that we have a network with a response function equal to the rational function 38 excited by a voltage equal to the time function given by Eq. 28. This network has a unit impulse response equal to the time function defined by Eq. 35. The net response for this altered situation is still given by Eq. 36 in which all coefficients A_ν are the same as before because the product of $Y(s)$ and $E(s)$ in Eq. 37 is the same.

Alternately we can assume a network with the response function $Y'(s) = Y(s) \cdot E(s)$ excited by a unit impulse. Equations 36 and 37 still yield the resultant time-domain behavior.

Since the resulting time response depends only upon the rational function given by the product of $Y(s)$ and $E(s)$, we can split this total rational function into factors and assign these to a pair of excitation and response functions in any way we wish. Many different physical situations (networks and excitation functions) are thus seen by inspection to have the same net response in the time domain.

In order to foster and exploit this attitude it is effective to suppress the separate identities of $Y(s)$ and $E(s)$ by introducing the total rational function

$$F(s) = Y(s) \cdot E(s) = \frac{N(s)}{D(s)} \tag{39}$$

in which $N(s)$ and $D(s)$ denote the resultant numerator and denominator polynomials; and since we no longer wish to identify specific poles of $F(s)$ with an excitation or response function, we lump them all together, let n denote their total number, and write

$$D(s) = (s - s_1)(s - s_2) \cdots (s - s_n) \tag{40}$$

Consistently, we denote the corresponding time function by $f(t)$, since it is not necessarily a current but could be either a voltage or a current response, and have more simply

$$f(t) = A_1 e^{s_1 t} + A_2 e^{s_2 t} + \cdots + A_n e^{s_n t} \tag{41}$$

with

$$A_\nu = [(s - s_\nu) \cdot F(s)]_{s=s_\nu} \tag{42}$$

$F(s)$ is the transform of $f(t)$, the latter is the inverse transform of $F(s)$, and formula 42 provides the means for carrying out the inverse transformation. The direct transformation from $f(t)$ to $F(s)$ is established by recognizing that the quantities A_ν are residues of the function $F(s)$ in its poles and that we have

$$F(s) = \frac{A_1}{s - s_1} + \frac{A_2}{s - s_2} + \cdots + \frac{A_n}{s - s_n} \tag{43}$$

from which this rational function can be constructed for any given time function in the form of equation 41, that is to say, for a sum of complex exponential time functions.

The assumption that $F(s)$ is a proper fraction is again implicit here. If this assumption is not fulfilled, then we reduce $F(s)$ to a proper form by the long-division process indicated in Eq. 31 and interpret the polynomial $\alpha_0 + \alpha_1 s$ as yielding an impulse and a doublet in the time domain. The following discussion elaborates this as well as numerous other interesting properties of the result summarized by Eqs. 39 through 43.

2. Elementary Properties of the Solution; TD-FD Correlations

In applying the preceding results to practical problems we frequently encounter situations where it is necessary to carry out elementary operations such as differentiation, integration, displacement of the origin or scaling in either the frequency or time domain. If such an operation is carried out on one of the two functions $f(t)$ or $F(s)$, it is useful to know the effect upon the other function. We refer to such inter-relationships as time-domain–frequency-domain (abbreviated TD–FD) correlations.

Regarding Eq. 41 we see by inspection that differentiation has the effect of multiplying each A_ν by the corresponding s_ν-value, while integration has the effect of dividing each A_ν by s_ν. Formula 42 shows that these effects upon the coefficients A_ν are alternately and respectively obtained if $F(s)$ is multiplied by or divided by s. We can state this result as an elementary property, namely:

If $f(t)$ is replaced by df/dt, then
$F(s)$ becomes replaced by $s\,F(s)$ (44)

and

If $f(t)$ is replaced by $\int f(t)\,dt$, then
$F(s)$ becomes replaced by $(1/s)\,F(s)$ (45)

The complementary pair of operations involve the differentiation of $F(s)$ and the resulting effect upon $f(t)$, but first we must consider the operation of displacement in the frequency domain. By this we mean replacing the variable s in $F(s)$ by $s - s_0$ where s_0 is any finite complex value. The effect upon $F(s)$ may best be seen by regarding the polynomials in Eq. 39 in their factored form as shown for $D(s)$ in Eq. 40. Each factor $s - s_\nu$ becomes $s - s_0 - s_\nu = s - (s_\nu + s_0)$. Thus each critical frequency (zero or pole) in $F(s)$ has an increment s_0 added to it; the entire pole-zero pattern of $F(s)$ is displaced by an amount s_0.

The effect upon $f(t)$ can be recognized from Eq. 41. If each s_ν is replaced by $s_\nu + s_0$, the entire expression is multiplied by an exponential factor $s^{s_0 t}$. Hence we can state the following property:

$$\begin{aligned} &\textit{If } F(s) \textit{ is replaced by } F(s - s_0), \textit{ then} \\ &f(t) \textit{ becomes replaced by } e^{s_0 t} \cdot f(t) \end{aligned} \tag{46}$$

In terms of this result we can formulate the derivative of $F(s)$ from its basic definition which reads

$$\frac{d\,F(s)}{ds} = \lim_{\Delta s \to 0} \left[\frac{F(s + \Delta s) - F(s)}{\Delta s}\right] \tag{47}$$

and since

$$\lim_{\Delta s \to 0} \left[\frac{e^{-\Delta st} f(t) - f(t)}{\Delta s}\right] = \frac{(1 - \Delta st) f(t) - f(t)}{\Delta s} = -t\,f(t) \tag{48}$$

we have the statements, complementary to 44 and 45:

$$\begin{aligned} &\textit{If } F(s) \textit{ is replaced by } dF/ds, \textit{ then} \\ &f(t) \textit{ is replaced by } -t\,f(t) \end{aligned} \tag{49}$$

and

$$\begin{aligned} &\textit{If } F(s) \textit{ is replaced by } \textstyle\int F(s)\,ds, \textit{ then} \\ &f(t) \textit{ is replaced by } (-1/t)\,f(t) \end{aligned} \tag{50}$$

An operation complementary to that expressed in 46 involves displacement of the time origin, which amounts to replacing t by $t - t_0$ in $f(t)$. The effect upon $F(s)$ can formally be recognized by noting in Eq. 41 that each coefficient A_ν becomes multiplied by an exponential factor $e^{-s_\nu t_0}$. In formula 42 this same result is obtained if $F(s)$ is replaced by $e^{-st_0}F(s)$. A disturbing feature here is the fact that this revised frequency function is no longer rational and the results summarized by Eqs. 39 through 43, which are derived for a lumped-parameter network, are strictly speaking not applicable.

However, if we are aware of the fact that a lossless transmission line is a means for producing a perfect time delay, we see that the transcendental function involved here is the limiting form of a rational transfer function since a lossless line may be approximated arbitrarily closely by a uniform lumped ladder network involving inductive series branches and capacitive shunt branches, simply by making the element values smaller and their number correspondingly larger. In other words, we may regard the function e^{-st_0} as approximatable by a rational function in a manner that is uniform over the entire finite s-plane and in the limit is still representative of a physical network which realizes this transcendental function exactly.*

In this sense we can accept the revised frequency function resulting from the change-of-variable $t \rightarrow (t - t_0)$ in $f(t)$ as belonging to the class of functions to which the derivation of results 39–43 pertain, and so we can make the statement:

$$\begin{aligned} &\textit{If } f(t) \textit{ is replaced by } f(t - t_0), \textit{ then} \\ &F(s) \textit{ is replaced by } e^{-st_0} \cdot F(s) \end{aligned} \tag{51}$$

Finally, we come to the operation of frequency scaling with which we are familiar from our elementary considerations of network theory but which we need to state more precisely in terms of our present results.

In Eq. 43 we observe that if we replace the frequency variable s by s/a, we get

$$F\frac{s}{a} = \sum_{\nu=1}^{n} \frac{aA_\nu}{s - as_\nu} \tag{52}$$

Each amplitude A_ν and each natural frequency s_ν is multiplied by a. The effect upon $f(t)$ is readily seen from Eq. 41. We get

$$\sum_{\nu=1}^{n} aA_\nu e^{as_\nu t} = af(at) \tag{53}$$

Hence we can state:

$$\begin{aligned} &\textit{If } F(s) \textit{ is replaced by } F(s/a), \textit{ then} \\ &f(t) \textit{ becomes replaced by } af(at) \end{aligned} \tag{54}$$

For $a > 1$, the s-plane pole-zero constellation of $F(s)$ is expanded; all s_ν-values are a times larger. The time scale is compressed since $t \rightarrow at$ makes the time pass faster. Also the time function is a times taller. Note that the area given by

$$\int_0^\infty f(t)\,dt \rightarrow \int_0^\infty f(at)\,d(at) \tag{55}$$

remains the same. *Net area enclosed by the time function is invariant to frequency scaling.*

These various elementary properties of the time- and frequency-domain functions are useful in that they often enable us to obtain many new

* In this connection see also the remarks in opening paragraphs of art. 4 below.

results from a given one with little effort. For example, in Eq. 24 the frequency function $E/(s - s_0)$ is recognized as being the transform of the time function defined by Eq. 2, and the latter becomes a unit step function if we let $s_0 = 0$ and $E = 1$. Hence we see that a unit step $u_{-1}(t)$ has the transform $1/s$. If we apply the properties 44 and 45 to this simple result, we see at once that the transform of singularity function $u_n(t)$ of order n is s^n. Specifically, the transform of a unit impulse $u_0(t)$ is constant, equal to unity.

Another example is provided by an elaboration of property 46 which reads:

$$\begin{aligned} &\textit{If } F(s) \rightarrow \tfrac{1}{2}[F(s - j\omega_0) + F(s + j\omega_0)], \\ &\textit{then } f(t) \rightarrow f(t) \cos \omega_0 t \end{aligned} \tag{56}$$

where the arrow is an abbreviation for the words "is replaced by." The resulting time function here is recognized as representing an amplitude-modulated carrier function. The result 56 immediately yields the corresponding pole-zero plot in the s-plane in terms of that pertinent to the modulation function $f(t)$. This result has many practical applications.

The scaling property 54 when applied to the singularity functions is seen by inspection to be correct since this process is equivalent to multiplication of the function by $1/a^n$. Thus for functions of order $n = -m = -1$ or less, which are $t^{m-1}/(m - 1)!$ for $t > 0$, the process $f(t) \rightarrow a f(at)$ clearly amounts to multiplication by $a^m = 1/a^n$. For the higher-order functions which are generated through a limit process $\delta \rightarrow 0$ from pulses of width δ and enclosed areas proportional to $1/\delta^n$, invariance of these areas to the transformation $t \rightarrow at$ or $\delta \rightarrow \delta/a$ again requires multiplication of the pertinent function by $1/a^n$. The effect upon the unit impulse incidentally is nil.

In a frequency-domain response function like the admittance 23, which may be written as a constant multiplier H times a quotient of frequency factors $s - s_k$, the transformation $s \rightarrow s/a$ may evidently be interpreted through replacing these factors by $s - as_k = s - s_k^*$ and the multiplier by $H^* = Ha^{(n-m)}$ where integers m and n denote the degrees of numerator and denominator polynomials respectively. The transformed admittance function Y^* has the property expressed by $Y^*(s) \equiv Y(s/a)$ or $Y^*(as) \equiv Y(s)$. If N^* and N denote the corresponding networks then these are alike except that values of all L's and C's in N^* are divided by the scale factor a.

3. Asymptotic Behavior and Moment Conditions

Additionally useful properties of the time and frequency functions are obtained by considering their behavior for very large and very small values of their independent variables. The first of these that we shall consider provides correlation between the behavior of $F(s)$ near $s = 0$ and that of $f(t)$ for large t. We begin by using Eq. 41 to form the following integral:

$$\int_0^\infty f(t)\,dt = \left[\sum_{\nu=1}^{n} \frac{A_\nu}{s_\nu} e^{s_\nu t}\right]_0^\infty \tag{57}$$

If $F(s)$ has only left half-plane poles (none on the j-axis) then this expression is zero for the upper limit. With the help of Eq. 43, we then get

$$\int_0^\infty f(t)\,dt = -\sum_{\nu=1}^{n} \frac{A_\nu}{s_\nu} = F(0) \tag{58}$$

In a situation for which $F(0) = 0$, this result tells us that $f(t)$ must cross the zero axis at least once since the net enclosed area could otherwise not be zero. It also tells us that for large values of t, $f(t)$ must become zero faster than $1/t$; otherwise the integral could not have a finite value, let alone the value zero. Hence we recognize that a correlation exists between the behavior of $F(s)$ near $s = 0$ and that of $f(t)$ near $t = \infty$.

By applying property 49 we can obtain a more general form of this result that reads

$$\int_0^\infty (-t)^n f(t)\,dt = \left(\frac{d^nF}{ds^n}\right)_{s=0} \tag{59}$$

If $F(s)$ has a zero of order $k + 1$ at $s = 0$ so that

$$F(0) = \left(\frac{dF}{ds}\right)_{s=0} = \cdots = \left(\frac{d^kF}{ds^k}\right)_{s=0} = 0 \tag{60}$$

then we have

$$\int_0^\infty t^n f(t)\,dt = 0 \qquad \text{for } n = 0, 1, 2, \ldots k \tag{61}$$

The quantity on the left-hand side of this relation is the *nth-order moment* of function $f(t)$, so called because of its similarity in form to moments encountered in mechanics. The results stated by 60 and 61, therefore, are referred to as the *moment conditions*. They are recognizable by inspection of the rational function $F(s)$, noting its behavior for $s \to 0$.

If $F(s)$ has a kth-order zero at $s = 0$ then all moments of $f(t)$ up to and including the one of order $k - 1$ are zero, which means that for large values of t, $f(t)$ vanishes at least as fast as $1/t^k$.

A result which is complementary to this one is also obtained from relations 41 and 43. Since the behavior of $F(s)$ for large s is involved here, it is helpful to put the expansion 43 into a form that is more appropriate to studying its high-frequency behavior. The Maclaurin expansion of a typical term about the point $s = \infty$ reads

$$\frac{A_\nu}{s - s_\nu} = \frac{A_\nu}{s}\left[1 + \frac{s_\nu}{s} + \frac{s_\nu^2}{s^2} + \cdots\right] \tag{62}$$

and 43 is correspondingly rewritten in the form

$$F(s) = \sum_{\nu=1}^{n} A_\nu\left[\frac{1}{s} + \frac{s_\nu}{s^2} + \frac{s_\nu^2}{s^3} + \cdots\right] \tag{63}$$

If it is known that $F(s)$ has a zero of order $k + 2$ at $s = \infty$ so that

$$F(s) \to \frac{K}{s^{k+2}} \qquad \text{for } s \to \infty \tag{64}$$

then the first $k + 1$ terms in the summand in Eq. 63 must vanish, as is indicated by writing

$$\sum_{\nu=1}^{n} s_\nu{}^n A_\nu = 0 \qquad \text{for } n = 0, 1, 2, \ldots k \tag{65}$$

Eq. 41 then yields

$$f(0) = \left(\frac{df}{dt}\right)_{t=0} = \cdots = \left(\frac{d^k f}{dt^k}\right)_{t=0} = 0 \tag{66}$$

which is the desired result. It enables one to see by inspection how the time function behaves near $t = 0$ when the rational function $F(s)$ is given; or if a desired behavior of $f(t)$ for $t \to 0$ is specified, it fixes the asymptotic behavior of the pertinent frequency function.

The singularity functions are readily seen to meet this asymptotic condition. The unit ramp $u_{-2}(t)$ with transform $1/s^2$ corresponds to $k = 0$ and hence 66 yields only $u_{-2}(0) = 0$, which is correct. For the parabolic ramp $u_{-3}(t)$ with transform $1/s^3$, 66 yields the initial value and initial derivative zero, which is also correct. Functions of higher order than the linear ramp are evidently not involved in this asymptotic criterion since they are either nonzero at $t = 0$ (like the step) or have a singular behavior there. This conclusion is also obvious from the fact that result 66 is meaningless if the integer k is negative or, according to Eq. 64, if $F(s)$ fails to vanish at least as fast as $1/s^2$ for large s.

4. Some Rational Approximants for the Delay Factor e^{-st_0}

In this article we wish to elaborate the contention that the transcendental delay factor e^{-st_0} can be regarded as the limit of a rational approximant as the number of its critical frequencies is indefinitely increased, and that we may, therefore, consider this factor as belonging to the same class of functions as those characterizing lumped networks. Stated in another way, this attitude recognizes that so-called *distributed-parameter* networks like transmission lines and waveguides are merely limiting forms of lumped-parameter networks, or that the latter are artificially simplified versions of the former, and that their describing functions in both time and frequency domains have essentially the same analytic properties whether the number of natural frequencies is finite or not.

The mere fact that networks like transmission lines and waveguides have an infinite number of natural frequencies does not destroy their kinship with lumped-parameter networks whose natural frequencies are finite in number. Rather it is logical for physical reasons to expect that the transcendental response functions characterizing distributed-parameter networks belong networkwise to the same class as response functions characterizing lumped-parameter networks. In fact, if we regard the subject of circuit theory as an approximation to field theory, then the so-called "lumped" network and its characterization in terms of rational functions is already playing the role of a discrete approximant to a distributed system, and any further distinction between general properties of these two embodiments is not only trivial but redundant as well.

In evaluating the transient response of waveguides and similar distributed-parameter networks by the Fourier methods discussed in following chapters, one is faced with the problem of evaluating integrals in which a combination of irrational and transcendental functions appear. The only way in which such integrals have successfully been evaluated is through replacing the integrand by an appropriate rational approximant. Such a method allows a careful control of error prediction and evaluation whereby the adequacy of the approximations can be measured. In such an approach we are in essence replacing the distributed-parameter network by a lumped-parameter approximation and accepting its response as a sufficiently close replica of what we set out to find.

In order to present a more concrete picture of what these remarks imply, we shall discuss some specific methods yielding rational approximants for the delay factor e^{-st_0} or, what amounts to the same thing, lumped-network approximations to a lossless transmission line.

For $s = j\omega$ this exponential factor evidently has a constant unit magnitude and a lag angle that is linearly increasing with frequency. The discussion in art. 3, Ch. IX shows that a pole-zero array yielding a transfer function having such a constant j-axis magnitude and continuously increasing phase lag is obtained by placing poles in the left half-plane and zeros at points in the right half-plane that are images of the pole positions about the j-axis. In particular, if we use a uniform spacing of poles, as shown in the s-plane sketch in Fig. 1, the resulting phase-lag function is linear except for a uniform superimposed ripple that becomes smaller with an increased spacing σ or a decreased spacing $\Delta\omega$. Since in a practical problem we are interested in simulating the delay factor e^{-ts_0} only over a finite frequency range, the pole pattern need not be infinite in extent. For any finite number of poles, however large, we obtain in this way a rational approximant to the transcendental delay factor. If β denotes the phase-lag function, we want it ideally to equal ωt_0; that is to say, we want the phase function to have a constant slope t_0. Assuming the pole array in Fig. 1 tentatively to be infinite in extent, we can evaluate the actual slope $d\beta/d\omega$ as a function of ω with the spacing $\Delta\omega$ and the quantity σ as parameters. We find in this way that the average slope is given by

$$t_0 = \left(\frac{d\beta}{d\omega}\right)_{\text{av}} = \frac{\pi}{\Delta\omega}\coth\frac{2\pi\sigma}{\Delta\omega} \approx \frac{\pi}{\Delta\omega} \tag{67}$$

and that the ripple factor becomes

$$\frac{(d\beta/d\omega)_{\max} - (d\beta/d\omega)_{\min}}{(d\beta/d\omega)_{\max} + (d\beta/d\omega)_{\min}} = \frac{1}{\cosh(2\pi\sigma/\Delta\omega)} \approx 2e^{-2\pi\sigma/\Delta\omega} \tag{68}$$

since it is rather obvious that the ratio $2\pi\sigma/\Delta\omega$ must be large.

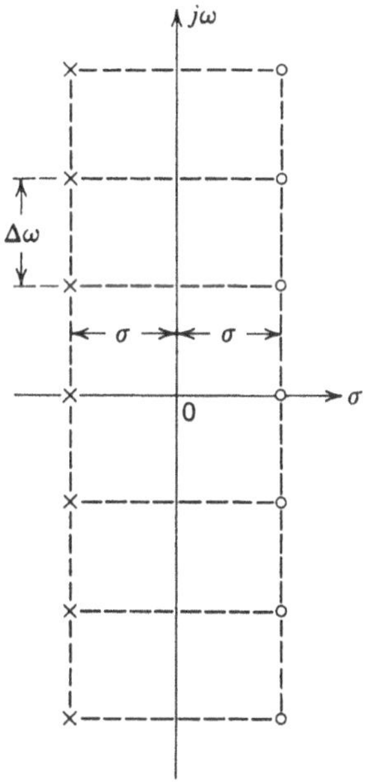

FIG. 1. Uniform pole-zero pattern yielding a rational approximant to the delay factor e^{-st_0}.

In terms of these relations we can construct a rational approximant to the delay factor e^{-st_0} for any desired delay and specified delay variation as determined by the ripple factor. Although formulas 67 and 68 are based upon an infinite uniform pole distribution, their accuracy for any finite array may be made arbitrarily good by considering a sufficiently large total number of poles.

Another approach to this problem, which is rather practical because it considers a lumped-network realization for the rational approximant at the outset, is based upon representing the transfer function by means of the so-called constant-resistance lattice network. This is the symmetrical lattice* in which the branch impedances z_a and z_b are reciprocal ($z_a z_b = 1$).

* See *ICT*, p. 164.

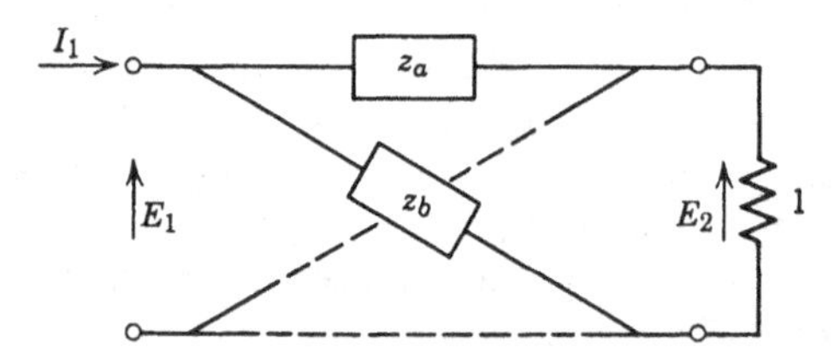

FIG. 2. Constant-resistance lossless lattice which may be used as a building block for the delay factor e^{-st_0}.

Under this condition the input impedance E_1/I_1 is a 1-ohm resistance when the lattice is terminated in 1 ohm, and the transfer function is given by

$$\frac{E_2}{I_1} = Z_{12} = \frac{1 - z_a}{1 + z_a} \tag{69}$$

and conversely

$$z_a = \frac{1 - Z_{12}}{1 + Z_{12}} \tag{70}$$

For $Z_{12} = e^{-s}$ this gives

$$z_a = \frac{1 - e^{-s}}{1 + e^{-s}} = \tanh \frac{s}{2} \tag{71}$$

By using the continued fraction expansion of the hyperbolic tangent function

$$\tanh \frac{s}{2} = \cfrac{1}{2/s + \cfrac{1}{6/s + \cfrac{1}{10/s + \cfrac{1}{14/s + \cdots}}}} \tag{72}$$

and identifying this with the driving-point impedance z_a, we can see by inspection that the ladder network in Fig. 3 is its realization. Theoretically this network is infinite in extent, but a rational approximant for e^{-s} is obtained if we break off the expansion after a finite number of terms, the approximation becoming better the larger the number of terms chosen. The network for z_b is readily found by duality, and any delay t_0 is accommodated by frequency scaling.

Finally we will show another practically useful method of constructing rational approximants for transcendental or other functions. Let the function be denoted by $H(s)$. The starting point in this method is the

Maclaurin expansion

$$H(s) = C_0 + C_1 s + C_2 s^2 + \cdots + C_k s^k + \cdots \tag{73}$$

Consider the two polynomials

$$P_m(s) = \alpha_0 + \alpha_1 s + \alpha_2 s^2 + \cdots + \alpha_m s^m \tag{74}$$

and

$$Q_n(s) = \beta_0 + \beta_1 s + \beta_2 s^2 + \cdots + \beta_n s^n \tag{75}$$

of degree m and n respectively. The pertinent approximants are defined as that set for which

$$H(s) - \frac{P_m(s)}{Q_n(s)} = 0 + 0 + \cdots + 0 + C_{m+n+1} s^{m+n+1} + C_{m+n+2} s^{m+n+2} + \cdots \tag{76}$$

for any combination of the integers m and n and a fixed value for their sum. For any such fixed value, all possible approximants can be arranged

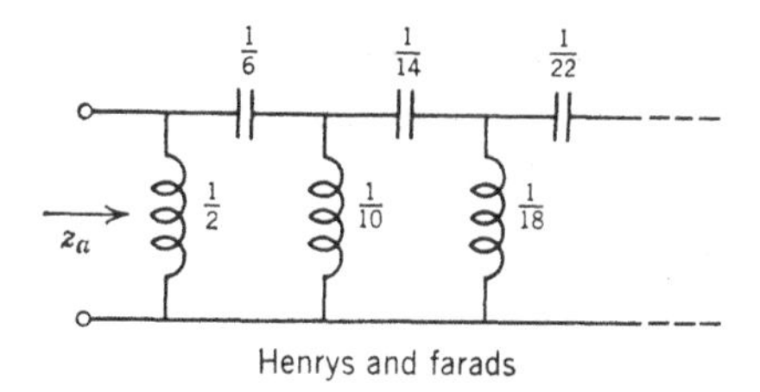

FIG. 3. Impedance realization in the a-arm of the lattice of Fig. 2.

in a rectangular table in which columns pertain to chosen m-values and rows to chosen n-values. Such an array is called a *Padé table* and the rational functions appearing therein are the *Padé approximants.*

The coefficients α and β for any of these are uniquely determined from the following equations which in essence assure that each Padé approximant shall have a Maclaurin expansion that coincides identically with that for the given function through $m + n$ terms

$$\begin{aligned}
&C_0\beta_0 = \alpha_0 \\
&C_1\beta_0 + C_0\beta_1 = \alpha_1 \\
&C_2\beta_0 + C_1\beta_1 + C_0\beta_2 = \alpha_2 \\
&\cdots\cdots\cdots\cdots \\
&C_m\beta_0 + C_{m-1}\beta_1 + \cdots + C_{m-n}\beta_n = \alpha_m \\
&C_{m+1}\beta_0 + C_m\beta_1 + \cdots + C_{m+1-n}\beta_n = 0 \\
&\cdots\cdots\cdots\cdots\cdots\cdots \\
&C_{m+n}\beta_0 + C_{m+n-1}\beta_1 + \cdots + C_n\beta_n = 0
\end{aligned} \tag{77}$$

Specifically for $m = n$, which leads to approximants located on the principal diagonal of the Padé table, we find for the function $H(s) = e^{-s}$ the polynomials

$$P_n(s) = 1 - \frac{s}{2} + \frac{n(n-1)}{2n(2n-1)} \cdot \frac{s^2}{2!} - \frac{n(n-1)(n-2)}{2n(2n-1)(2n-2)} \cdot \frac{s^3}{3!} + \cdots \quad (78)$$

and

$$Q_n(s) = 1 + \frac{s}{2} + \frac{n(n-1)}{2n(2n-1)} \cdot \frac{s^2}{2!} + \frac{n(n-1)(n-2)}{2n(2n-1)(2n-2)} \cdot \frac{s^3}{3!} + \cdots \quad (79)$$

The resultant transfer function may again be realized by the constant-resistance lattice in Fig. 2. It is interesting to find that for this particular Padé approximant one obtains precisely the same network as shown in Fig. 3 for the preceding method.

Although there are many other variants to these procedures for constructing rational approximants* for the delay factor e^{-s}, our purpose is served sufficiently by these examples which amply demonstrate that this transcendental may be regarded as the limit toward which the transfer function of an appropriate lumped-parameter network tends as the number of its elements is made larger and larger.

Problems

1. A unit voltage step applied to a series RLC circuit produces a current $i(t)$ for $t > 0$. A damped sinusoid

$$e(t) = e^{-\alpha t} \cos \omega_d t$$

applied to a different circuit produces the same current $i(t)$ for $t > 0$. Determine this circuit and the common response $i(t)$. Show that the asymptotic conditions discussed in art. 3 are fulfilled.

2. A circuit has the input admittance function

$$Y(s) = \frac{s + \frac{1}{2}}{s^2 + s + 1}$$

At $t = 0$ the voltage

$$e(t) = \cos t$$

* In this connection see *SPN*, pp. 632–644 and pp. 687–728. Also "An Application of Modern Network Synthesis to the Design of Constant-Time-Delay Networks with Low-Q Elements," Leo Storch, *1954 IRE Convention Record, Part 2, Circuit Theory*, pp. 105–117; other references will be found there.

is suddenly applied. Find the resulting current $i(t)$ for $t > 0$: (*a*) by using formulas 24 and 25; (*b*) by using formulas 36 and 37, showing that the results are identical.

3. Determine and plot the pole-zero patterns for transforms of the following time functions and indicate values of their constant multipliers:

(*a*) $f(t) = \cos \omega t$ (*b*) $f(t) = \sin \omega t$

(*c*) $f(t) = e^{-t} \cos \omega t$ (*d*) $f(t) = e^{-t} \sin \omega t$

(*e*) $f(t) = \cos (\omega t + 30°)$ (*f*) $f(t) = e^{-t} \sin (\omega t - 30°)$

Choose $\omega = 10$.

4. With reference to the adjacent circuit, the box N contains any finite combination of linear passive elements. With the switch S open, the capacitance

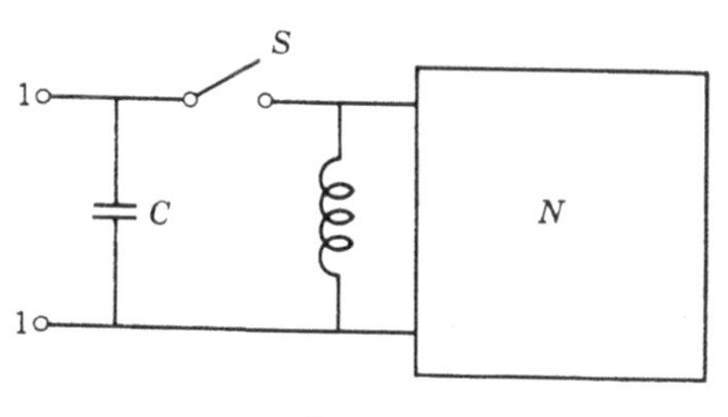

PROB. 4.

C is charged. At $t = 0$ the switch closes. Sketch the resulting voltage at terminals 1-1 for $t > 0$ showing its behavior for small t (whether or not it is continuous at $t = 0$) and the general character of the following response. Is it oscillatory, and if so, how many zero crossings does it display?

5. A unit current impulse is applied to a linear passive network and its voltage response is observed at (*a*) the driving point, (*b*) some other terminal pair. Can the response exhibit one or more zero crossings at (*a*) or at (*b*)? Can the response be continuous through $t = 0$ with one or more zero derivatives at (*a*) or at (*b*)? Explain in detail.

6. Given the adjacent pole-zero plot for a function $F(s) = F_1(s) \times F_2(s)$, consider time functions $f_1(t)$ and $f_2(t)$ which respectively are the inverse transforms of $F_1(s)$ and $F_2(s)$. Further, let N_1 and N_2 be networks realizing $F_1(s)$ and $F_2(s)$. Assume $f_1(t) = \cos t$ for $t > 0$.

(*a*) Find $f_2(t)$.

(*b*) If $f_1(t)$ for $t > 0$ excites network N_2 having the frequency response $F_2(s)$, find the response in the time domain for $t > 0$.

(*c*) If $f_2(t)$ for $t > 0$ excites network N_1 having the frequency response $F_1(s)$, find the response in the time domain for $t > 0$.

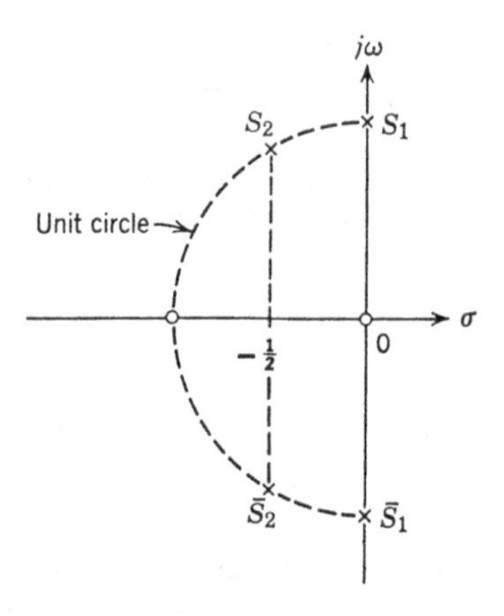

PROB. 6.

7. Relative to the circuit shown here, find the transfer impedance

$$Z_{12}(s) = \frac{E_2}{I_1}$$

Expand it in partial fractions, and, making use of property 49 (this chapter), determine the unit current impulse response of this network.

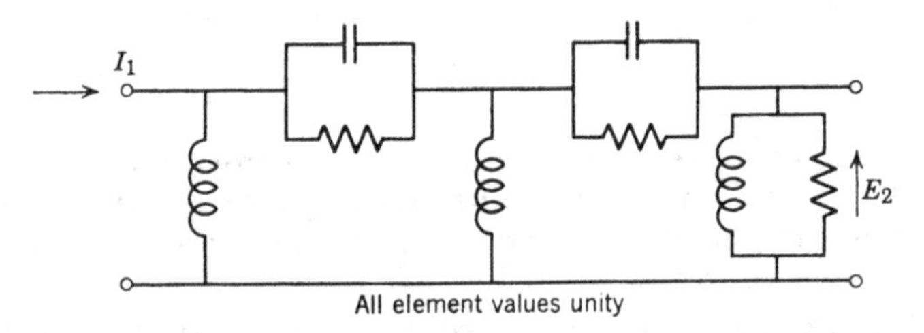

PROB. 7.

8. Consider the double-tuned circuit in Fig. 8, Ch. VIII for $R_1 = L_1 = \infty$. Choose the source and load resistance R equal to 1 ohm and the midband frequency equal to 1 radian per second. Fix the remaining circuit elements so that the frequency response is that of a second-order Butterworth function with a bandwidth yielding a ratio of center frequency to bandwidth equal to 100 (the filter $Q = 100$). Find the voltage output for $t > 0$ (*a*) for a unit current step input, (*b*) for an input current $i(t) = \cos t$ applied at $t = 0$.

Now scale the network to be appropriate to a midband frequency of 1 megacycle and $R = 10{,}000$ ohms. What is the effect of this operation upon the responses (*a*) and (*b*)? (Don't forget that a constant multiplier is important.) Indicate all element values in the circuit for this situation.

9. Given:

$$F_1(s) = \frac{1}{(s+1)(s+2)(s+3)} \quad \text{and} \quad f_2(t) = e^{-t}\cos t$$

Using only the methods developed in this chapter, find the inverse transform of

$$F(s) = F_1(s) \times F_2(s)$$

where $F_2(s)$ is the transform of $f_2(t)$.

10. Given:

$$f_1(t) = e^{-t}\cos t \quad \text{and} \quad f_2(t) = \sin t$$

Find the pole-zero plot and the appropriate multiplier of the transform $F(s)$ of the time function

$$f(t) = f_1(t) \times f_2(t)$$

Use only the methods developed in this chapter.

CHAPTER XI

Time- and Frequency-Domain Correlation for Periodic Functions through Fourier Analysis

1. Boundary-Value Problems and the Expansion of an Arbitrary Function in a Series of Normal Functions

It is an interesting fact that the kind of physical problem which led to the discovery of the Fourier series is not at all like that with which Fourier theory is commonly associated today, except perhaps in some of its purely mathematical aspects. Jean Baptiste Joseph Fourier (1768–1830), generally recognized as a distinguished French mathematician, while at Grenoble in the early nineteenth century, carried on elaborate investigations on the conduction of heat in metallic rods—a diffusion problem belonging to the general class of boundary-value problems, so called because the mathematical function representing a solution not only must fulfill the required equilibrium conditions within the medium but must also be able to assume arbitrarily specified boundary conditions.

In the heat-conduction problem a solution must permit the specification of fixed conditions at the ends of the rod for all values of t, plus a distribution throughout the length of the rod at $t = 0$, a so-called initial distribution function. The situation is very similar to that of an electrical transmission line or cable which not only imposes certain fixed volt-ampere relations at its terminals but also calls for the solution to meet the requirement of an arbitrary initial charge distribution on the line, such as a bound charge held there by an overhead storm cloud, the problem being to determine what happens when the cloud suddenly discharges and frees this charge distribution.

Another example is that of a violin string being plucked. The ends of the string are held immobile by its supports while the string's displacement from an equilibrium position prior to the moment of release by the musician's finger is describable by some initial-distribution function, in this case a triangular displacement-versus-distance function with its apex at the plucking point.

The functions satisfying equilibrium conditions on the rod (line or string) plus terminal conditions, are sinusoids involving the natural frequencies of this system, which is a distributed-parameter system and hence is characterized by an infinite number of natural frequencies. If the terminal conditions are simple constraints it turns out that the natural frequencies are integer multiples of a so-called fundamental, like the fundamental vibration frequency of a violin string and its overtones. The formal solution consists of an infinite sum of sinusoids having these integrally related natural frequencies and amplitudes as yet unspecified. The initial distribution must be met by adjustment of these amplitudes alone; and the question is: Can values for these amplitudes always be found such that the formal solution equals an arbitrarily specified initial distribution?

Euler and other contemporary mathematicians who studied this problem before Fourier were of the opinion that such a representation was not possible except for continuous distribution functions. Fourier's discovery was that a representation by continuous trigonometric functions is valid also for *discontinuous* distribution functions. Although he himself did not offer a rigorous mathematical proof, the mathematician Lejeune Dirichlet in 1837 gave such a demonstration.

Fourier's work not only was the motivation for detailed studies by numerous later mathematicians of this particular problem, but it was also the origin of the theory of orthogonal functions which later were further developed principally by Fredholm in his work on integral equations.

The way in which this aspect of the Fourier theory comes about is in the specific manner in which the amplitudes of the sinusoids (in general normal functions) are determined. Suppose in a certain boundary-value problem the normal functions are the set $\phi_1(x)$, $\phi_2(x)$, . . . , and the specified distribution is given by the function $f(x)$. The problem is to determine coefficients $a_1, a_2, \ldots$ in the infinite series

$$f(x) = a_1\phi_1(x) + a_2\phi_2(x) + a_3\phi_3(x) \cdots \tag{1}$$

such that this series becomes uniformly convergent within a specified range $a \leq x \leq b$ and hence may there be regarded as a valid representation for the function $f(x)$.

If we consider a partial sum

$$s_n(x) = a_1\phi_1(x) + a_2\phi_2(x) + \cdots + a_n\phi_n(x) \tag{2}$$

of the infinite series 1, and divide the interval a to b into n equal subintervals whose midpoints are the x-values indicated by writing $a < x_1 < x_2 < \cdots < x_n < b$, then the coefficients $a_1 \cdots a_n$ found by solving the set of simultaneous algebraic equations

$$\begin{aligned}
a_1\phi_1(x_1) + a_2\phi_2(x_1) + \cdots + a_n\phi_n(x_1) &= f(x_1) \\
a_1\phi_1(x_2) + a_2\phi_2(x_2) + \cdots + a_n\phi_n(x_2) &= f(x_2) \\
\cdot \quad \cdot \quad \cdot \quad \cdot \quad \cdot \quad \cdot \quad \cdot \quad \cdot \quad \cdot & \quad \cdot \quad \cdot \\
a_1\phi_1(x_n) + a_2\phi_2(x_n) + \cdots + a_n\phi_n(x_n) &= f(x_n)
\end{aligned} \tag{3}$$

yield a function $s_n(x)$ which equals $f(x)$ at the selected points $x_1 \cdots x_n$. Although not much can be said about the manner in which $s_n(x)$ approximates $f(x)$ at any other points, we may expect that by taking n sufficiently large and assuming $f(x)$ to be a smooth function, the resulting partial sum 2 will ultimately become identical with the specified function $f(x)$.

Such a process of solution requires, for a finite n, inversion of the matrix

$$\begin{bmatrix}
\phi_1(x_1) & \phi_2(x_1) & \cdots & \phi_n(x_1) \\
\phi_1(x_2) & \phi_2(x_2) & \cdots & \phi_n(x_2) \\
\cdot & \cdot & \cdot & \cdot \\
\phi_1(x_n) & \phi_2(x_n) & \cdots & \phi_n(x_n)
\end{bmatrix} \tag{4}$$

which in general is a laborious task, and becomes hopeless if not impossible as n is increased without limit. However, if this matrix is an orthogonal one, its inverse is simply the transpose of 4, and the solution can then be written down immediately.

We recall that the vector set of an orthogonal matrix of order n is a set of n mutually orthogonal unit vectors, and that if the rows of a matrix satisfy this condition then the columns automatically satisfy a similar condition. That is to say, the transposed vector set of an orthogonal matrix is likewise a set of mutually orthogonal unit vectors.

The orthogonality conditions for the matrix 4 can, therefore, be written either

$$\sum_{k=1}^{n} \phi_r(x_k)\phi_s(x_k) = \begin{cases} 1 & \text{for } r = s \\ 0 & \text{for } r \neq s \end{cases} \tag{5}$$

or

$$\sum_{k=1}^{n} \phi_k(x_r)\phi_k(x_s) = \begin{cases} 1 & \text{for } r = s \\ 0 & \text{for } r \neq s \end{cases} \tag{6}$$

If conditions 5 are fulfilled then 6 automatically follow, and vice versa. In this event, multiplying Eqs. 3 respectively by $\phi_s(x_1)$, $\phi_s(x_2)$, $\ldots$ $\phi_s(x_n)$ and adding yields the desired solutions

$$a_s = \sum_{k=1}^{n} \phi_s(x_k)f(x_k) \tag{7}$$

An equivalent method of solution can be used for the determination of coefficients in the infinite series 1 if the normal functions satisfy an orthogonality condition which reads

$$\int_a^b \phi_r(x)\,\phi_s(x)\,dx = \begin{cases} 1 & \text{for } r = s \\ 0 & \text{for } r \neq s \end{cases} \tag{8}$$

For then we can multiply Eq. 1 by $\phi_s(x)$ and integrate over the range $a \leq x \leq b$ to get

$$a_s = \int_a^b \phi_s(x)\,f(x)\,dx \tag{9}$$

which is the desired solution.

The question that naturally arises at this point is: Suppose the functions $\phi_1(x)$, $\phi_2(x)$, . . . do not satisfy the orthogonality conditions 8; what does one do then? The answer is: One *constructs* an orthogonal set, expressing these as appropriate linear combinations of the given functions, just as one can always construct a set of mutually orthogonal unit vectors from any given set of arbitrary vectors. The detailed procedure can take a variety of forms; the following is a possible one. Let $\psi_1(x)$, $\psi_2(x)$, . . . be the orthogonal set of functions that we wish to construct, and write

$$\begin{aligned} \psi_1(x) &= a_{11}\phi_1(x) \\ \psi_2(x) &= a_{21}\phi_1(x) + a_{22}\phi_2(x) \\ \psi_3(x) &= a_{31}\phi_1(x) + a_{32}\phi_2(x) + a_{33}\phi_3(x) \\ &\cdots\cdots\cdots\cdots \end{aligned} \tag{10}$$

Then the integral

$$\int_a^b \psi_1^2(x)\,dx = 1 \tag{11}$$

determines a_{11}. The two integrals

$$\int_a^b \psi_1(x)\,\psi_2(x)\,dx = 0 \quad \text{and} \quad \int_a^b \psi_2^2(x)\,dx = 1 \tag{12}$$

determine the constants a_{21} and a_{22}; and so forth.

Since the functions $\phi_1(x)$, $\phi_2(x)$, . . . satisfy the equilibrium equations of the system (which is linear) it is clear that the functions $\psi_1(x)$, $\psi_2(x)$, . . . also satisfy these same equilibrium equations and in addition satisfy the orthogonality conditions. They are referred to as an *orthonormal-function set*.

Fortunately the trigonometric functions which are normal functions appropriate to the problem of heat conduction in rods, or voltage or current propagation on electrical transmission lines, or the vibrating string, are an orthogonal set to start with if the boundary conditions are simple (open- or short-circuit constraints on a transmission line, immobile supports for the vibrating string, etc.). In these situations the natural frequencies (as mentioned above) are integer multiples of a lowest one called the *fundamental*. The normal functions are simply sin nx and cos nx for integer values of n, and the fundamental range can always be regarded as defined by $a \leq x \leq a + 2\pi$ where a can be any finite number. (If it is not defined this way to start with, we can always achieve this result through a suitable change of variable.)

The orthogonality conditions for the trigonometric functions stem from the result that

$$\frac{1}{2\pi}\int_a^{a+2\pi} e^{jkx}\,dx = \frac{e^{jka}(e^{jk2\pi} - 1)}{2\pi jk} = \begin{cases} 1 & \text{for } k = 0 \\ 0 & \text{for } k \neq 0 \end{cases} \tag{13}$$

Expressing sin nx and cos nx in terms of the exponential function, we thus find specifically that

$$\frac{1}{\pi}\int_a^{a+2\pi} \sin mx \sin nx\,dx = \begin{cases} 1 & \text{for } m = n \neq 0 \\ 0 & \text{for } m \neq n \end{cases} \tag{14}$$

$$\frac{1}{\pi}\int_a^{a+2\pi} \cos mx \cos nx\,dx = \begin{cases} 1 & \text{for } m = n \neq 0 \\ 0 & \text{for } m \neq n \end{cases} \tag{15}$$

and additionally that

$$\int_a^{a+2\pi} \sin mx \cos nx\,dx = 0 \tag{16}$$

for *all* integer values of m and n.

Suppose an infinite series of these normal functions is to be found such that an arbitrary function $f(x)$ specified over the interval $a \leq x \leq a + 2\pi$ may be represented. Formally, we require that

$$\begin{aligned} f(x) = \frac{a_0}{2} &+ a_1 \cos x + a_2 \cos 2x + a_3 \cos 3x + \cdots \\ &+ b_1 \sin x + b_2 \sin 2x + b_3 \sin 3x + \cdots \end{aligned} \tag{17}$$

Multiplying both sides of this equation by cos nx and integrating over the fundamental range, we observe in view of orthogonality condition 16 that all of the sine terms in 17 yield nothing, and that all but one of the cosine terms also yield nothing because of condition 15. From the one nonzero term we obtain the formula

$$a_n = \frac{1}{\pi}\int_a^{a+2\pi} f(x) \cos nx\,dx \tag{18}$$

which applies to all coefficients of the cosine terms. In this connection the constant term may be regarded as a cosine term for $n = 0$, and by writing it $a_0/2$ in 17, formula 18 is also valid for the determination of a_0.

Analogously the orthogonality conditions 14 and 16 yield a formula for coefficients of the sine terms. Thus, if we multiply both sides of Eq. 17 by $\sin nx$ and integrate over the fundamental range, all terms yield nothing except the pertinent sine term, and we find

$$b_n = \frac{1}{\pi} \int_a^{a+2\pi} f(x) \sin nx \, dx \tag{19}$$

In this derivation of formulas 18 and 19 we have tacitly assumed that the infinite series in 17 converges uniformly over the fundamental range $a \leq x \leq a + 2\pi$; otherwise it cannot there be identified with the function $f(x)$, and the term-by-term integration which is involved in this process becomes meaningless. In other words, the validity of these formulas remains to be established through a study of the approximation and convergence properties of this infinite series. Some aspects of this topic are discussed in the immediately following articles; a more thorough study is presented in Ch. XIV.

2. Representation of a Periodic Time Function by a Fourier Series

The theory of the Fourier series is important in network analysis and synthesis because it provides the means for approximating to an arbitrary function* over a given time interval by a sum of sinusoids. The additive property of linear networks, which permits the superposition of solutions pertinent to the individual terms in such a sum, immediately suggests that one may apply sinusoidal steady-state circuit analysis to situations where the associated time functions are arbitrary, and thus preserve for the analysis and study of these more general problems the simplicity and circumspection that is characteristic of network theory in the sinusoidal steady state.

A full exploitation of these thoughts leads ultimately to an invaluable explicit relationship between transient and steady-state circuit response, providing a whole new philosophy with regard to linear system behavior that is indispensable to the solution of problems in the communications and electronic-control art. For example, it provides the basis upon which one can judge the adequacy of a circuit component for the transmission of a transient signal through use of sinusoidal steady-state measurements or calculations alone. It suggests means for separating messages transmitted through a common medium (the problem of radio or carrier telephony),

* This statement implies the extension to aperiodic functions given in the following chapters.

for making studies of noise and interference and devising means for minimizing their effects, for formally reducing the problem of transient synthesis to an equivalent one involving steady-state response functions alone, and ways of dealing with an endless variety of other problems of a similar or related nature. It is the purpose of this and the following chapters to present the mathematical basis upon which this point of view rests.

A Fourier series is a sum of sinusoids (also called a *trigonometric polynomial*) whose arguments all are integer multiples of a smallest one characterizing the sinusoid known as the *fundamental*; the others are referred to as harmonics and the pertinent integer in each of these determines the *order* of that harmonic. If the independent variable is the time, as it is in our discussions in this volume, then the fundamental has associated with it a certain frequency, and the frequencies of the harmonics are integer multiples of this one. The function of time represented by the series is periodic, with a period equal to the reciprocal of the fundamental (cyclic) frequency.

The fundamental and each harmonic, like any sinusoid, is characterized by three things: an amplitude, a frequency, and a time phase. However, once the period of the resultant time function is chosen, all frequencies are fixed, and only the amplitudes and time phases remain arbitrary. In essence the Fourier theorem states that appropriate values for these amplitudes and time phases can always be found such that the resultant series approximates arbitrarily closely to any decently behaved function that one may specify throughout a time interval which becomes the period. A "decently behaved" function may have a finite number of discrete discontinuities within this period (also called the fundamental range) and it may become infinite at a finite number of isolated points provided only that the function is integrable; that is to say, that the area enclosed by the function over the fundamental range is finite. Practically speaking, the restrictions involved in the Fourier theorem are not serious.

In the solution of circuit problems involving Fourier series it is usually found expedient to express sinusoids in complex form. In general, for a complex amplitude $E = |E|\, e^{j\psi}$, we have two alternate complex forms to consider, as indicated in the following:

$$|E| \cos(\omega t + \psi) = \text{Re}\,[Ee^{j\omega t}] = \tfrac{1}{2}[Ee^{j\omega t} + \bar{E}e^{-j\omega t}] \qquad (20)$$

Wherever we encounter operations that are commutable with respect to taking the real part, the first of these forms is chosen since one term then serves as well as the sum of two. In our present discussions, however, one finds the second form in Eq. 20 preferable since it avoids the Re-sign and any restrictions that attach to its manipulation, while not adding noticeably to the awkwardness of the resulting expression which must involve the summation of many sinusoids (theoretically an infinite number) in any case.

If we define a *complex amplitude of order k* as given by

$$\alpha_k = |\alpha_k| \, e^{-j\psi_k} \tag{21}$$

and use for the designation of its conjugate value the notation

$$\alpha_{-k} = \bar{\alpha}_k = |\alpha_k| \, e^{j\psi_k} \tag{22}$$

then we have the identity

$$2 \, |\alpha_k| \cos (k\omega_0 t - \psi_k) = \alpha_k e^{jk\omega_0 t} + \alpha_{-k} e^{-jk\omega_0 t} \tag{23}$$

In this way an infinite sum of sinusoids with frequencies equal to an integer multiple of the fundamental radian frequency ω_0 may compactly be written as

$$f(t) = \sum_{\nu=-\infty}^{\infty} \alpha_\nu e^{j\nu\omega_0 t} \tag{24}$$

in which the summation index ν assumes all positive and negative integer values, including the value zero.

The pair of terms for $\nu = k$ and $\nu = -k$ yield the sinusoid 23 of order k, having the peak amplitude $2\,|\alpha_k|$, the radian frequency $k\omega_0$, and the time-phase angle $-\psi_k$. For $k = 1$, Eq. 23 represents the fundamental, which can alternately be regarded as a harmonic of first order. For $k = 2$, this same expression represents the second harmonic; for $k = 3$, it is the third harmonic; and so on. Pairs of terms in the sum 24 involving the same absolute value of the integer ν represent sinusoids just as the pair of conjugate terms in Eq. 20 represents the sinusoid given there. Since the factor $\frac{1}{2}$ in Eq. 20 is not included in Eq. 24, it turns out that the amplitude of the pertinent sinusoid, as given by Eq. 23, is twice the absolute value of the associated complex coefficient α_k.

It should be noted that the sum in Eq. 24 includes a term for $\nu = 0$, which is the constant α_0. This term may be regarded as a harmonic of zero order (having zero frequency), or more simply as an additive constant that may be needed in the representation of the given periodic function $f(t)$. Its period τ is the reciprocal of a fundamental cyclic frequency f_0, related to the fundamental radian frequency ω_0 by the expression

$$\tau = \frac{1}{f_0} = \frac{2\pi}{\omega_0} \tag{25}$$

The constant α_0 must evidently be equal to the average value of $f(t)$ over this period since the average values of all the sinusoidal terms in the sum 24 are zero.

An interpretation of the compact expression 24 may be indicated through writing out the terms in the following way:

$$f(t) = \alpha_0 + \begin{cases} \alpha_1 e^{j\omega_0 t} + \alpha_2 e^{j2\omega t} + \alpha_3 e^{j3\omega t} + \cdots \\ \alpha_{-1} e^{-j\omega_0 t} + \alpha_{-2} e^{-j2\omega t} + \alpha_{-3} e^{-j3\omega t} + \cdots \end{cases} \tag{26}$$

All terms except the one for $\nu = 0$ are regarded as grouped into pairs involving opposite signs of the index ν. The pair for $\nu = \pm 1$ yields the fundamental sinusoid, that for $\nu = \pm 2$ yields the second harmonic, and so forth. The amplitude of each sinusoid is twice the absolute value of its associated complex coefficient, and the time-phase angle of each sinusoid (expressed as a *lag* angle) equals the angle of this coefficient, as indicated in Eq. 21. Equation 26 may, therefore, be written in the equivalent trigonometric form

$$f(t) = \alpha_0 + 2\,|\alpha_1| \cos(\omega_0 t - \psi_1) + 2\,|\alpha_2| \cos(2\omega_0 t - \psi_2) + \cdots \tag{27}$$

Here the phase angles appear explicitly while in Eq. 26 they are implied through the complex character of the coefficients α_k. Equation 24 is referred to as the *complex form* of the Fourier series; Eq. 27 is one possible trigonometric form. Other trigonometric forms may be obtained through writing sines instead of cosines, which is done by adding $\pi/2$ to each phase angle, or through replacing each cosine by the sum of a sine and a cosine. The result of carrying out this scheme is to obtain the sum of a sine and a cosine series as in Eq. 17. We will have more to say about this form later on. Our more immediate objective is to study the manner in which the sum of sinusoids in Eq. 24 or 27 approximates a given function $f(t)$ over the fundamental range when the coefficients have values determined by the method discussed in art. 1 above.

3. The Mean-Square-Approximation Property

This topic is important in dealing with practical problems because we can seldom work with the infinite series, but must for obvious reasons content ourselves with a finite number of terms; that is, with the so-called *partial sum*

$$s_n(t) = \sum_{\nu=-n}^{n} \alpha_\nu e^{j\nu\omega_0 t} \tag{28}$$

In a particular problem not only do we need to know how $s_n(t)$ approximates $f(t)$ over the fundamental range for a given value of n, but we must know how the character of this approximation changes with an increase in n, in order that we may make an intelligent estimate regarding the sufficiency of any chosen value of n.

Since the nature of the approximation depends entirely upon the values of the coefficients α_ν, one can attempt to control this aspect of the problem through using the desired approximating property as a means for determining these coefficients from the given $f(t)$. To carry out this thought we form the error between $f(t)$ and the partial sum

$$\epsilon(t) = f(t) - s_n(t) \tag{29}$$

and then see whether it is possible to find expressions for the α_ν which conform with an acceptable behavior of $\epsilon(t)$ over the fundamental range. In regard to this behavior, a number of specifications may be considered. For example, we may wish to stipulate that the maximum absolute value of the error remain constant or vary in some particular way over the fundamental range. In general it is very difficult to determine the α_ν-values appropriate to such given error specifications, and one is forced to resort to a cut-and-try procedure to obtain a solution.*

A particular specification for the error function that does lend itself to an analytic determination of the appropriate coefficients α_ν, and coincidentally has some reason for being acceptable otherwise, is expressed by stating that the squared error integrated over the fundamental range shall become a minimum. According to this so-called mean-square-error criterion, our problem is to find an expression for α_ν in the partial sum 28 such that the integral

$$\frac{1}{\tau}\int_{t_0}^{t_0+\tau} \epsilon^2(t)\, dt = \overline{\epsilon^2} \tag{30}$$

becomes a minimum. Surprisingly enough, the solution to this problem presents no serious difficulties.

By substituting from Eqs. 28 and 29, the integral to be minimized is made to read

$$F(\alpha_\nu) = \int_{-\tau/2}^{\tau/2} \left[f(t) - \sum_{-n}^{n} \alpha_\nu e^{j\nu\omega_0 t} \right]^2 dt \tag{31}$$

where we have chosen the limits of integration in a symmetrical fashion as we are at liberty to do. Since the variables in this minimization problem are the coefficients α_ν, the integral 31 is regarded for the moment as a function of these coefficients.

Assuming that the integrand is a continuous function and that it has finite values throughout the fundamental range, we may form the partial derivative of 31 with respect to α_k by differentiating under the integral sign. Setting this derivative equal to zero reads

$$-2\int_{-\tau/2}^{\tau/2} \left[f(t) - \sum_{-n}^{n} \alpha_\nu e^{j\nu\omega_0 t} \right] e^{jk\omega_0 t}\, dt = 0 \tag{32}$$

* These matters are more fully considered in Ch. XIV.

where it should be noted that the term in 31 for $\nu = k$ alone involves the variable α_k.

If we interchange the order of summation and integration, this result yields

$$\sum_{\nu=-n}^{n} \alpha_\nu \int_{-\tau/2}^{\tau/2} e^{j(\nu+k)\omega_0 t}\, dt = \int_{-\tau/2}^{\tau/2} f(t) e^{jk\omega_0 t}\, dt \tag{33}$$

Carrying out the integration on the left, noting Eq. 25, we get

$$\sum_{\nu=-n}^{n} \alpha_\nu \left[\frac{e^{j(\nu+k)\pi} - e^{-j(\nu+k)\pi}}{j(\nu + k)\omega_0} \right] = \int_{-\tau/2}^{\tau/2} f(t)\, e^{jk\omega_0 t}\, dt \tag{34}$$

Since $\nu + k$ is an integer, we see that every term in the above sum is zero except the one for $\nu = -k$ which evaluates to

$$\frac{2\pi}{\omega_0} \alpha_{-k} = \int_{-\tau/2}^{\tau/2} f(t) e^{jk\omega_0 t}\, dt \tag{35}$$

or, replacing k by $-\nu$ and dividing both sides by $\tau = 2\pi/\omega_0$, we have the coefficient formula

$$\alpha_\nu = \frac{1}{\tau} \int_{-\tau/2}^{\tau/2} f(t) e^{-j\nu\omega_0 t}\, dt \tag{36}$$

We may conclude that if the coefficients in series 24 are computed for the given periodic function $f(t)$ by this formula, then the partial sum 28 for any finite n approximates $f(t)$ throughout the fundamental range in a manner that renders the mean value of the squared error 29 a minimum. It turns out that formulas 18 and 19 obtained earlier using the orthogonal properties of the trigonometric functions agree precisely with formula 36.

In order to see that this is so, write

$$\alpha_\nu = \tfrac{1}{2}(a_\nu - jb_\nu), \quad \alpha_{-\nu} = \tfrac{1}{2}(a_\nu + jb_\nu) \tag{37}$$

and replace the exponential in 24 by its sine and cosine equivalent so that this equation can be written in the alternate equivalent form

$$f(t) = \alpha_0 + \sum_{\nu=1}^{\infty} (\alpha_\nu + \alpha_{-\nu}) \cos \nu\omega_0 t + \sum_{\nu=1}^{\infty} j(\alpha_\nu - \alpha_{-\nu}) \sin \nu\omega_0 t \tag{38}$$

In view of 37 this yields the sine and cosine series

$$f(t) = \alpha_0 + \sum_{\nu=1}^{\infty} a_\nu \cos \nu\omega_0 t + \sum_{\nu=1}^{\infty} b_\nu \sin \nu\omega_0 t \tag{39}$$

while substitution of 37 into 36 and separation into real and imaginary parts leads to the coefficient formulas

$$a_\nu = \frac{2}{\tau}\int_{-\tau/2}^{\tau/2} f(t)\cos\nu\omega_0 t\,dt \tag{40}$$

and

$$b_\nu = \frac{2}{\tau}\int_{-\tau/2}^{\tau/2} f(t)\sin\nu\omega_0 t\,dt \tag{41}$$

which agree respectively with 18 and 19 if we recognize that $x = \omega_0 t$ and $\omega_0/\pi = 2/\tau$.

It should carefully be observed that the coefficients α_ν are Fourier coefficients only when evaluated in conformance with formula 36. Only then is the series 24 a Fourier series. Otherwise it is a trigonometric series of some other sort with whatever approximation properties pertain to the specific coefficient determination used. That is to say, a Fourier series is a particular trigonometric series of the form given by Eq. 24 or 39 that results if one applies the mean-square-error criterion to the solution of the pertinent approximation problem. Sometimes this criterion is undesirable, and appropriate modifications of the coefficient values from those obtained by the Fourier formula must be carried out. In most problems of this kind, the Fourier coefficients are useful as a starting point, and sometimes the desired result can be obtained by a simple cut-and-dry procedure.

When the function $f(t)$ has a finite number of discrete discontinuities or integrable infinite values within the fundamental range, then the above procedure leading to the result expressed by Eq. 36 can still be carried out if in the integration 31 we exclude these points. We may then conclude that the series 24 with coefficients determined by 36 approximates $f(t)$ in the mean-square sense throughout the fundamental range except at these isolated points. There its behavior must be investigated separately (as is done in Ch. XIV).

4. Even and Odd Time Functions

A function $f(t)$ is said to be *even* when it satisfies the condition

$$f(-t) = f(t) \tag{42}$$

A plot of the function then is symmetrical about the vertical axis at the origin. On the other hand, the function is *odd* if

$$f(-t) = -f(t) \tag{43}$$

A plot of the function is then antisymmetrical about a vertical axis at the origin.

Since the trigonometric sine is odd and the cosine even, it follows that the series representation for a periodic function that is even contains only cosine terms while that for a periodic function that is odd contains only sine terms. For the former, all coefficients b_ν in the series 39 vanish, and for the latter, all coefficients a_ν become zero. In the first of these situations (t) is represented by a *cosine series*, in the second by a *sine series*.

When $f(t)$ is neither even nor odd, but arbitrary, it can always be expressed as the sum of two component functions of which one is odd and the other even. Denoting these components by $f_1(t)$ and $f_2(t)$ respectively, we can express this statement by the relation

$$f(t) = f_1(t) + f_2(t) \tag{44}$$

According to the definitions of even and odd functions given by Eqs. 42 and 43 it follows that

$$f(-t) = -f_1(t) + f_2(t) \tag{45}$$

Addition and subtraction of Eqs. 44 and 45 yield respectively

$$\begin{aligned} f_2(t) &= \tfrac{1}{2}[f(t) + f(-t)] \\ f_1(t) &= \tfrac{1}{2}[f(t) - f(-t)] \end{aligned} \tag{46}$$

By means of these relations the even and odd components of a given function may be determined either analytically or graphically.

Since the component function $f_1(t)$ has a sine series and the component $f_2(t)$ a cosine series representation, it becomes clear that the series representation of an arbitrary function $f(t)$ must consist of a sine and a cosine part as in Eq. 39.

5. The Phase Angles of the Harmonic Components

In view of the trigonometric identities

$$\begin{aligned} c_\nu \cos(\nu\omega_0 t - \psi_\nu) &\equiv c_\nu \sin\left(\nu\omega_0 t + \frac{\pi}{2} - \psi_\nu\right) \\ &\equiv c_\nu \cos\psi_\nu \cos\nu\omega_0 t + c_\nu \sin\psi_\nu \sin\nu\omega_0 t \end{aligned} \tag{47}$$

it is possible to recombine the sine and cosine series in Eq. 39 by letting

$$a_\nu = c_\nu \cos\psi_\nu \quad \text{and} \quad b_\nu = c_\nu \sin\psi_\nu \tag{48}$$

These relations and those in Eqs. 37 show that

$$c_\nu = \sqrt{a_\nu^2 + b_\nu^2} = 2\,|\alpha_\nu| \tag{49}$$

so that the equivalence of 27 and 39 is once more evident while for the harmonic phase angles in the series 27 we have

$$\psi_\nu = \tan^{-1}\frac{b_\nu}{a_\nu} \tag{50}$$

or

$$\tan\psi_\nu = \frac{b_\nu}{a_\nu}$$

On account of the periodic nature of the function $f(t)$ it is usually possible in practical problems to select the origin for the independent variable t at any convenient point. In subsequent manipulations it may be desirable to shift the origin to a new location. If the variable t in $f(t)$ is replaced by $t - t_0$, the effect is to shift the origin *back* by an amount t_0 or to retard the sequence of values of $f(t)$ by t_0. Graphically this change corresponds to shifting the entire plot of $f(t)$ *forward* by an amount t_0.

A typical harmonic component as given by the left-hand side of identity 47 becomes

$$c_\nu \cos\left[\nu\omega_0(t - t_0) - \psi_\nu\right] = c_\nu \cos\left[\nu\omega_0 t - (\psi_\nu + \nu\omega_0 t_0)\right] \tag{51}$$

showing that the harmonic phase-lag angles are now given by

$$\psi_\nu' = \psi + \nu\omega_0 t_0 \tag{52}$$

This same result may also be obtained from the complex series 24 noting that

$$t \to t - t_0 \quad \text{yields} \quad \alpha_\nu \to \alpha_\nu e^{-j\nu\omega_0 t_0} \tag{53}$$

We conclude that shifting the origin so as to retard all values of $f(t)$ by t_0 has the effect of decreasing each harmonic phase angle (or increasing each harmonic phase-lag angle) by an amount proportional to its order ν, the proportionality factor being $\omega_0 t_0 = 2\pi t_0/\tau$. Such a proportionate change in the harmonic phase angles is, therefore, seen to leave the form of the function $f(t)$ intact except for a translation in its entirety.

6. Even and Odd Harmonics

It frequently happens that a given periodic function satisfies the condition

$$f\left(t + \frac{\tau}{2}\right) = -f(t) \tag{54}$$

which means that the sequence of values of the function throughout any half-period are negatives of the values encountered throughout a succeeding half-period. The particular form which a Fourier series assumes under this condition may be seen from Eqs. 40 and 41 for the cosine and sine coefficients which may in general be written in the form

$$a_\nu = \frac{\omega_0}{\pi}\int_0^{\pi/\omega_0}\left[f(t)\cos\nu\omega_0 t + f\left(t+\frac{\tau}{2}\right)\cos\nu\omega_0\left(t+\frac{\tau}{2}\right)\right]dt \quad (55)$$

$$b_\nu = \frac{\omega_0}{\pi}\int_0^{\pi/\omega_0}\left[f(t)\sin\nu\omega_0 t + f\left(t+\frac{\tau}{2}\right)\sin\nu\omega_0\left(t+\frac{\tau}{2}\right)\right]dt \quad (56)$$

Noting that $\nu\omega_0\tau/2 = \nu\pi$, and hence that

$$\cos\nu\omega_0\left(t+\frac{\tau}{2}\right) = \cos\nu\pi\cos\nu\omega_0 t \quad (57)$$

and

$$\sin\nu\omega_0\left(t+\frac{\tau}{2}\right) = \cos\nu\pi\sin\nu\omega_0 t \quad (58)$$

we see that condition 54 leads to the following special forms for the cosine and sine coefficients:

$$a_\nu = \frac{\omega_0}{\pi}(1-\cos\nu\pi)\int_0^{\pi/\omega_0} f(t)\cos\nu\omega_0 t\,dt \quad (59)$$

$$b_\nu = \frac{\omega_0}{\pi}(1-\cos\nu\pi)\int_0^{\pi/\omega_0} f(t)\sin\nu\omega_0 t\,dt \quad (60)$$

The factor $1-\cos\nu\pi$ is zero for all *even*-integer values of ν and equal to 2 for all *odd*-integer values. Hence it becomes clear that *when the given periodic function has the property expressed by Eq. 54, its Fourier series contains odd harmonics only.* The average value or constant component α_0 is evidently zero also. The coefficients are then given by

$$a_\nu = \frac{4}{\tau}\int_0^{\tau/2} f(t)\cos\nu\omega_0 t\,dt \quad (61)$$

and

$$b_\nu = \frac{4}{\tau}\int_0^{\tau/2} f(t)\sin\nu\omega_0 t\,dt \quad (62)$$

for odd-integer values of ν only.

When the given periodic function satisfies a condition complementary to 54, namely one that reads

$$f\left(t+\frac{\tau}{2}\right) = f(t) \quad (63)$$

the effect upon the Fourier coefficients may be seen by recognizing that the factor $1-\cos\nu\pi$ in Eqs. 59 and 60 is changed to $1+\cos\nu\pi$. However, a moment's reflection reveals that condition 63 merely states that

$f(t)$ has the period $\tau/2$ instead of τ. The result of retaining τ as the fundamental period is that the corresponding Fourier series contains only *even* harmonics, which is just another way of saying that the period of $f(t)$ is actually half as large.

On the basis of these thoughts it appears that any periodic function may be assumed to consist of two components of which one satisfies the condition 54 while the other has twice the fundamental frequency. The

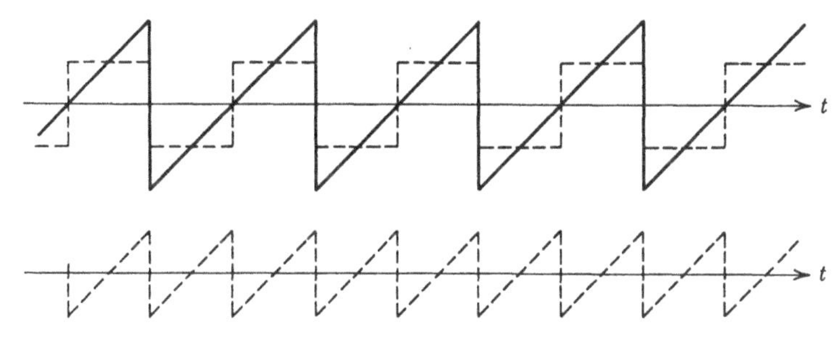

FIG. 1. Illustrating that the decomposition of a sawtooth wave into even and odd harmonics yields a square wave and a double-frequency sawtooth respectively.

first of these contributes the odd harmonic components to the resulting Fourier series while the second contributes the even harmonics.

An interesting example is illustrated by the sawtooth wave in Fig. 1. Here the square wave (shown dotted in the upper sketch) represents the odd harmonic content, while the even harmonics yield the sawtooth wave of double frequency (shown dotted in the lower sketch) which remains after the square-wave component is subtracted from the original function.

7. Amplitude and Phase Spectra

A plot of the magnitude of the complex coefficient α_ν in the series 24 versus the harmonic frequency $\nu\omega_0$ is called the *amplitude spectrum* of the periodic function $f(t)$. A plot of the phase angle ψ_ν of α_ν versus $\nu\omega_0$ is called the *phase spectrum* of $f(t)$. Since the index ν assumes only integer values, $\nu\omega_0$ is a discrete variable. The amplitudes and phase spectra are not continuous curves but are given by a set of ordinates erected at intervals corresponding to integer values of ν. They are, therefore, referred to as *line* spectra.

The pair of line spectra determined by α_ν as a function of the discrete variable $\nu\omega_0$ are in every way equivalent to a plot of the periodic function $f(t)$ versus the continuous variable t in the sense that either representation completely characterizes the function. The function α_ν versus $\nu\omega_0$ specifies the periodic function in the so-called *frequency* domain just as $f(t)$ versus t specifies it in the *time domain.*

The one form of characterization is convertible into the other by means of Eqs. 24 and 36. The latter may be regarded as *transforming* one characterization of the periodic function into the other equivalent one. In this sense, α_ν as a function of $\nu\omega_0$ is regarded as the *Fourier transform* of $f(t)$, and the latter as the *inverse* transform of α_ν. Equations 24 and 36 form a pair of mutually inverse relations in that one *undoes* what the other *does.*

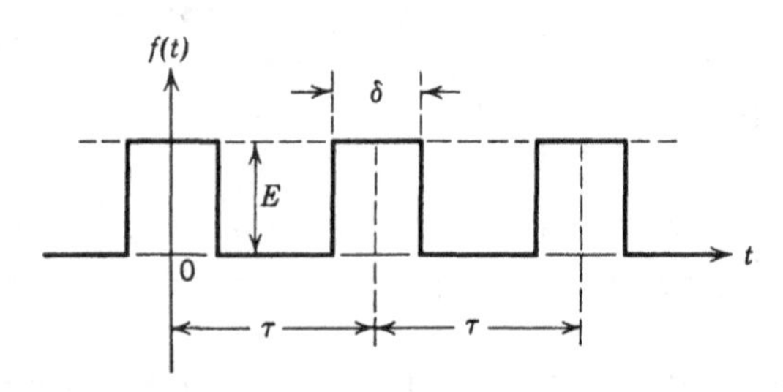

FIG. 2. Periodically repeated rectangular pulse yields Fourier coefficients given by Eq. 64.

Relation 36 is also said to perform Fourier analysis since it analyzes $f(t)$ into its equivalent spectrum; and 24 is said to perform Fourier synthesis because it synthesizes the spectrum so as to regain the function $f(t)$. In a broader sense, Fourier transformation may be looked upon as a process whereby the independent variable may be changed from t to $\nu\omega_0$ and back again. The usefulness of such a change of variable so far as circuit theory is concerned lies in the fact that the process of solving circuit equations is much simpler in the frequency than in the time domain, as we have seen in the discussions of Chs. IX and X.

The detailed procedure for finding the spectrum for a given periodic function will be illustrated for the one shown in Fig. 2 which consists of a succession of identical rectangular pulses of magnitude E and duration δ, with the origin located at the center of one of them. Equation 36 yields the following coefficients in the complex series 24:

$$\alpha_\nu = \frac{E}{\tau}\int_{-\delta/2}^{\delta/2} e^{-j\nu\omega_0 t}\,dt = \frac{E}{\tau}\cdot\frac{e^{j\nu\omega_0\delta/2} - e^{-j\nu\omega_0\delta/2}}{j\nu\omega_0}$$

or

$$\alpha_\nu = \frac{E\delta}{\tau}\cdot\frac{\sin \nu\omega_0\delta/2}{\nu\omega_0\delta/2} \tag{64}$$

These coefficients are real, as may be expected because $f(t)$ is an even function and hence representable by a cosine series so that the sine coefficients b_ν in relations 37 are zero. The angles ψ_ν are, therefore, identically zero and hence, so is the phase spectrum. The amplitude spectrum is obtained by plotting 64 versus the discrete variable $\nu\omega_0$. This process is facilitated through first plotting a dotted curve for the function 64 considering $\nu\omega_0$ as a continuous variable and then erecting ordinates at frequencies corresponding to integer values of ν. The result is shown in Fig. 3.

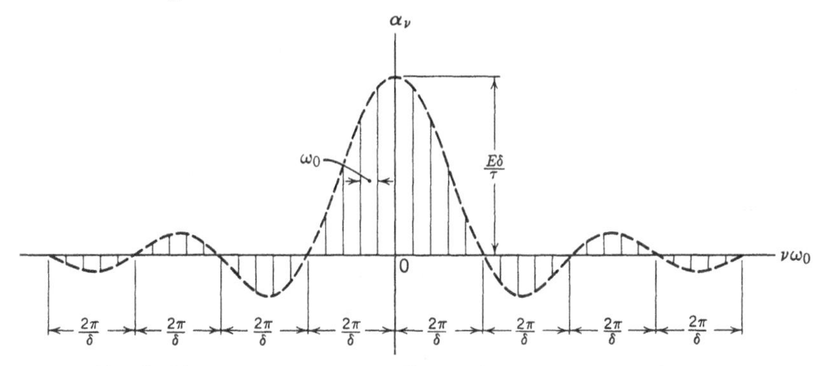

FIG. 3. Line spectrum corresponding to the periodic wave of Fig. 2.

The dotted curve, which acts as an envelope for the lines in this amplitude spectrum, passes through zero at the values of $\nu\omega_0$ for which $\nu\omega_0\delta/2$ equals integer multiples of π. It is significant to note that this envelope curve does not depend upon the ratio δ/τ (the duration of the pulse relative to the fundamental period) except that all ordinates are linearly proportional to it, that is, except for a scale factor. The plot of this envelope may be carried out before a numerical choice for the ratio δ/τ is made. This choice merely determines the *spacing* of the lines in the spectrum. In Fig. 3 the ratio

$$\frac{\delta}{\tau} = \frac{1}{5} \tag{65}$$

is chosen. Since

$$\frac{\nu\omega_0\delta}{2} = \nu\pi\frac{\delta}{\tau} \tag{66}$$

the first zero of the envelope, which falls at $\nu\omega_0\delta/2 = \pi$, corresponds to $\nu = 5$. The uniform spacing of lines then equals $\frac{1}{5}$ the distance from the origin to the first envelope zero. Any number of lines are thus readily drawn.

It should be emphasized again that the phase spectrum in this example is zero because of the symmetry of the rectangular pulses in Fig. 2 about a vertical center line and because of the particular location chosen for the origin. Suppose the origin is shifted back (the plot in Fig. 2 is shifted forward) by t_0; that is, $f(t)$ is replaced by $f(t - t_0)$. Then, as shown by the relations 53 in art. 5, α_ν is replaced by $\alpha_\nu e^{-j\nu\omega_0 t_0}$. The amplitude spectrum is not affected by this shift of the origin, but the phase spectrum (angle of the complex Fourier coefficient) is now equal to $-\nu\omega_0 t_0$ radians. As a function of the harmonic frequency $\nu\omega_0$, the phase-lag function is determined by a straight line through the origin having a slope equal to t_0 seconds. As also pointed out in art. 5, *a delay in the function $f(t)$ by t_0 seconds is seen to correspond to adding a linear increment to the phase-lag function, the slope of which numerically equals the delay.*

The converse is also true, and forms an important basis for judging the effect of modifications in the phase characteristic produced by a network through which the given function may be transmitted. The details of this topic are discussed in Ch. XVI.

8. Useful Artifices in the Evaluation of Fourier Coefficients.*

In many practical situations where the Fourier series representation for a given periodic function is called for, one may achieve the desired result in a way that avoids most of the disagreeable manipulation encountered in a straightforward evaluation of Fourier coefficients by means of the integral 36. For convenience we repeat the pertinent relations 24 and 36 here:

$$f(t) = \sum_{\nu=-\infty}^{\infty} \alpha_\nu e^{j\nu\omega_0 t} \tag{67}$$

$$\alpha_\nu = \frac{1}{\tau}\int_{-\tau/2}^{\tau/2} f(t)e^{-j\nu\omega_0 t}\,dt \tag{68}$$

In many situations, as we shall show, it is easier to carry out the integration in Eq. 68 if $f(t)$ is replaced by its first or higher derivative, especially where the derivative function is given by a sum of impulses, a result that can be brought about in any case if we are willing to settle for an approximate result. Although the effect upon α_ν of replacing $f(t)$ by its first or higher derivative may be found formally by simply differentiating the series 67 term by term, one has at this stage no assurance that the infinite series thus found actually represents the pertinent derivative of $f(t)$ even though the infinite series 67 is known to represent this function itself. A procedure which yields the desired result and does not involve

* Based upon *Tech. Rept. 268, Mass. Inst. Technol. Res. Lab. Electron.*, Sept. 2, 1953, presented at the National Electronics Conference (Chicago) of that date, and published in their *Transactions*, Vol. 9, Feb. 1954.

this difficulty makes use of the well-known formula for integration by parts, which reads

$$\int u\, dv = uv - \int v\, du \tag{69}$$

Suppose we apply this formula to the integral 68 letting

$$u = f(t) \quad \text{and} \quad dv = e^{-j\nu\omega_0 t}\, dt \tag{70}$$

so that

$$du = f^{(1)}(t)\, dt \quad \text{and} \quad v = \frac{e^{-j\nu\omega_0 t}}{-j\nu\omega_0} \tag{71}$$

where the superscript (1) on $f(t)$ indicates its first derivative. For the integral 68 we then get

$$\alpha_\nu = \frac{1}{\tau}\left\{\left[\frac{f(t)e^{-j\nu\omega_0 t}}{-j\nu\omega_0}\right]_{-\tau/2}^{\tau/2} + \frac{1}{j\nu\omega_0}\int_{-\tau/2}^{\tau/2} f^{(1)}(t)e^{-j\nu\omega_0 t}\, dt \right. \tag{72}$$

The first term in the bracket evaluates to

$$\frac{f(\tau/2)e^{-j\nu\pi} - f(-\tau/2)e^{j\nu\pi}}{-j\nu\omega_0} = 0 \tag{73}$$

and so we have

$$\alpha_\nu = \frac{1}{\tau(j\nu\omega_0)}\int_{-\tau/2}^{\tau/2} f^{(1)}(t)e^{-j\nu\omega_0 t}\, dt \tag{74}$$

We can repeat the same process k times and get

$$\alpha_\nu = \frac{1}{\tau(j\nu\omega_0)^k}\int_{-\tau/2}^{\tau/2} f^{(k)}(t)e^{-j\nu\omega_0 t}\, dt \tag{75}$$

in which the superscript (k) on $f(t)$ indicates its kth derivative. This derivation is valid so long as the integral 75 exists (the integrand is integrable).

If we denote the complex coefficients in a Fourier series representation for the function $d^k f/dt^k$ by γ_ν; that is to say, if

$$\gamma_\nu = \frac{1}{\tau}\int_{-\tau/2}^{\tau/2} f^{(k)}(t)e^{-j\nu\omega_0 t}\, dt \tag{76}$$

then we have the result

$$\alpha_\nu = \frac{\gamma_\nu}{(j\nu\omega_0)^k} \tag{77}$$

which allows us to calculate the Fourier coefficient for the function $f(t)$ from that for the kth derivative of this function. Hence if the kth derivative of the given function is one for which the integration involved in its

Fourier coefficient determination is easier to evaluate, we may find the coefficient γ_ν pertinent to $f^{(k)}(t)$ first and then get the desired α_ν through division by $(j\nu\omega_0)^k$ as indicated in Eq. 77.

As an example of the application of this modified procedure for Fourier coefficient evaluation, consider again the periodic succession of rectangular pulses shown in Fig. 2. The derivative of this function consists of two impulses per period. Specifically for the period defined by $-\tau/2 < t < \tau/2$, we have

$$f^{(1)}(t) = Eu_0\left(t + \frac{\delta}{2}\right) - Eu_0\left(t - \frac{\delta}{2}\right) \tag{78}$$

that is, an impulse of value E occurring at $t = -\delta/2$ followed by an impulse of value $-E$ occurring at $t = \delta/2$.

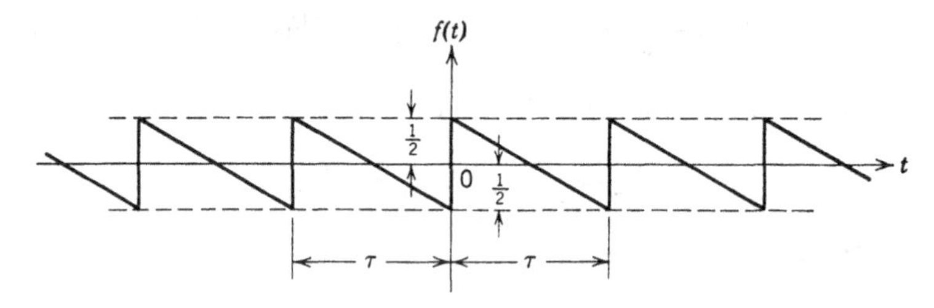

FIG. 4. Example to show how easily the Fourier series for a sawtooth wave may be gotten.

Regarding the evaluation of integral 76 for this time function consisting of a sum of impulses, we note first of all that

$$\int_{-\tau/2}^{\tau/2} u_0(t - t_0)e^{-j\nu\omega_0 t}\,dt = e^{-j\nu\omega_0 t_0} \qquad \text{for } -\tau/2 < t_0 < \tau/2 \tag{79}$$

because $u_0(t - t_0)$ is zero everywhere except at $t = t_0$ where it is infinite and encloses unit area. Hence the result of substituting the time function 78 into integral 76 may be written down by inspection. It reads

$$\gamma_\nu = \frac{E}{\tau}\{e^{j\nu\omega_0\delta/2} - e^{-j\nu\omega_0\delta/2}\} = \frac{2jE}{\tau}\sin\frac{\nu\omega_0\delta}{2} \tag{80}$$

The desired coefficient α_ν is found (according to Eq. 77) through division by $j\nu\omega_0$, thus:

$$\alpha_\nu = \frac{2E}{\tau}\cdot\frac{\sin \nu\omega_0\delta/2}{\nu\omega_0} \tag{81}$$

which checks the result given by Eq. 64.

As a second example, suppose we determine the Fourier coefficients for the sawtooth wave in Fig. 4. Over the period $-\tau/2 < t < \tau/2$ the derivative of this function is given by

$$f^{(1)}(t) = u_0(t) - \frac{1}{\tau} \tag{82}$$

Disregarding the constant term for the moment, Eqs. 76 and 79 yield

$$\gamma_\nu = \frac{1}{\tau} \tag{83}$$

indicating that all harmonic coefficients for the function $f^{(1)}(t)$ in this example are alike and that the one for $\nu = 0$ (the constant term or dc

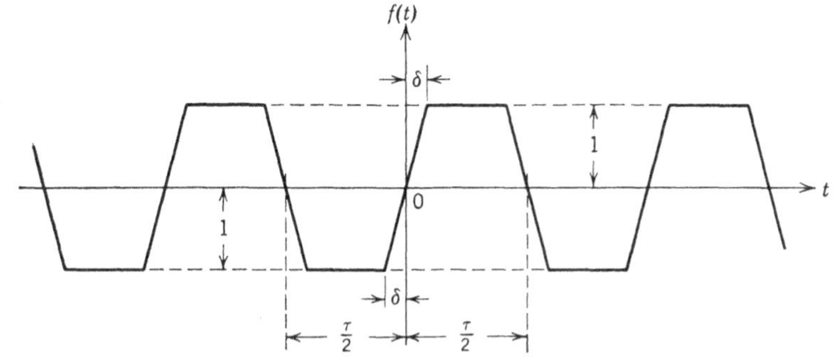

FIG. 5. A second example illustrating the impulse-train method of evaluating Fourier coefficients.

coefficient) is zero because this term in the Fourier series cancels the $-1/\tau$ term in Eq. 82.

The complex coefficient in the Fourier series for the sawtooth wave in Fig. 4 is thus given by

$$\alpha_\nu = \frac{1}{j\nu\omega_0\tau} = \frac{1}{2\pi j\nu} \tag{84}$$

It is purely imaginary and leads to the sine series

$$f(t) = \frac{1}{\pi}\left\{\frac{\sin \omega_0 t}{1} + \frac{\sin 2\omega_0 t}{2} + \frac{\sin 3\omega_0 t}{3} + \cdots\right\} \tag{85}$$

as might have been expected since $f(t)$ is an odd function.

Next we shall derive the Fourier series for the trapezoidal wave having a rise interval δ as shown in Fig. 5. If we differentiate this function once, the result is an alternation of identical positive and negative rectangular pulses of height $1/\delta$ and duration 2δ, with a positive pulse centered at

$t = 0$. The second derivative consists of a sum of impulses. Specifically, for the interval $-\tau/2 < t < \tau/2$ we have

$$f^{(2)}(t) = \frac{1}{\delta}\left[u_0\left(t + \frac{\tau}{2} - \delta\right) + u_0(t + \delta) - u_0(t - \delta) - u_0\left(t - \frac{\tau}{2} + \delta\right)\right] \tag{86}$$

Use of Eqs. 76 and 79 enables us to write at once

$$\gamma_\nu = \frac{1}{\delta\tau}\{e^{j\nu\omega_0[(\tau/2)-\delta]} + e^{j\nu\omega_0\delta} - e^{-j\nu\omega_0\delta} - e^{-j\nu\omega_0[(\tau/2)-\delta]}\} \tag{87}$$

Noting that $\omega_0\tau/2 = \pi$, and combining conjugate exponentials, we have

$$\gamma_\nu = \frac{2j \sin \nu\omega_0\delta}{\delta\tau}(1 - e^{j\nu\pi}) \tag{88}$$

The factor $1 - e^{j\nu\pi}$ is zero for $\nu = 0, 2, 4, \ldots$ and equal to 2 for $\nu = 1, 3, 5, \ldots$. Hence the trapezoidal wave of Fig. 5 contains only odd harmonics, as may have been predicted according to the discussion in art. 6. For ν equal to odd-integer values, Eqs. 77 and 88 thus yield

$$\alpha_\nu = \frac{4j \sin \nu\omega_0\delta}{\delta\tau(-\nu^2\omega_0^2)} = \frac{2}{j\pi\nu}\left\{\frac{\sin \nu\omega_0\delta}{\nu\omega_0\delta}\right\} \tag{89}$$

As the rise interval δ is chosen smaller and smaller, the trapezoidal function of Fig. 5 becomes the familiar square wave; the bracket factor in Eq. 89 becomes unity, and the corresponding series 67 is given by

$$f(t) = \frac{4}{\pi}\left(\frac{\sin \omega_0 t}{1} + \frac{\sin 3\omega_0 t}{3} + \frac{\sin 5\omega_0 t}{5} + \cdots\right) \tag{90}$$

It should be clear from these examples that any function consisting of straight-line segments will have its second derivative given by a sum of impulses as in Eq. 86. The pertinent Fourier coefficient can then be written down by inspection without the necessity of evaluating any integrals. Thus many practical situations are easily taken care of, especially since one can often use a broken-line approximation to some given function that does not initially consist of straight-line segments.

In order to illustrate another technique that may be used to obtain Fourier coefficients without formal integration, consider the so-called "half-sine" wave shown in the top sketch of Fig. 6. Below this one are drawn respectively the first and second derivatives of this function. The impulses contained in the latter all have the value ω_0 which equals the common value of the discontinuities in the first derivative.

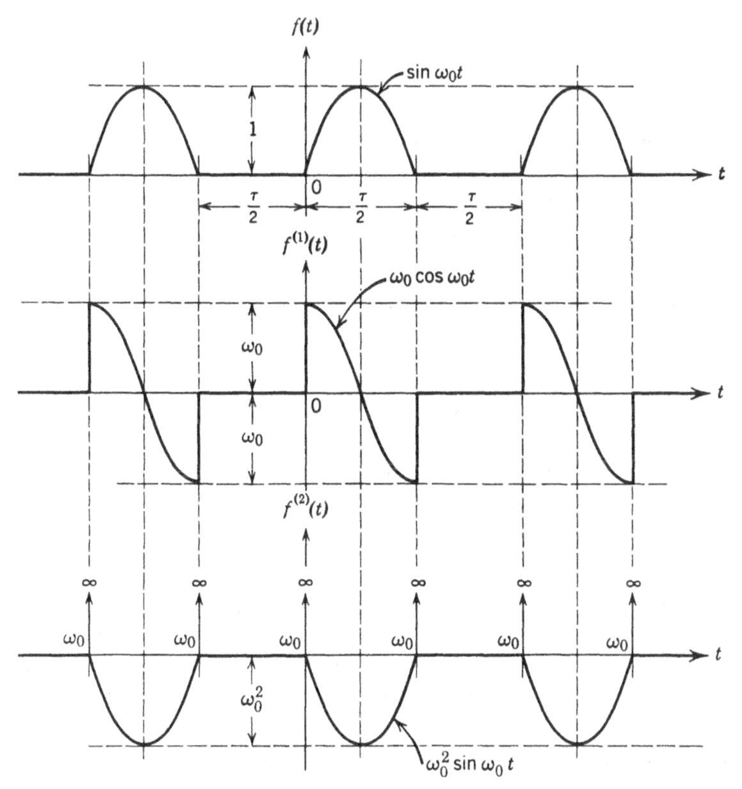

FIG. 6. Illustrating a second technique for obtaining Fourier coefficients without integration.

According to these results, one can evidently write for the period $0 \leq t < \tau$

$$f^{(2)}(t) = -\omega_0^2 f(t) + \omega_0 u_0(t) + \omega_0 u_0\left(t - \frac{\tau}{2}\right) \tag{91}$$

or

$$\omega_0^2 f(t) + f^{(2)}(t) = \omega_0\left[u_0(t) + u_0\left(t - \frac{\tau}{2}\right)\right] \tag{92}$$

If α_ν denotes the Fourier coefficient for $f(t)$, then that for the time function on the left of Eq. 92 is given by

$$\omega_0^2 \alpha_\nu (1 - \nu^2) \tag{93}$$

while the time function on the right of Eq. 92 yields the coefficient

$$\frac{\omega_0}{\tau}(1 + e^{-j\nu\pi}) = \frac{\omega_0^2}{2\pi}(1 + e^{-j\nu\pi}) \tag{94}$$

Equating these two we have

$$\alpha_\nu = \frac{1 + e^{-j\nu\pi}}{2\pi(1 - \nu^2)} \tag{95}$$

which is the desired Fourier coefficient for $f(t)$ in Fig. 6. Note that for $\nu = \pm 1$, this expression has the finite nonzero value $\mp j/4$, but for all other odd-integer values of ν, $\alpha_\nu = 0$. Except for the fundamental, the "half-sine" wave of Fig. 6 contains no odd harmonics.

9. A Criterion Regarding Asymptotic Convergence

Let us suppose that a given function $f(t)$ and all of its successive derivatives $f^{(1)}(t), f^{(2)}(t), \ldots f^{(k-1)}(t)$ are continuous, but that the kth derivative $f^{(k)}(t)$ has n discontinuities $\delta_1, \delta_2, \ldots \delta_n$ at the points $t_1, t_2, \ldots t_n$ within the fundamental range $-\tau/2 < t < \tau/2$. If we denote the saw-tooth wave in Fig. 4 by $h(t)$, then the function

$$f^{(k)}(t) - \sum_{i=1}^{n} \delta_i h(t - t_i) \tag{96}$$

is still continuous and hence it can be differentiated once more before impulses are encountered. If the procedure leading to the integral 75 is carried one step further we see, therefore, that the Fourier coefficient α_ν must decrease in magnitude for large values of ν at least as rapidly as the ratio $1/\nu^{k+1}$. Since the analysis integral 68 applied to an impulse in the fundamental range yields a constant, as expressed by Eq. 79, it is likewise clear that the coefficient α_ν in this case cannot decrease faster than this ratio for large values of ν.

This result provides a simple and often useful means for predicting the asymptotic rate of convergence of the Fourier series representation for a given periodic function. Thus if this function contains discontinuities like the rectangular pulse wave of Fig. 2 or the sawtooth wave of Fig. 4, then the harmonics can at most be inversely proportional to their order, as the formulas 64 and 84 show. The trapezoidal wave of Fig. 5, on the other hand, yields harmonics that vary in amplitude inversely as the square of their order. Formula 89 bears out this fact and also shows that as δ becomes smaller this convergence rate reverts to $1/\nu$ as it should, for the trapezoidal wave becomes a square wave for $\delta = 0$.

Each time the Fourier series 67 is integrated term by term, an additional factor ν is introduced into the denominator of the expression for α_ν, and each time it is differentiated, a factor ν is canceled out of the denominator. It is readily seen, therefore, that integration improves the rate of convergence of the Fourier series, whereas differentiation makes the series converge more slowly. This fact may have been expected since integrating a function makes it smoother whereas differentiating it accentuates any irregularities.

10. Computation of Power and Effective Values

When periodic voltages and currents occur in circuit theory the calculation of average power involves evaluation of the integral of a product of two Fourier series over their common fundamental range. Suppose the functions

$$f_1(t) = \sum_{\nu=-\infty}^{\infty} \alpha_\nu e^{j\nu\omega_0 t} \tag{97}$$

and

$$f_2(t) = \sum_{\mu=-\infty}^{\infty} \beta_\mu e^{j\mu\omega_0 t} \tag{98}$$

represent the voltage and current at a driving point, and we are interested in the so-called *power product*

$$P_{\text{av}} = \frac{1}{\tau} \int_{-\tau/2}^{\tau/2} f_1(t) f_2(t)\, dt \tag{99}$$

The integrand involved here is given by the double sum

$$f_1(t) f_2(t) = \sum_{\nu,\mu=-\infty}^{\infty} \alpha_\nu \beta_\mu e^{j(\nu+\mu)\omega_0 t} \tag{100}$$

Substituting this result into the integral 99 and integrating term by term (which the discussion in the two foregoing articles has shown to be permissible even when $f_1(t)$ and $f_2(t)$ contain discontinuities), we get

$$P_{\text{av}} = \frac{1}{\tau} \sum_{\nu,\mu=-\infty}^{\infty} \alpha_\nu \beta_\mu \int_{-\tau/2}^{\tau/2} e^{j(\nu+\mu)\omega_0 t}\, dt \tag{101}$$

Evaluation of the integral involved here reads

$$\int_{-\tau/2}^{\tau/2} e^{j(\nu+\mu)\omega_0 t}\, dt = \frac{e^{j(\nu+\mu)\pi} - e^{-j(\nu+\mu)\pi}}{j(\nu+\mu)\omega_0} = \begin{cases} \tau & \text{for } \nu = -\mu \\ 0 & \text{for } \nu \neq -\mu \end{cases} \tag{102}$$

Hence all terms in the double sum 101 vanish except those for which $\nu = -\mu$, and we have the simple result

$$P_{\text{av}} = \sum_{\nu=-\infty}^{\infty} \alpha_\nu \beta_{-\nu} \tag{103}$$

If we let

$$\alpha_\nu = |\alpha_\nu|\, e^{-j\psi_\nu}, \quad \beta_\nu = |\beta_\nu|\, e^{-j\phi_\nu} \tag{104}$$

Then 103 may be written in the alternate form

$$P_{\text{av}} = \alpha_0\beta_0 + 2 \sum_{\nu=1}^{\infty} |\alpha_\nu \beta_\nu| \cos(\psi_\nu - \phi_\nu) \tag{105}$$

which illustrates the interesting fact that average power depends only upon products of harmonics of the same order. In other words, none of the cross-product terms resulting from the product of the two series for $f_1(t)$ and $f_2(t)$ contributes to the average value of this product, a result that again demonstrates the orthogonal property discussed in the introductory article of this chapter.

The coefficient $\cos(\psi_\nu - \phi_\nu)$ appearing in the sum 105 is the *power factor* of the harmonic of order ν; and the factor 2 ahead of the sum is needed because the magnitudes of α_ν and β_ν are *half* the peak values of the pertinent sinusoids, so that $2\,|\alpha_\nu\beta_\nu|$ equals the product of their *effective* or *rms* values.

Incidentally, the effective value of either function 97 or 98 is the value of integral 99 for $f_1 \equiv f_2$, and hence is given by Eq. 103 if we identify β_ν with α_ν. Thus we have

$$(f_1)_{\text{eff}} = \sum_{\nu=-\infty}^{\infty} |\alpha_\nu|^2 = \alpha_0 + 2\sum_{\nu=1}^{\infty} |\alpha_\nu|^2 \tag{106}$$

From Eqs. 99 and 106 we then have the interesting result

$$\frac{1}{T}\int_{-T/2}^{T/2} f_1^2(t)\,dt = \sum_{\nu=-\infty}^{\infty} |\alpha_\nu|^2 \tag{107}$$

which states that the square of a time function integrated over its fundamental range in the time domain equals the square of the corresponding spectrum function summed over its infinite range in the frequency domain, except for a proportionality constant.

Problems

1. Four arbitrary vectors are defined by the rows of the following matrix:

$$[A] = \begin{bmatrix} 1 & -1 & 2 & 1 \\ 2 & 1 & -1 & 1 \\ 1 & -2 & 1 & 2 \\ -1 & 1 & -1 & 1 \end{bmatrix}$$

From these construct an orthonormal set, the first of which coincides with a_1 of $[A]$, the second of which lies in the plane determined by a_1 and a_2, the third of which is a linear combination of a_1, a_2 and a_3, etc. Determine also the linear transformation connecting the given vectors with the orthonormal set.

2. A periodic square-wave voltage source is connected to a series *RLC* circuit.

(*a*) Using only the known unit step response of this circuit, obtain an expression for the resulting periodic current when the circuit is tuned to a harmonic of order ν, making approximations consistent with the assumption that the Q of the circuit is very large.

(*b*) Using Fourier analysis for the applied voltage wave, find the expression for current response under the same circuit conditions, making analogous approximations compatible with the high-Q property of the circuit.

Are these two results necessarily the same? Discuss with regard to the minimum-square-error criterion implicit in the Fourier coefficient formula.

3. A class-C amplifier has a sinusoidal voltage applied to its grid with peak value E. The grid has a negative bias voltage equal to $0.9E$. The tube characteristic may be assumed given by two linear segments such that for a net negative grid voltage the I_p versus E_g slope is zero while for a net positive grid voltage the slope is g_m. The plate-cathode load is a simple tank circuit (C in parallel with series RL) with $Q = 100$ and $R = 50$ ohms at the operating frequency. Compute the essentially sinusoidal voltage amplitude across the tank circuit for $g_m = 1000$ micromhos and thus obtain the voltage gain of this amplifier.

4. By substituting Eq. 37 for α_ν into Eq. 24 obtain Fourier representation

$$f(t) = \frac{1}{\tau}\int_{-\tau/2}^{\tau/2} f(\xi)\,d\xi + f_1(t) + f_2(t) + \cdots$$

where

$$f_\nu(t) = \frac{2}{\tau}\int_{-\tau/2}^{\tau/2} f(\xi)\cos\nu\omega_0(t-\xi)\,d\xi$$

Show that this integral may alternately be written

$$f_\nu(t) = \frac{2}{\tau}\int_{-\tau/2}^{\tau/2} \cos\nu\omega_0\xi\, f(t-\xi)\,d\xi$$

and that its evaluation may be interpreted graphically by considering the function $f(t)$ in the form of a template which, for variable t, is slid across the function $\cos\nu\omega_0\xi$, the value of $f_\nu(t)$ for any given t being the net area enclosed by the product of the cosine function and $f(t-\xi)$ evaluated over a period τ. In terms of this interpretation, which is described by saying that the cosine function is being "scanned" by $f(t)$—the scanning function, show that $f_\nu(t)$ is a harmonic function of time having the form

$$f_\nu(t) = c_\nu \cos(\nu\omega_0 t - \psi_\nu)$$

Thus the net area, during the scanning process, varies harmonically and reaches a maximum value c_ν at the time $t_{\max} = \psi_\nu/\nu\omega_0$. In situations where one can recognize $t_{\max}$ by inspection, this interpretation yields a direct evaluation for the phase angle ψ_ν and the harmonic amplitude c_ν.

5. Apply the graphical process suggested in the preceding problem to the evaluation of the Fourier series for a square wave, making sketches appropriate to the evaluation of pertinent harmonic amplitudes and phase angles.

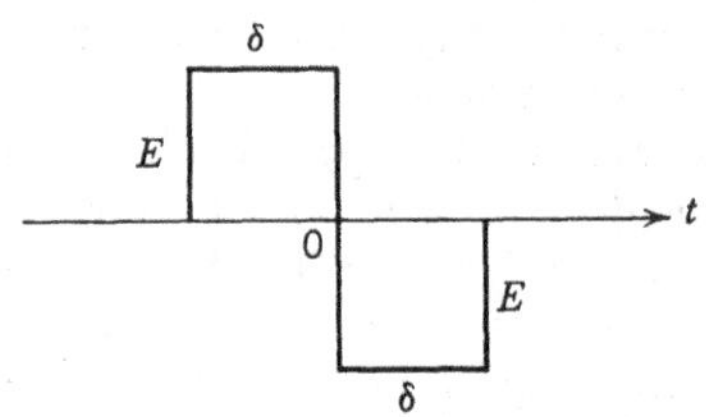

(a) Pulse doublet periodically repeated

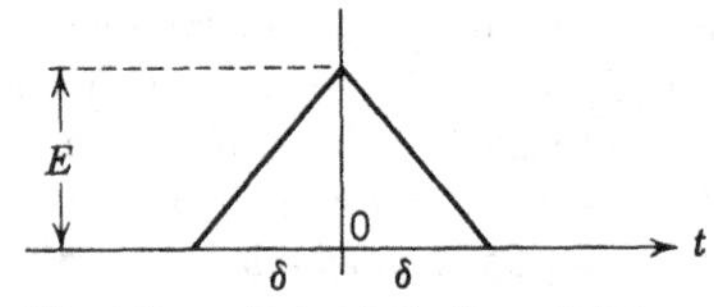

(b) Triangular pulse periodically repeated

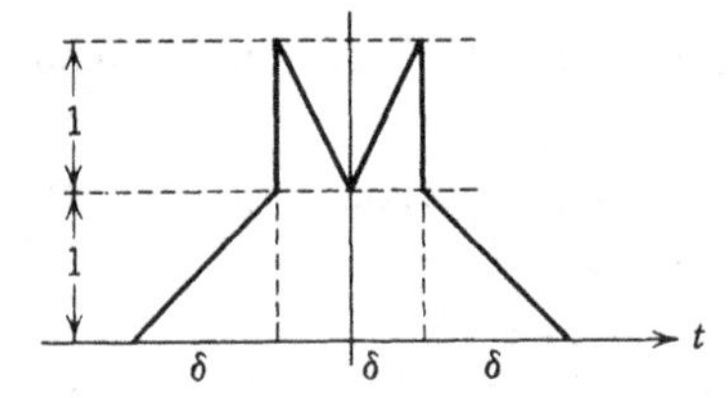

(c) Queer pulse periodically repeated

PROB. 9.

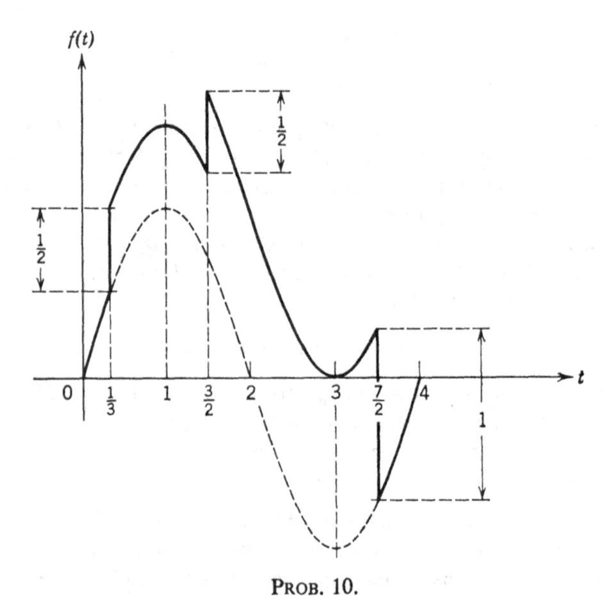

PROB. 10.

6. In a given communication system, a sinusoidal carrier is keyed off and on periodically. During a period τ let us say that the carrier is on for a time t_1 and off for a time t_2. During the period $\tau = t_1 + t_2$ the carrier completes a whole number of cycles so that the carrier frequency ω_c is a harmonic of the resulting periodic function.

The question is: What are the amplitude and phase angle of this harmonic frequency relative to the amplitude and phase of the carrier when it is not mutilated by the keying process? Assume that t_1 or t_2 equals an integer number of carrier half-cycles.

7. A frequency-modulated wave consists of the sinusoid $-A \sin 2\omega_0 t$ during the half-period interval $-\tau/4 < t < \tau/4$, continuously joined to the sinusoid $A \sin 4\omega_0 t$ during the remainder of a fundamental period centered at $t = 0$. The problem is to evaluate the amplitudes and phase angles of the particular harmonics having the two given frequencies. Do this in a simple, effortless manner.

8. By means of the relations 24 and 36, prove directly the following properties of the time-domain and frequency-domain functions $f(t)$ and α_ν:

(*a*) If $f(t)$ is replaced by its derivative or integral, then α_ν is multiplied or divided by $j\nu\omega_0$.

(*b*) If the time function is delayed or advanced by t_0, then α_ν is multiplied or divided by $e^{-j\nu\omega_0 t_0}$.

(*c*) If $f(t)$ is multiplied by $e^{jn\omega_0 t}$, then α_ν is replaced by $\alpha_{\nu-n}$, both ν and n being integers.

(*d*) The complex Fourier coefficient of the periodically amplitude-modulated carrier function $f(t) \times \cos n\omega_0 t$ is $\frac{1}{2}(\alpha_{\nu-n} + \alpha_{\nu+n})$.

9. Using the result 64 for the pulse function in Fig. 2 and appropriate properties in prob. 8, write down by inspection the complex Fourier coefficients for the periodic functions shown in the foregoing sketches, page 332.

10. The function illustrated on page 332 consists of a fundamental sinusoid (shown dotted) with several discontinuities inserted as indicated. The whole is repeated periodically. The problem is to determine with a minimum of effort the amplitude and phase of the resultant fundamental.

CHAPTER XII

Time- and Frequency-Domain Correlation for Aperiodic Functions Through Fourier Analysis

1. Frequency Groups and Interference Patterns

Discussions in the foregoing chapter are aimed at studying a given periodic function to determine its representation in terms of simple harmonic components. Let us consider instead the reverse problem in which a finite or infinite number of discrete frequency components are given and the object is to determine the function resulting from their linear superposition.

To begin with a simple example let three steady sinusoids with any finite amplitudes and fixed relative phase angles be given, having the frequencies: 100, 125 and 150 cycles per second. The resultant function given by their sum is periodic, the fundamental frequency being 25 cycles per second. This is clear from the fact that 25 is the highest common factor (HCF) of the group of numbers 100, 125, and 150. The fundamental period is $\frac{1}{25}$th of a second. Throughout this interval the 100-cycles-per-second component completes 4 cycles, the 125-cycles-per-second component completes 5 cycles, and the 150-cycles-per-second component completes 6 cycles. The original state is then re-established because each component has completed a *whole* number of cycles. This statement is true no matter what the relative phase angles of the components are.

In the language of Fourier series, the resultant periodic function is represented by its fourth, fifth and sixth harmonic components alone. All other components, including the fundamental, are absent. Note that the linear superposition of a group of harmonic components yields a periodic function *if and only if the component frequencies are commensurable.* This requirement is met if the component frequencies as a group of numbers have a common factor. If an irrational number like $\sqrt{2}$ or a transcendental number like π or the Napierian base e is included with some rational numbers to form a frequency group, then the resultant function *never exactly repeats its sequence of values*; that is, its period becomes infinite.

For example, the function

$$f(t) = \cos 100t + \cos (100 + \pi)t \tag{1}$$

never repeats its sequence of values. This situation should not be confused with the well-known result that the function given by Eq. 1 can be interpreted as a *beat phenomenon* by applying a familiar trigonometric identity to the right-hand side so as to express it in the equivalent form

$$f(t) = 2 \cos \frac{\pi}{2} t \cdot \cos \left(100 + \frac{\pi}{2}\right) t \tag{2}$$

which is customarily plotted by considering the slowly varying function 2 $\cos (\pi/2)t$ as an envelope enclosing the rapidly varying simple harmonic function $\cos [100 + (\pi/2)]t$. Each half period of the envelope function $\cos (\pi/2)t$ is commonly referred to as a "beat period," but this is a true period only if the difference between the two frequencies involved is at the same time their highest common factor. Otherwise the resultant function is referred to as being *quasi-periodic.*

Even when the two frequencies are rational numbers the beat period is not necessarily a true period. For example, the frequencies 100 cycles per second and 103 cycles per second give rise to a beat frequency of 3 cycles per second, but the true fundamental frequency is 1 cycle per second. In other words, the exact pattern of the resultant function does not repeat until *three* beat periods have lapsed. The well-known experimental fact that the human ear (under proper circumstances) appears to recognize the beat frequency as though it were actually present in the form of a frequency component is a physiological phenomenon due in part to a nonlinear characteristic in the response mechanism of the ear, a situation that is not to be confused with the strictly *linear* superposition of component frequencies considered here.

It is interesting as well as instructive to generalize the problem of the simple beat phenomenon by inquiring how the resultant pattern, commonly referred to as an *interference pattern,* looks when more than two frequency components are superimposed. Incidentally, it is essential to recognize that the familiar beat pattern in the case of two frequencies is pronounced only when the increment between these frequencies is small compared with either one. When the two frequencies and their difference are of the same order of magnitude, the trigonometric conversion from Eq. 1 to Eq. 2 is, of course, still valid, but the two functions in the right-hand product of Eq. 2 then vary at about the same rate and the beat character of the resultant is lost.

In generalizing the problem of the beat phenomenon it is therefore essential to assume a group of simple harmonic components whose spacing in the frequency spectrum is small compared with the mean frequency of the group. The line spectrum of such a group containing seven components is illustrated in Fig. 1. The mean angular frequency is designated as ω_m. The adjacent frequencies are $\omega_m + \omega_0$, $\omega_m - \omega_0$, $\omega_m + 2\omega_0$, $\omega_m - 2\omega_0$, etc., the uniform spacing being ω_0. In general let there be n components in the group, and for simplicity let all their amplitudes be equal and all their phases be zero. It is expedient to set the common

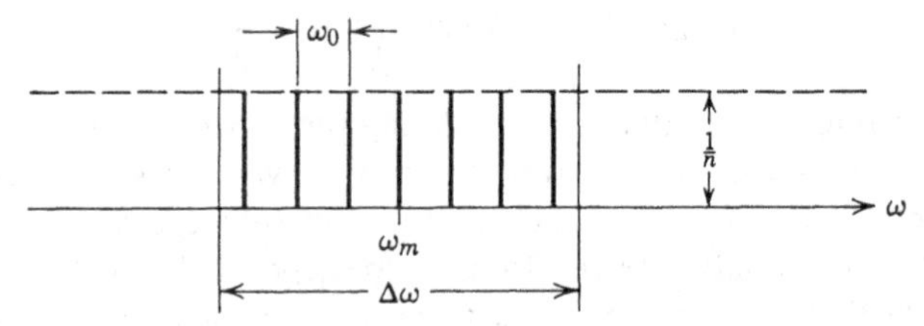

FIG. 1. Uniform line spectrum yielding the periodic interference pattern shown in Fig. 2.

amplitude equal to $1/n$, and to define the "width" of the group as

$$\Delta\omega = n\omega_0 \tag{3}$$

The analytic expression for this group is

$$f(t) = \frac{1}{n}\cos\omega_m t + \frac{1}{n} \times \begin{Bmatrix} \cos(\omega_m + \omega_0)t + \cos(\omega_m + 2\omega_0)t + \cdots + \cos\left(\omega_m + \dfrac{n-1}{2}\,\omega_0\right)t \\ \cos(\omega_m - \omega_0)t + \cos(\omega_m - 2\omega_0)t + \cdots + \cos\left(\omega_m - \dfrac{n-1}{2}\,\omega_0\right)t \end{Bmatrix} \tag{4}$$

By making use of the trigonometric identity for the cosine of the sum of two angles, this expression can be rewritten in the form

$$f(t) = \frac{1}{n}\cos\omega_m t\left\{1 + 2\cos\omega_0 t + 2\cos 2\omega_0 t + \cdots + 2\cos\frac{n-1}{2}\,\omega_0 t\right\} \tag{5}$$

The trigonometric polynomial within the curved brackets may be put into a closed form by making use of the well-known expressions

$$\frac{1 - e^{jkx}}{1 - e^{jx}} = 1 + e^{jx} + e^{j2x} + \cdots + e^{j(k-1)x} \tag{6}$$

$$\frac{1 - e^{jkx}}{1 - e^{jx}} \cdot e^{-j(k-1)x} = 1 + e^{-jx} + e^{-j2x} + \cdots + e^{-j(k-1)x} \tag{7}$$

Addition of the left-hand sides gives

$$\frac{e^{jkx/2} - e^{-jkx/2}}{e^{jx/2} - e^{-jx/2}} (e^{j(k-1)x/2} + e^{-j(k-1)x/2}) = \frac{2 \sin \dfrac{kx}{2} \cos \dfrac{(k-1)x}{2}}{\sin \dfrac{x}{2}} \tag{8}$$

which, through use of another well-known identity, becomes

$$\frac{\sin \dfrac{(2k-1)x}{2} + \sin \dfrac{x}{2}}{\sin \dfrac{x}{2}} = 1 + \frac{\sin (k - \frac{1}{2})x}{\sin \dfrac{x}{2}} \tag{9}$$

while addition of the right-hand sides yields

$$2 + 2 \cos x + 2 \cos 2x + \cdots + 2 \cos (k-1)x \tag{10}$$

If we let $k = (n + 1)/2$ this becomes the trigonometric polynomial in Eq. 5, and we thus obtain the following expression for the frequency group in Eq. 4:

$$f(t) = \frac{\sin \dfrac{n\omega_0}{2} t}{\sin \dfrac{\omega_0}{2} t} \cdot \cos \omega_m t \tag{11}$$

Here the function

$$F(t) = \frac{\sin \dfrac{n\omega_0}{2} t}{\sin \dfrac{\omega_0}{2} t} \tag{12}$$

is slowly varying compared with the mean-frequency component $\cos \omega_m t$ and hence may be regarded as an envelope enclosing the latter. The beat phenomenon is placed in evidence by the envelope function just as in the simple case of two interfering sinusoids.

This envelope function is plotted versus time t for the case $n = 7$ in Fig. 2. It illustrates the interference pattern for the group of frequencies whose line spectrum is shown in Fig. 1. The regions of constructive interference, which are the main humps of the envelope, occur at intervals of

$$\tau_g = \frac{2\pi}{\omega_0} \text{ seconds} \tag{13}$$

The duration of each main hump is $4\pi/\Delta\omega$ seconds. The group period τ_g is inversely proportional to the frequency increment (the line spacing) ω_0, and the duration of a region of constructive interference is inversely proportional to the width $\Delta\omega$ of the group. If this width is kept constant while more components are added to the group, regions of constructive interference occur at longer intervals, but their duration remains the same. The number of small humps between the large ones increases; but for a large number of component frequencies, the amplitudes of these smaller humps become insignificant (of the order $1/n$) midway between the large

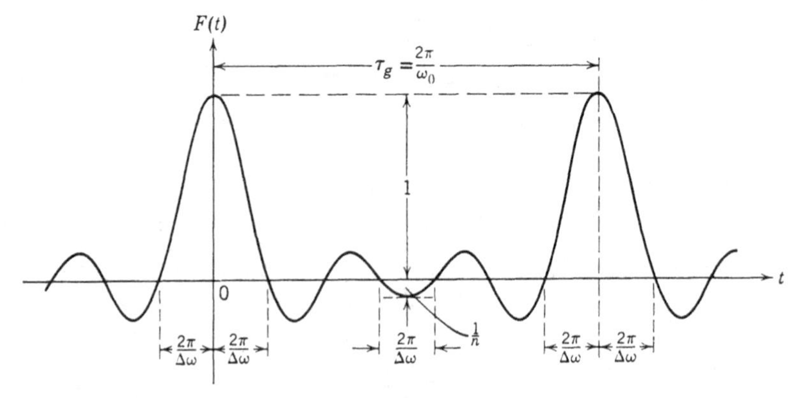

FIG. 2. Periodic time function corresponding to the line spectrum in Fig. 1.

ones and in this vicinity which is referred to as a region of *destructive* interference since the sinusoidal components there are very nearly canceling or annulling each other.

The fact that the interval between regions of constructive interference is inversely proportional to the frequency increment ω_0 leads to the conclusion that, as this increment is allowed to approach zero, the beat period grows without limit. The frequency group finally becomes continuous, and the resulting function has only one region of constructive interference in the entire time scale from minus to plus infinity, notwithstanding that each frequency component individually maintains a constant amplitude throughout the entire time scale as a true steady-state component should.

This limiting form for the envelope function 12 is readily evaluated. Since the width $\Delta\omega$ of the group remains constant, n and ω_0 are inversely proportional to each other as the limit process indicated by $n \to \infty$ and $\omega_0 \to 0$ is carried to completion. The trigonometric sine in the denominator of Eq. 12 may be replaced by its argument, and the limiting form of this function is seen to be

$$F(t) = \frac{\sin \dfrac{\Delta\omega}{2} t}{\dfrac{\Delta\omega}{2} t} \tag{14}$$

This time function is plotted in Fig. 3 where the dotted curve, a rectangular hyperbola, shows the way in which amplitudes decrease for large positive or negative values of t. Essentially this limiting form of the envelope function is given by a single large hump like those in Fig. 2 pertinent to a finite nonzero spacing ω_0 in the associated line spectrum.

The resultant group function $f(t)$, which is the mean frequency component $\cos \omega_m t$ bounded by this envelope, is now entirely transient or

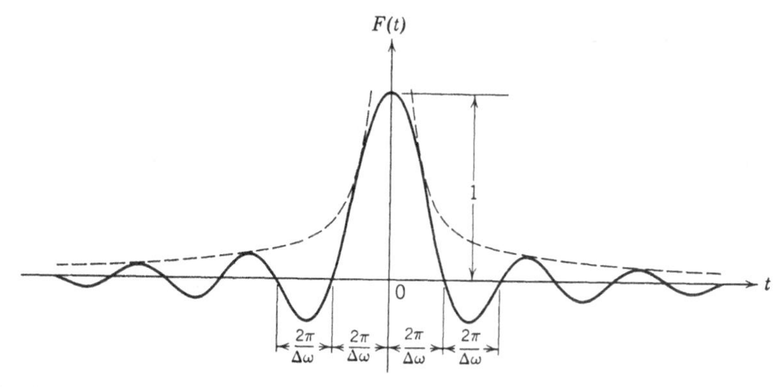

FIG. 3. Limiting form of the interference pattern in Fig. 2 resulting from letting the line spectrum of Fig. 1 become continuous.

aperiodic in character; that is, it never repeats but has only one major region of constructive interference. Beyond this region the components in this group (infinite in number and infinitesimal in size) interfere more or less destructively forever.

Since the phases of all components are chosen equal to zero, the region of constructive interference lies at the time origin. If in the argument of each cosine term in Eq. 4 a phase $-\psi$ having the frequency dependence

$$\psi(\omega) = (\omega - \omega_m)t_0 \tag{15}$$

is inserted, then the variable t in the above formulas for the envelope function is replaced by $t - t_0$. The region of constructive interference then occurs in the vicinity of the point $t = t_0$. If the phases of the components in the group have a random distribution, the form of the resulting pattern is not determinable and has no well-defined region of constructive interference.

These examples strongly suggest that Fourier analysis for periodic functions may be extended to include aperiodic functions as well. We shall discuss the details of such an extension in a moment. Meanwhile we shall pursue further the thought that Fourier analysis may be regarded as a mechanism for determining the relative amplitudes and phases of steady sinusoidal components whose linear superposition yields an interference pattern approximating a given function more and more closely as the number of components is increased.

In this sense the infinite line spectrum shown in Fig. 3 of the previous chapter may be regarded as yielding the periodic interference pattern shown there in Fig. 2. If the period τ of this function is increased more and more, keeping the dimensions E and δ of the pulse unchanged, the dotted envelope of the line spectrum or frequency group in Fig. 3 changes only by a scale factor for its ordinates, while the line spacing $\omega_0 = 2\pi/\tau$ becomes smaller and smaller. For a sufficiently large value of τ, we are for all practical purposes left with just a single rectangular pulse centered at the time-domain origin. The pattern created by this limiting frequency group has one region of constructive interference during which the resultant equals a constant value E for a duration δ, while for all time before and after this interval the interference is completely and uniformly destructive.

This situation is obviously an extreme test of what a group of steady sinusoids can produce by linear superposition or interference. Since Fourier theory is essentially a theory of interference patterns creatable by the superposition of simple harmonic functions, it is extremely important that one should develop an understanding of the graphical aspects of this process, as afforded by examples like the ones discussed here. The fact that steady (everlasting) sinusoids may be regarded as building blocks out of which one can construct aperiodic functions (nonzero over a finite interval only) is a truely remarkable result. Even more astounding is the fact (discussed in art. 3 below) that these building blocks can be generalized to include exponentially growing or decaying sinusoids notwithstanding that the resulting interference pattern may remain bounded throughout the entire time scale. We shall, however, first consider the formal process of extending Fourier analysis and synthesis in terms of steady sinusoids to include aperiodic as well as periodic time functions.

2. The Fourier Integral

An important aspect of the foregoing discussion may be summarized by stating that: whereas the linear superposition of a group of discrete frequency components (finite or infinite line spectrum) gives rise to an interference pattern of a periodic or quasi-periodic nature, this pattern becomes aperiodic in character when the frequencies in the group are assumed continuously distributed. The resulting so-called *continuous* spectrum should be thought of as a line spectrum in which the spacing of lines has become infinitesimal; and the aperiodic character of the resulting time function may be thought of as the limiting form of a periodic function for which the period has become infinite.

That these two limit processes are consistent is evident from the fact that the spacing of lines in the spectrum of a periodic function equals its fundamental radian frequency, which is 2π times the inverse of its period. It may further be helpful in this connection to recognize that the period equals the line *density* expressed as the *number of lines per cycle per second.* As the period is allowed to grow without limit, the line density grows without limit, so that finally the spectrum becomes a continuous one and the function never repeats; that is, it becomes an aperiodic function.

By carrying out this limit process in terms of the Fourier series and its spectrum function or coefficient formula, Eqs. 67 and 68 in the preceding chapter, the analytic means are obtained for representation in closed form of an arbitrary aperiodic function. The value of such a mathematical tool in connection with communication problems is readily appreciated.

Repeating, for convenience, the mathematical statement of the periodic case

$$f(t) = \sum_{\nu=-\infty}^{\infty} \alpha_\nu e^{j\nu\omega_0 t} \tag{16}$$

$$\alpha_\nu = \frac{1}{\tau}\int_{-\tau/2}^{\tau/2} f(t)e^{-j\nu\omega_0 t}\,dt \tag{17}$$

it should be observed that the limit process indicated by $\tau \to \infty$ and $\omega_0 \to 0$ evidently is accompanied by $\alpha_\nu \to 0$; that is to say, as the line spacing in the spectrum is allowed to become infinitesimal, the amplitudes of the harmonic components also become infinitesimal. However, the ratio α_ν/ω_0 approaches a finite nonzero limit. Before carrying out the limit process it is, therefore, expedient to write Eqs. 16 and 17 in the form

$$f(t) = \frac{1}{2\pi}\sum_{\nu\omega_0=-\infty}^{\infty}\left(\frac{2\pi\alpha_\nu}{\omega_0}\right)e^{j\nu\omega_0 t}\,\Delta(\nu\omega_0) \tag{18}$$

$$\left(\frac{2\pi\alpha_\nu}{\omega_0}\right) = \int_{-\tau/2}^{\tau/2} f(t)e^{-j\nu\omega_0 t}\,dt \tag{19}$$

where $\Delta(\nu\omega_0) = \omega_0$ is the uniform increment in the discrete variable $\nu\omega_0$, or the line spacing in the spectrum.

The limit process is formally indicated by the transitions

$$\left.\begin{aligned} \tau &\to \infty \\ \omega_0 &\to 0 \\ \nu\omega_0 &\to \omega \\ \Delta(\nu\omega_0) &\to d\omega \\ \left(\frac{2\pi\alpha_\nu}{\omega_0}\right) &\to F(\omega) \end{aligned}\right\} \tag{20}$$

Just as the discrete variable $\nu\omega_0$ refers to a particular harmonic frequency in the line spectrum of the periodic function 18, so the continuous variable ω refers to a particular harmonic frequency in the continuous spectrum of the aperiodic function resulting from the transitions 20. Consistent with the definition of this continuous frequency variable, the differential $d\omega$ replaces the uniform increment $\Delta(\nu\omega_0)$.

Introduction of the variable ω is really only a convenience in thought and in writing the limiting forms of the various functions. It is suggested by the fact that the constituent parts of $\nu\omega_0$, namely ν and ω_0, become improper in the limit. It should be observed, however, that $\nu\omega_0$ remains a proper variable, and still refers to the frequency of any harmonic component in the limit precisely as it does before this limit is carried to completion. If this limit process is thought of as being carried out in steps by doubling and redoubling the period τ, then it becomes clear that each time τ is doubled, any specific harmonic component doubles its order ν. If attention is fixed upon a specific harmonic frequency, ν and ω_0 must vary inversely as the period is increased. The frequency increment $\Delta(\nu\omega_0)$ evidently becomes the differential increment $d\omega$, and the ratio $2\pi\alpha_\nu/\omega_0 = \alpha_\nu/\tau$ becomes a new spectrum function $F(\omega)$ of the continuous frequency variable ω, expressing the variation of harmonic amplitudes in the limit in which these amplitudes are infinitesimal yet have a spectral distribution that uniquely determines the resulting time function.

As the limit process is carried to completion, the sum in Eq. 18 becomes replaceable by an integral. The final forms of the pair of relations 18 and 19 read

$$f(t) = \frac{1}{2\pi}\int_{-\infty}^{\infty} F(\omega)\, e^{j\omega t}\, d\omega \tag{21}$$

$$F(\omega) = \int_{-\infty}^{\infty} f(t)\, e^{-j\omega t}\, dt \tag{22}$$

While the above heuristic derivation of these forms does not establish their validity on a rigorous mathematical basis, our interest from an engineering point of view lies primarily in their interpretation and use, to which we devote the remaining chapters of this book.

In connection with the transitions 20 it is rather significant to point out the additional one

$$2\pi\alpha_\nu \rightarrow F(\omega)\, d\omega \tag{23}$$

which is obvious since $\omega_0 = \Delta(\nu\omega_0)$. It thus becomes clear that harmonic amplitudes in the integral 21 are given, not by $F(\omega)$, but by $F(\omega)\, d\omega$, which places their infinitesimal character in evidence since it has been seen that $F(\omega)$ is finite. Just as $\Delta(\nu\omega_0)$ is a uniform spacing in the line spectrum, so $d\omega$ is also to be thought of as uniform or constant. Hence $F(\omega)$ is proportional to $F(\omega)\, d\omega$, so that a plot of $F(\omega)$ versus ω shows the correct variation of harmonic amplitudes with frequency even though these amplitudes are all infinitely small.

Equation 21 is called the Fourier integral representation for the function $f(t)$, and Eq. 22 is its mate. The pair of relations are also called Fourier transformations. The second transforms a time function $f(t)$ into its equivalent frequency function $F(\omega)$, and the first integral reverses the process. The second integral *analyzes* the time function into a spectrum, and the first integral *synthesizes* the spectrum to regain the time function. Just as with periodic functions, we assume that any given function has two equivalent modes of representation: one in the time domain and one in the frequency domain. The integrals 21 and 22 convert one of these representations into the other and vice versa. In this connection $F(\omega)$ is referred to as the *direct Fourier transform of* $f(t)$ and the latter as the *inverse Fourier transform* of $F(\omega)$.

Graphical interpretation of the process of Fourier transformation by means of the integrals 21 and 22 remains exactly the same as for periodic functions where one of the pertinent relations involves an infinite series. The essential difference is that the spectrum function is continuous and that synthesis of the spectrum is accomplished by means of an integral instead of a sum. This latter point is actually an advantage because there are more formulas available for the evaluation of integrals than there are for the evaluation of sums. The principal advantage, however, lies in the fact that the Fourier integral is capable of handling transient functions.

Conditions under which Fourier transformation of aperiodic functions is possible are essentially the same as already discussed for periodic functions. Since the integrals 21 and 22 must exist, it is sufficient if $f(t)$ or $F(\omega)$ are integrable over the infinite domain of the variable t or ω. Subsequent discussions which are aimed at developing techniques for evaluating these integrals under a variety of particular conditions and with a minimum of computational effort will also consider a number of special devices for evaluating direct and inverse Fourier transformations when the integrability condition is not fulfilled but physical or graphical interpretation indicates that the pertinent function nevertheless exists.

3. Interpretation in the Complex Frequency Plane; the Laplace Transform

In the practical use of Fourier methods it is usually the evaluation of integral 21 involving the inverse transformation that presents considerable difficulty, especially when the analysis of a fairly complicated system is contemplated. Although the integrand here is in general complex, the variable of integration, which is ω, is real, and hence we are dealing with a problem in real integration. A rather well-known artifice that is widely used in difficult cases of real integration is to replace the real independent variable by a complex one and thus make available the technique of complex integration.

In our situation this expedient is almost self-suggestive inasmuch as the concept of a complex frequency variable is already established. In fact, we can interpret 21 as a complex integral along the j-axis as it stands. All that we need do in order to make complex integration completely available is to replace the function $F(\omega)$ by its analytic continuation in the complex plane.

In order to see how this is done, suppose we consider a specific example, namely the spectrum function resulting from applying the transitions 20 to the periodic succession of rectangular pulses shown in Fig. 2 in the preceding chapter. Starting with the pertinent expression for α_ν as given there by Eq. 64, we readily find for the spectrum function of a single pulse centered at $t = 0$

$$F(\omega) = E\delta \frac{\sin \omega\delta/2}{\omega\delta/2} = E\delta \frac{\sinh j\omega\delta/2}{j\omega\delta/2} = \hat{F}(j\omega) \tag{24}$$

Our main purpose here is to show that the function $F(\omega)$ can always be rewritten as a function of $j\omega$ if we wish. The latter is functionally not the same as $F(\omega)$, and so for the moment we put a distinguishing mark over it. Analytic continuation into the complex frequency plane is now accomplished through the simple expedient of replacing $j\omega$ by the complex variable s. In the complex plane the transform of the single rectangular pulse, therefore, reads

$$\hat{F}(s) = E\delta \cdot \frac{\sinh s\delta/2}{s\delta/2} \tag{25}$$

Since this function is identical with $F(\omega)$ in Eq. 24 on the j-axis of the s-plane, the identity theorem* establishes the fact that 25 is the desired analytic continuation of $F(\omega)$.

We can apply this simple procedure to the pair of integrals 21 and 22 and thus obtain an extended version of the Fourier transformation that involves a spectrum function of the complex frequency variable s directly. In 21 we multiply and divide by j so as to get $d(j\omega)$ instead of $d\omega$. The function $F(\omega)$ may be replaced by $F(j\omega)$ as in Eq. 24; and the integral 22 can just as well be regarded as yielding this function of $j\omega$. If we then write s for $j\omega$ and drop the distinguishing mark over the function F since it is no longer needed, we have a pair of mutually inverse transformations

* See *MCA*, p. 290.

in the form

$$f(t) = \frac{1}{2\pi j} \int_{-j\infty}^{j\infty} F(s)\, e^{st}\, ds \tag{26}$$

and

$$F(s) = \int_{-\infty}^{\infty} f(t)\, e^{-st}\, dt \tag{27}$$

where the limits on the first of these two integrals indicate that integration is along the j-axis as in 21. We will see shortly, however, that this path of integration is not so rigidly fixed.

In this extended form, the integrals are referred to as Laplace rather than as Fourier transformations. $F(s)$ is designated the direct Laplace transform of $f(t)$ and the latter as the inverse Laplace transform of $F(s)$. The Fourier transform $F(j\omega)$ then is simply the Laplace transform evaluated along the j-axis. We will elaborate and illustrate this statement later on.

The particular form of the Laplace transformation obtained from an extension of the Fourier integrals into the complex plane is spoken of as a "double-ended" Laplace transformation in order to distinguish it from a somewhat more restricted form (the "single-ended" Laplace transformation) which has the built-in tacit assumption that $f(t) \equiv 0$ for $t < 0$. The double-ended form given here obviously includes the other as a special case, and we shall, therefore, consider only this more general form which is so simply related to the Fourier situation. In fact our only reason for introducing the Laplace transformation at all is to extend the spectrum function to the complex frequency plane so that all existing s-plane techniques and interpretations as well as the methods of complex integration will become available to us.

Regarding the method of complex integration we are reminded that it is applicable only to the evaluation of integrals extended around a *closed* path or contour. For this reason the process is also known as *contour integration.* According to Cauchy's *residue theorem** the value of such an integral equals $2\pi j$ times the sum of the residues of the integrand in all its poles enclosed by the contour. It is obvious that this method is applicable only if the integrand has no singularities other than poles within the pertinent region.

The integrand in 26 contains the exponential function which has no singularities in the entire finite s-plane. So long as $F(s)$ is rational or meromorphic, the integrand will have no singularities other than poles in the entire finite s-plane. If $F(s)$ does not fulfill this condition then it must be approximated by a rational function. Although this may be a difficult task in some cases, there is no other satisfactory way of evaluating this

* See *MCA*, p. 302.

integral. We need not become too concerned about this situation, however, since in most practical problems $F(s)$ is either rational or meromorphic or an entire transcendental.

The fact that the integrand has an essential singularity at $s = \infty$ does not disturb us since this point is not involved in the integration. To be sure, the infinite limits on the integral seem to contradict this statement, but this is only apparently so. As in the Fourier series (Eq. 67, Ch. XI) it is actually more appropriate to regard these limits as arbitrarily large but finite. The difference between the integral and the function $f(t)$ which is an error function like $\epsilon(t)$ in Eq. 29, Ch. XI then becomes smaller and smaller as the finite limits are made larger and larger. Convergence of the integral (which is discussed in detail in Ch. XIV) depends upon the fact that *some sufficiently large but finite values for these limits exist for which the error becomes smaller than an arbitrarily small but nonzero assigned value*. Stated more simply, we can achieve an arbitrarily small nonzero error with limits that are still finite. We can in any relevant situation obtain a negligible error without actually extending the limits of integration to the point $s = \infty$.

As the integral 26 stands, therefore, the path of integration is open-ended. This fact is very important, for without it the methods of complex integration would do us little good because of the essential singularity of the integrand in the point at infinity. As mentioned before, we cannot apply Cauchy's residue theorem unless the contour of integration is closed; that is to say, we must close the open-ended path in integral 26 by a link that detours around the point $s = \infty$ in such a way that makes no contribution to the integral. If this can be done then the integral around the closed contour will be the same as it is along the open-ended one, and hence we can find the value along the open-ended path by applying the methods of contour integration (Cauchy's residue theorem) to an integral around the appropriately closed path.

It is not difficult to see that the path increment that detours around the point $s = \infty$ cannot be fixed if it is to be appropriate for both $t < 0$ and $t > 0$; it must lie in the right half-plane for $t < 0$ and in the left half-plane for $t > 0$ so that the product st in the exponent of the exponential function has a consistently negative real part.*

A collaterally useful principle in complex integration is known as *Cauchy's integral law* which states that the integral extended between two fixed points a and b is independent of the path joining them if any two alternate paths enclose a region in which the function is analytic (has no singularities) and if no singularities lie upon either path. Since the difference between the integrals along two such paths equals the integral

* For a more thorough consideration of this point see *MCA*, pp. 547–551.

around the closed contour formed by both, we see that this law may be stated by saying that the contour integral surrounding a region of analyticity is zero. This statement is obviously in agreement with Cauchy's residue theorem.

We observe that unless the function $F(s)$ in integral 26 has singularities immediately adjacent to the j-axis, the path of integration which presumably is indicated to follow this axis is not so rigidly fixed, as we have already mentioned. In fact we can deform this path into any shape and push it partly into the right and partly into the left half-plane in any fashion, so long as none of these distortions causes the path to pass over a singularity of $F(s)$.

This result is more remarkable than we might off-hand suppose, especially if we interpret this integral as constructing an interference pattern by superimposing everlasting sinusoids. If we choose the path of integration along the j-axis, we are superimposing sinusoids having constant amplitudes. If the path veers into the right half-plane these sinusoids become exponentially increasing; if the path veers into the left half-plane they become exponentially decreasing (i.e. increasing for negative values of t). Cauchy's integral law in effect tells us the truly remarkable fact that we can build the same interference pattern from constant-amplitude sinusoids or growing or decaying sinusoids, or from endless mixtures of these kinds of sinusoids! This fact is almost unbelievable but nevertheless true.

4. Some Interesting Interference Patterns

It is useful at this point to consider a number of examples illustrative of the kinds of interference patterns constructible with the help of Fourier or Laplace integrals; but first we shall present a completely independent way of obtaining these integrals which is interesting from a collateral viewpoint. This approach follows the thought introduced in art. 1, Ch. XI in connection with Fourier series and ways of obtaining a coefficient formula by making use of the orthogonal properties of sinusoids.

To begin with we may write, purely on a heuristic basis, the following representation for a time function

$$f(t) = \frac{1}{2\pi}\int_{-\infty}^{\infty} F(\omega)\, e^{j\omega t}\, d\omega \tag{28}$$

simply because this is in a sense a sum of sinusoids and hence a logical extension of the complex Fourier series (Eq. 24, Ch. XI) to a continuous distribution rather than a discrete one. Having written down this form of representation for $f(t)$, the important question is: How do we find the distribution function $F(\omega)$? This problem is analogous to the determination of a coefficient formula for the Fourier series.

Being reminded of the orthogonality conditions discussed in art. 1, Ch. XI, we try multiplying $f(t)$ by $e^{j\mu t}$ and integrating over the infinite time domain. This gives

$$\begin{aligned}\int_{-\infty}^{\infty} f(t)\, e^{j\mu t}\, dt &= \int_{-\infty}^{\infty} \frac{1}{2\pi} \int_{-\infty}^{\infty} F(\omega)\, e^{j(\omega+\mu)t}\, d\omega\, dt \\ &= \frac{1}{2\pi} \int_{-\infty}^{\infty} F(\omega)\, d\omega \int_{-\infty}^{\infty} e^{j(\omega+\mu)t}\, dt \\ &= \frac{1}{2\pi} \int_{-\infty}^{\infty} F(\omega)\, d\omega \left[\frac{e^{j(\omega+\mu)t}}{j(\omega+\mu)}\right]_{-\infty}^{\infty} \end{aligned} \tag{29}$$

where rearrangement in the order of integration is evidently permitted. Evaluation of the function within square brackets for infinite limits is straightforward if we interpret these tentatively as having some finite value T that approaches infinity as a limit but never actually gets there. We thus have for this bracket expression

$$\left.\frac{e^{j(\omega+\mu)t}}{j(\omega+\mu)}\right]_{-T}^{T} = \frac{2T \sin{(\omega+\mu)T}}{(\omega+\mu)T} \tag{30}$$

which we regard as a function of $(\omega + \mu)$ for fixed T.

It is essentially the function $\sin x/x$ like $F(t)$ in Eq. 14 shown plotted in Fig. 3. This function equals unity for $x = 0$; it passes through zero for x equal to integer multiples of π; and it encloses a net area* equal to π. The function 30 plotted versus $\omega + \mu$ passes through zero at integer multiplies of π/T; it has the value $2T$ for $\omega + \mu = 0$; and it encloses a net area equal to 2π. Since this area is independent of T, and the plot of function 30 becomes more and more like a tall slim spike as T becomes larger and larger, it is clear that this function approaches an impulse of value 2π occurring at $\omega + \mu = 0$. Hence Eq. 29 yields

$$\int_{-\infty}^{\infty} f(t)\, e^{j\mu t}\, dt = \int_{-\infty}^{\infty} u_0(\omega+\mu)\, F(\omega)\, d\omega = F(-\mu) \tag{31}$$

because the integrand is zero everywhere except at the point $\omega = -\mu$ where the unit impulse occurs. Replacing μ by $-\mu$, we now have

$$F(\omega) = \int_{-\infty}^{\infty} f(t)\, e^{-j\omega t}\, dt \tag{32}$$

* It is interesting to note that this is the area of a triangle inscribed within the main hump of the $\sin x/x$ function, its base being 2π and altitude unity. In this connection see Jahnke and Emde, *Tables of Functions with Formulae and Curves*, Dover Publications, 1943, p. 3 *et seq.*

which is the desired mate to integral 28, obtained by the same methods that yield a coefficient formula for the Fourier series.

Now let us consider a number of resultant time functions (interference patterns) and determine the distributions of steady or growing or decaying sinusoids that produce them. The first of these we define as

$$\begin{aligned} f(t) &\equiv 0 \qquad \text{for } t < 0 \\ f(t) &= e^{-t} \qquad \text{for } t > 0 \end{aligned} \tag{33}$$

By means of integral 32 we get

$$F(\omega) = \int_0^\infty e^{-(j\omega+1)t}\,dt = \frac{e^{-(j\omega+1)t}}{-(j\omega+1)}\Bigg]_0^\infty = \frac{1}{j\omega+1} \tag{34}$$

Analytic continuation of this function into the complex frequency plane yields the so-called Laplace transform of time function 33 given by

$$F(s) = \frac{1}{s+1} \tag{35}$$

Substituting this result into integral 26 and using the method of complex integration discussed above, we can readily regain the time function 33. Thus, for $t < 0$, for which the point at $s = \infty$ is avoided by a semicircular detour in the right half-plane, the closed contour consisting of the j-axis plus this semicircular arc encloses no singularities in the finite s-plane and $f(t) = 0$ as it should. For $t > 0$, the contour consists of the j-axis plus a semicircular detour to the left of the point at $s = \infty$. This contour may be regarded as surrounding the entire finite left half-plane, the direction of traversal being counterclockwise.

The function $F(s)$, Eq. 35, has a simple pole at $s = -1$ with residue equal to $+1$; the integrand in 26, therefore, has the residue e^{-t} in this pole, and since it has no other singularities within the pertinent contour, Cauchy's residue theorem is seen to yield the result expressed by the second of Eqs. 33.

It is significant to note that Cauchy's integral law allows us to deviate from the j-axis in forming the appropriate contours for $t < 0$ and $t > 0$ so long as such deformations do not cause the path to sweep across the pole at $s = -1$. In other words, we can construct the time function defined by Eqs. 33 from the interference of exponentially growing or decaying sinusoids as well as from sinusoids having constant amplitudes.

For our second example we consider the time function

$$\begin{aligned} f(t) &= e^{t} \qquad \text{for } t < 0 \\ f(t) &\equiv 0 \qquad \text{for } t > 0 \end{aligned} \tag{36}$$

Here integral 32 gives

$$F(\omega) = \int_{-\infty}^{0} e^{-(j\omega-1)t}\,dt = \frac{e^{-(j\omega-1)t}}{-(j\omega-1)}\Bigg]_{-\infty}^{0} = \frac{-1}{j\omega - 1} \tag{37}$$

and the corresponding Laplace transform is

$$F(s) = \frac{-1}{s-1} \tag{38}$$

The time function defined by Eqs. 36 is readily regained by substituting 38 into the integral 26 and again making use of complex integration. Here the integrand has a simple pole in the right half-plane at $s = 1$ with residue equal to $-e^t$. Therefore the path for $t < 0$ encloses this pole and hence yields a nonzero result while that for $t > 0$ encloses no singularities and yields nothing. Regarding algebraic signs we observe that the path enclosing the pole for $t < 0$ is traversed in the clockwise direction, and since Cauchy's residue theorem implies a *counter*clockwise encirclement, we must form the negative of $2\pi j$ times the residue. The latter being $-e^t$ in this case, it is clear that time function 36 is regained.

Let us elaborate upon the significance of this result lest its true meaning be lost in the formality of complex integration. Suppose we substitute the spectrum function 38 into integral 26 and contemplate carrying out the integration along the j-axis. We have to perform the real integration

$$f(t) = \frac{1}{2\pi}\int_{-\infty}^{\infty} \frac{e^{j\omega t}}{1 - j\omega}\,d\omega \tag{39}$$

But

$$\frac{e^{j\omega t}}{1 - j\omega} = \frac{\cos\omega t + j\sin\omega t}{1 - j\omega} = \frac{\cos\omega t - \omega\sin\omega t + j(\omega\cos\omega t + \sin\omega t)}{1 + \omega^2} \tag{40}$$

from which it is clear that the imaginary part contributes nothing because it is an odd function, and so we have

$$f(t) = \frac{1}{\pi}\int_{0}^{\infty} \frac{\cos\omega t}{1 + \omega^2}\,d\omega - \frac{1}{\pi}\int_{0}^{\infty} \frac{\omega\sin\omega t}{1 + \omega^2}\,d\omega \tag{41}$$

where we integrate over positive frequencies only and multiply by 2 since the integrands are even functions.

These two integrals are respectively even and odd functions of t, and hence may be interpreted as the even and odd components of $f(t)$. According to 36, these are

$$\underset{\text{even}}{f(t)} = \tfrac{1}{2}e^{-|t|} \qquad \text{for all } t \tag{42}$$

and

$$f_{\text{odd}}(t) = \begin{cases} \frac{1}{2}e^{t} & \text{for } t < 0 \\ -\frac{1}{2}e^{-t} & \text{for } t > 0 \end{cases} \tag{43}$$

which evidently cancel for $t > 0$ and yield e^t for $t < 0$ as they should. We then have

$$f_{\text{even}}(t) = \frac{1}{\pi}\int_0^\infty \frac{\cos \omega t}{1 + \omega^2}\, d\omega \tag{44}$$

and

$$f_{\text{odd}}(t) = -\frac{1}{\pi}\int_0^\infty \frac{\omega \sin \omega t}{1 + \omega^2}\, d\omega \tag{45}$$

Thus we may conclude that cosine functions having a continuous amplitude distribution given by the spectral function $1/(1 + \omega^2)$ yield an interference pattern given by 42; and that sine functions having a continuous amplitude distribution given by the spectral function $\omega/(1 + \omega^2)$ yield an interference pattern given by 43 (apart from the factor $1/\pi$). The upper limits in integrals 44 and 45 may be replaced by some finite value T, and the resulting interference patterns thought of as approximating more and more closely the time functions 42 and 43 as T is made larger and larger, the error becoming smaller than any assigned nonzero value for a sufficiently large but still finite value of T.

The paths in these integrals are very definitely open-ended, and the correct behavior for $f(t)$ in 41 for *both* $t < 0$ and $t > 0$ comes to pass because of the particular spectral pattern of the interfering sinusoids.

We keep emphasizing these things because the use of complex integration with its stipulation of distinct paths for $t < 0$ and $t > 0$ is apt to suggest that the proper behavior of $f(t)$ for $t < 0$ and for $t > 0$ is achieved through some trick and that the Fourier or Laplace integral by itself does not yield the correct behavior of $f(t)$ for *all* t.

Nothing could be further from the truth; and any "doubting Thomas" can convince himself of this fact by approximating the integrals 41 graphically with a summation process. By thus actually constructing the interference pattern as we did in art. 1 for the discrete frequency group in Fig. 1, he will see for himself that Fourier theory (and, of course, Laplace theory also) is fundamentally a mathematical scheme whereby we can approximate arbitrarily closely to any continuously or discontinuously defined function by adding together an appropriate group of sinusoids.

The next example we shall consider involves a time function defined by

$$\begin{aligned} f(t) &= e^{-t} && \text{for } t < 0 \\ f(t) &\equiv 0 && \text{for } t > 0 \end{aligned} \tag{46}$$

which differs fundamentally from the time functions assumed in the two previous examples in that it does not remain bounded for all values of t. Again using integral 32, we have

$$F(\omega) = \int_{-\infty}^{0} e^{-(j\omega+1)}\, dt = \frac{e^{-(j\omega+1)t}}{-(j\omega+1)}\Big]_{-\infty}^{0} \tag{47}$$

which does not yield a finite result; and so we see that the Fourier integral cannot be applied in a situation in which the time function is exponentially unbounded. The meaning of this result is simply that it is not possible to construct the exponentially unbounded function 46 from sinusoids with constant amplitudes. It does not mean that we cannot construct *any* unbounded functions from such steady sinusoids, for we can, as we shall see later on.

In order to construct exponentially growing time functions we shall see that we must use exponentially growing sinusoids, which is rather logical. As we have already seen, it does *not* follow that exponentially growing sinusoids necessarily yield exponentially growing time functions; however, the latter cannot be constructed unless at least some exponentially growing sinusoids contribute to the interference pattern.

In order to obtain an appropriate spectral function in a situation of this kind we must resort to a rather obvious special device which avoids the divergent integral 47. Thus if we compare the time functions defined by 36 and 46, we see that the former (let us call this the "old" one) is the latter (or "new" one) multiplied by the exponential factor e^{2t}. In terms of the synthesis integral 28 and the result 37 for the "old" function we can, therefore, write

$$\underset{\text{old}}{f(t)} = e^{2t} \underset{\text{new}}{f(t)} = \frac{1}{2\pi} \int_{-\infty}^{\infty} \frac{-1}{j\omega - 1}\, e^{j\omega t}\, d\omega \tag{48}$$

We can now multiply on both sides by e^{-2t} (certainly for all finite t however large) and get

$$\underset{\text{new}}{f(t)} = \frac{-1}{2\pi} \int_{-\infty}^{\infty} \frac{1}{j\omega - 1}\, e^{(j\omega-2)t}\, d\omega \tag{49}$$

whereupon continuation of the integrand into the complex plane yields

$$\underset{\text{new}}{f(t)} = \frac{-1}{2\pi j} \int_{-j\infty}^{j\infty} \frac{1}{s - 1}\, e^{(s-2)t}\, ds \tag{50}$$

It is now expedient to make a frequency shift through replacing $(s - 2)$ by s or s by $(s + 2)$ and obtain this result in the form

$$\underset{\text{new}}{f(t)} = \frac{-1}{2\pi j} \int_{-j\infty-2}^{j\infty-2} \frac{1}{s + 1}\, e^{st}\, ds \tag{51}$$

where the effect of this shift upon the path of integration is most important; in fact the spectral function 35 obtained for the time function defined by 33, except for a change in algebraic sign, is the same as that involved here.

The appropriate spectral function in this case is found by evaluating the rational function $-1/(s + 1)$ along a line parallel to, and two units to the left of, the j-axis. Clearly then the interfering sinusoids are exponentially growing for negatively increasing time. This same rational function (except for reversal of its algebraic sign) evaluated along the j-axis is the spectral function 34 characterizing steady sinusoids that yield an interference pattern approximating the time function defined by 33.

Making use of Cauchy's integral law, we see of course that the paths that yield the time functions defined by either 33 or 46 are not so rigidly fixed. A spectral function yielding the time function 33 is $F(s)$ in Eq. 35 evaluated along a path that passes on the right of the pole at $s = 1$, while one yielding the time function defined by 46 is the negative of this same $F(s)$ evaluated along a path that passes on the left of this pole. In so doing, the spectral function is assured of including *at least some* sinusoids that increase faster than the exponential factor e^{-t} for negatively increasing time. This much is essential to the construction of an interference pattern approximating the time function defined by 46.

Now that we know how to deal with exponentially growing time functions, let us consider a fourth example with

$$\begin{aligned} f(t) &\equiv 0 \qquad \text{for } t < 0 \\ f(t) &= e^{t} \qquad \text{for } t > 0 \end{aligned} \tag{52}$$

which is the same as function 33 in our first example multiplied by e^{2t}. In view of the spectral function given by Eq. 35 for the first of the above examples, the synthesis integral 28 yields in this case

$$f(t) = \frac{1}{2\pi j} \int_{-j\infty}^{j\infty} \frac{1}{s + 1}\, e^{(s+2)t}\, ds \tag{53}$$

Replacing $(s + 2)$ by s or s by $(s - 2)$, we get

$$f(t) = \frac{1}{2\pi j} \int_{-j\infty+2}^{j\infty+2} \frac{1}{s - 1}\, e^{st}\, ds \tag{54}$$

The function $F(s)$ involved here is the negative of 38 which is associated with the time function defined by 36 (our second example). However, the path of integration now passes to the right of the pole at $s = 1$ instead of to the left. Therefore the responses for $t < 0$ and $t > 0$ become interchanged, as is evident from a comparison of 36 and 52.

These four examples are summarized in the sketches of Fig. 4 where the s-plane for each example is shown on the left and the time function on the right. Each s-plane sketch shows the pole location of the pertinent transform $F(s)$ as well as a path along which the spectral function yielding the corresponding interference pattern $f(t)$ must be evaluated. Thus, for the time function 33 shown in part (*a*) the path may be any one passing to the right of the pole, and hence the j-axis is a possible one. For the time function 36 shown in part (*b*) the pole of the pertinent transform $F(s)$ lies

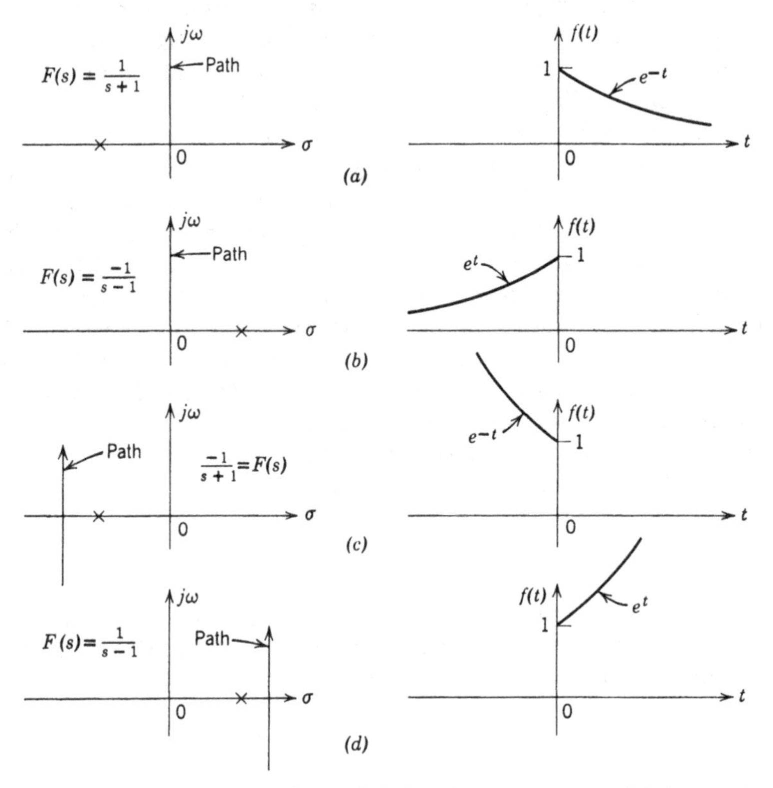

FIG. 4. A series of spectrum functions and the interference patterns which they produce.

in the right half of the s-plane and an appropriate path must pass on the left. Hence the j-axis is again a possibility.

For either time function 46 or 52, shown in parts (*c*) and (*d*) respectively, the j-axis is not a possible path since, as pointed out for these examples, the pertinent interference patterns require at least some exponentially or growing sinusoids. Possible path locations here are as indicated in the pertinent s-plane sketches.

By combining these examples in various ways, we can find other spectral functions and their resulting interference patterns. We can, however, combine only such functions for which a common path is possible. For example, we can add the functions in parts (*a*) and (*b*) in Fig. 4 to obtain the Fourier transform pair

$$f(t) = e^{-|t|} \tag{55}$$

$$F(\omega) = \frac{2}{\omega^2 + 1} \tag{56}$$

Similarly we can add functions (*a*) and (*d*) since the path for (*a*) can be displaced so as to coincide with the path for (*d*) without affecting the resulting interference pattern for (*a*). We thus obtain the time function

$$\begin{aligned} f(t) &\equiv 0 \qquad &\text{for } t < 0 \\ f(t) &= \cosh t \qquad &\text{for } t > 0 \end{aligned} \tag{57}$$

and its corresponding integral representation

$$f(t) = \int_{-j\infty+2}^{j\infty+2} \frac{s}{s^2 - 1}\, e^{st}\, ds \tag{58}$$

Functions (*b*) and (*c*) can likewise be added by displacing the path for (*b*) until it coincides with that for (*c*). We get

$$\begin{aligned} f(t) &= \cosh t \qquad &\text{for } t < 0 \\ f(t) &\equiv 0 \qquad &\text{for } t > 0 \end{aligned} \tag{59}$$

and

$$f(t) = \int_{-j\infty-2}^{j\infty-2} \frac{-s}{s^2 - 1}\, e^{st}\, ds \tag{60}$$

One cannot similarly add functions (*c*) and (*d*) because the pertinent paths cannot be made to coincide. Although the same comment appears to apply also to functions (*a*) and (*c*) or (*b*) and (*d*), we can combine these functions in pairs by cascading their paths. For example, after changing the algebraic sign of $F(s)$ in (*c*) and traversing the path in the opposite direction, its cascade with the path in (*a*) evidently amounts simply to a counterclockwise encirclement of the pole at $s = -1$; and the resulting time function, which is the sum of functions $f(t)$ in (*a*) and (*c*), is e^{-t} for all t. In like manner we can combine (*b*) and (*d*) to get an integral representation for the time function e^{t}; but these results are not very useful in a Fourier sense.

5. Elementary Properties of Fourier or Laplace Transforms

Consider the Laplace integrals 26 and 27 which are Fourier integrals for $s = j\omega$. The time function in general has even and odd components as placed in evidence by writing

$$f(t) = f_e(t) + f_o(t) \tag{61}$$

where the subscripts e and o refer to *even* and *odd* respectively. The synthesis integral 27 is then decomposable into two parts as follows

$$F(j\omega) = \int_{-\infty}^{\infty} f_e(t)\, e^{-j\omega t}\, dt + \int_{-\infty}^{\infty} f_o(t)\, e^{-j\omega t}\, dt \tag{62}$$

If the exponential function is replaced by its sine and cosine equivalent

$$e^{-j\omega t} = \cos \omega t - j \sin \omega t \tag{63}$$

and we recognize the even and odd character of the cosine and sine functions, it becomes clear that the sine term contributes nothing to the first integral while the cosine term contributes nothing to the second, since the integrand in either integral must be an even function of t in order to contribute a net area. It follows that the two integrals in Eq. 62 yield respectively the real and imaginary parts of $F(j\omega)$. If we write

$$F(j\omega) = F_1(\omega) + jF_2(\omega) \tag{64}$$

it follows that

$$F_1(\omega) = \int_{-\infty}^{\infty} f_e(t)\, e^{-j\omega t}\, dt = 2\int_0^{\infty} f_e(t) \cos \omega t\, dt \tag{65}$$

and

$$F_2(\omega) = \int_{-\infty}^{\infty} f_o(t)\, e^{-j\omega t}\, dt = -2\int_0^{\infty} f_o(t) \sin \omega t\, dt \tag{66}$$

The real part $F_1(\omega)$, which incidentally turns out to be an even function, is seen to be determined by the even component of $f(t)$ alone; and the imaginary part $F_2(\omega)$, which is an odd function of ω, is determined by the odd component of $f(t)$. Thus we recognize an important property of Fourier transforms, namely if a given time function $f(t)$ is even, then its transform $F(j\omega)$ is even and purely real, whereas if $f(t)$ is odd, then $F(j\omega)$ is also odd and in this case it is purely imaginary. This property is the equivalent of a similar one involving periodic functions and their spectrum representation as pointed out in the previous chapter.

By substituting the decompositions 61 and 64 into analysis integral 26, a similar manipulation leads to a pair of relations that are inverse to 65 and 66, namely

$$f_e(t) = \frac{1}{2\pi}\int_{-\infty}^{\infty} F_1(\omega)\, e^{j\omega t}\, d\omega = \frac{1}{\pi}\int_0^{\infty} F_1(\omega) \cos \omega t\, d\omega \tag{67}$$

$$f_o(t) = \frac{j}{2\pi}\int_{-\infty}^{\infty} F_2(\omega)\, e^{j\omega t}\, d\omega = -\frac{1}{\pi}\int_0^{\infty} F_2(\omega) \sin \omega t\, d\omega \tag{68}$$

Another interesting mutual property of the time function $f(t)$ and its Fourier transform is more effectively expressed in terms of the integrals 21 and 22. If in these we replace ω by t' and t by $-\omega'$, take proper account of the effect upon limits of integration, and then drop the prime on ω and t, we get

$$f(-\omega) = \frac{1}{2\pi} \int_{-\infty}^{\infty} F(t)\, e^{-j\omega t}\, dt \tag{69}$$

$$F(t) = \int_{-\infty}^{\infty} f(-\omega)\, e^{j\omega t}\, d\omega \tag{70}$$

Comparison with the integrals 21 and 22 shows that if we now regard $F(t)/2\pi$ as a time function then $f(-\omega)$ is its transform or spectrum representation. If the original function $f(t)$ is even, then $f(-\omega) = f(\omega)$, and we have the result that we may interchange the identities of the time and spectrum functions except for a factor of 2π. If we have found the Fourier transform $F(\omega)$ of an even time function $f(t)$ then we may turn about and regard $1/2\pi$ times the transform as a given time function and the original one as its transform.

Another fundamental property that involves similar reasoning follows either from the pair of Fourier integrals 21 and 22 or from the extended form given by 26 and 27. Either pair remains unchanged if we replace s by $-s$ (or ω by $-\omega$) *and* t by $-t$. We can express this property by saying that if $f(t)$ and $F(s)$ are a transform pair, then so are $f(-t)$ and $F(-s)$.

In Chapter X we derived a closed form for the impulse response of a network (initially at rest) in terms of its frequency response (or *system function*) for given points of entry, and interpreted this pair of functions, denoted by $f(t)$ and $F(s)$ respectively, as time-domain and frequency-domain representations of the pertinent linear system. In terms of these closed-form relations (as expressed by Eqs. 41, 42 and 43 of Ch. X) we were able to derive a number of interesting mutual properties of the impulse-response and system functions (Eqs. 44 through 56).

Later on (Ch. XVI) when we discuss the use of Fourier theory in formulating network response, we shall see that a time function and its Fourier transform may, under certain circumstances, be identified respectively with the impulse-response and system functions of a physical network. Properties of the latter should, therefore, also be properties of the Fourier transform pair, although the reverse is not necessarily true since the impulse response of a physical network is more restricted than a time function possessing a spectrum representation. At the moment we wish to show that all of the mutual properties of $f(t)$ and $F(s)$ obtained in Ch. X can readily be gotten from the Fourier integrals 21 and 22 or their extended forms as given by 26 and 27.

Let us consider first the derivative properties which we can get by differentiation under the integral sign. This operation is permitted here for the same reason that justifies it in the case of periodic functions (as discussed in art. 8, Ch. XI) since we regard the aperiodic case as a limiting form of the periodic.

If we thus form the derivative of $f(t)$ given by integral 26, we see that $F(s)$ in the integrand becomes multiplied by s; and if we similarly form the derivative of $F(s)$ by differentiating with respect to s under the integral 27, $f(t)$ in the integrand is multiplied by $-t$. We can evidently do or undo either of these operations repeatedly so long as the resultant time function remains Fourier representable. Hence we can state the complementary properties

If $f(t)$ is replaced by d^nf/dt^n, then
$F(s)$ becomes replaced by $s^nF(s)$ (71)

and

If $F(s)$ is replaced by d^nF/ds^n, then
$f(t)$ becomes replaced by $(-t)^nf(t)$ (72)

In either of these statements, the integer n may be positive or negative; if negative, the operation is interpreted as indefinite integration.

The effect of a shift in time or in frequency may likewise be seen by inspection. If in the time function given by integral 26 we replace t by $t - t_0$, $F(s)$ becomes multiplied by the exponential factor e^{-st_0}; and if in the spectral function given by integral 27 we replace s by $s - s_0$, $f(t)$ becomes multiplied by the exponential factor e^{s_0t}. Hence we have the results

If $f(t)$ is replaced by $f(t - t_0)$, then
$F(s)$ becomes replaced by $e^{-st_0}\,F(s)$ (73)

If $F(s)$ is replaced by $F(s - s_0)$, then
$f(t)$ becomes replaced by $e^{s_0t}\,f(t)$ (74)

It is incidentally worth noting that these properties may be used to obtain, in an alternate manner, the derivative properties expressed by statements 71 and 72. Thus, according to the basic definition of a derivative we get the following results:

$$\frac{f(t+\Delta t) - f(t)}{\Delta t} \rightarrow \frac{e^{s\Delta t}\,F(s) - F(s)}{\Delta t} \rightarrow s\,F(s) \tag{75}$$

$$\frac{f(s+\Delta s) - F(s)}{\Delta s} \rightarrow \frac{e^{-\Delta st}f(t) - f(t)}{\Delta s} \rightarrow -t\,f(t) \tag{76}$$

where the limit process $\Delta t \rightarrow 0$ or $\Delta s \rightarrow 0$ is carried out by writing the first two terms in a Maclaurin expansion for the exponential function.

From 73 and 74 we can also obtain the following relations which are frequently found useful:

$$f(t) \cdot \cos \omega_0 t = \tfrac{1}{2} f(t)[e^{j\omega_0 t} + e^{-j\omega_0 t}] \rightarrow \tfrac{1}{2}[F(s - j\omega_0) + F(s + j\omega_0)] \tag{77}$$

$$f(t) \cdot \sin \omega_0 t = \frac{1}{2j} f(t)[e^{j\omega_0 t} - e^{-j\omega_0 t}] \rightarrow \frac{1}{2j}[F(s - j\omega_0) - F(s + j\omega_0)] \tag{78}$$

$$F(j\omega) \cdot \cos \omega t_0 = \tfrac{1}{2} F(j\omega)[e^{j\omega t_0} + e^{-j\omega t_0}] \rightarrow \tfrac{1}{2}[f(t + t_0) + f(t - t_0)] \tag{79}$$

$$F(j\omega) \cdot j \sin \omega t_0 = \tfrac{1}{2} F(j\omega)[e^{j\omega t_0} - e^{-j\omega t_0}] \rightarrow \tfrac{1}{2}[f(t + t_0) - f(t - t_0)] \tag{80}$$

The first two of these demonstrate a principle that is fundamental to many communication systems, namely, that if a time function $f(t)$ is used to modulate the amplitude of a carrier (the function $\cos \omega_0 t$ in 77 or $\sin \omega_0 t$ in 78) then the effect upon the associated spectrum function is to displace it along the frequency axis by the carrier frequency ω_0 in both positive and negative directions. A lowpass spectrum is thus converted into a bandpass centered at $\omega = \omega_0$ with the same bandwidth as the lowpass centered at $\omega = 0$.

This result is illustrated in Fig. 5 for the rectangular pulse, part (*a*), for which integral 22 yields the spectral function

$$F(\omega) = E \int_{-\delta/2}^{\delta/2} e^{-j\omega t}\, dt = E\delta \cdot \frac{\sin \omega\delta/2}{\omega\delta/2} \tag{81}$$

plotted in part (*b*). For the amplitude modulated carrier function

$$\hat{f}(t) = f(t) \cos \omega_0 t \tag{82}$$

drawn in part (*c*), the result 77 yields the spectral function

$$\hat{F}(\omega) = \frac{E\delta}{2} \times \left\{ \frac{\sin (\omega - \omega_0)\delta/2}{(\omega - \omega_0)\delta/2} + \frac{\sin (\omega + \omega_0)\delta/2}{(\omega + \omega_0)\delta/2} \right\} \tag{83}$$

shown in part (*d*). In practical situations the bandwidth of the spectral function (which is essentially $2\pi/\delta$) is usually small compared with the carrier frequency ω_0 so that overlap of the two displaced spectral components in (*d*) is negligible; each is simply the lowpass function (*b*) reoriented in the spectrum.

In this connection, if $F(s)$ is a rational function, as it might be if it represents the frequency response of a network, and if it has poles at the points $s = s_\nu$ with residues k_ν, then $F(s \mp j\omega_0)$ has poles at the points $s = s_\nu \pm j\omega_0$ with residues k_ν also. The shift has no effect upon the residues. For a Butterworth function of order 6 (see Ch. IX, art. 4) the pole patterns of the lowpass and of the corresponding bandpass according to 77 are shown in sketches (*a*) and (*b*) of Fig. 6.

It should be noted that this lowpass-bandpass relationship is not the same as that discussed in art. 5 of Ch. IX, although for a narrow band-

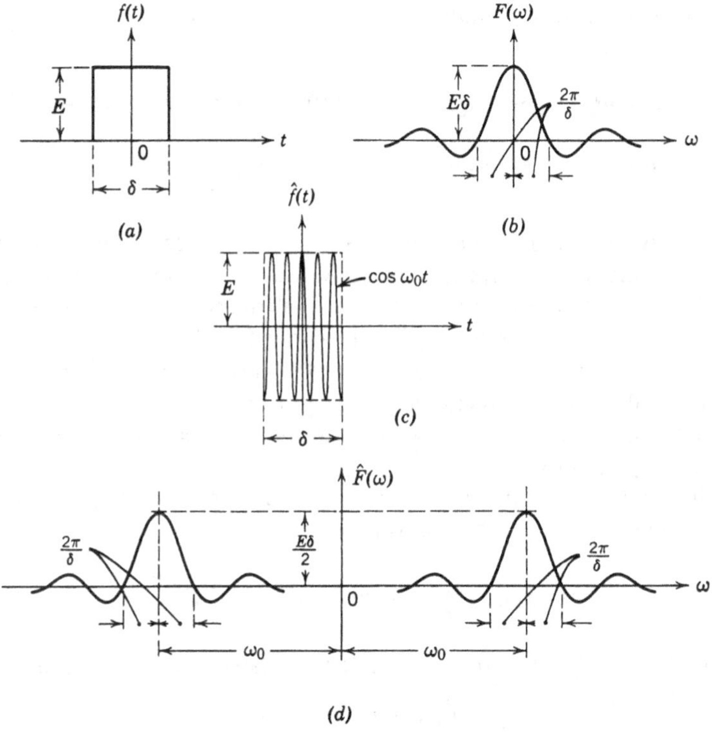

FIG. 5. Illustrating an important elementary property of Fourier transforms.

width w relative to ω_0 the difference is inappreciable. More precisely, the bandpass here has the same arithmetic symmetry about its center frequency as does the lowpass, while the change of variable discussed in Ch. IX yields a bandpass with geometric symmetry.

Finally, the effect of frequency and time scaling is also gotten with little effort from the integrals 26 and 27. If we replace s by s/a and t by at so that the product st is unchanged, these integrals assume the form

$$a\,f(at) = \frac{1}{2\pi j}\int_{-j\infty}^{j\infty} F(s/a)e^{st}\,ds \tag{84}$$

and

$$F(s/a) = \int_{-\infty}^{\infty} a\,f(at)e^{-st}\,dt \tag{85}$$

from which it is clear that

$$\begin{array}{l}\text{If } F(s) \text{ is replaced by } F(s/a)\text{, then}\\ f(t) \text{ becomes replaced by } a\,f(at)\end{array} \tag{86}$$

Frequency and time are affected inversely by the scale factor a, and the net area enclosed by the time function, which is $F(0)$, remains invariant.

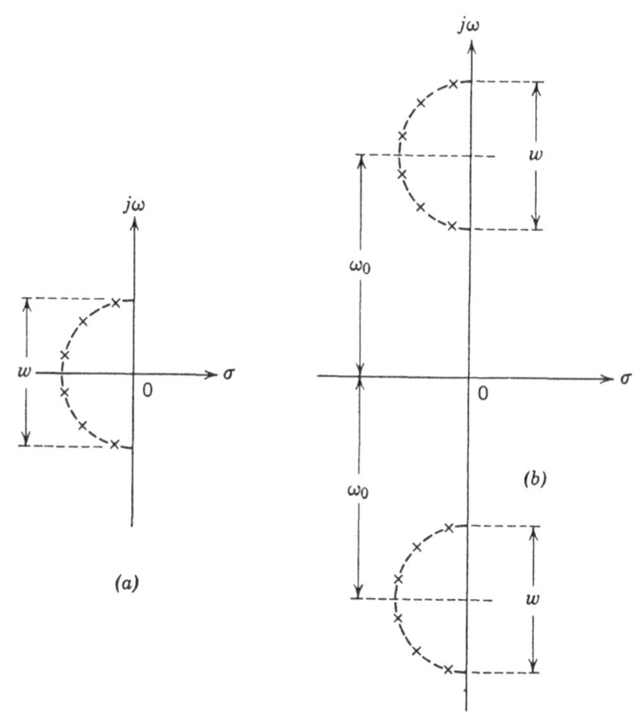

FIG. 6. Frequency-domain result corresponding to the time-domain property illustrated in Fig. 5.

If $F(s)$ is a rational function, its pole-zero pattern is uniformly expanded or contracted by this scaling process; all critical frequencies become multiplied by a.

6. Asymptotic Behavior and Moment Conditions

From the integral 26 and property 71 we have

$$\frac{d^n f}{dt^n} = \frac{1}{2\pi j} \int_{-j\infty}^{j\infty} s^n F(s)\, e^{st}\, ds \tag{87}$$

which, for $t = 0$, yields the result

$$\left(\frac{d^n f}{dt^n}\right)_{t=0} = \frac{1}{2\pi j}\int_{-j\infty}^{j\infty} s^n\, F(s)\, ds \tag{88}$$

If $f(t)$ has a zero of order $k + 1$ at $t = 0$, then

$$f(0) = \left(\frac{df}{dt}\right)_{t=0} = \cdots = \left(\frac{d^k f}{dt^k}\right)_{t=0} = 0 \tag{89}$$

and Eq. 88 yields

$$\int_{-j\infty}^{j\infty} s^n\, F(s)\, ds = 0 \qquad \text{for } n = 0, 1, 2, \ldots k \tag{90}$$

This result requires that $F(s)$ become zero faster than $1/s^{k+1}$ for large values of s. If $F(s)$ is rational it requires an asymptotic behavior described by

$$F(s) \rightarrow \frac{1}{s^{k+2}} \qquad \text{for } s \rightarrow \infty \tag{91}$$

Conversely, if $F(s)$ is known to have this asymptotic behavior, then we may conclude that the pertinent time function has a zero of order $k + 1$ at $t = 0$.

A complementary result is obtained from integral 27 which, through use of property 72, may be written in the form

$$\frac{d^n F(s)}{ds^n} = \int_{-\infty}^{\infty} (-t)^n f(t)\, e^{-st}\, dt \tag{92}$$

For $s = 0$, this gives

$$\left[\frac{d^n F(s)}{ds^n}\right]_{s=0} = \int_{-\infty}^{\infty} (-t)^n\, f(t)\, dt \tag{93}$$

If $F(s)$ has a zero of order $k + 1$ at $s = 0$, then

$$F(0) = \left(\frac{dF}{ds}\right)_{s=0} = \cdots = \left(\frac{d^k F}{ds^k}\right)_{s=0} = 0 \tag{94}$$

and Eq. 93 yields

$$\int_{-\infty}^{\infty} t^n\, f(t)\, dt = 0 \qquad \text{for } n = 0, 1, 2, \ldots k \tag{95}$$

This result requires that $f(t)$ become zero faster than $1/t^{k+1}$ for large values of t. Hence the asymptotic behavior of $f(t)$ is seen to be linked with the behavior of $F(s)$ at $s = 0$.

The integral in Eq. 95 is referred to as the *nth-order moment* of the function $f(t)$ about the time origin. The zero-order moment equals the net area under $f(t)$; the first-order moment equals this area multiplied by the distance of its "center of gravity" from the origin, which is the way this moment is calculated in mechanics. The conditions 95 for various values of n are, therefore, referred to as *moment conditions.*

7. The Family of Singularity Functions

The most useful members of this family of functions are the unit impulse and the unit step which are familiar to us from our consideration of the transient response of simple circuits consisting of two- and three-element combinations. We denote any member in this family by $u_n(t)$ and refer to it as the singularity function of order n. It is expressible as the derivative of a function of next lower order or as the indefinite integral of a function of next higher order. For example the unit impulse $u_0(t)$ equals the derivative of the step $u_{-1}(t)$ or the integral of the doublet $u_1(t)$.

We have also seen (in art. 2, Ch. X) that the frequency-domain representation of $u_n(t)$ is simply s^n. We shall now show that this result is consistent with the Fourier method of correlating time and frequency domains; and in the process we will find some collaterally useful generalizations.

Let us begin with showing that the spectrum function of a unit impulse is the constant *unity*. To this end consider integral 27 for $f(t) = u_0(t)$:

$$F(s) = \int_{-\infty}^{\infty} u_0(t)\, e^{-st}\, dt \tag{96}$$

Since the integrand is zero everywhere except at $t = 0$ where the exponential equals 1, the net area enclosed by the integrand is unity and we have the expected result

$$F(s) = 1 \tag{97}$$

We can regain the unit impulse function by substituting this result into integral 26, which we write in the form

$$f(t) = \frac{1}{2\pi} \int_{-\Omega}^{\Omega} e^{j\omega t}\, d\omega \tag{98}$$

giving

$$f(t) = \frac{e^{j\omega t}}{2\pi j t}\Bigg]_{-\Omega}^{\Omega} = \frac{\Omega}{\pi} \cdot \frac{\sin \Omega t}{\Omega t} \tag{99}$$

This is the familiar sin x/x function having the form shown in sketch (*b*) of Fig. 5; its zero ordinate is Ω/π, and its zeros occur at integer multiples of π/Ω, the net enclosed area being unity.* Hence, as Ω is allowed to

* See Jahnke and Emde, *Tables of Functions with Formulae and Curves*, Dover Publications, 1943, p. 3.

become larger and larger, this function becomes taller and slimmer while the enclosed area, which is independent of Ω, becomes more and more concentrated at $t = 0$. It is clear, therefore, that $f(t)$ in Eq. 99 approaches the unit impulse $u_0(t)$ as the limit $\Omega \to \infty$ is carried to completion.

It was pointed out in art. 5 that when $f(t)$ and $F(j\omega)$ are even functions, as they are in this example, then their identities may be interchanged (apart from the constant multiplier 2π). Thus if we assume a constant for $f(t)$ we should get an impulse for $F(j\omega)$. Suppose we let $f(t) = 1$ in the integral 27 and write it in the form

$$F(j\omega) = \int_{-T}^{T} e^{-j\omega t}\, dt \tag{100}$$

where T is some large but finite value of time. We get

$$F(j\omega) = \frac{e^{-j\omega t}}{-j\omega}\bigg]_{-T}^{T} = \frac{2T \sin \omega T}{\omega T} \tag{101}$$

which is the $\sin x/x$ function again, this time with a zero ordinate equal to $2T$, zeros at integer multiples of $\omega = \pi/T$, and a net enclosed area equal to 2π independent of T. As T becomes larger and larger, the function 101 approaches an impulse of value 2π; that is to say, the Fourier transform of $f(t) = 1$ is

$$F(j\omega) = 2\pi u_0(\omega) \tag{102}$$

which bears out the property mentioned above. We can readily regain $f(t)$ by substituting this result into integral 26. The evaluation is straightforward.

These same results may be recognized directly from the graphical sketches (*a*) and (*b*) in Fig. 5. Thus in sketch (*a*) let $E = 1$ and let δ become larger and larger so that $f(t)$ approaches the constant unity. The $\sin x/x$ function in sketch (*b*) becomes taller and slimmer; the enclosed area equals 2π independent of δ. In the limit $\delta \to \infty$, $F(\omega) = 2\pi u_0(t)$. Conversely, in sketch (*b*) let the zero ordinate $E\delta = 1$ and consider the limit process $\delta \to 0$. $F(\omega)$ now approaches unity, while $f(t)$ in sketch (*a*) is a rectangular pulse enclosing area $= E\delta = 1$, becoming taller and slimmer as δ becomes smaller and smaller. In the limit $\delta \to 0, f(t) = u_0(t)$. We see, incidentally, that the impulse function can be approached not only as a limiting form of $\sin x/x$ but also as a limiting form of the square pulse (or of a triangular pulse or of many other pulse shapes as we shall appreciate more clearly as we go along).

Suppose we next consider members of the family of singularity functions formed through successive integration of the impulse; that is to say, the unit step, the unit linear ramp, and so forth, for which the frequency-domain representations are $1/s$, $1/s^2$, ... Assuming a unit step for $f(t)$ in the integral 27, we get

$$F(s) = \int_0^\infty e^{-st}\, dt \tag{103}$$

which must be interpreted as the integral

$$F(s) = \int_0^T e^{-st}\, dt = \frac{1 - e^{-sT}}{s} \tag{104}$$

for an arbitrarily large value of T. This interpretation is aided by the sketch in Fig. 7 in which, for any finite T, $f(t)$ is regarded as the difference

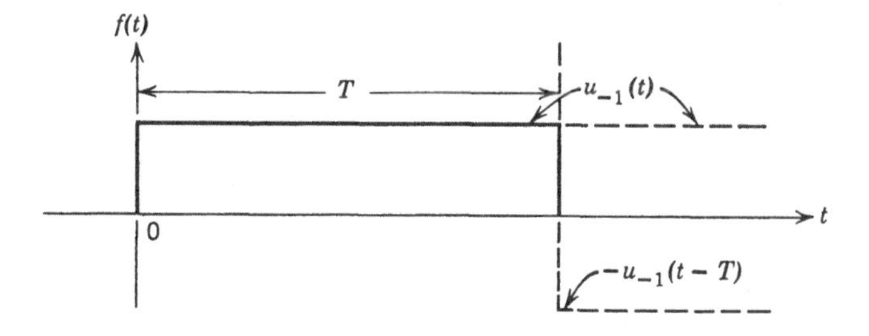

FIG. 7. Function which becomes the unit step in the limit $T \to \infty$.

between two unit step functions, one occurring at $t = 0$ and the other at $t = T$. That is to say, 104 is the transform of a time function

$$f(t) = u_{-1}(t) - u_{-1}(t - T) \tag{105}$$

which clearly becomes the desired unit step function for an arbitrarily large value of T. Being mindful of the property expressed in statement 73, it is thus evident that the desired frequency domain representation of the unit step is given by

$$F(s) = \frac{1}{s} \tag{106}$$

The same approach may be taken to evaluate the transform of $u_{-n}(t)$ for $n > 1$. Since

$$u_{-n}(t) = \frac{t^{n-1}}{(n-1)!} \cdot u_{-1}(t) \qquad \text{for } n \geq 1 \tag{107}$$

the desired transform is given by the limit of the integral

$$F_{-n}(s) = \frac{1}{(n-1)!} \int_0^T t^{n-1} e^{-st}\, dt \tag{108}$$

for an arbitrarily large value of T. Through repeatedly applying the method of integration by parts, we find

$$F_{-n}(s) = \frac{1}{s^n} - \left[\frac{T^{n-}}{s(n-1)!} + \frac{T^{n-2}}{s^2(n-2)!} + \cdots + \frac{1}{s^n}\right]e^{-sT} \tag{109}$$

which is regarded as the transform of

$$u_{-n}(t) - u_{-n}(t) \times u_{-1}(t - T) \tag{110}$$

The Taylor expansion of the second term in this expression for $t > T$ is

$$u_{-n}(t) \cdot u_{-1}(t - T) = \sum_{k=0}^{n-1} \frac{1}{k!} \left[\frac{d^k u_{-n}}{dt^k}\right]_{t=T} \cdot (t - T)^k \cdot u_{-1}(t - T) \tag{111}$$

In view of 107 we have

$$\left[\frac{d^k u_{-n}}{dt^k}\right]_{t=T} = (u_{-n+k})_{t=T} = \frac{T^{n-k-1}}{(n-k-1)!} \tag{112}$$

and

$$(t - T)^k u_{-1}(t - T) = k!\, u_{-k-1}(t - T) \tag{113}$$

so that 111 becomes

$$u_{-n}(t) \times u_{-1}(t - T) = \sum_{k=0}^{n-1} \frac{T^{n-k-1}}{(n-k-1)!} u_{-k-1}(t - T) \tag{114}$$

The time function 110, of which 109 is the transform, may thus be written

$$u_{-n}(t) - \left[\frac{T^{n-1}}{(n-1)!} u_{-1}(t - T) + \frac{T^{n-2}}{(n-2)!} u_{-2}(t - T) + \cdots + u_{-n}(t - T)\right] \tag{115}$$

Comparison with 109 and consideration of property 73 shows at once that the desired transform of $u_{-n}(t)$, which we shall denote by $v_{-n}(s)$, is given by

$$v_{-n}(s) = \frac{1}{s^n} \tag{116}$$

which agrees with our expectation.

To complete the picture we now consider members of the family of singularity functions formed through successive differentiation of the impulse, namely the unit doublet $u_1(t)$, the triplet $u_2(t)$, and so forth. We shall do this by considering initially a family of functions of nonzero duration which approaches the desired family of singularity functions as some parameter, like δ for the pulse of part (*a*) in Fig. 5, is allowed to become smaller and smaller. The function of zero order and nonzero duration in this family must, moreover, be differentiable n times if the function of nth order is to exist in a conservative sense.

We arrive at this family of nonzero duration in two steps. First we construct the following auxiliary family:

$$h_0(t) = \frac{u_{-1}(t) - u_{-1}(t - \delta)}{\delta} \tag{117}$$

which is a rectangular pulse of duration δ and unit enclosed area, independent of δ. The next member is

$$h_1(t) = \frac{h_0(t) - h_0(t - \delta)}{\delta} = \frac{u_{-1}(t) - 2u_{-1}(t - \delta) + u_{-1}(t - 2\delta)}{\delta^2} \tag{118}$$

where 117 is substituted to obtain the final result. Next we form

$$h_2(t) = \frac{h_1(t) - h_1(t - \delta)}{\delta} = \frac{h_0(t) - 2h_0(t - \delta) + h_0(t - 2\delta)}{\delta^2}$$

$$= \frac{u_{-1}(t) - 3u_{-1}(t - \delta) + 3u_{-1}(t - 2\delta) + u_{-1}(t - 3\delta)}{\delta^3} \tag{119}$$

It is not difficult to see that the function of order n in this family is given by

$$h_n(t) = \sum_{k=0}^{n+1} \frac{(-1)^k}{\delta^{n+1}} \binom{n+1}{k} u_{-1}(t - k\delta) \tag{120}$$

in which

$$\binom{m}{k} = \frac{m(m-1)(m-2)\cdots(m-k+1)}{k!} \tag{121}$$

is the familiar binomial coefficient.

It is significant to observe that for the limit $\delta \to 0$, $h_1(t)$ in 118 is the conventionally defined derivative of $h_0(t)$, $h_2(t)$ in 119 is the conventionally defined derivative of $h_1(t)$, and so forth. Although strictly speaking, none of these functions are differentiable, we may regard the relations 117, 118, 119, for a *nonzero* δ as a sequence of functions related to one another by imperfect or approximate differentiation which is a clearly defined and rigorously justified procedure.

The nth order function 120, which is thus derived through an n-fold succession of imperfect differentiations from $h_0(t)$ in 117, we now subject to n successive steps of perfect integration and get

$$f_0(t) = \sum_{k=0}^{n+1} \frac{(-1)^k}{\delta^{n+1}} \binom{n+1}{k} \times u_{-n-1}(t - k\delta) \tag{122}$$

This zero-order function may conservatively be differentiated successively n times. In view of the result expressed in Eq. 116, the property 73, and the binomial expansion

$$(1 - e^{-s\delta})^{n+1} = \sum_{k=0}^{n+1} (-1)^k \binom{n+1}{k} e^{-ks\delta} \tag{123}$$

it follows that the transform of $f_0(t)$ in Eq. 122 is given by

$$F_0(s) = \left\{\frac{1 - e^{-s\delta}}{s\delta}\right\}^{n+1} \tag{124}$$

Since $F_0(0) = 1$, and this (according to integral 27) is the net area under the associated time function, we see that the zero-order function 122 encloses unit area, independent of δ, becoming more and more concentrated at $t = 0$ as δ is made smaller and smaller. $f_0(t)$ and its n successive derivatives, all of which exist on a conservative basis, are, therefore, the desired family since they become identical with the singularity functions $u_0(t)$ through $u_n(t)$ as the limit $\delta \to 0$ is carried out.

The function of order ν in this family is given by the expression 122 with u_{-n-1} replaced by $u_{-n+\nu-1}$, and its transform, according to property 71, is the function $F_0(s)$ in 124 multiplied by s^ν. For $\delta \to 0$, this yields the transform of $u_\nu(t)$ for $\nu \geq 1$ in the form

$$v_\nu(s) = s^\nu \tag{125}$$

which is what we wished to demonstrate.

There is another way to approach this problem that is of some collateral interest. The following manipulation of a time function:

$$\int_{-\infty}^{\infty} \frac{f[t + (\delta/2)] - f[t - (\delta/2)]}{\delta} dt \tag{126}$$

may be interpreted as a smoothing operation since, for a small but nonzero δ, it represents imperfect differentiation followed by perfect integration. If carried out on an impulse, for example, it converts it into a rectangular pulse; and if carried out on the square pulse, a triangular one results. Properties 71 and 73 show that the frequency-domain equivalent of this operation is to multiply the associated $F(s)$ by

$$\frac{e^{s\delta/2} - e^{-s\delta/2}}{s\delta} = \frac{\sinh s\delta/2}{s\delta/2} \tag{127}$$

Starting with the unit impulse $u_0(t)$ and its transform $v_0(s) = 1$, $(n + 1)$ successive operations as defined by 126 yield a pulse with the transform

$$\left(\frac{\sinh s\delta/2}{s\delta/2}\right)^{n+1} \tag{128}$$

This pulse is conservatively differentiable n times, for only on the nth differentiation are discontinuities encountered. Except for a displacement of $(n + 1)\delta/2$ seconds to the left, it is the same as the pulse $f_0(t)$ in Eq. 122 with transform 124. In the limit $\delta \to 0$ this difference becomes negligible, and so we again arrive at the conclusion 125 as before.

8. Evaluation of Laplace Transforms on the j-Axis

In art. 3 we extended the Fourier transform to the complex plane by applying to the function $F(\omega)$ defined by integrals 21 and 22 the process of analytic continuation and the identity theorem, well known in the theory of functions of a complex variable. The resulting extended forms of the Fourier integrals, given by Eqs. 26 and 27 and designated as Laplace integrals, involve the function $F(s)$ which must revert to the Fourier transform when evaluated on the j-axis.

In this evaluation process we must be careful to observe the principles expressed in art. 4 of Ch. VIII to the effect that if a rational function has j-axis poles, then the evaluation of its real and imaginary parts along this axis must be done with a clear understanding of their behavior in the immediate vicinity of such singularities, as illustrated there by the sketches in Figs. 1 and 2. More specifically, if a simple pole with positive real residue k is involved, the real part exhibits a j-axis impulse of value $k\pi$ at the pole position. Let us illustrate with several examples.

We have just seen that the unit step function in the time domain has a Laplace transform

$$F(s) = \frac{1}{s} \tag{129}$$

Evaluation of this function on the j-axis is not done by simply writing $j\omega$ for s. The more careful considerations given in art. 4, Ch. VIII show that the Fourier transform corresponding to 129 reads

$$F(j\omega) = F_1(\omega) + jF_2(\omega) = \pi u_0(\omega) + \frac{1}{j\omega} \tag{130}$$

As pointed out in art. 5, the real and imaginary parts of this transform correspond to even and odd parts of the unit step in the time domain. If, as in Eq. 61, we write

$$u_{-1}(t) = f_e(t) + f_0(t) \tag{131}$$

then evidently

$$f_e(t) = \tfrac{1}{2} \tag{132}$$

$$f_0(t) = \begin{cases} -\tfrac{1}{2} & \text{for } t < 0 \\ \tfrac{1}{2} & \text{for } t > 0 \end{cases} \tag{133}$$

Thus the term $1/j\omega$ in 130, which we might erroneously regard as the j-axis evaluation of 129, actually is the Fourier transform of the odd part 133 alone. The even part of the unit step function, Eq. 132, is contributed by the real part of 130 which is an impulse of value π. The importance of correctly evaluating the j-axis behavior of a function like $F(s)$ in Eq. 129 is rather evident.

A second example is the steady cosine function for $t > 0$ given by

$$f(t) = u_{-1}(t) \cdot \cos \omega_0 t \tag{134}$$

Making use of the property expressed in 77 we have the corresponding Laplace transform

$$F(s) = \frac{1}{2}\left[\frac{1}{s - j\omega_0} + \frac{1}{s + j\omega_0}\right] = \frac{s}{s^2 + \omega_0^2} \tag{135}$$

The proper j-axis evaluation here yields the Fourier transform

$$F(j\omega) = \frac{\pi}{2}\left[u_0(\omega - \omega_0) + u_0(\omega + \omega_0)\right] + \frac{j\omega}{\omega_0^2 - \omega^2} \tag{136}$$

Similarly for

$$f(t) = u_{-1}(t) \cdot \sin \omega_0 t \tag{137}$$

The property expressed in 78 gives

$$F(s) = \frac{1}{2j}\left[\frac{1}{s - j\omega_0} - \frac{1}{s + j\omega_0}\right] = \frac{\omega_0}{s^2 + \omega_0^2} \tag{138}$$

and its proper j-axis evaluation reads

$$F(j\omega) = \frac{\omega_0}{\omega_0^2 - \omega^2} - j\frac{\pi}{2}\left[u_0(\omega - \omega_0) - u_0(\omega + \omega_0)\right] \tag{139}$$

Note that here the impulses are in the imaginary part which is an odd function of ω (although u_0 is even) whereas in 136 they are in the real part which is even.

In contrast let us consider the sine or cosine functions for all t rather than for $t > 0$ only. If we start from the transform pair $f(t) = 1$, $F(j\omega) = 2\pi u_0(\omega)$ in Eq. 102 and apply property 77, we get the Fourier transform pair

$$f(t) = \cos \omega_0 t \tag{140}$$

$$F(j\omega) = \pi\left[u_0(\omega - \omega_0) + u_0(\omega + \omega_0)\right] \tag{141}$$

Applying property 78, on the other hand, yields

$$f(t) = \sin \omega_0 t \tag{142}$$

$$F(j\omega) = -j\pi\left[u_0(\omega - \omega_0) - u_0(\omega + \omega_0)\right] \tag{143}$$

It is interesting to observe that in these last two examples no Laplace transform exists. The steady sine or cosine function for all t possesses a Fourier transform only. We can obtain a check on this result by applying the property (pointed out in art. 5) that a transform pair remains appropriate if we replace t by $-t$ and s by $-s$. If in 134 we replace t by

$-t$ the result is $\cos \omega_0 t$ for $t < 0$; and if we add this to $\cos \omega_0 t$ for $t > 0$, we get $\cos \omega_0 t$ for all t as in Eq. 140. The resulting Laplace transform is 135 plus the same function with s replaced by $-s$, which yields zero. Replacing ω by $-\omega$ in 136, on the other hand, and adding this result to 136 (noting the even character of u_0) yields 141 as it should.

Similarly $f(t)$ in 137 minus $f(-t)$ is $\sin \omega_0 t$ for all t as in 142. $F(s)$ in 138 minus $F(-s)$ yields zero; whereas $F(j\omega)$ in 139 minus $F(-j\omega)$ yields 143.

It is collaterally useful to observe that 140 and 142 are periodic functions and that 141 and 143 are line spectra, as one should expect in the case of periodic functions. The "lines" in these spectra are impulses since comparison with the Fourier coefficients for periodic functions (as dealt with in Ch. XI) requires, according to the transitions 20 and Eq. 23 in art. 2, that we regard $F(j\omega) \cdot d\omega/2\pi$ as the comparable quantity. On this basis the lines are a pair of ordinates of height $\frac{1}{2}$ since the limit process discussed in art. 2 yields the interpretation

$$[u_0(\omega - \omega_0)\, d\omega]_{\omega=\omega_0} = 1 \tag{144}$$

We may conclude more generally that any periodic function defined for the entire time domain, $-\infty < t < \infty$, possesses a Fourier transform but no Laplace transform; and that the Fourier transform consists entirely of j-axis impulses which are the equivalent of coefficients in a Fourier series.

A periodic function defined only for $t > 0$ (identically zero for $t < 0$), on the other hand, has a Laplace transform consisting exclusively of simple j-axis poles. The Fourier transform in this case includes j-axis impulses, the evaluation of which requires an understanding of the interpretations given in art. 4, Ch. VIII.

9. Power Spectra and Parseval's Theorem

In applying Fourier methods to practical problems we may encounter a situation where time functions $f_1(t)$ and $f_2(t)$ represent voltage and current at some terminal pair and we are interested in evaluating the net energy involved, as expressed by the integral

$$\int_{-\infty}^{\infty} f_1(t) \cdot f_2(t)\, dt \tag{145}$$

For the separate time functions we have the Fourier integral representations

$$f_1(t) = \frac{1}{2\pi} \int_{-\infty}^{\infty} F_1(x)\, e^{jxt}\, dx \tag{146}$$

and

$$f_2(t) = \frac{1}{2\pi}\int_{-\infty}^{\infty} F_2(y)\, e^{jyt}\, dy \tag{147}$$

where we have used the symbols x and y to denote angular frequency, the variable of integration.

Substitution into 145 and rearrangement of the order of integration gives

$$\int_{-\infty}^{\infty} f_1(t) f_2(t)\, dt = \frac{1}{(2\pi)^2}\iint F_1(x)\, F_2(y)\, dx\, dy \int_{-\infty}^{\infty} e^{j(x+y)t}\, dt \tag{148}$$

The integration with respect to time may be carried out separately; and proper interpretation of this integral, as has previously been seen, requires that we tentatively replace the infinite limits by T and subsequently consider the limit $T \to \infty$. We thus have

$$\int_{-T}^{T} e^{j(x+y)t}\, dt = \left.\frac{e^{j(x+y)t}}{j(x+y)}\right]_{-T}^{T} = \frac{2T \sin (x+y)T}{(x+y)T} \tag{149}$$

Regarded as a function of $x + y$, the zero ordinate here is $2T$, zeros occur at integer multiples of π/T, and the net enclosed area, which becomes more and more concentrated at $x + y = 0$ as T tends toward larger and larger values, equals 2π independently of T. We therefore have the interpretation

$$\int_{-\infty}^{\infty} e^{j(x+y)t}\, dt = 2\pi u_0(x+y) \tag{150}$$

The remaining double integration over x and y in 148 evidently becomes curtailed to a single integration since the integrand is zero except for $x = -y$. The end result can evidently be written

$$\int_{-\infty}^{\infty} f_1(t) f_2(t)\, dt = \frac{1}{2\pi}\int_{-\infty}^{\infty} F_1(-\omega)\, F_2(\omega)\, d\omega \tag{151}$$

which is equivalent to the result expressed by Eqs. 99 and 103 in Ch. XI where the analogous problem involving periodic time functions is dealt with. In the present situation, however, the result is more striking since integrals appear on both sides of the relation instead of there being an integral on the left and a summation on the right. In this connection it may also be borne in mind that the minus sign in the argument of F_1 may be transferred to F_2 without affecting the value of the integral.

An even more symmetrical result is obtained if $f_1(t)$ and $f_2(t)$ are identical. Since (according to discussion in art. 5)

$$F(-\omega)F(\omega) = |F(\omega)|^2 \tag{152}$$

we get in this case

$$\int_{-\infty}^{\infty} f^2(t)\, dt = \frac{1}{2\pi}\int_{-\infty}^{\infty} |F(\omega)|^2\, d\omega \tag{153}$$

Here 152 is referred to as the *power spectrum* of the function $f(t)$, and the result expressed in Eq. 153 is known as *Parseval's theorem*.

The result 151 can be put into another form by making use of relations 69 and 70 which state that if $f(t)$ and $F(\omega)$ are a pair, then so are $F(t)/2\pi$ and $f(-\omega)$. Accordingly in 151 we make the replacements

$$f_1(t) \to \frac{F_1(t)}{2\pi} \quad \text{and} \quad F_1(-\omega) \to f_1(\omega) \tag{154}$$

and get the very symmetrical result

$$\int_{-\infty}^{\infty} F_1(t)\,f_2(t)\,dt = \int_{-\infty}^{\infty} f_1(\omega)\,F_2(\omega)\,d\omega \tag{155}$$

Here each integrand is the product of a direct and an inverse Fourier transform, and one should not be misled into thinking that a function of ω is necessarily the direct transform of a function of t. For this reason it is perhaps better to write the result 155 as

$$\int_{-\infty}^{\infty} F_1(x)\,f_2(x)\,dx = \int_{-\infty}^{\infty} f_1(x)\,F_2(x)\,dx \tag{156}$$

where x is any real variable and a capital letter designates the direct Fourier transform of a function denoted with a lowercase letter while the latter is the inverse transform of a function denoted by the corresponding capital letter.

Problems

1. For the time function shown in the adjacent sketch, obtain and plot the Fourier transform $F(\omega)$ by:

(*a*) Use of integral 22 directly.
(*b*) Differentiating to get impulses and then using properties 71 and 72.
(*c*) Considering first the periodic repetition of $f(t)$ and then carrying through the transitions 20 on the resulting Fourier series.

Compare results.

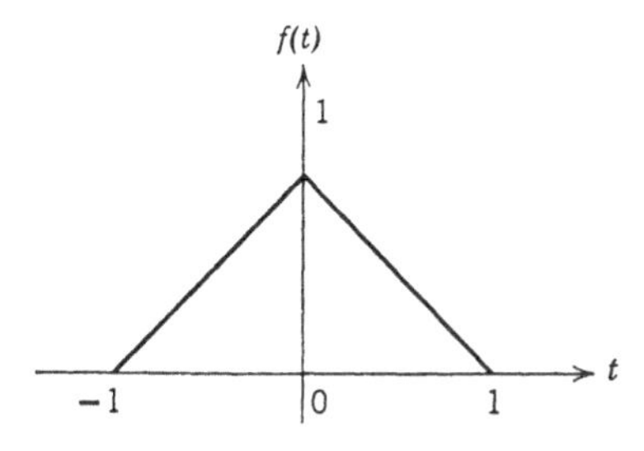

PROB. 1.

2. With reference to $f(t)$ in prob. 1, consider the function $f(t) \cos 10\,\pi t$. Determine and plot the corresponding spectrum function $F(\omega)$.

3. Consider the rectangular pulse in part (*a*) of Fig. 5 and its Laplace transform as given by Eq. 25. By integrating along the right half-plane path shown in the adjacent s-plane sketch, show that this pulse can be constructed as the

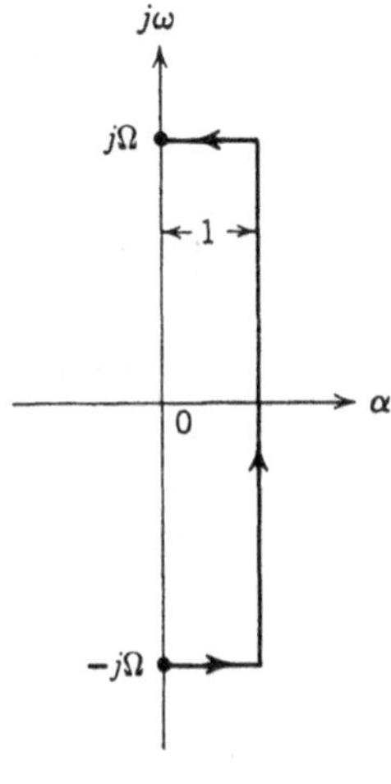

PROB. 3.

interference pattern resulting from exponentially growing sinusoids. Carry out the integration from $-j\Omega$ to $j\Omega$ as indicated, and plot the resulting approximation for $\Omega = 10\pi/\delta$. Make use of any known properties of functions of a complex variable that aid in obtaining the desired result, but justify in detail all pertinent steps.

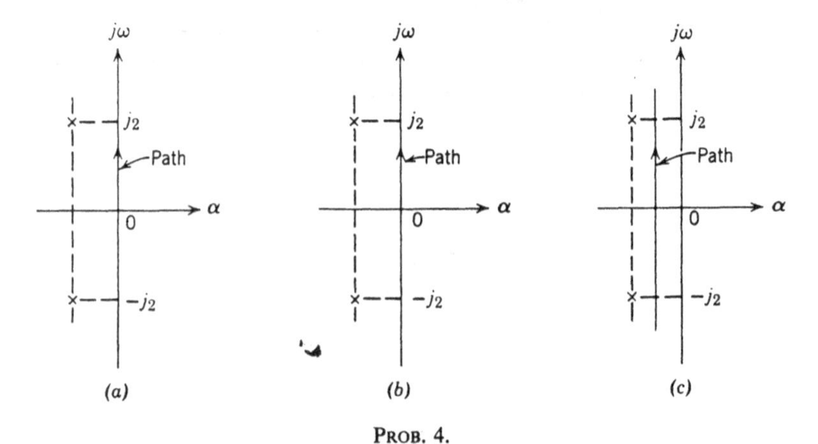

PROB. 4.

4. Evaluate the synthesis integral 26 along indicated paths for rational $F(s)$ functions having the pole-zero patterns shown. Their constant multipliers are unity.

5. Prove the statement: Necessary and sufficient conditions that $f(t)$ be real are that $F(s)$ be real for real s.

6. Evaluate $f(t)$ for the following pole-zero plots and paths of integration:

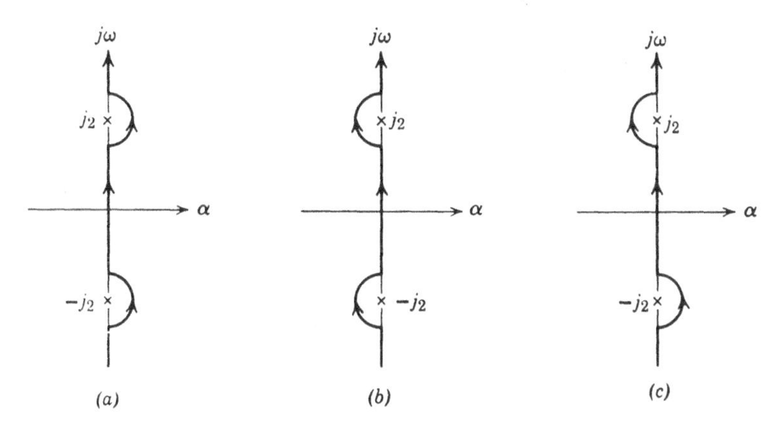

PROB. 6.

7.

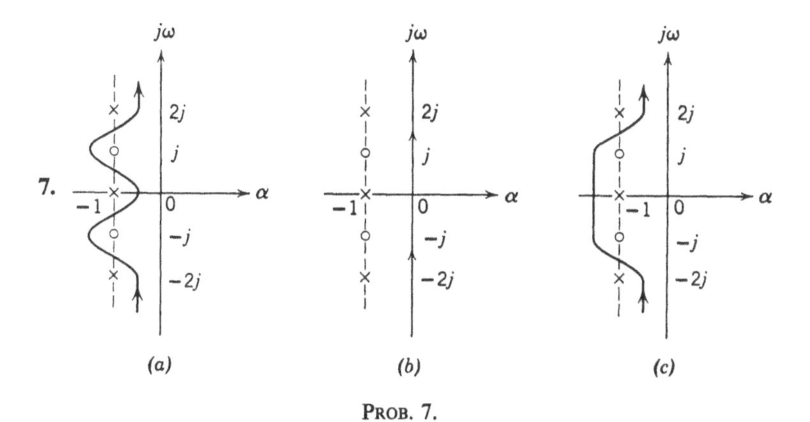

PROB. 7.

For the above pole-zero patterns and paths of integration, evaluate the pertinent Laplace integrals for $f(t)$.

8. The $F(s)$ function in the preceding problem is replaced by $F(s - s_0)$ with $s = 1 + j$. If the path of integration is the j-axis with semicircular detours into the right half-plane for any poles that fall upon this axis, evaluate $f(t)$.

9. (*a*) Determine and plot to scale the $f(t)$ corresponding to $F(s) = 1/(s + 1)^5$.

(*b*) From this result obtain and plot to scale the $f(t)$ corresponding to $F(s) = A/(s + 4)^5$ with $F(0) = 1$ as in part (*a*).

(*c*) If N_a and N_b are networks whose response functions are the $F(s)$ functions in parts (*a*) and (*b*) respectively, how are these networks related?

10. Given:

$$F(s) = \frac{a_0 + a_1 s + a_2 s^2 + \cdots + a_m s^m}{b_0 + b_1 s + b_2 s^2 + \cdots + b_n s^n}$$

(*a*) The corresponding time function $f(t)$ and its first three derivatives are zero at $t = 0$. What special forms do the polynomials in $F(s)$ have?

(*b*) If in $F(s)$, $a_0 = a_1 = a_2 = a_3 = a_4 = 0$, what can you say about the behavior of $f(t)$?

11. Consider the periodic function $f(t)$ shown in Fig. 2 of Ch. XI, having the line spectrum shown there in Fig. 3.

(*a*) Determine the corresponding Fourier transform $F(\omega)$.

(*b*) Determine the Fourier transform corresponding to $f[t - (\tau/2)]$.

(*c*) Determine the Fourier transform corresponding to $u_{-1}(t)f[t - (\tau/2)]$.

(*d*) Determine the Laplace transform corresponding to $u_{-1}(t)f[t - (\tau/2)]$.

12. Using only the properties of Fourier transforms and integrals, evaluate the following:

(*a*)
$$\int_{-\infty}^{\infty} \frac{\sin x}{x}\, dx$$

(*b*)
$$\int_{-\infty}^{\infty} \left(\frac{\sin x}{x}\right)^2 dx$$

13. Plot carefully and to scale the time function resulting from applying the smoothing operation 126 to the unit impulse once, twice, . . . four times, and for each, plot also the corresponding spectrum function.

CHAPTER XIII

Convolution

1. The Convolution Integral

In a study of the approximation properties of Fourier series as well as in many related problems arising in network and communication theory, an understanding of the following integral relationship:

$$g(t) = \int_{-\infty}^{\infty} f(t - \xi)\, K(\xi)\, d\xi \tag{1}$$

is essential. It is, therefore, necessary that we become familiar with its detailed characteristics before proceeding further with the discussion of our main topic.

This integral is variously regarded either as expressing a relationship between the functions $f(t)$ and $g(t)$ that is determined by the nature of the function $K(t)$, or as a kind of interaction of the functions $f(t)$ and $K(t)$ that generates a new function $g(t)$. In the latter instance, no particular distinction is made regarding the roles played by $f(t)$ and $K(t)$, and indeed, this point may be emphasized through pointing out that one can just as well write integral 1 in the form

$$g(t) = \int_{-\infty}^{\infty} K(t - \xi)\, f(\xi)\, d\xi \tag{2}$$

The equivalence of 1 and 2 is readily demonstrated through introducing in 1 the change of variable

$$t - \xi = \eta, \quad \xi = t - \eta, \quad d\xi = -d\eta \tag{3}$$

whence, noting that $\eta = -\infty$ where $\xi = \infty$ and vice versa, we can transform the integral 1 into

$$g(t) = -\int_{\infty}^{-\infty} f(\eta)\, K(t - \eta)\, d\eta \tag{4}$$

Since an interchange of the limits of integration reverses the algebraic sign of the integral, and since η may now be replaced by the letter ξ or any other symbol with which we wish to designate the variable of integration, it is clear that 2 and 4 are equivalent.

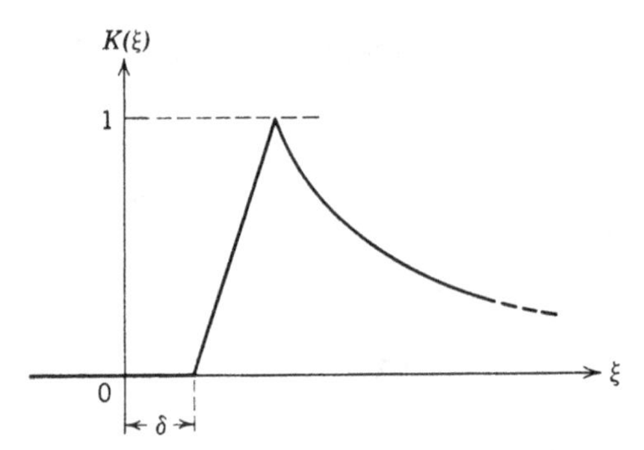

FIG. 1. One of the functions whose convolution is considered.

Either form is designated a *convolution integral* expressing the so-called *convolution* of functions $f(t)$ and $K(t)$. We also say that $f(t)$ and $K(t)$ are *convolved*, and $g(t)$ is the result of this process of convolution. As the equivalence of 1 and 2 shows, it is a mutual relationship so far as $f(t)$ and $K(t)$ are concerned; it is immaterial whether we convolve $f(t)$ with $K(t)$ or vice versa.

A graphical interpretation of the process defined by this integral not only places in evidence its interesting and useful characteristics but also suggests several more meaningful names both for the mechanism portrayed by it and for the participating functions $f(t)$ and $K(t)$. The plots in Figs. 1 through 4 are such graphical aids in the interpretation of integral 2. The first two of these figures show arbitrary forms for the functions $K(\xi)$ and $f(\xi)$ whose convolution is to be evaluated. Since most time functions encountered in practical problems start at some finite instant which may be arbitrary, $K(\xi)$ is chosen to be zero for $\xi < \delta$ and $f(\xi)$ for $\xi < t_0$.

Reference to the integrand in Eq. 2 shows that we should next form the function $K(t - \xi)$. Somewhat easier to visualize is the function $K(\xi - t)$,

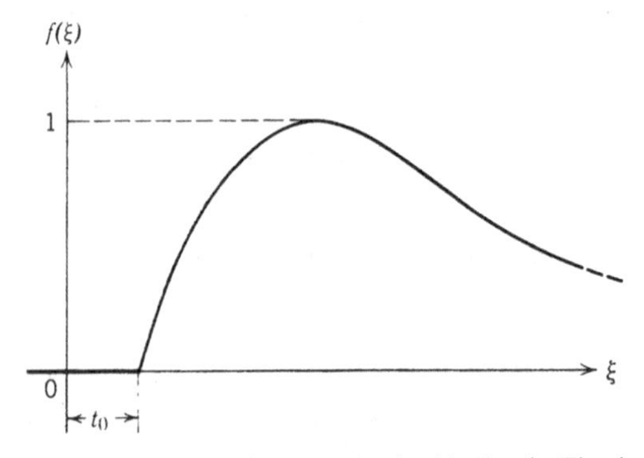

FIG. 2. Function to be convolved with that in Fig. 1.

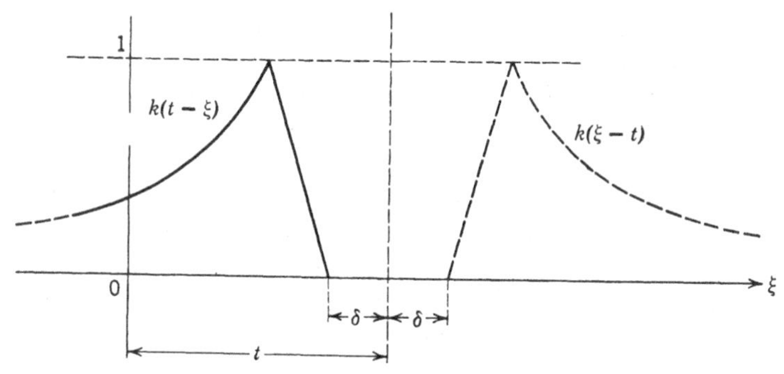

Fig. 3. Function in Fig. 1 and its image about the ordinate at $\xi = t$.

since on a ξ-scale, this is simply the function $K(\xi)$ delayed by t seconds. Graphically one obtains $K(\xi - t)$ by displacing $K(\xi)$ to the right (the positive ξ-direction) by an amount t, as shown by the dotted curve in Fig. 3. Since the argument $t - \xi$ is the negative of $\xi - t$, we next see that function $K(t - \xi)$ is the mirror image of $K(\xi - t)$ about a vertical line erected at the point $\xi = t$. The solid curve in Fig. 3, which is $K(t - \xi)$, and the dotted one are such a pair of images.

It should specifically be observed that ξ is the independent variable in these plots and t is a parameter. For different values of t we get different plots for the function $K(t - \xi)$, but these can all be drawn with a single template which is simply displaced along the ξ-axis appropriate to any chosen value of t.

In Fig. 4 are shown dotted the function $f(\xi)$ as in Fig. 2 and the function $K(t - \xi)$ as in Fig. 3. The integrand appearing in the integral 2 is the product of these two functions and is depicted by the solid curve in Fig. 4. The value of the integral is the area under this solid curve.

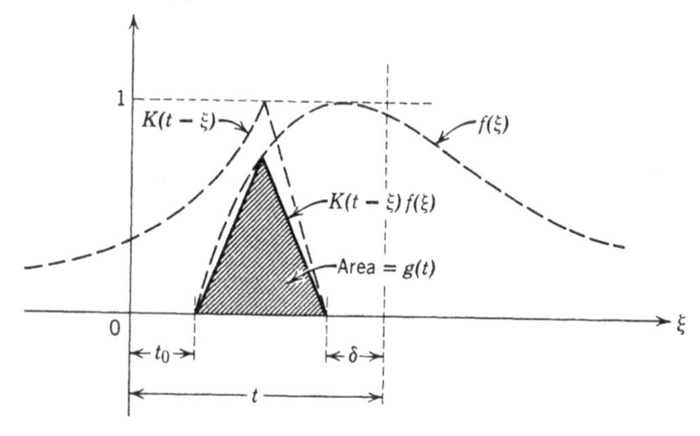

Fig. 4. Illustrating the process of convolving the functions in Figs. 1 and 2.

Since, for different values of t, the curve for $K(t - \xi)$ assumes various positions along the ξ-axis, it is clear that the solid curve enclosing the pertinent area changes both in shape and duration as t takes on these values, and thus the area becomes a continuous function of time. This function is $g(t)$.

From the plots in Fig. 4 it is also clear that $g(t)$ is zero for $t < (t_0 + \delta)$ and starts to have nonzero values at $t = t_0 + \delta$ since the curves for $K(t - \xi)$ and $f(\xi)$ have overlapping nonzero portions only when $t > t_0 + \delta$.

Interpretation of the resulting area as a function of t is facilitated if we visualize $f(\xi)$ as a fixed curve drawn on the paper, but $K(t - \xi)$ as being the cut-away portion of a cardboard template as shown in Fig. 5. Placed

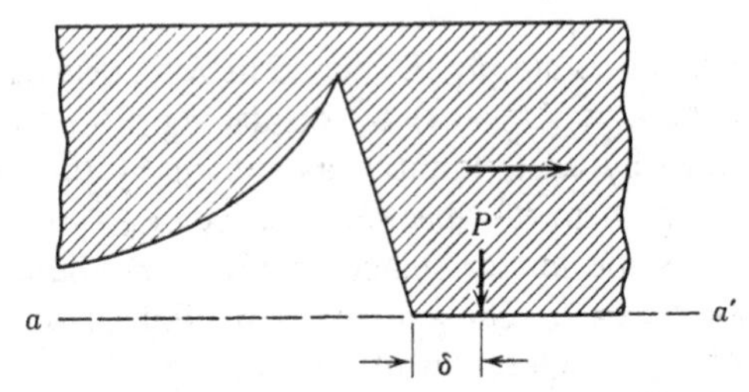

FIG. 5. Template cut for the function of Fig. 1.

upon the plot of Fig. 2, the line a-a' coinciding with the ξ-axis, this template is moved into positions appropriate to various values of t by displacing it to the right or left, a particular t-value being equal to the distance that the point P on the template is to the right of the origin 0 in Fig. 2.

For $t < t_0 + \delta$ the plot of $f(\xi)$ in Fig. 2 is completely covered by the solid portion of the template. As the latter is moved to the right and the distance 0–P becomes larger than $t_0 + \delta$, the cut-away portion in the template uncovers more and more of the plot of $f(\xi)$. Through visualizing the product $K(t - \xi) \cdot f(\xi)$, whose nonzero portion is confined to the range uncovered by the template, we obtain an appreciation of the enclosed area $g(t)$ and how it varies with the time t.

Since the area evaluated by the integral 2 extends only from $\xi = t_0$ to $\xi = t - \delta$, the reader may be wondering why the limits on this integral are chosen to be $-\infty$ and ∞. The answer is that the functions $K(\xi)$ and $f(\xi)$ may in some cases be infinite in extent. The infinite limits are appropriate in any case, since no contribution to the value of the integral is made by those portions of its range that extend either to the right or to the left of the interval in which the integrand is nonzero. That is to say, the infinite limits as indicated in Eq. 2 provide an adequate range, and one that automatically becomes abridged in a situation like that shown in Fig. 4

since the process of integration (viewed as an area accumulation process) ceases of itself where the integrand is zero.

2. Convolution with a Narrow Slit; the Scanning Concept

Keeping the same $f(\xi)$ function as in Fig. 2, let us now suppose that $K(\xi)$ is the tall narrow pulse shown in Fig. 6, to which the superimposed plots of $K(t - \xi)$ and $f(\xi)$ in Fig. 7 correspond. If we think of $K(t - \xi)$ in this figure as being in the form of a template, then the only portion of $f(\xi)$ uncovered is the small segment visible through the slitlike opening. Since the integrand $K(t - \xi) \cdot f(\xi)$ is nonzero only over this narrow range in which $f(\xi)$ may with reasonable accuracy be regarded as equal to the constant value $f(t - \delta)$ pertaining to the center of the slit, the integral 2 in this case is seen to yield substantially

$$g(t) \approx A \cdot f(t - \delta) \tag{5}$$

where A equals the area of the slit (Fig. 6).

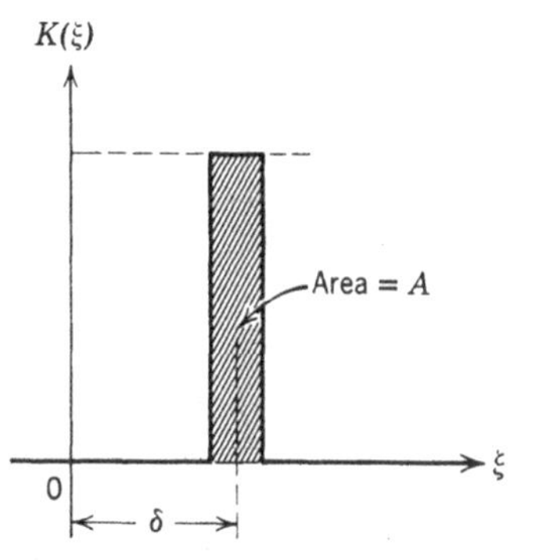

FIG. 6. A particular form of scanning function.

This example illustrates a very important property of the convolution integral, namely, that if one of the two functions being convolved has the form of a narrow slit, then the result is substantially the other function except for a delay equal to the time of occurrence of the slit (the time δ in Fig. 6). The area of the slit appears as a constant multiplier.

Graphical interpretation involving this process of viewing the function (ξ) through the opening of a horizontally displaceable template which

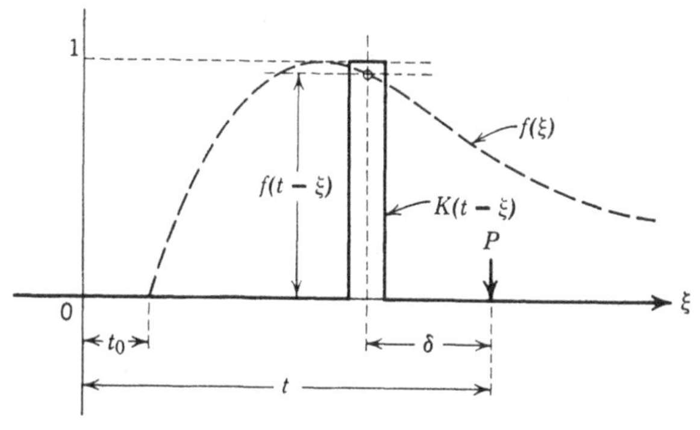

FIG. 7. The process of scanning the function in Fig. 2 with the function shown in Fig. 6.

characterizes $K(t - \xi)$ is called *scanning*. The template is the *scanning function* (also called a *window function* because the function being scanned is viewed through the opening provided by this function as if it were a window) and the process which we shall refer to simply as the scanning of $f(\xi)$ with this function is the equivalent of evaluating the convolution integral as described above.

What we have shown by means of the last example is that the scanning of a function through a narrow slit yields essentially that same function

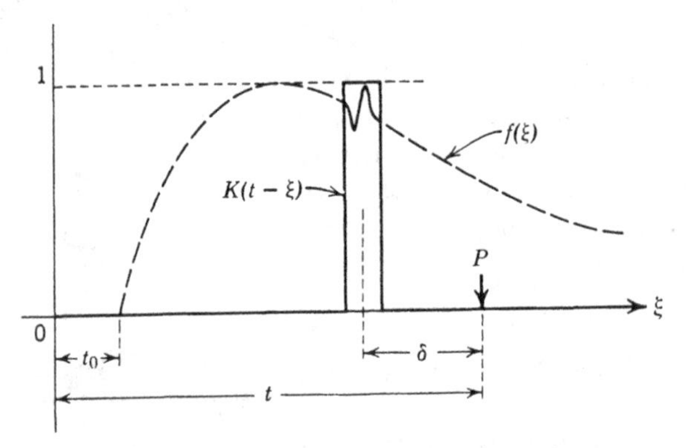

FIG. 8. The scanning function of Fig. 6, viewed as a narrow slit, is not narrow enough to reproduce a fast wiggle in $f(t)$.

except for a constant multiplier equal to the slit area and a time delay equal to that of the slit function or scanning function. The correct use of this result involves, of course, an interpretation of the term *narrow* as applied to the slit function.

What we are tempted to state in this regard is that the slit or pulse (as in Fig. 6) is "narrow" if its width is small compared with its height. This conclusion, however, is wholly fallacious since the narrowness of the slit in connection with our present problem is not a characteristic that can be judged from the appearance of the slit alone but only by comparing the slit width with the character of the function $f(\xi)$ to be scanned. The height of the slit relative to its width has no bearing whatever upon this question, as the following discussion shows.

Essential to the result 5 in the graphical interpretation of Fig. 7 is the conclusion that *the variation of $f(\xi)$ throughout the ξ-range uncovered by the slit is negligible* so that one can without excessive error replace $f(\xi)$ by an average constant value over this range. In order to appreciate the importance of this condition, consider, for example, the $g(t)$ function resulting from a convolution of this same $K(\xi)$ function and an $f(\xi)$

containing a fast wiggle as shown in Fig. 8. It is evident that $g(t)$ is substantially the same as it is for the $f(\xi)$ function without this wiggle because the resulting area measured by integral 2 for the slit position shown as well as for neighboring positions is almost the same whether the wiggle in $f(\xi)$ is there or not. In order that the resulting $g(t)$ function may contain a reasonably close replica of this wiggle in $f(\xi)$, the slit must be considerably narrower, so narrow in fact, that even for the vicinity of this wiggle, the function $f(\xi)$ has a negligible variation throughout the ξ-range uncovered by the slit.

The criterion for narrowness is thus seen to involve the rate of time variation in the function to be scanned by the slit; and, of course, also involved is the degree of exactness with which we are interested in having $g(t)$ reproduce the shape of $f(t)$. These matters will engage our attention more completely when we apply the convolution integral as a tool in judging the adequacy of a given transmission network for the reproduction of an input signal waveform which is one of the many problems in which the results of our present discussions prove valuable.

These thoughts suggest that the unit impulse function would make a perfect window or scanning function in the sense that any function scanned by it must be reproduced perfectly. Recalling that the notation for the unit impulse occurring at $t = 0$ is $u_0(t)$, we can express this conclusion by writing

$$f(t) = \int_{-\infty}^{\infty} u_0(t - \xi) \cdot f(\xi)\, d\xi \tag{6}$$

which is true because the scanning function $u_0(t)$ has zero width and encloses unit area. It is also obviously true because we can graphically regard the differential area element $f(\xi)\, d\xi$ as an impulse of this value occurring at $t = \xi$. The integrand in 6 is thus seen to be a typical element or *sample* of the function $f(t)$ and the sum of all such samples yields this function itself. *Convolution of any function with the unit impulse leaves that function unchanged.*

When a scanning function does not have the shape of a narrow slit, but is arbitrary like the one shown in Fig. 1, we may consider its decomposition into additive components of which each is a narrow slit. This idea is indicated in Fig. 9 where the function of Fig. 1 is shown dotted and a typical component or sample in the form of a slit of width $d\xi$ is drawn with solid lines. Since the instant of occurrence of this sample is at the time ξ, and its enclosed area is $K(\xi)\, d\xi$, the corresponding contribution to the net result, according to Eq. 5, is

$$dg = K(\xi)\, d\xi \cdot f(t - \xi) \tag{7}$$

and the whole of the function $g(t)$ is obtained through summing such typical

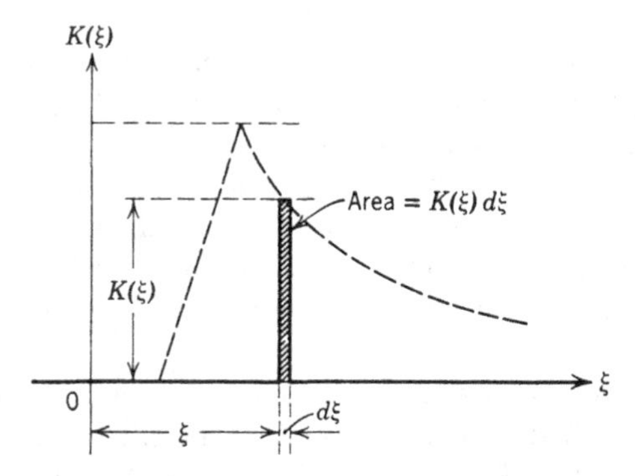

FIG. 9. Interpreting the integral representation of a function as in Eq. 6 as a scanning process with the impulse function.

contributions due to all the additive components comprising $K(\xi)$. This summation evidently takes the form of the following integral:

$$g(t) = \int_{-\infty}^{\infty} K(\xi)\, d\xi \cdot f(t - \xi) \tag{8}$$

whereupon we have succeeded in regaining Eq. 1 with which the discussion of this chapter began.

3. Convolution as a Superposition of Delayed Replicas

It is interesting and helpful for purposes of visualization to consider in detail the solution of a specific example according to the interpretation just given. Suppose, therefore, that we wish to determine the result of convolving a triangular and a square pulse like the $f(\xi)$ and $K(\xi)$ functions shown in Fig. 10. We can evaluate the result through assuming either that K scans f or that f scans K, since as shown above, an interchange of the roles played by these two functions has no effect upon the answer (the integrals 1 and 2 are equivalent). If we consider decomposition of the scanning function into additive components as indicated in Fig. 9, we conclude that $K(\xi)$ is best used in this way since all components have the same height. If we choose a uniform width for the components, then all the slit areas become equal.

In Fig. 10 we choose a uniform slit width of $\frac{1}{2}$-second, which is not as narrow as might be thought proper in view of the rate-of-change of $f(\xi)$, but smaller increments make the figure difficult to draw, and the example will illustrate the ideas involved just as well with a $\frac{1}{2}$-second slit width. The slit area, which in Eq. 7 is $K(\xi)\, d\xi$, is thus in our present problem equal to the finite value $\frac{1}{2}$, and so each contribution dg is a triangular pulse like $f(\xi)$ multiplied by a factor $\frac{1}{2}$. The total $g(t)$ function, therefore, is given by the sum of eight such half-sized triangular pulses displaced from each other by $\frac{1}{2}$-second intervals. This resultant as well as the eight component triangles are also shown in Fig. 10, where no attempt is made to indicate the relative positions in time of the three functions f, K and g, but only to take account of their shapes.

Actually each of the components shown dotted in the $g(t)$ plot is not a perfect triangle because the $\frac{1}{2}$-second slit width chosen for the elements of

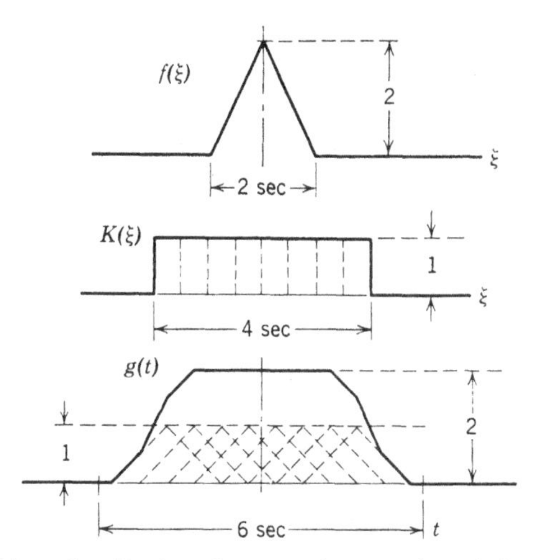

FIG. 10. The scanning of a triangle and a rectangle approximately interpreted according to the method illustrated in Fig. 7.

K is not small enough to reproduce f with such a degree of exactness, and so the resultant plot for $g(t)$ is correspondingly approximate. Nevertheless the example in Fig. 10 illustrates the thought that evaluation of the convolution process may be regarded as a superposition of delayed replicas which become individually more exact as their spacing in time is decreased.

It should also be noted in this connection that the triangular shape of each of the components has nothing to do with $K(\xi)$ in this example being a flat-topped pulse. The latter circumstance results in all component triangles having the same size; their triangular shape would be the same if $K(\xi)$ had any functional variation—only their relative sizes would then correspond to this variation.

4. A Generalized Form of the Convolution Integral

We are now in a position to make another important observation about a property of the convolution integral. Since we have seen that $g(t)$ may be regarded as a sum of $f(\xi)$ functions uniformly displaced in time and multiplied by appropriate amplitude factors (which happen in the preceding example to be equal but in general are not), it is evident that if $f(\xi)$ were replaced by its time derivative, then $g(t)$ would become replaced by its time derivative because each of the components whose sum equals $g(t)$ would be so replaced, and we know that the sum of derivatives equals the derivative of a sum of functions (i.e., differentiation and summation are interchangeable operations).

This statement is just another way of saying that we can form the derivative dg/dt in Eq. 8 by differentiating $f(t - \xi)$ with respect to t since t is a parameter so far as the integration is concerned.

By the same reasoning we can say that if $f(\xi)$ is replaced by the integral

$$f^{(-1)}(t) = \int_{-\infty}^{t} f(\xi)\, d\xi \tag{9}$$

then $g(t)$ becomes correspondingly replaced by its integral

$$g^{(-1)}(t) = \int_{-\infty}^{t} g(\eta)\, d\eta \tag{10}$$

By the integral of a function, as expressed by the forms 9 and 10, we here mean what we have consistently meant by this term, namely the accumulated area under that function since the beginning of its existence.

Now comes the important conclusion. Since $K(\xi)$ and $f(\xi)$ play equivalent roles, the effect upon $g(t)$ is the same whether we replace K or f by its derivative or by its integral and, therefore, the effect upon $g(t)$ is nil if we replace f by its derivative and K by its integral or vice versa. Generalization of this result permits us to write the convolution integral in an extremely flexible form which capitalizes upon the fact that the above reasoning remains true no matter how often the differentiations or integrations are repeated.

If a superscript as in $K^{(m)}(\xi)$ denotes the mth successive derivative if m is a positive integer, and the mth successive integration (as in Eqs. 9 and 10) if m is a negative integer, then we can write the convolution integral in the very general form

$$g^{(m+n)}(t) = \int_{-\infty}^{\infty} K^{(m)}(t - \xi) \cdot f^{(n)}(\xi)\, d\xi \tag{11}$$

in which m and n are any positive or negative integers.

In applying the convolution integral in this general form, one must make use of the singularity functions if K or f contains discontinuities, for the derivative functions will then contain impulses and doublets, etc. So long as these are correctly taken into account, one will encounter no difficulty which might otherwise arise on account of the presence of such discontinuities.

Through choosing the scanning function $K(\xi)$ equal to the unit impulse $u_0(\xi)$, and noting that $K^{(m)}$ then becomes the singularity function $u_m(\xi)$ of order m, Eq. 11 for $n = 0$ yields a generalization of the relation expressed by Eq. 6 that reads

$$f^{(m)}(t) = \int_{-\infty}^{\infty} u_m(t - \xi) f(\xi) \, d\xi \tag{12}$$

For $m = 0$, this states the aforementioned result that scanning a function with the unit impulse yields that function. We now see additionally that scanning a function with the doublet $u_1(t) = du_0/dt$ differentiates that function; scanning with the unit step $u_{-1}(t)$ integrates it; and so forth.

5. Impulse-Train Methods of Evaluation

Usefulness of the generalized form 11 resides in the fact that it may frequently be easier to visualize the result expressed by the convolution integral if K or f is replaced by its derivative or integral. Thus in the problem of Fig. 10 suppose we replace $K(\xi)$ by its derivative. Recalling that the derivative of a unit step is the unit impulse, we see by inspection that the derivative of $K(\xi)$ in this example yields a positive unit impulse followed 4 seconds later by a negative unit impulse.

If we now make use of the fact that a function scanned by the unit impulse is unchanged, we see that scanning $f(\xi)$ in Fig. 10 with the derivative of $K(\xi)$ yields a positive triangle followed 4 seconds later by a negative triangle, as shown in the top part of Fig. 11. Integration then yields $g(t)$ as given in the bottom part of the same figure. This result, which incidentally is exact, checks well with the approximate result in Fig. 10, and shows how one may make effective use of form 11 of the convolution integral.

More specifically this example illustrates the fact that the generalized form of the convolution integral given in Eq. 11 is particularly effective if some derivative of one of the functions yields only impulses, for then the corresponding derivative of $g(t)$ can be written down by inspection, and all that remains so far as evaluation of the convolution process is concerned is to perform one or more successive integrations. Let us illustrate this idea further by carrying out the convolution of the pair of functions shown in parts (a) and (b) of Fig. 12. Here $f(t)$ in part (a) is the rectangular hyperbola

$$f(t) = \frac{1}{t} \tag{13}$$

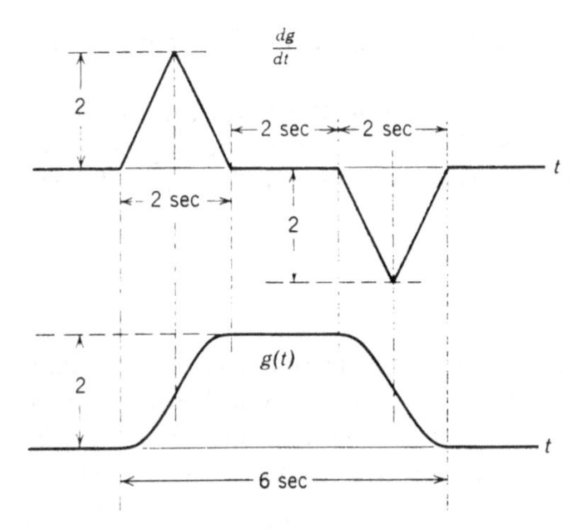

Fig. 11. Exact convolution of the functions in Fig. 10 making use of impulse-train methods or the generalized form of convolution shown in Eq. 11.

while $K(t)$ consists of straight-line segments defined by

$$K(t) = t \qquad \text{for } -1 < t < 1$$

$$K(t) = \begin{cases} -1 & \text{for } t < -1 \\ 1 & \text{for } t > 1 \end{cases} \tag{14}$$

It is not difficult to see that the first derivative of $K(t)$ is a rectangular

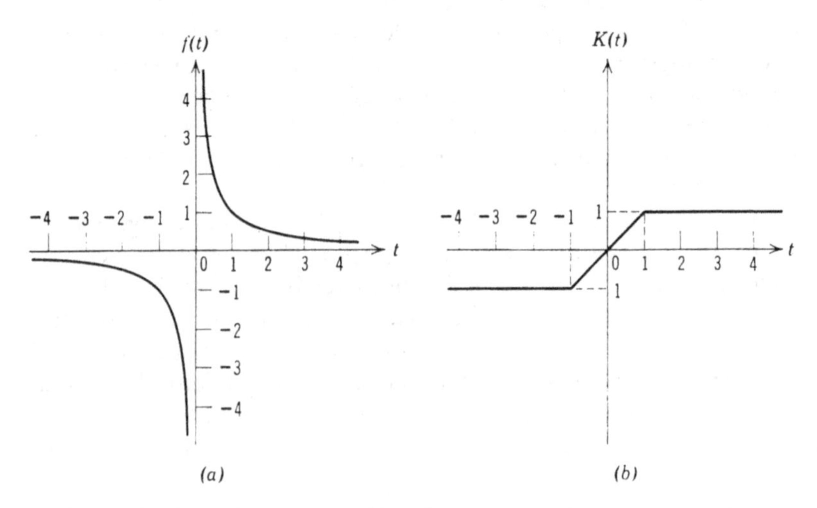

Fig. 12. A pair of functions to be used in a further illustration of the generalized convolution integral 11.

pulse and that the second derivative is a pair of impulses as given by the equation

$$K^{(2)}(t) = u_0(t + 1) - u_0(t - 1) \tag{15}$$

The second derivative of $g(t)$ can, therefore, be written down at once. It reads

$$g^{(2)}(t) = \frac{1}{t + 1} - \frac{1}{t - 1} \tag{16}$$

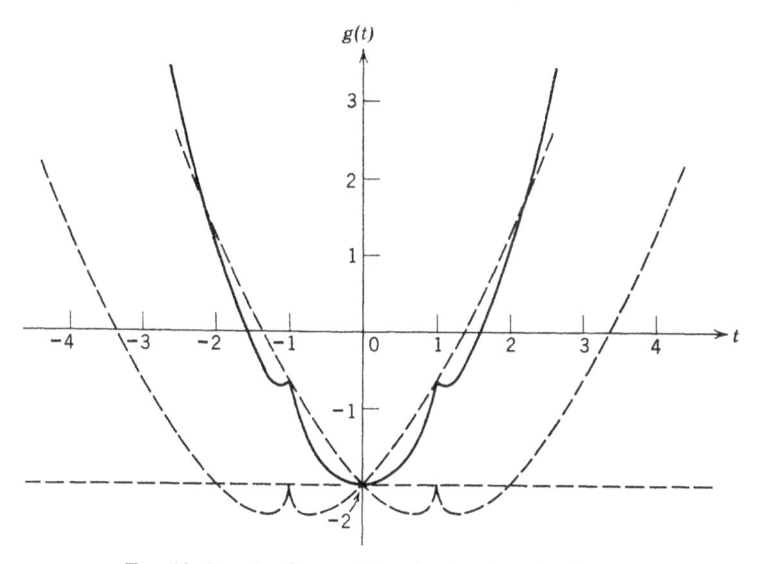

FIG. 13. Result of convolving the functions in Fig. 12.

Integrating once we have

$$g^{(1)}(t) = \ln |t + 1| - \ln |t - 1| \tag{17}$$

where the absolute-value signs are needed because we know that the result is real. Integrating once more we find the desired convolution of $f(t)$ and $K(t)$ in Fig. 12 to read

$$g(t) = |t + 1| \ln |t + 1| - |t - 1| \ln |t - 1| - 2 \tag{18}$$

This function is shown plotted in Fig. 13 where the dotted curves are the separate terms in 18. Without the technique used here the convolution in this example would be rather difficult to evaluate.

At this point we may well raise the question as to how we can capitalize upon this method when both functions to be convolved are smooth so that differentiation does not produce impulses. For example, suppose we wish to convolve the same $f(t)$ in Eq. 13 with $K(t)$ given by the smooth curve in the upper sketch of Fig. 14 which incidentally may be a graphically given function having no analytic representation. In this case we must resort to an approximate numerical method.

We can proceed by approximating $K(t)$ with a broken line as indicated by the confluent linear segments shown dotted in the same sketch. The

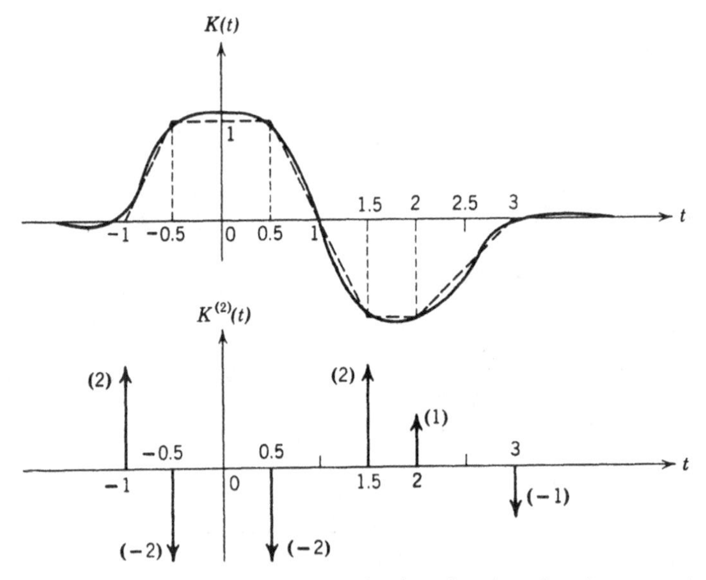

FIG. 14. Representation of some higher derivative of a given function as an impulse train.

second derivative of this broken-line approximation is given by the impulses whose values and locations are indicated in the bottom sketch of Fig. 14. This so-called impulse-train approximation to the second derivative of $K(t)$ has the form

$$K^{(2)}(t) = \sum_{k=1}^{n} a_k u_0(t - t_k) \tag{19}$$

in which $n = 6$ in this example and we have

$$\begin{aligned} a_1 &= 2 & t_1 &= -1 \\ a_2 &= -2 & t_2 &= -\tfrac{1}{2} \\ a_3 &= -2 & t_3 &= \tfrac{1}{2} \\ a_4 &= 2 & t_4 &= \tfrac{3}{2} \\ a_5 &= 1 & t_5 &= 2 \\ a_6 &= -1 & t_6 &= 3 \end{aligned} \tag{20}$$

The desired result is given by the convolution of this impulse train with the second integral of $f(t)$. For a typical impulse this yields

$$g_k(t) = a_k(\,|t - t_k| \ln |t - t_k| - |t - t_k|\,) \tag{21}$$

and the net result is expressed by the sum

$$g(t) = \sum_{k=1}^{n} g_k(t) \tag{22}$$

Graphically this function can be constructed straightforwardly in the manner that the convolution of functions $f(t)$ and $K(t)$ of Fig. 12 are constructed in Fig. 13 according to the expression given by Eq. 18.

This result is, of course, approximate to the extent that the broken line is an approximation to the given $K(t)$ function. By increasing the number of straight-line segments we can reduce the error as much as we like, but this expedient also increases the number of break points and hence the number of impulses or the terms in the sums 19 and 22. The computational labor involved in the final graphical construction of the desired result is correspondingly increased.

A preferable scheme is to approximate the given function with a set of confluent parabolic arcs, which is equivalent to approximating its first derivative by a broken line. One can construct the parabolic-arc approximation directly from the given function. The principles involved are illustrated in Fig. 15 where the upper sketch shows a function $K(t)$ constructed from confluent parabolic arcs and the lower one is its first derivative.

The first step in fitting parabolic arcs to a given smooth curve is to draw lines connecting the maxima and minima like the dotted ones in the top sketch of Fig. 15. Next draw the vertical dotted lines at maxima and minima so as to locate all corresponding abscissae like t_1, t_3, t_5 and t_7. Now consider the pair of confluent parabolic arcs connecting the origin t_1 with the first maximum occurring at t_3. Since these arcs result from integrating the triangle with base $t_3 - t_1$ and height h_1 in the lower sketch, the area of this triangle must equal the ordinate increment a_1 corresponding to the abscissa increment $t_3 - t_1$. Hence we have

$$2h_1 = \frac{a_1}{t_3 - t_1} \tag{23}$$

Observe that the point t_2 at which the tip of this triangle occurs is not yet fixed. In the upper sketch this is the point of confluency of the two parabolic arcs. If we want this point to lie upon the given curve $K(t)$ then we fix t_2 where the dotted line joining the origin t_1 with the first maximum intersects this curve. The first triangle in the bottom sketch is then determined in such a way that its integral yields confluent parabolic arcs that fit $K(t)$ at points t_1, t_2 and t_3.

The next triangle in the bottom sketch, with base $t_5 - t_3$ and altitude h_2, must have an area equal to the total ordinate increment in the $K(t)$ curve between t_3 and t_5. This condition fixes the height h_2 as

$$2h_2 = \frac{a_1 + a_2}{t_5 - t_3} \tag{24}$$

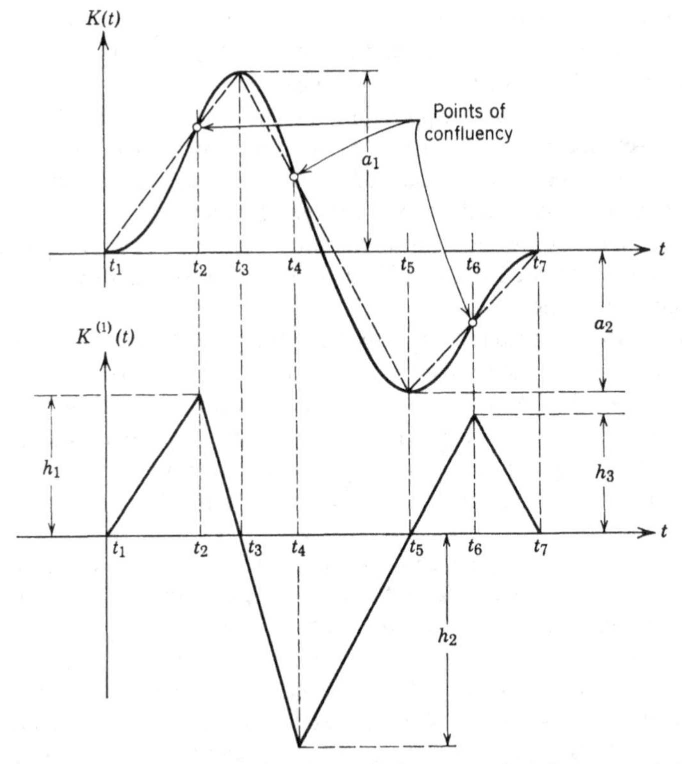

FIG. 15. Approximating a given function $K(t)$ by a set of confluent parabolic arcs.

and having the point of confluency of the next pair of arcs fall upon the $K(t)$ curve at t_4 locates the tip of the second triangle.

The third triangle with base $t_7 - t_5$ and height h_3 is similarly determined. We have

$$2h_3 = \frac{a_2}{t_7 - t_5} \tag{25}$$

and t_6 is fixed by again locating the point of confluency of the parabolic arcs on the given $K(t)$ curve.

In this example, we chose the $K(t)$ curve in such a way that the straight-line segments of triangles meeting at t_3 and t_5 have the same slope, so that the next derivative $K^{(2)}(t)$ has no discontinuities at these points. For an arbitrary curve $K(t)$, this condition may not be fulfilled, but if the discrepancy is small we can revise the locations of points t_3 and t_5 slightly without significantly impairing the net approximation and thus simplify the resulting numerical work. It may, of course, happen that the given $K(t)$ curve has a long flat maximum or minimum, in which case additional break points are unavoidable, but the same methods for constructing an appropriate confluent parabolic-arc approximation still apply, and we will in any case get better accuracy with fewer total break points than would be possible with a broken-line approximation.

The third derivative of the parabolic-arc approximation is an impulse train for which the third derivative of the desired function $g(t)$ can be written down by inspection. Successive integration is then all that is needed to evaluate the convolution process.

Whether or not it is useful to extend these ideas to an approximation by third-degree parabolic arcs (or even higher-order ones) depends upon the nature of the given $K(t)$ curve. This question, which gets rather deeply involved in purely geometrical matters will not be further pursued at this time.*

6. Considerations When One or Both Time Functions Are Periodic

Additional remarks are pertinent when one or both of the functions $f(t)$ and $K(t)$ are periodic. Since the integral 1 or 8 represents $g(t)$ as a sum of equally displaced f-functions multiplied by various amplitude factors (as exemplified by the dotted triangles in the lower portion of Fig. 10), it is graphically obvious that the resulting $g(t)$ is periodic if $f(t)$ is periodic. Analytically this statement may be seen to follow if in Eq. 1 we replace the variable t by $t + k\tau$, thus:

$$g(t + k\tau) = \int_{-\infty}^{\infty} f(t + k\tau - \xi)K(\xi)\,d\xi \tag{26}$$

and note that the periodicity of $f(t)$ is expressed by

$$f(t + k\tau - \xi) = f(t - \xi) \tag{27}$$

in which τ is the period and k an integer. Substitution of 27 into 26, and reference to Eq. 1, then gives

$$g(t + k\tau) = g(t) \tag{28}$$

establishing the periodicity of $g(t)$.

* It is discussed in detail in Ch. XVII, art. 3.

The interchangeability of the roles of $f(t)$ and $K(t)$ shows that $g(t)$ is likewise periodic if $K(t)$ instead of $f(t)$ is periodic. Important in this connection is the fact that only one of the functions f or K need be periodic in order to establish the periodic character of $g(t)$. If *both* f and K are periodic, it is in general not sensible to attempt drawing conclusions as to the character of $g(t)$ without first considering its possible existence.

It has been taken for granted in the preceding discussion that the area enclosed by the integrand in either integral 1 or 2 throughout the infinite time domain remains finite; otherwise the resulting function $g(t)$ does not exist. A sufficient condition insuring the existence of $g(t)$ is that $f(t)$ and $K(t)$ remain finite or have only a finite number of integrable infinities, and that at least one of them be nonzero over only a *finite* portion of the t-domain. Functions like those considered in the problem illustrated in Fig. 10 are typical examples. Since a periodic function is nonzero over the *infinite* t-domain, the existence of $g(t)$ must be considered more carefully when such functions are involved in the convolution.

If only one of the functions $f(t)$ or $K(t)$ is periodic and integrable over a period, and if the other is integrable and nonzero over only a finite portion of the same domain, then there is obviously no question about the existence of $g(t)$, for the pertinent area that this function represents for any t is surely finite. On the other hand, when both $f(t)$ and $K(t)$ are periodic then the total area yielding $g(t)$ is just as surely infinite since the product of f and K forming the integrand is nonzero over the infinite time domain and (barring degenerate cases) the net area enclosed is correspondingly infinite. Moreover, if $f(t)$ and $K(t)$ have different periods τ_1 and τ_2 respectively, and if these periods are incommensurable, then the integrand is not periodic and a convolution of $f(t)$ and $K(t)$ cannot be defined. However, if τ_1 and τ_2 are commensurable so that there exists a finite number divisible by both (the well-known LCM), then the integrand is periodic and the smallest number into which both τ_1 and τ_2 are divisible is its period.

Even so, the total area enclosed by the integrand extended over the infinite time domain is in general infinite, and a $g(t)$ function as defined by Eq. 1 or 2 still does not exist. In view of the periodic character of the integrand in this case, we can nevertheless obtain a unique and meaningful result through restricting the range of integration to a single period. Since the integration over additional periods merely yields integer multiples of the contribution due to one period, nothing is gained through integrating over such additional whole periods. The uniqueness of $g(t)$, on the other hand, is lost if the range of integration extends over a fractional period, and everything is lost if we integrate over an infinite range.

We shall be interested primarily in situations where $f(t)$ and $K(t)$ have the same period τ. In this case their convolution must be defined by restricting the integration in Eqs. 1 and 2 to a single period. Hence in this important periodic case, the convolution of $f(t)$ and $K(t)$ is defined by the integral

$$g(t) = \int_{-\tau/2}^{\tau/2} f(t - \xi)K(\xi)\, d\xi \tag{29}$$

or by the equivalent one

$$g(t) = \int_{-\tau/2}^{\tau/2} K(t - \xi)f(\xi)\, d\xi \tag{30}$$

Although we might be tempted to regard these integrals as restricted forms of 1 and 2, it is actually more proper to do just the reverse and look upon the latter as special forms that result from 29 and 30 when the period τ becomes infinite. This alternate view, in fact, is frequently useful, as some of the later applications (particularly those in the next chapter) will show.

7. Interpretation in the Frequency Domain

Since every operation in the time domain implies a corresponding one in the frequency domain, it is appropriate at this point to determine the frequency-domain equivalent of convolution in the time domain. We can do this straightforwardly by applying the Fourier analysis integral

$$G(j\omega) = \int_{-\infty}^{\infty} g(t)e^{-j\omega t}\, dt \tag{31}$$

to the function resulting from the convolution integral 1. After interchanging the order of integration we get

$$G(j\omega) = \int_{-\infty}^{\infty} K(\xi)\, d\xi \int_{-\infty}^{\infty} f(t - \xi)e^{-j\omega t}\, dt \tag{32}$$

Here

$$\int_{-\infty}^{\infty} f(t - \xi)e^{-j\omega t}\, dt = F(j\omega)e^{-j\omega\xi} \tag{33}$$

because of the property expressed by statement 73 in Ch. XII. Hence Eq. 32 becomes

$$G(j\omega) = F(j\omega)\int_{-\infty}^{\infty} K(\xi)e^{-j\omega\xi}\, d\omega \tag{34}$$

If we denote the Fourier transform of $K(t)$ by $H(j\omega)$, this yields

$$G(j\omega) = F(j\omega)H(j\omega) \tag{35}$$

and so we see that *multiplication in the frequency domain corresponds to convolution in the time domain.*

This simple result places the mutual character of convolution in evidence since an interchange of the roles played by $f(t)$ and $K(t)$ amounts in Eq. 35 to an interchange of the transforms $F(j\omega)$ and $H(j\omega)$, the effect of which is obviously nil. In the time domain where convolution is expressed by integral 1, this property (although readily determined as we did in art. 1 above) is not so obvious. This situation illustrates an important fact, namely that certain properties or relationships are often much easier to recognize in one domain than in the other. Fourier transformation has this collaterally useful feature.

8. Convolution in the Frequency Domain

Here we begin with multiplication in the time domain by considering the product

$$g(t) = f(t) \cdot K(t) \tag{36}$$

Replacing $f(t)$ and $K(t)$ by their respective Fourier integrals, we have

$$g(t) = \frac{1}{2\pi}\int_{-\infty}^{\infty} F(ju)e^{jut}\,du \cdot \frac{1}{2\pi}\int_{-\infty}^{\infty} H(jv)e^{jvt}\,dv \tag{37}$$

which we can rewrite in the form

$$g(t) = \frac{1}{2\pi}\int_{-\infty}^{\infty} F(ju)\,du \cdot \frac{1}{2\pi}\int_{-\infty}^{\infty} H(jv)e^{j(u+v)t}\,dv \tag{38}$$

For the integration with respect to v we introduce the change of variable

$$u + v = \omega \quad \text{or} \quad v = \omega - u \tag{39}$$

and since u is a constant parameter so far as this integration is concerned we have

$$dv = d\omega \tag{40}$$

The second integral in Eq. 38 is then rewritten as

$$\frac{1}{2\pi}\int_{-\infty}^{\infty} H(jv)e^{j(u+v)t}\,dv = \frac{1}{2\pi}\int_{-\infty}^{\infty} H(j\omega - ju)e^{j\omega t}\,d\omega \tag{41}$$

Substituting this form back into Eq. 38 and rearranging things slightly we have

$$g(t) = \frac{1}{2\pi}\int_{-\infty}^{\infty}\left\{\frac{1}{2\pi}\int_{-\infty}^{\infty} F(ju)H(j\omega - ju)\,du\right\}e^{j\omega t}\,d\omega \tag{42}$$

If for the expression within curved brackets we write

$$G(j\omega) = \frac{1}{2\pi}\int_{-\infty}^{\infty} F(ju)H(j\omega - ju)\,du \tag{43}$$

then Eq. 42 reads

$$g(t) = \frac{1}{2\pi} \int_{-\infty}^{\infty} G(j\omega) e^{j\omega t} \, d\omega \tag{44}$$

which is the normal Fourier integral representation for $g(t)$.

Equation 43 is thus seen to express the Fourier transform of $g(t)$ in terms of the Fourier transforms of $f(t)$ and $K(t)$, and since the right-hand side of Eq. 43 (apart from the factor 2π) is a convolution integral involving the convolution of spectral functions $F(j\omega)$ and $H(j\omega)$, we see that *convolution in the frequency domain corresponds to multiplication in the time domain*, a result that is complementary to that found in the foregoing article.

Through replacing $j\omega$ and ju by complex variables s and λ respectively, we can convert 43 into the complex integral

$$G(s) = \frac{1}{2\pi j} \int_{-j\infty}^{j\infty} F(\lambda) H(s - \lambda) \, d\lambda \tag{45}$$

which is sometimes said to express complex convolution. As in the Laplace integral, the path of integration may here deviate from the j-axis of the λ-plane in any way so long as such modifications do not sweep the path over any singularities of the integrand, and it is tacitly understood that the same path applies to the evaluation of $G(s)$.

It is interesting to apply this result to the convolution of a pair of rational functions $F(s)$ and $H(s)$ having the partial fraction expansions

$$F(s) = \frac{A_1}{s - s_1} + \frac{A_2}{s - s_2} + \cdots + \frac{A_m}{s - s_m} \tag{46}$$

and

$$H(s) = \frac{A_{m+1}}{s - s_{m+1}} + \frac{A_{m+2}}{s - s_{m+2}} + \cdots + \frac{A_{m+n}}{s - s_{m+n}} \tag{47}$$

The corresponding time functions are readily found by noting that the unit step $u_{-1}(t)$ and $1/s$ are a transform pair (Eq. 106, Ch. XII) and that property 74 of Ch. XII converts these into the form

$$u_{-1}(t) e^{s_k t} \quad \text{and} \quad \frac{1}{s - s_k} \tag{48}$$

Hence we have

$$f(t) = [A_1 e^{s_1 t} + \cdots + A_m e^{s_m t}] u_{-1}(t) \tag{49}$$

$$K(t) = [A_{m+1} e^{s_{m+1} t} + \cdots + A_{m+n} e^{s_{m+n} t}] u_{-1}(t) \tag{50}$$

The product of these two yields

$$g(t) = u_{-1}(t) \sum_{j=1}^{m} \sum_{k=m+1}^{m+n} A_j A_k e^{(s_j + s_k) t} \tag{51}$$

and the convolution of 46 and 47 is seen to be

$$2\pi G(s) = 2\pi \sum_{j=1}^{m} \sum_{k=m+1}^{m+n} \frac{A_j A_k}{s - (s_j + s_k)} \tag{52}$$

which is a rational function having $m \times n$ poles equal to sums of poles of $F(s)$ and $H(s)$ in all combinations.

9. Commutability of Convolution and Multiplication by an Exponential

At times we wish to consider the convolution of time functions $f(t)$ and $K(t)$ that do not remain bounded for large values of t but that can be converted into bounded functions through multiplication by an appropriate exponential. In other situations, convolution can be carried out more easily by inspection after the given functions are multiplied by an exponential. It is useful to know in such cases that if $f(t)$ and $K(t)$ are each multiplied by the same exponential $e^{s_0 t}$ then the effect upon $g(t)$ is for it also to become multiplied by this exponential. The demonstration is straightforward.

In integral 2, replace $f(t)$ and $K(t)$ by $f(t)e^{s_0 t}$ and $K(t)e^{s_0 t}$ respectively so as to get

$$\int_{-\infty}^{\infty} f(\xi)e^{s_0 \xi}K(t - \xi)e^{s_0(t-\xi)}\, d\xi = e^{s_0 t}\int_{-\infty}^{\infty} f(\xi)K(t - \xi)\, d\xi \tag{53}$$

But this is simply $g(t)e^{s_0 t}$, which agrees with the statement made above.

As an example, suppose we wish to convolve

$$f(t) = e^{-2t}u_{-1}(t) \tag{54}$$

and

$$K(t) = e^{-6t}u_{-1}(t) \tag{55}$$

If we multiply both of these by e^{2t}, the first is just a unit step and its convolution with the second yields

$$g(t)e^{2t} = \int_0^t e^{-4\xi} = \tfrac{1}{4}(1 - e^{-4t})u_{-1}(t) \tag{56}$$

Hence the solution to this example reads

$$g(t) = \tfrac{1}{4}(1 - e^{-4t})e^{-2t}u_{-1}(t) \tag{57}$$

Next consider the problem of convolving the following:

$$f(t) = e^{t}u_{-1}(t) \tag{58}$$

$$K(t) = e^{-|3t|} \tag{59}$$

Multiplying both by e^{-3t} we get the revised functions

$$f^*(t) \;= e^{-2t}u_{-1}(t) \tag{60}$$

$$K^*(t) = u_{-1}(-t) + e^{-6t}u_{-1}(t) \tag{61}$$

The first of these is 54 and the second term in 61 is 55. Hence 57 is part of the answer to the convolution of 60 and 61. For the other part we note first that

$$u_{-1}(-t) = 1 - u_{-1}(t) \tag{62}$$

Convolving unity with the function 60 gives the constant

$$\int_0^\infty e^{-2t}\,dt = \tfrac{1}{2} \tag{63}$$

Convolving the unit step with 60 yields the integral

$$\int_0^t e^{-2\xi}\,d\xi = \tfrac{1}{2}(1 - e^{-2t})u_{-1}(t) \tag{64}$$

Hence the convolution of 60 and 61 reads

$$\tfrac{1}{2} - \tfrac{1}{2}(1 - e^{-2t})u_{-1}(t) + \tfrac{1}{4}(1 - e^{-4t})e^{-2t}u_{-1}(t) \tag{65}$$

and the convolution of 58 and 59 is given by this expression multiplied by e^{3t}.

Problems

1. Consider $f(t) = \sin \omega_0 t$ convolved with the scanning function $K(t)$ shown in the adjacent sketch, and think of the evaluation as being done by the process of decomposing $K(t)$ into rectangular elements as indicated in Figs. 9 and 10. Since $K(t)$ is itself a rectangular window having a width 2δ, it is interesting to consider it as a single element and evaluate the result for different values of this width. In particular, it is interesting to express this width in terms of the period $\tau = 2\pi/\omega_0$ of the sinusoid $f(t)$. One has the intuitive feeling that a reasonable reproduction of $f(t)$ requires a width 2δ at least smaller than τ.

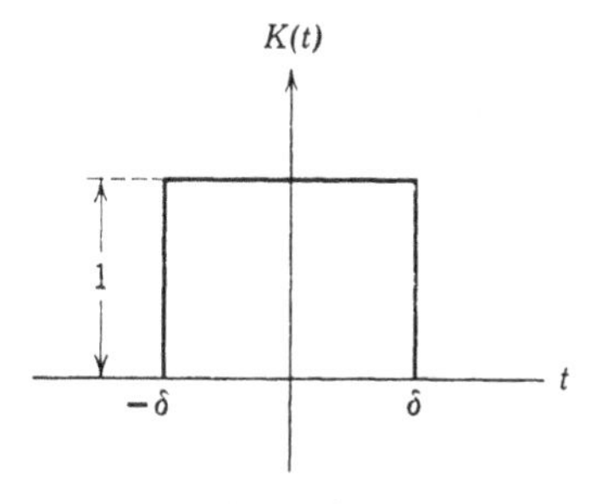

PROB. 1.

Study this problem graphically and see what you can predict as to the form and amplitude of the result. Then check your prediction making use of the principle that convolution in the time domain corresponds to multiplication in the frequency domain.

2. The sinusoid $f(t) = \sin \omega_0 t$ is scanned by the window function

$$K(t) = \frac{\sin \delta t}{\delta t}$$

having a nominal width (the main hump) equal to $2\pi/\delta$. Study and discuss the result for values of this width in the vicinity of the period $\tau = 2\pi/\omega_0$. Compare with the situation in prob. 1.

3. Evaluate and plot the Fourier transform corresponding to

$$K(t) \cos \omega_0 t$$

for $K(t)$ as defined in (*a*) prob. 1, (*b*) prob. 2. Assume $\omega_0 = \pi/\delta$ for part (*a*) and $\omega_0 = \delta$ for part (*b*).

4. Prove that the convolution of a sinusoid with any function integrable over the infinite domain is again a sinusoid. In particular, consider the convolution of $\sin t$ with any of the singularity functions $u_n(t)$.

5. Evaluate the convolution of the following two functions:

$$f(t) = \frac{\delta}{\pi} \times \frac{\sin \delta t}{\delta t} \quad \text{and} \quad h(t) = \frac{\alpha}{\pi} \times \frac{\sin \alpha t}{\alpha t}$$

for

$$\alpha < \delta; \quad \alpha = \delta; \quad \alpha > \delta$$

What conclusions can you draw from these results?

6. Show that the periodic function in Fig. 2 of Ch. XI can be obtained by convolving the single pulse in Fig. 5, part (*a*) of Ch. XII with the uniform impulse train

$$f(t) = \sum_{k=-\infty}^{\infty} u_0(t - k\tau)$$

From the Fourier transform of this impulse train and that of the single pulse, obtain the Fourier transform for the periodic function mentioned above and thus gets its Fourier series representation.

7. Find the convolution of the functions

$$e^{-t} \text{ and } e^{-2t} \text{ for } t > 0; \qquad \text{zero for } t < 0$$

8. Find the convolution of the following two Laplace transforms:

$$\frac{1}{s+1} \quad \text{and} \quad \frac{1}{s+2}$$

Choose the j-axis as a common path of evaluation.

9. (*a*) Convolve $1/(1 + \omega^2)$ with itself.
(*b*) Convolve $-\omega/(1 + \omega^2)$ with itself.
(*c*) Convolve $1/(1 + \omega^2)$ with $-\omega/(1 + \omega^2)$.
(*d*) Make an independent check of the foregoing results.

10. Find the time function resulting from the convolution of

$$f(t) = e^{-t} \sin t \quad \text{and} \quad g(t) = \tfrac{1}{2}e^{-t} - 2e^{-2t} + \tfrac{5}{2}e^{-3t}$$

for $t > 0$. Both are zero for $t < 0$.

11. Convolve the trigonometric polynomial

$$p(t) = \cos t - 9.5 \sin 3t + 6.4 \cos 5t - 2 \sin 7t$$

for all t, with the function

$$f(t) = \frac{\omega_0}{\pi} \times \frac{\sin \omega_0 t}{\omega_0 t}$$

choosing for ω_0 the values:

(*a*) $\omega_0 = 10$; (*b*) $\omega_0 = 6$ radians per second

12. Convolve the following frequency domain functions:

$$\frac{\sinh 2s}{2s} \quad \text{and} \quad \frac{1}{(s + 1)(s + 2)}$$

Choose the j-axis as a common path of evaluation and express the solution as a function of ω.

13. Given the two time functions (impulse trains)

$$f_1(t) = \sum_{k=0}^{m} a_k u_0(t - kt_0) \quad \text{and} \quad f_2(t) = \sum_{k=0}^{n} b_k u_0(t - kt_0)$$

Show that their transforms are given respectively by the polynomials

$$F_1(s) = a_0 + a_1 x + a_2 x^2 + \cdots + a_m x^m$$

and

$$F_2(s) = b_0 + b_1 x + b_2 x^2 + \cdots + b_n x^n$$

with $x = e^{-st_0}$; and thus show that convolution of $f_1(t)$ and $f_2(t)$ is equivalent in this situation to polynomial multiplication.

CHAPTER XIV

Approximation and Convergence Properties of the Fourier Series or Integral

1. The Dirichlet Kernel and the Gibbs Phenomenon

We are now in a position to study in some detail the manner in which a Fourier series or Fourier integral approximates the periodic or aperiodic function which it represents. This question is particularly important in dealing with Fourier series since it is concerned with the detailed appearance of the function represented by a finite number of terms, the so-called *partial sum* of a given infinite series. In most practical work we must, for reasons of computational feasibility, be content with the approximation afforded by a finite number of harmonics, and we cannot gauge the adequacy of such a partial sum unless we evaluate graphically its appearance in comparison with the function which the series represents.

A straightforward evaluation involving computation, plotting and summation of the individual harmonics becomes excessively tedious, especially if many terms are involved and their number must be varied in order to determine the smallest number that is still adequate in view of a stipulated tolerance requirement. In the following we shall develop a method of dealing with this problem that obviates such tedious computations entirely and is equally simple to apply whether a large or small number of terms is involved.

Consider the partial sum (Eq. 28, Ch. XI) given by

$$s_n(t) = \sum_{\nu=-n}^{n} \alpha_\nu e^{j\nu\omega_0 t} \tag{1}$$

and the coefficient formula

$$\alpha_\nu = \frac{1}{\tau} \int_{-\tau/2}^{\tau/2} f(\xi) e^{-j\nu\omega_0\xi}\, d\xi \tag{2}$$

derived on the basis that the mean square error (Eq. 30, Ch. XI) becomes a minimum. The manner in which this partial sum approximates $f(t)$ may be studied by substituting formula 2 into the expression 1 for $s_n(t)$. After reversing the order of summation and integration, the result may be written

$$s_n(t) = \int_{-\tau/2}^{\tau/2} f(\xi) D_n(t - \xi)\, d\xi \tag{3}$$

where we have introduced the notation

$$D_n(t) = \frac{1}{\tau} \sum_{\nu=-n}^{n} e^{j\nu\omega_0 t} \tag{4}$$

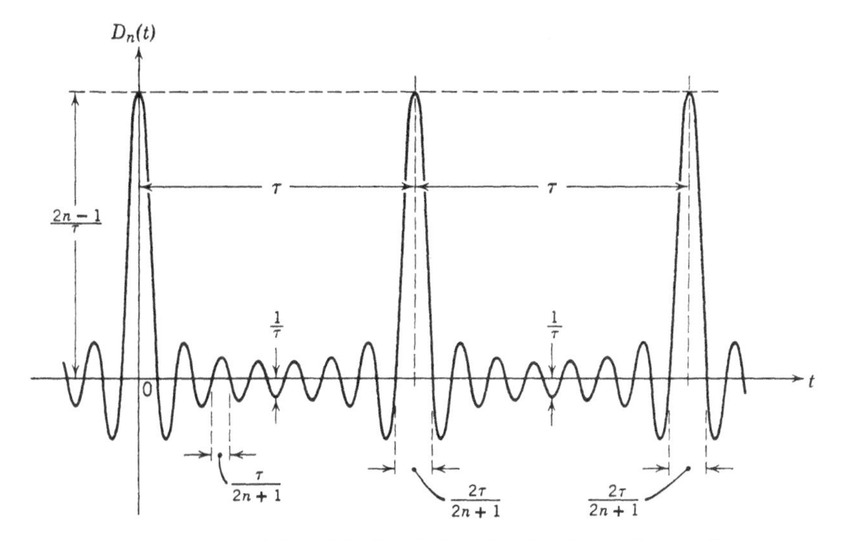

FIG. 1. The periodic Dirichlet window function drawn for $n = 7$.

Like $f(t)$, this is a periodic function having the period τ. It may be written in the equivalent trigonometric form

$$D_n(t) = \frac{1}{\tau} \{1 + 2 \cos \omega_0 t + 2 \cos 2\omega_0 t + \cdots + 2 \cos n\omega_0 t\} \tag{5}$$

This same expression occurs in the discussion of frequency groups in art. 1 of Ch. XII. Specifically, the sum within the curved brackets here is the same as in Eq. 5 of Ch. XII except that $(n - 1)/2$ there is replaced by n here, or n there by $2n + 1$ here. Making this change in Eq. 11 of Ch. XII and noting also the difference in the factors multiplying the expression within the curved brackets, we see that

$$D_n(t) = \frac{(n + \frac{1}{2})\omega_0}{\pi} \left\{ \frac{\sin (n + \frac{1}{2})\omega_0 t}{(2n + 1) \sin \omega_0 t/2} \right\} \tag{6}$$

Its appearance is shown in Fig. 1 which is drawn for $n = 7$. The height of each main spike is $(2n + 1)/\tau$ and the width of its base is $2\tau/(2n + 1)$.

The net area enclosed per period is seen by inspection of Eq. 5 to be unity independent of n. In a ripply sort of way, $D_n(t)$ becomes more and more like a periodic succession of unit impulses as n increases.

The partial sum according to Eq. 3 is given by the convolution of $f(t)$ and $D_n(t)$ extended over one period, as the discussion in art. 6 of Ch. XIII shows to be appropriate in a situation of this sort. Each period of $f(t)$ thus becomes scanned by one period of the function $D_n(t)$ which in this connection is called the *periodic Dirichlet kernel* or *window function.* As n is chosen larger and larger, the scanning process reproduces $f(t)$ with greater accuracy except for a residual effect owing to the ripply character of the scanning function.

Before discussing these details, let us try a somewhat different approach to this problem using the Fourier integrals to express the periodic function $f(t)$ in terms of itself. Thus an equivalent of the expression 1 for a partial sum is obtained which reads

$$s_n(t) = \frac{1}{2\pi}\int_{-\infty}^{\infty} f(\xi)\int_{-n\omega_0}^{n\omega_0} e^{j\omega(t-\xi)}\,d\omega\,d\xi \tag{7}$$

This expression can be written

$$s_n(t) = \int_{-\infty}^{\infty} f(\xi)D_a(t-\xi)\,d\xi \tag{8}$$

where

$$D_a(t) = \frac{1}{2\pi}\int_{-n\omega_0}^{n\omega_0} e^{j\omega t}\,d\omega = \frac{e^{j\omega t}}{2\pi jt}\Bigg]_{-n\omega_0}^{n\omega_0}$$

$$= \frac{\sin n\omega_0 t}{\pi t} = \frac{n\omega_0}{\pi}\cdot\frac{\sin n\omega_0 t}{n\omega_0 t} \tag{9}$$

In contrast with the periodic Dirichlet kernel 6, this one is aperiodic; and in contrast with the convolution integral 3, the alternate representation for the partial sum in 8 involves convolution over the infinite time domain. It is indeed interesting that the partial sum of a Fourier series may thus be evaluated by convolving the periodic function $f(t)$ either with the periodic Dirichlet kernel 6 over a single period or with the aperiodic Dirichlet kernel 9 over the infinite domain.*

In the following detailed study of the approximation properties we will use the convolution integral 8 and the aperiodic Dirichlet kernel 9 since the results are then appropriate not only to the Fourier series but also to a Fourier integral representation inasmuch as the formulation presented by

* Further discussion of this item is given in art. 7 of the next chapter.

Eqs. 7, 8 and 9 applies as well to an aperiodic function $f(t)$. It is merely necessary to interpret $n\omega_0$ as some top frequency Ω at which the spectrum synthesis terminates. As in the periodic case, $s_n(t)$ is regarded as an approximation to the desired $f(t)$.

Although we are already familiar with the function 9 which seems to appear quite persistently in Fourier analysis, it is plotted once again in

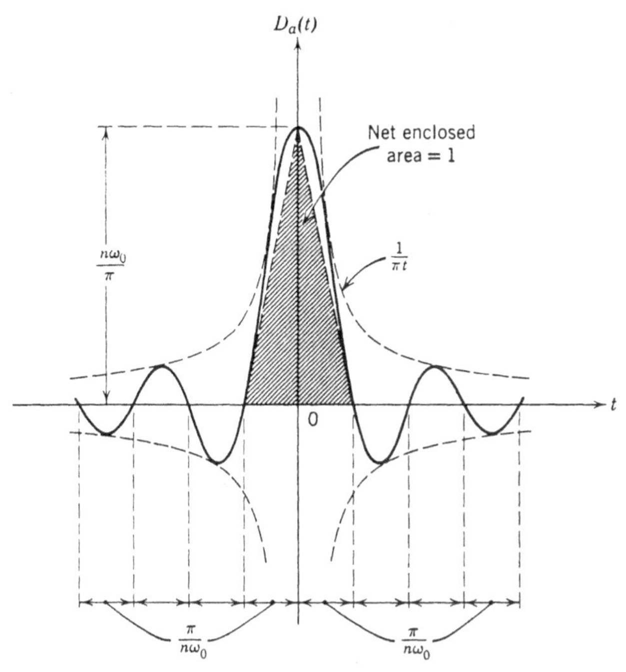

FIG. 2. The aperiodic Dirichlet window function. Shaded triangle equals net enclosed area between $-\infty$ and $+\infty$.

Fig. 2 in which certain features particularly pertinent to the present discussion are emphasized. The net enclosed area, incidentally, is equal to the triangular shaded region inscribed within the main hump. Of major significance is the fact that the subsidiary humps on either side are contained within the hyperbolic envelope function $1/\pi t$ *which is independent of the parameter n*. As a result, the amplitudes of the adjacent humps relative to the amplitude of the main hump are likewise independent of n.

As this parameter becomes larger and larger, the aperiodic Dirichlet kernel $D_n(t)$ changes by becoming compressed in the abscissa direction while remaining bounded by the invariant envelope $1/\pi t$. All ripple amplitudes including that of the main hump increase but retain fixed ratios to one another; all humps become narrower and taller while the area enclosed by each remains unaltered and the net area enclosed by the total function retains the fixed value unity.

As n increases without limit, this function approaches the unit impulse $u_0(t)$; and it is clear, therefore, that in the limit $n = \infty$ the partial sum $s_n(t)$ in Eq. 3 or 8 becomes equal to the pertinent periodic or aperiodic function $f(t)$. However, because of the peculiar ripply way in which $D_a(t)$ approaches $u_0(t)$, the approximation to $f(t)$ for any finite n however large may (in certain situations to be described presently) possess amplitude deviations that do not become smaller as n becomes larger and are present even in the limit $n = \infty$ in spite of the fact that the mean square error tends toward zero.

It can readily be seen that this state of affairs obtains whenever the given function $f(t)$ possesses discontinuities. By choosing for $f(t)$ the unit step $u_{-1}(t)$ and noting the detailed manner in which it is altered when scanned by the Dirichlet window in Eq. 9 (Fig. 2) the approximation properties in the vicinity of a typical discontinuity can be studied in detail.

The convolution of $u_{-1}(t)$ with $D_a(t)$ is more easily visualized if we assume that $D_a(t)$ is scanned by $u_{-1}(t)$ as expressed by the integral

$$g(t) = \int_{-\infty}^{\infty} D_a(\xi) u_{-1}(t - \xi)\, d\xi \tag{10}$$

According to Eqs. 9 and 12 in the previous chapter, we then get

$$g(t) = \int_{-\infty}^{t} D_a(\xi)\, d\xi \tag{11}$$

Substituting from Eq. 9 this yields the function

$$g(t) = \frac{1}{\pi} \int_{-\infty}^{\Omega t} \frac{\sin u}{u}\, du \tag{12}$$

in which $\Omega = n\omega_0$ and u is a dummy variable.

The calculation and plotting of this result is aided by noting that the function

$$Si(x) = \int_0^x \frac{\sin u}{u}\, du = \tfrac{1}{2} \int_{-x}^{x} \frac{\sin u}{u}\, du \tag{13}$$

called the "sine-integral of x" is tabulated* for various values of the independent variable x. Since the integrand is even, $Si(x)$ is odd about

* See Jahnke and Emde, *Tables of Functions with Formulas and Curves*, Dover Publications, 1943, pp. 1–9.

$x = 0$. In terms of the constant value

$$\int_{-\infty}^{0} \frac{\sin u}{u}\, du = \int_{0}^{\infty} \frac{\sin u}{u}\, du = \frac{\pi}{2} \tag{14}$$

we, therefore, have

$$g(t) = \frac{1}{2} + \frac{1}{\pi}\int_{0}^{\Omega t} \frac{\sin u}{u}\, du = \frac{1}{2} + \frac{1}{\pi}\, Si\Omega t \tag{15}$$

This result is shown plotted in Fig. 3.

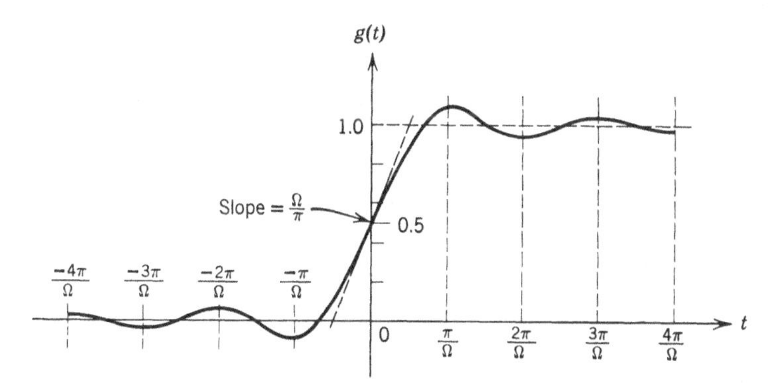

FIG. 3. Reproduction of a unit step by means of a finite portion of its spectrum.

Of primary significance is the undulatory manner in which this function approximates the unit step and the fact that the amplitudes of the ripples are independent of the parameter Ω. Thus the amplitude of the pair of ripples nearest $t = 0$ is about 9%, that of the next adjacent pair 5%, then 3%, and so forth. As the parameter Ω becomes larger, the abscissa scale contracts; the slope of $g(t)$ at $t = 0$ becomes proportionately steeper and the ripple frequency which varies inversely with Ω becomes larger. All ripples of significant amplitude become more and more concentrated at $t = 0$.

For an infinitely large value of Ω, all ripples occur at $t = 0$ and endure for a zero time interval, so that in a sense we can say that the result is a unit step, for any deviations therefrom are of zero duration.

In the periodic case, $\Omega \to \infty$ corresponds to $n \to \infty$; the partial sum 1 becomes an infinite series. In the aperiodic case the integral 7 is identified with that yielding $f(t)$, in this example, the unit step function. In either case the ripples in Fig. 3 are compressed into a single vertical line at the point of discontinuity. Even in this limit the Fourier series or integral is still observed to yield the overswing of 9% which is characteristic of the function $Si(x)$.

Although in the limit $\Omega = \infty$ this overswing together with significant adjacent ripples occupies zero space so that practically speaking one may say that no residual discrepancy remains, nevertheless, this peculiarity in the approximation property of a Fourier series or integral is noteworthy from a mathematical standpoint inasmuch as it illustrates the ultimate effect of the failure of the scanning function $D_a(t)$ properly to approach its ideal form (the unit impulse) in the limit. It is, moreover, of practical concern also since it reveals the disheartening fact that, by means of a finite portion of a Fourier series, a given function can never be approximated in the vicinity of a discontinuity with a tolerance less than the characteristic 9% overswing, no matter how many terms one may be willing to use.

This peculiar characteristic of Fourier analysis, known as the *Gibbs phenomenon*, is a direct consequence of the mean-square-error criterion implied in the formula for coefficient evaluation in the series (discussed in art. 3, Ch. XI), for it is this formula that leads inevitably to the present result which is a very real disadvantage in some practical problems. Because of this phenomenon, it is necessary in connection with certain approximation problems to use other types of trigonometric series which are appropriate modifications of the Fourier series. This aspect of Fourier theory is presented in the following article.

2. Cesàro and Other Types of Summation or Truncation

Evaluation of the partial sum $s_n(t)$ in the periodic case, or of the approximation to $f(t)$ in the aperiodic one, is evidently equivalent to determining the effect upon $f(t)$ of multiplying the pertinent spectrum function by

$$\begin{aligned} H(\omega) &= 1 \qquad \text{for } -\Omega < \omega < \Omega \\ H(\omega) &\equiv 0 \qquad \text{for } |\omega| > \Omega \end{aligned} \tag{16}$$

This so-called cut off function has the form of a rectangle when plotted versus the variable ω. Its inverse transform or time-domain aspect is the aperiodic Dirichlet window function $D_a(t)$ given by Eq. 9 and plotted in Fig. 2. Since multiplication in the frequency domain corresponds to convolution in the time domain, the effect of multiplying the spectrum of $f(t)$ by $H(\omega)$ is to convolve $f(t)$ with $D_a(t)$. The foregoing approach arrives at this same result in a slightly different manner.

Regarding the process as a cutoff operation with the rectangular cutoff function 16 allows us to make use of pertinent elementary transform properties and thus see significant results more easily. For example, since the value of $H(\omega)$ for $\omega = 0$ equals the net area enclosed by $D_a(t)$, we see at once that this must be unity regardless of the value of Ω. Variation of

Ω is equivalent to frequency scaling and hence it is immediately clear that this manipulation leaves the area enclosed by $D_a(t)$ invariant. Increasing Ω compresses the time scale and makes all humps in Fig. 2 taller and narrower but does not change the enclosed areas, as is pointed out above.

Convolution with the ripply $\sin x/x$ function (the inverse transform of the rectangular cutoff function 16) introduces ripples in the result which are most noticeable at discontinuities of $f(t)$ for any finite Ω and yield the Gibbs phenomenon, an error whose magnitude persists undiminished regardless of how large Ω may become.

This view of the problem not only makes it easier to understand the results just presented, but also suggests how we can diminish the undesirable ripply effect in a situation where truncation of the spectrum function is unavoidable. Thus truncation merely means that the spectrum function shall forcibly be made to equal zero for all ω larger than some finite cutoff frequency Ω; it leaves the form of the cutoff function completely arbitrary within the interval $-\Omega < \omega < \Omega$. By choosing different forms for $H(\omega)$ here, we can obtain a large variety of inverse transforms $h(t)$ to take the place of $D_a(t)$ in the convolution integral 8 and produce approximation properties of a correspondingly varied nature, some of which can avoid the Gibbs phenomenon altogether while others can reduce its unwanted features to any desired extent.

Such methods of summation or truncation applied to the infinite Fourier spectrum of a given time function yield results that are no longer appropriately designated as Fourier representations. In the periodic case, for example, we speak of the resulting series as a trigonometric polynomial of a certain type depending upon the kind of truncation function involved. One such type that has received considerable attention from mathematicians involves the triangular cutoff function shown in the upper part of Fig. 4. The sketch directly below it illustrates the corresponding time function which is found straightforwardly to be given by

$$h(t) = \frac{\Omega}{2\pi}\left(\frac{\sin \Omega t/2}{\Omega t/2}\right)^2 \tag{17}$$

Essentially, this is the Dirichlet function $D_a(t)$ squared.

The net enclosed area, equal to $H(\Omega)$, is again unity, and the function also has a ripply character but with a very important difference, namely that it remains positive for all t. In sharp contrast with the function in Fig. 3, the behavior of $s_n(t)$ in the vicinity of a discontinuity of $f(t)$ now approaches its asymptote monotonically instead of in an oscillatory manner. At time $t = 2\pi/\Omega$ the pertinent $g(t)$ in a plot analogous to that in Fig. 3 is still about 10% *below* its final value, at $t = 4\pi/\Omega$ it is 5% below, and so forth.

The Gibbs phenomenon is completely absent. The present truncation function, which is referred to as a Cesàro summation* of the Fourier series, is actually ultraconservative in this regard. Although an infinite Fourier series no longer converges uniformly at discontinuities of $f(t)$, the Cesàro sum which also becomes an infinite series if we let Ω approach infinity, does converge uniformly even at points of discontinuity notwithstanding the fact that all harmonic amplitudes of finite order are

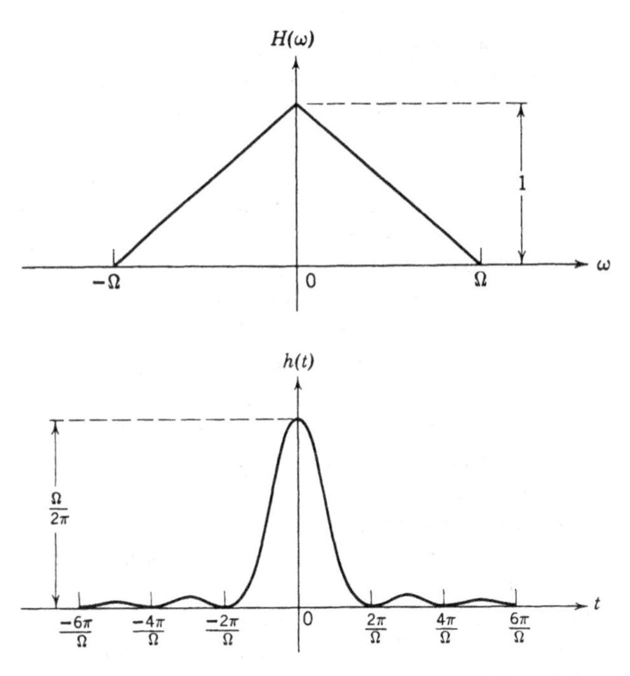

FIG. 4. Cesàro summation applied to the spectrum of the unit impulse.

exactly the same as for the Fourier series since the broken line in Fig. 4 becomes indistinguishable from a horizontal line for $\Omega \rightarrow \infty$. This situation involves one of the many subtleties with which Fourier theory abounds.

In comparing the rectangular and triangular truncation functions for which the respective scanning functions in the time domain are shown in Figs. 2 and 4, it is significant to note that the width of the main hump (the so-called *aperture* of the window function) is half as large for rectangular truncation and the same cutoff frequency Ω. For a given aperture, the

* For further detailed discussion see *MCA*, pp. 496–501.

cutoff frequency in a Cesáro sum must be twice that in the partial sum of a Fourier series. Suppression of the Gibbs phenomenon is, therefore, achieved at the cost of having to consider twice as many harmonics for roughly the same degree of approximation since the latter depends essentially upon the width of the slitlike scanning function.

For practical purposes, a trapezoidal cutoff function, somewhere between the rectangular and the triangular one, may yield an effective compromise. This thought and its logical elaboration are further discussed in the following article.

3. The Problem of Optimum Truncation; Uniqueness of the Fourier Transform

Suppose we consider the trapezoidal truncation function shown in Fig. 5. We will use the impulse-train method for obtaining its time-domain counterpart by observing that the second derivative of this spectrum function reads

$$\frac{d^2H}{d\omega^2} = \frac{1}{2\Delta}[u_0(t+\Omega+\Delta) - u_0(t+\Omega-\Delta) - u_0(t-\Omega+\Delta) + u_0(t-\Omega-\Delta)] \tag{18}$$

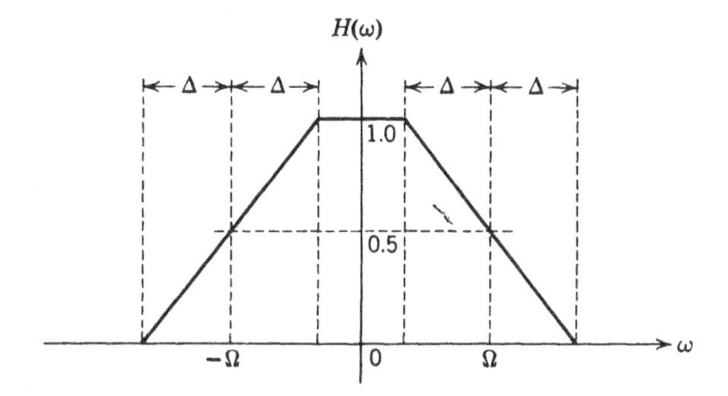

FIG. 5. A trapezoidal type of truncation function.

Since differentiation with respect to $s = j\omega$ corresponds to multiplication of $f(t)$ by $-t$, according to property 72 in Ch. XII, we see that the operation in Eq. 18 multiplies the desired time function $h(t)$ by $(-jt)^2$. Evaluation of the synthesis integral (Eq. 21, Ch. XII) for the function within the square bracket can be written down by inspection; we have

$$(-jt)^2h(t) = \frac{1}{4\pi\Delta}[e^{j\Omega t}(e^{j\Delta t} - e^{-j\Delta t}) - e^{-j\Omega t}(e^{j\Delta t} - e^{-j\Delta t})] \tag{19}$$

Straightforward manipulation yields the result

$$h(t) = \frac{\Omega}{\pi} \cdot \frac{\sin \Omega t}{\Omega t} \cdot \frac{\sin \Delta t}{\Delta t} \tag{20}$$

which is shown plotted in Fig. 6 for the choice $\Delta/\Omega = \frac{2}{3}$. The separate factors in Eq. 20 are shown dotted and the resultant function given by their product is drawn with a solid line. The chosen ratio Δ/Ω is seen to

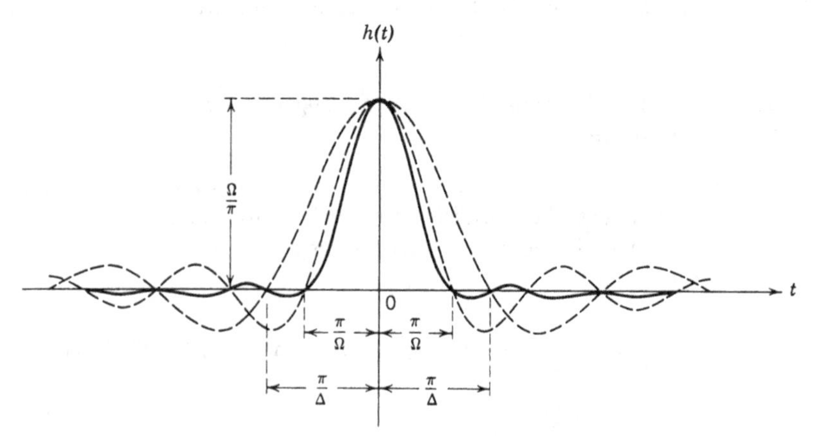

FIG. 6. Time function resulting from the trapezoidal truncation function shown in Fig. 5.

yield a resultant function $h(t)$ that is almost zero for $t > \pi/\Omega$. With this window function, discontinuities in $f(t)$ will cause practically no ripples. The trapezoidal cutoff function in Fig. 5 is very nearly ideal in this respect since the scanning function $h(t)$ which it produces approaches a unit impulse for $\Omega \to \infty$ in an almost-smooth manner.

What one might regard as ideal in this connection would be a truncation function $H(\omega)$ such that it produces an $h(t)$ having one large hump and nothing else. In other words, we would like an $h(t)$ function which, like the cutoff function $H(\omega)$, is non-negative and nonzero over a finite interval. Unfortunately, this result is not possible since it can readily be shown that the Fourier transformation cannot yield a pair of time-domain and frequency-domain functions that are both nonzero over a finite interval.

We can show, for example, that a contrary assumption leads to a contradiction. Thus, suppose we assume that $H(\omega)$ and $h(t)$ are nonzero over finite intervals and that $H(\omega)$ is identically zero for $|\omega| > \Omega$. In this case $H(\omega)$ is unaffected if we multiply it by a rectangular cutoff function having unit amplitude and any finite cutoff larger than Ω. In the time domain, however, this operation implies convolution with a $\sin x/x$ function, which not only changes the form of $h(t)$ but renders it nonzero over the infinite time domain.* Since the Fourier transform or inverse transform is unique, this conclusion is untenable, and so the original supposition is likewise untenable.

Uniqueness of the Fourier transformation, incidentally, may be seen from the following argument: If $f(t)$ had two different transforms $H_1(\omega)$ and $H_2(\omega)$ then their nonzero difference $\delta(\omega)$ would be required to have an identically zero inverse transform $d(t)$ as the resulting interference pattern of steady sinusoids with spectral distribution $\delta(\omega)$. The impulse function $u_0(t)$ comes closest to being an identically zero interference pattern for a nonzero spectrum, for it is nonzero at only one point. $d(t)$ is required to be zero *everywhere* and yet have a nonzero transform $\delta(\omega)$. This is impossible since the analysis integral yielding $\delta(\omega)$ has an identically zero integrand, even to the exclusion of impulses, which in this connection are spoken of as *functions of zero measure* because they occupy no space in the time domain.

In this connection it is significant to point out that if we allow the spectrum function to include exponentially growing or decaying sinusoids as well as steady ones, then an infinite variety of such functions may be found to yield the same interference pattern; and in this broader sense the Fourier transform is by no means unique. Uniqueness follows only if we fix the path in the s-plane along which the spectrum function is to be evaluated. In the usual Fourier transform evaluation this path is the j-axis.

It is interesting to observe that whereas we cannot find a truncation function $H(\omega)$, nonzero only for the finite range $-\Omega < \omega < \Omega$, whose time-domain counterpart $h(t)$ has only one large hump and nothing else, it is nevertheless possible to approximate this situation arbitrarily closely. To show that this is so, let us consider the family of functions developed in the latter part of art. 7, Ch. XII, the functions whose transforms are given there by Eq. 128. Here we will write the time functions in the form

$$h_n(t) = \sum_{k=0}^{n} \frac{(-1)^k}{\delta^n} \binom{n}{k} u_{-n}\left(t + \frac{n\delta}{2} - k\delta\right) \tag{21}$$

Their transforms are

$$H_n(\omega) = \left(\frac{\sin \omega\delta/2}{\omega\delta/2}\right)^n \tag{22}$$

and the integer n is regarded as designating the *order* of any specific pair.

* Visualize the $h(t)$ of finite duration displaced with respect to the $\sin x/x$ function. If the result of this convolution were to be nonzero for a finite interval, then the product for *any* displacement greater than some fixed value would have to enclose zero area, not only for a particular $\sin x/x$ function but for *all* of these having apertures less than $2\pi/\Omega$.

The time-domain functions for several orders are

$$h_0(t) = u_0(t) \tag{23}$$

$$h_1(t) = \frac{1}{\delta}\left[u_{-1}\left(t + \frac{\delta}{2}\right) - u_{-1}\left(t - \frac{\delta}{2}\right)\right] \tag{24}$$

$$h_2(t) = \frac{1}{\delta^2}[u_{-2}(t + \delta) - 2u_{-2}(t) + u_{-2}(t - \delta)] \tag{25}$$

$$h_3(t) = \frac{1}{\delta^3}[u_{-3}(t + \tfrac{3}{2}\delta) - 3u_{-3}(t + \tfrac{1}{2}\delta) + 3u_{-3}(t - \tfrac{1}{2}\delta) - u_{-3}(t - \tfrac{3}{2}\delta)] \tag{26}$$

$$h_4(t) = \frac{1}{\delta^4}[u_{-4}(t + 2\delta) - 4u_{-4}(t + \delta) + 6u_{-4}(t) - 4u_{-4}(t - \delta) + u_{-4}(t - 2\delta) \tag{27}$$

The corresponding frequency-domain functions are the familiar $\sin x/x$ function to the power n, the one for $n = 0$ being the constant unity. The time-domain function of zero order is the unit impulse. The first-order function 24 is a rectangular pulse; the second-order function 25 is a triangular pulse; and so forth. All time-domain functions are nonzero over a finite interval; their transforms are not, but for large orders their nonzero region also becomes essentially restricted to a finite interval.

Figure 7 illustrates the five time functions 23 through 27 together with their respective transforms. Since these are all even functions their identities in the time and frequency domains may be interchanged. Specifically, if we multiply all the frequency functions in Fig. 7 by 2π and then interchange ω and t, we again have transform pairs. In this case the time function of order 1 is the aperiodic Dirichlet kernel; the one of order 2 corresponds to Cesàro summation; and those of higher order represent the logical continuation of this type of summation process.

4. The Fourier Series as a Special Form of the Laurent Expansion

It is of collateral interest to recognize that the Fourier series may be regarded as a special form of the Laurent expansion and hence derivable from Cauchy's integral formula* which is so useful in connection with functions of a complex variable.

A given function $f(z)$ of the complex variable $z = x + jy$ is assumed to be regular and continuous within the annular region circumscribed by a pair of concentric circles with radii r_1 and r_2 and upon these circles themselves, as shown in the sketch of Fig. 8. Singularities of $f(z)$ lie either within the smaller circle or outside of the larger one or both. The point

* See *MCA*, pp. 290–295 and 459–460.

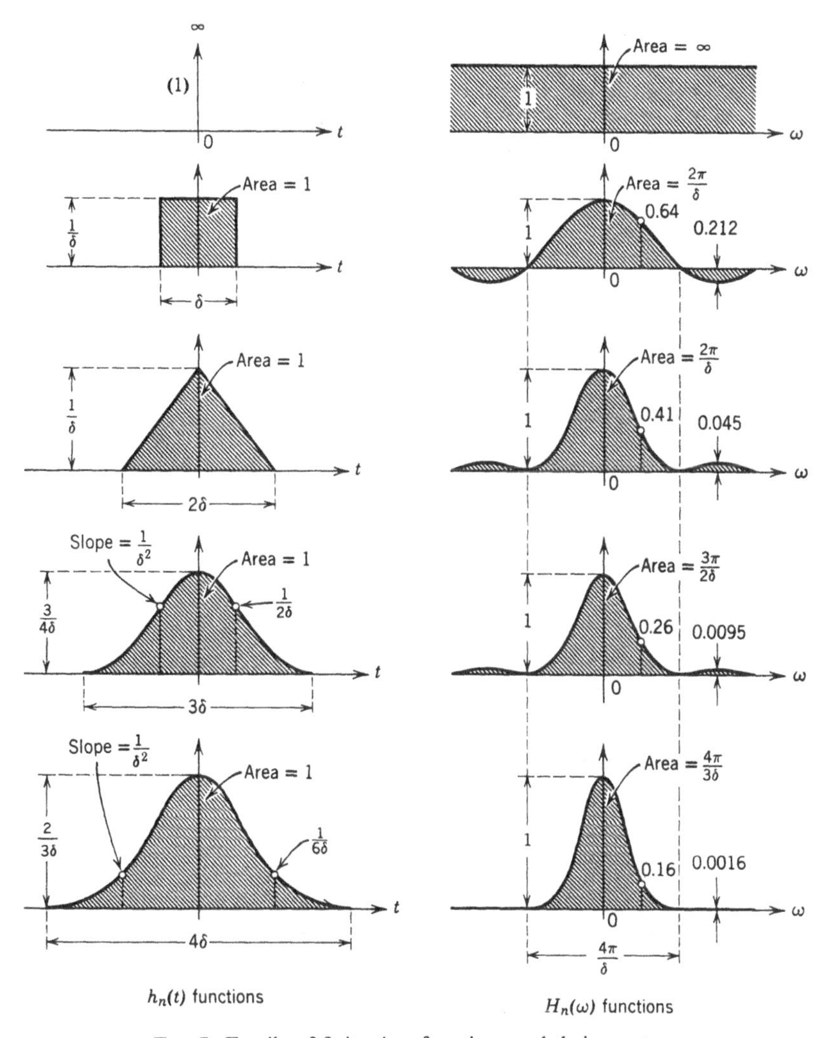

FIG. 7. Family of finite time functions and their spectra.

z_0 which is their common center, may or may not be a singular point. The Laurent series, which is uniformly convergent for all points z within the annular space, reads

$$f(z) = \sum_{\nu=-\infty}^{\infty} \alpha_\nu (z - z_0)^\nu \tag{28}$$

with coefficients determined by the formula

$$\alpha_\nu = \frac{1}{2\pi j} \oint \frac{f(\zeta)\, d\zeta}{(\zeta - z_0)^{\nu+1}} \tag{29}$$

where the integration extends over any closed contour within the annular space or coincident with either boundary circle, and the variable ζ refers to points on this contour. The concentric circle C is a possible contour, and the variable point z as well as ζ may lie upon it, although more generally z may also be any other point within the annular ring-shaped space.

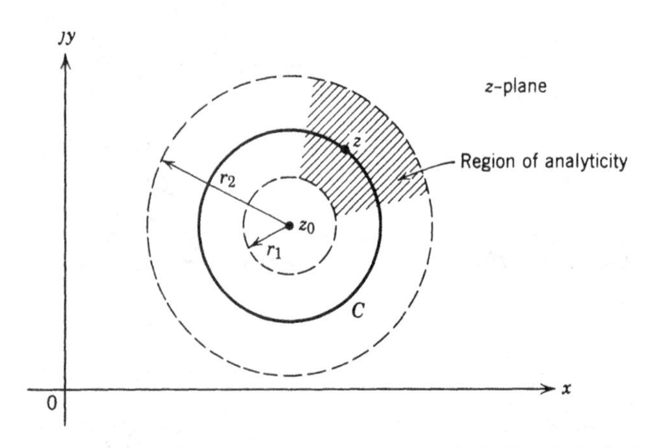

FIG. 8. Region to which Cauchy's integral formula is applied to obtain the Fourier series.

Since, through an appropriate change of variable, we can always make z_0 fall upon the origin, and can fulfill the condition

$$r_1 < 1 < r_2 \tag{30}$$

so that we may choose for C the unit circle, we can without loss of generality assume from now on that these special choices have been made. The relations 28 and 29 then have the simpler forms

$$f(z) = \sum_{\nu=-\infty}^{\infty} \alpha_\nu z^\nu \tag{31}$$

and

$$\alpha_\nu = \frac{1}{2\pi j} \int_C \frac{f(\zeta)\, d\zeta}{\zeta^{\nu+1}} \tag{32}$$

where C now is the unit circle about the origin. We assume, moreover, that both z and ζ lie upon this circle and write

$$z = e^{j\phi} \tag{33}$$

$$\zeta = e^{j\psi} \tag{34}$$

The expansion 31 then takes the form

$$f(\phi) = \sum_{\nu=-\infty}^{\infty} \alpha_\nu e^{j\nu\phi} \tag{35}$$

in which $f(\phi)$ may be regarded as a real function of the real variable ϕ. From 34 we have

$$\frac{d\zeta}{\zeta} = j\, d\psi \tag{36}$$

and hence the coefficient formula 32 becomes

$$\alpha_\nu = \frac{1}{2\pi} \int_a^{a+2\pi} f(\psi) e^{-j\nu\psi}\, d\psi \tag{37}$$

The infinite series 35 we recognize as the complex form of a Fourier series and 37 as its coefficient formula. The Laurent expansion is thus seen to yield the Fourier series as a special case. A number of additional comments are pertinent to this interesting result.

Since the function $f(z)$ is required to be analytic within the annular region, $f(\phi)$ in the Fourier series 35 is understood to be regular and continuous throughout the fundamental range $0 < \phi < 2\pi$. Actually we have seen that this condition is not necessary for the validity of the Fourier series. In fact, as pointed out in art. 1 of Ch. XI, the essence of Fourier's contribution was to point out that the function $f(\phi)$ need not be continuously defined over the fundamental range. Hence we must interpret the analyticity requirement of the Laurent expansion as a sufficient but not necessary condition for obtaining the Fourier representation 35.

In most applications of Fourier analysis the function $f(\phi)$ arises as a real function of a real variable, and so the existence of a first derivative does not imply anything with regard to the existence of higher derivatives. On the other hand, if $f(\phi)$ arises from the complex function $f(z)$, its analyticity which requires only the existence of a first derivative implies the existence of all higher derivatives. The uniform convergence, which is guaranteed for the Laurent expansion, automatically applies to the Fourier series 35 *and to any of its term-by-term derivatives.* This is a remarkable result that cannot be had if the Fourier series is approached from a given real function $f(\phi)$. In other words, the analyticity requirement which must be met for the Fourier series 35 derived from a complex function $f(z)$ seems at first to be rather exacting, but the remarkable properties of the resulting series are correspondingly extraordinary.

We can readily construct Fourier series with such extraordinary properties by starting from any complex function $f(z)$ which is analytic on the unit circle. Rational functions offer simple and versatile possibilities.

If $f(z)$ is to be real on the unit circle then poles must come in pairs that are images with respect to this circle; and with each such pair of poles there must be associated either a double-order zero at $z = -1$ or a simple zero at the origin, as well as an appropriate constant multiplier.

The sketch in Fig. 9 illustrates the pertinent geometry for such a pair of image points P and P'. The corresponding z-values are conjugate reciprocals. A double-order zero at the point $z = -1$ or a simple zero at the origin 0, together with these poles and an appropriate multiplier, yields a rational function that is real on the unit circle.

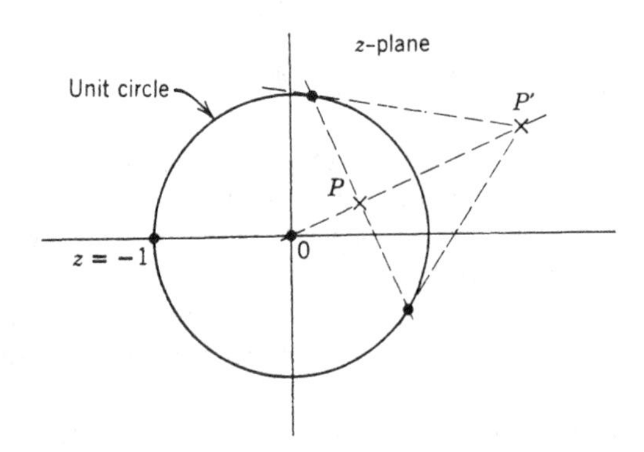

FIG. 9. Illustrating image points P and P' about the unit circle.

In this connection the change of variable

$$z = \frac{1 - w}{1 + w} \tag{38}$$

or

$$w = \frac{1 - z}{1 + z} \tag{39}$$

may be helpful. We recall that this familiar transformation maps the interior of a unit circle in the z-plane upon the right half of the w-plane, the j-axis of the latter corresponding to the unit circle itself, as is shown in Fig. 10 where cross-hatching is used to indicate corresponding regions. Points like P and P' (Fig. 9) that are images about the unit circle become points symmetrical about the j-axis in the w-plane, and vice versa. If w_1 and w_2 are such a pair of points, the function

$$\frac{1}{(w - w_1)(w - w_2)} \tag{40}$$

evidently is real on the j-axis of the w-plane. Since this function has a double-order zero at the point $w = \infty$, corresponding to the point $z = -1$, the pertinent function in the z-plane, apart from a constant multiplier, has the form

$$\frac{(z+1)^2}{(z-z_0)(z-z_0^*)} \tag{41}$$

In the w-plane w_2 is the negative conjugate of w_1:

$$w_2 = -\bar{w}_1 \tag{42}$$

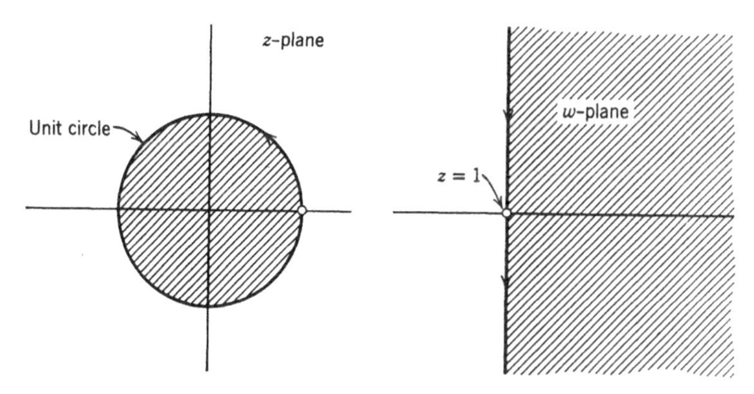

FIG. 10. Illustrating mapping properties of the transformation 38 or 39.

so that Eq. 38 gives

$$z_0 = \frac{1-w_1}{1+w_1}, \quad z_0^* = \frac{1+\bar{w}_1}{1-\bar{w}_1} = \frac{1}{\bar{z}_0} \tag{43}$$

as pointed out above.

Since 40 multiplied by $w^2 - 1$ is also real on the j-axis and corresponds, apart from a constant multiplier, to

$$\frac{z}{(z-z_0)(z-z_0^*)} \tag{44}$$

we can construct a rational function from a product of either or both 41 and 44, and provide a constant multiplier that renders the function real at some arbitrary point on the unit circle. A good choice is the point $z = 1$.

As an example let us consider the function

$$f(z) = \frac{z(1-z_0)(1-z_0^*)}{(z-z_0)(z-z_0^*)} \tag{45}$$

with poles on the real axis close to the point $z = 1$ so that the function

$f(\phi)$ having the Fourier series 35 is a sharp pulse centered at $\phi = 0$. The partial fraction expansion of this function reads

$$f(z) = \frac{z_0 - 1}{z_0 + 1}\left\{\frac{1}{[1 - (z/z_0)]} + \frac{1/zz_0}{[1 - (1/zz_0)]}\right\} \tag{46}$$

If z_0 has a value slightly larger than unity then the first term represents a pole lying outside the unit circle and the second term represents one inside. The Maclaurin expansion of the first term, which reads

$$\frac{z_0 - 1}{z_0 + 1}\left\{1 + \frac{z}{z_0} + \frac{z^2}{z_0^2} + \frac{z^3}{z_0^3} + \cdots\right\} \tag{47}$$

is therefore the ascending branch of the pertinent Laurent series 31. Coefficients in the Fourier series 35 are thus seen to be given by

$$\alpha_\nu = \frac{z_0 - 1}{z_0 + 1} \cdot \frac{1}{z_0^\nu} \qquad \text{for } \nu = 0, 1, 2, \ldots \tag{48}$$

For

$$z_0 = 1 + \delta \tag{49}$$

and a value of δ that is small compared with unity we have

$$\alpha_\nu \approx \frac{\delta}{2} e^{-\nu\delta} \tag{50}$$

Thus we see that the Fourier coefficients for series constructed in this manner need not be determined by the conventional formula 37 involving integration of $f(\phi)$. Since these coefficients are those in the pertinent Laurent expansion of $f(z)$, we can find them by algebraic means; and we need determine those for only one branch of the expansion because the ascending and descending branches have conjugate complex coefficients. In this connection the ascending branch is simply a Maclaurin expansion for that part of $f(z)$ contributed by poles lying outside the unit circle. Hence we can find all coefficients by the familiar Taylor series formula and avoid integration altogether.

Interpretation of $f(\phi)$ along the j-axis of the w-plane is made easier by substitution of the relation 33 into 39 which gives

$$w = -j \tan \phi \tag{51}$$

We can apply to the w-plane the methods discussed in arts. 3 and 4 of Ch. IX for determining pole distributions yielding desired j-axis behavior. For example, use of the Butterworth distribution plus its right half-plane image is a good way to produce a rectangular pulse-like function that may be made to approximate as closely as desired to a perfect rectangle. We can in this way construct a Fourier series that approximates arbitrarily closely to a square wave and converges uniformly for all points, and still converges uniformly for all points if differentiated term by term arbitrarily often! Such a result is not easily achievable in any other way.

This method of constructing a Fourier series approximation for a given function is, of course, not feasible when some arbitrary function $f(\phi)$ is given in analytic or piecewise analytic or graphical form, for one must first determine a rational function approximating $f(\phi)$ on the j-axis, and this step presents a problem in itself that is in general quite formidable.* However, there are situations where an appropriate rational function can be written down by inspection or where the approximation of $f(\phi)$ by a rational function is known or available in the pertinent literature. In such cases the relationship between rational functions (or other complex functions) and Fourier series discussed here can be very useful. And so we conclude that the discussions of the present article are not only of academic but also of practical interest.

Problems

1. A desirable scanning function, in place of the Dirichlet window, that produces a linear ramp with no overswing or rippling in contrast with the function $g(t)$ in Fig. 3 is a rectangular slit of finite width δ. In the frequency domain, however, this yields a $\sin x/x$ function and hence it fails to truncate the spectrum at a finite point. If we consider artificially terminating this $\sin x/x$ function at a finite point, then the discontinuities in the rectangular window become modified in the manner shown in Fig. 3, but in the function scanned by this modified window the effect is almost negligible.

Study this approach to the truncation problem and compare detailed results with those obtained for the triangular truncation function in Fig. 4 and the trapezoidal one in Fig. 5.

2. If (in connection with the discussion in art. 4) we want a polynomial in z that is real on the unit circle of the z-plane, consider the transformation

$$z = \frac{1 + js}{1 - js}; \qquad s = \frac{1 - z}{1 + z}$$

which maps the unit circle of the z-plane upon the real axis of the s-plane (the upper half of the s-plane is mapped upon the interior of the unit circle in the z-plane). Any rational function in s involving polynomials with real coefficients is real for real s and hence yields a rational function in z that is real on the unit circle.

* For its discussion see *SPN*, Ch. XIV.

If we want a peak near $z = 1$, we place a pair of poles on the j-axis near $s = 0$. This gives

$$f(s) = \frac{1}{s^2 + a^2} \rightarrow \frac{1}{a^2 - \left(\dfrac{1 - z}{1 + z}\right)^2} = f(z); \qquad a \ll 1$$

Determine and plot the resulting function $f(\phi)$ in the Fourier series 35 for $a = 0.1$ and $a = 0.01$ and find the Fourier coefficient formula.

From this result find a Fourier series representing a correspondingly close approximation to a square wave. Plot the result for $a = 0.1$ and $a = 0.01$.

3. A triangular wave of unit amplitude is approximated by the partial sum of its Fourier series. Determine the form of the resultant error function, the points of maximum deviation, and corresponding values of the error for harmonics through: (*a*) the fifth; (*b*) the fifteenth.

If the maximum error should not exceed 1%, how many harmonics must be considered in the partial sum?

4. Construct a Dirichlet window function like the one shown in Fig. 1 except that the main spikes are alternately positive and negative, so that its integral yields an approximation to the square wave. Is this approximation the same as that given by the partial sum of a Fourier series for this wave function?

5. The partial sum of a Fourier series for a square wave exhibits a ripply approximation. Find an expression for the exact locations of all maxima and minima.

6. A finite trigonometric polynomial involving only odd harmonics is to be found which approximates the square wave in such a way that for one term (the fundamental alone), the flat portion of a half-period is divided into three equal parts; for two terms (the fundamental and third harmonic) it is divided into five equal parts; for three terms (a fundamental, a third, and a fifth harmonic) the flat portion is divided into seven equal parts; etc. Determine a formula for the coefficients in terms of the order n of the highest harmonic.

7. Consider the trapezoidal truncation function in Fig. 5 yielding the scanning function $h(t)$ in Fig. 6. Estimate the ripple amplitudes it causes in scanning a square wave as contrasted with the ripples caused by a Dirichlet window involving the same number of harmonics in the partial sum. Compare also the corresponding rise intervals and comment upon the practicality of the trapezoidal versus the rectangular truncation function.

8. Essentially, a Dirichlet window function has the form

$$D_n(x) = \frac{\sin nx}{\sin x}$$

Plot this function for several values of n and note that it repeats in an even way for n odd and in an odd way for n even. Its alternate form is given by the expression

$$D_n(x) = a_0 + a_1 \cos x + a_2 \cos 2x + \cdots + a_n \cos nx$$

and the problem is to determine the coefficients $a_0, a_1, \ldots, a_n$. A simple way to do this is to make use of some well-known trigonometric identities and write

$$\begin{aligned}\sin nx &= (a_0 + a_1 \cos x + a_2 \cos 2x + \cdots + a_n \cos nx) \sin x \\ &= \left(a_0 - \frac{a_2}{2}\right) \sin x + \frac{a_1 - a_3}{2} \sin 2x + \frac{a_2 - a_4}{2} \sin 3x + \cdots\end{aligned}$$

By equating coefficients of like sine terms on the two sides of this equation, one can determine the a_k's sequentially. Do this and then by inspection normalize the window so that the net enclosed area is unity per period for n odd and per half-period for n even.

CHAPTER XV

Sampled and Band-Limited Functions

1. The Concept of a Sampled Function

A so-called sampled function is one whose values are known only at discrete values of the independent variable; and a *uniformly* sampled function is one for which the discrete values of the independent variable occur at equal intervals. If a time function $f(t)$ is sampled at uniformly spaced t-values with the interval t_0, then t is replaced by the discrete variable kt_0 where k assumes only integer values.

A function characterized in this manner is known to us from our study of Fourier series in Ch. XI; namely, the Fourier coefficient α_ν, alternately designated as the spectrum function $\alpha(\nu\omega_0)$, as a function of the discrete frequency variable $\nu\omega_0$, is an example of a uniformly sampled function, the samples being represented by lines in the pertinent spectrum like the one illustrated in Fig. 3 of Ch. XI for the periodic time function shown there in Fig. 2.

If we are to work with functions in sampled form then we must have an analytic expression which represents them. The following reasoning leads us to an effective solution to this requirement: The spectrum function associated with the single steady sinusoid $\cos \omega_\nu t$ consists of just two lines, each of height $\frac{1}{2}$, at the points $\omega = \pm\omega_\nu$. The "continuous" spectrum function $F(\omega)$ for this time function $f(t)$, according to Eqs. 140 and 141 of Ch. XII, is given by

$$F(\omega) = \pi[u_0(\omega - \omega_\nu) + u_0(\omega + \omega_\nu)] \tag{1}$$

Hence we may conclude that a single line of unit height at a point ω_k in a line spectrum has the representation

$$2\pi u_0(\omega - \omega_k) \tag{2}$$

That is to say, a sampled function is equivalent to a sum of impulses.

This conclusion is supported by the relationship expressed by Eq. 6 of Ch. XIII where a given time function $f(t)$ is represented in terms of itself by the convolution integral

$$f(t) = \int_{-\infty}^{\infty} u_0(t - \xi) f(\xi)\, d\xi \tag{3}$$

The sketch in Fig. 1 yields a simple graphical interpretation for this integral. The dotted rectangle of height $f(\xi)$ and width $d\xi$ occurring at $t = \xi$ is a typical element of the function $f(t)$ which is obviously expressible as a sum of all such elements. Since each element may be regarded as an

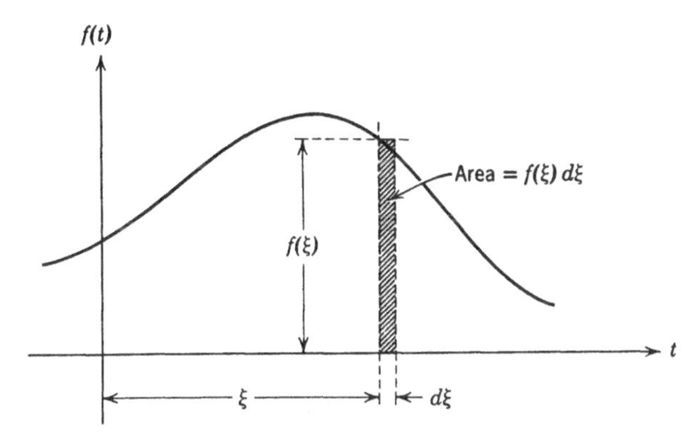

FIG. 1. A given function is represented in terms of itself by the convolution integral 3.

impulse having a value equal to the pertinent elementary area, the integrand in Eq. 3 is seen to be the typical element occurring at $t = \xi$; the integration sums all of these to form $f(t)$.

We may regard the elementary areas $f(\xi)\, d\xi$ as samples of the function $f(t)$ and the integral 3 as a continuously sampled representation of this function. Discrete sampling differs from continuous sampling only in that the sampling interval has a finite nonzero value t_0 taking the place of a differential increment $d\xi$. The total function is given by a sum of impulses whose values are the samples of $f(t)$ expressed as elementary areas of width t_0. The continuous variable ξ is replaced by the discrete variable kt_0 in which k is an integer. This sampled version of $f(t)$, therefore, has the representation

$$f_s(t) = \sum_{k=-\infty}^{\infty} \beta_k u_0(t - kt_0) \tag{4}$$

in which coefficients

$$\beta_k = t_0 f(kt_0) \tag{5}$$

are the samples. The sum 4 evidently reverts to the integral 3 if $t_0 \to d\xi$ and consistently kt_0 becomes the continuous variable ξ, whereupon $f_s(t)$ is identified with $f(t)$.

The sum 4 is spoken of as an *impulse train* and $f_s(t)$ is alternately referred to as an *impulse-train representation* of the function $f(t)$. Such a representation is admittedly a pretty ragged approximation of the function $f(t)$, but it is surprising how rapidly it becomes smooth through repeated integration.

This fact is illustrated in Fig. 2 where the bottom sketch shows the impulse-train representation $f_s(t)$ of some function $f(t)$ given by the dotted

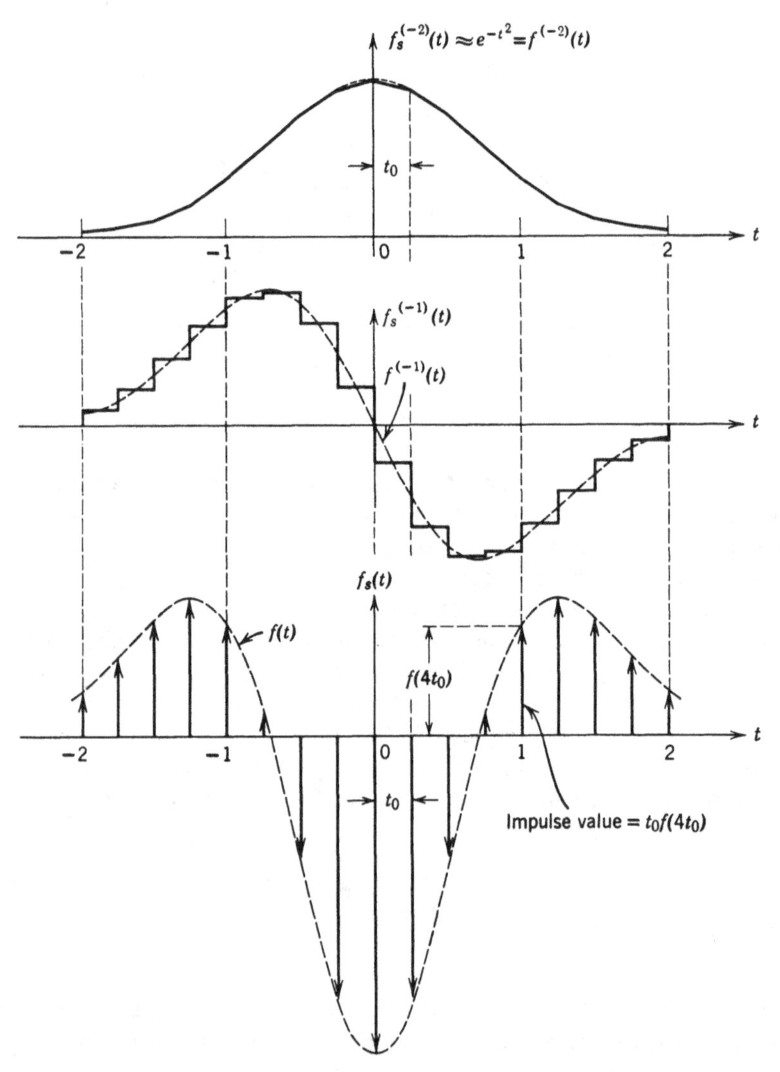

FIG. 2. Showing how integration rapidly smooths an impulse train for some function.

curve. In the center are shown the first integrals of $f_s(t)$ and of $f(t)$, again the dotted curve; and at the top of the figure are the second integrals. The first integral of the impulse train is a step function; the second integral is a broken-line or piecewise-linear approximation to the correspondingly twice-integrated smooth function $f(t)$. Here there is almost no noticeable difference between the continuous function and its approximation, notwithstanding the fact that a rather large sampling interval t_0 is involved. Further integration will for all practical purposes wipe out any remaining discrepancy between the smooth function and its sampled version. This remarkable result is useful in several practical applications to be discussed later on.

2. The Uniform Impulse Train and Its Spectrum

If in the sum 4, all coefficients β_k are equal, the sampled function $f_s(t)$ is referred to as an *infinite uniform impulse train*. It is evidently a periodic function with the period t_0. We may regard it as a limiting form of the function $f(t)$ in Fig. 2 of Ch. XI consisting of a succession of rectangular pulses of height E and duration δ if we consider letting $E\delta$ equal unity while δ becomes smaller and smaller. The corresponding line spectrum given there by the function $\alpha(\nu\omega_0)$ in Fig. 3 is seen in the limit to yield lines of equal height $1/\tau$ (which in our present considerations is $1/t_0$) with uniform spacing ω_0. Making use of the representation 2 for a typical line of unit height, and noting that $2\pi/t_0 = \omega_0$, we recognize that an infinite uniform impulse train in the time domain given by

$$f_u(t) = \sum_{k=-\infty}^{\infty} u_0(t - kt_0) \tag{6}$$

has the spectrum function

$$F_u(\omega) = \omega_0 \sum_{k=-\infty}^{\infty} u_0(\omega - k\omega_0) \tag{7}$$

Thus, the infinite uniform impulse train is its own transform, apart from a constant multiplier.

It is interesting to check this result independently. First we determine the Fourier series for the periodic function $f_u(t)$ given by Eq. 6. The Fourier coefficient formula yields

$$\alpha_\nu = \frac{1}{t_0}\int_{-t_0/2}^{t_0/2} f_u(t)\, e^{-j\nu\omega_0 t}\, dt = \frac{1}{t_0} \tag{8}$$

since only the unit impulse at $t = 0$ contributes to this integral. The Fourier series equivalent for the periodic function 6 therefore reads

$$f_u(t) = \frac{1}{t_0} \sum_{\nu=-\infty}^{\infty} e^{j\nu\omega_0 t} \tag{9}$$

Using this result to compute the pertinent spectrum function we have

$$F_u(\omega) = \int_{-\infty}^{\infty} \frac{1}{t_0} \sum e^{j\nu\omega_0 t} e^{-j\omega t}\, dt \tag{10}$$

or, interchanging the order of summation and integration,

$$F_u(\omega) = \frac{1}{t_0} \sum_{\nu=-\infty}^{\infty} \int_{-\infty}^{\infty} e^{-j(\omega-\nu\omega_0)t} dt \tag{11}$$

Replacing the infinite limits in this integral by some finite value T, we find

$$\int_{-T}^{T} e^{-j(\omega-\nu\omega_0)t}\, dt = 2\pi \cdot \frac{T}{\pi} \cdot \frac{\sin(\omega - \nu\omega_0)T}{(\omega - \nu\omega_0)T} \tag{12}$$

which we regard as a function of $(\omega - \nu\omega_0)$ with T as a parameter. Apart from the factor 2π, its value at the origin is T/π, its zeros occur at integer multiples of π/T, and the net enclosed area is unity, independent of T. Evidently the unit impulse is approached as T becomes larger and larger. Since $2\pi/t_0 = \omega_0$, we thus see that 11 yields the result

$$F_u(\omega) = \omega_0 \sum_{\nu=-\infty}^{\infty} u_0(\omega - \nu\omega_0) \tag{13}$$

in agreement with 7.

3. The Spectrum of a Sampled Time Function or Vice Versa

Now let us return to Eq. 4 representing the sampled version of some arbitrary time function $f(t)$ and determine the associated continuous spectrum. Substitution into the Fourier analysis integral gives

$$\begin{aligned} \tilde{F}(\omega) &= \int_{-\infty}^{\infty} \sum \beta_\nu u_0(t - \nu t_0) e^{-j\omega t}\, dt \\ &= \sum_{-\infty}^{\infty} \beta_\nu \int_{-\infty}^{\infty} u_0(t - \nu t_0) e^{-j\omega t}\, dt = \sum_{\nu=-\infty}^{\infty} \beta_\nu e^{-j\nu\omega t_0} \end{aligned} \tag{14}$$

This result has the form of a complex Fourier series in the variable ω. The spectrum function corresponding to a uniformly sampled time function, therefore, is periodic with the period

$$\Omega = \frac{2\pi}{t_0} \tag{15}$$

On the other hand, for a sampled spectrum function given by

$$F_s(\omega) = \sum_{\nu=-\infty}^{\infty} \alpha_\nu u_0(\omega - \nu\omega_0) \tag{16}$$

with samples

$$\alpha_\nu = \omega_0 F(\nu\omega_0) \tag{17}$$

the Fourier synthesis integral yields a periodic function

$$\begin{aligned}\tilde{f}(t) &= \frac{1}{2\pi}\int_{-\infty}^{\infty} \sum \alpha_\nu u_0(\omega - \nu\omega_0)e^{j\omega t}\,d\omega \\ &= \frac{1}{2\pi}\sum_{-\infty}^{\infty} \alpha_\nu \int_{-\infty}^{\infty} u_0(\omega - \nu\omega_0)e^{j\omega t}\,d\omega = \frac{1}{2\pi}\sum_{\nu=-\infty}^{\infty} \alpha_\nu e^{j\nu\omega_0 t}\end{aligned} \tag{18}$$

with the period

$$\tau = \frac{2\pi}{\omega_0} \tag{19}$$

In connection with these results it is important to observe that ω_0 is here not equal to $2\pi/t_0$ as it is in the pair of relations 6 and 7 since the sampling interval used to sample an arbitrary function in the time domain is in general completely independent of the sampling interval used in the frequency domain. In 6 and 7, the sampling intervals t_0 and ω_0 are at the same time the periods of the respective functions, but this is a very special situation. Whereas the period in one domain is always determined by the uniform sampling interval in the other, sampling in one domain does not even imply sampling in the other, and hence there is no reason why the sampling intervals in these domains should be related in any way.

Although the results given by Eqs. 14 and 18 are moderately interesting, they fail to reveal the detailed character of the pertinent periodic functions since the Fourier series by themselves tell us little else beyond the fact that these functions are periodic. The following interpretations, by contrast, yield a more explicit result.

We begin by pointing out that uniform sampling of an arbitrary time function $f(t)$ may be regarded as a process of amplitude modulation applied to the uniform impulse train. That is to say, if we amplitude-modulate the uniform impulse train 6 with an arbitrary time function $f(t)$ we obtain the sampled version of this time function expressed in a form that is essentially the same as, and entirely equivalent to, that given by Eqs. 4 and 5. This form reads

$$f_s(t) = t_0 f(t)\cdot f_u(t) = t_0 f(t)\cdot \sum_{k=-\infty}^{\infty} u_0(t - kt_0) \tag{20}$$

In art. 8, Ch. XIII we learned that multiplication in the time domain corresponds to convolution in the frequency domain, times a factor $1/2\pi$ as given there by Eq. 43. Hence, if $F(\omega)$ denotes the transform of $f(t)$, then the transform of $f_s(t)$ in Eq. 20 is $t_0/2\pi$ times the convolution of $F(\omega)$ with the uniform impulse train given by Eq. 7 which must here be rewritten in the form

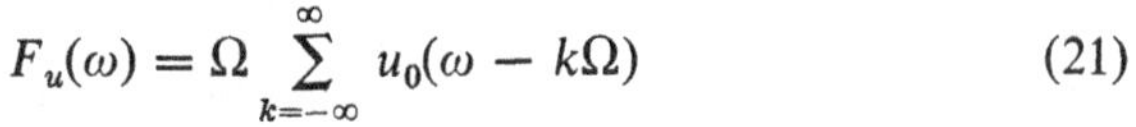

$$F_u(\omega) = \Omega \sum_{k=-\infty}^{\infty} u_0(\omega - k\Omega) \tag{21}$$

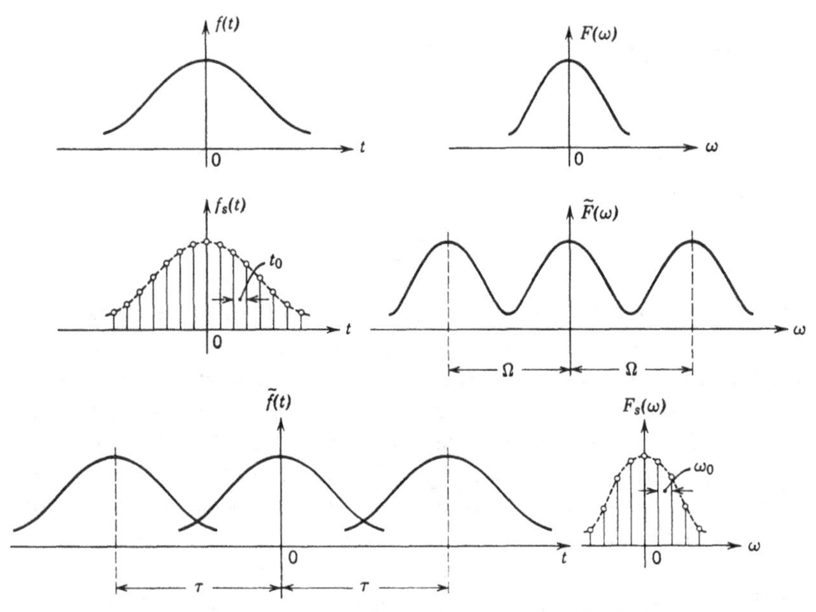

FIG. 3. Sampling in the time domain yields periodic repetition in the frequency domain, and periodic repietition in the time domain yields sampling in the frequency domain.

in view of remarks made in the paragraph following Eq. 19. Since convolution of $F(\omega)$ with the impulse $u_0(\omega - k\Omega)$ yields $F(\omega - k\Omega)$, we thus see that the spectrum function corresponding to $f_s(t)$ in Eq. 20 is given by

$$\tilde{F}(\omega) = \sum_{k=-\infty}^{\infty} F(\omega - k\Omega) \tag{22}$$

and that this expression is completely equivalent to the Fourier series in Eq. 14.

Both results 14 and 22 are useful, depending upon the particular application to be made of them. An interesting feature of 14 is the fact that the samples β_k of the time function $f(t)$ are at the same time the Fourier coefficients in the resulting periodic spectrum function. The relation 22, on the other hand, expresses this spectrum function as a periodic repetition of the spectrum $F(\omega)$ corresponding to the continuous time function $f(t)$. This result enables us to visualize the desired periodic spectrum function easily if we know the spectrum of $f(t)$.

The sketches in Fig. 3 are relevant to this interpretation. At the top are shown a pair of time- and frequency-domain functions $f(t)$ and $F(\omega)$. Directly below the sketch for $f(t)$ is drawn its sampled version with a sampling interval t_0, and to the right is the resulting periodic spectrum function in the form of a periodically repeated version of the spectrum $F(\omega)$. The latter is assumed essentially zero outside a frequency band of width Ω so that the periodically repeated components of $F(\omega)$ have little if any overlapping portions. This condition may, of course, not always be fulfilled, but even if it is not, the form of the resultant periodic spectrum is still fairly readily visualizable.

In the bottom sketches in Fig. 3 is shown the complementary situation in which the frequency function $F(\omega)$ is replaced by its uniformly sampled version $F_s(\omega)$, whereupon the associated time function becomes a periodic repetition of $f(t)$. To formalize this result analytically, we point out first that $F_s(\omega)$ in Eq. 16 may alternately be represented in the form of the uniform impulse train 7, amplitude-modulated by the continuous spectrum function $F(\omega)$, thus:

$$F_s(\omega) = F(\omega) \cdot F_u(\omega) = F(\omega) \cdot \omega_0 \sum_{k=-\infty}^{\infty} u_0(\omega - k\omega_0) \tag{23}$$

The corresponding operation in the time domain is the convolution of $f(t)$ with the uniform impulse train 6 rewritten in the form

$$f_u(t) = \sum_{k=-\infty}^{\infty} u_0(t - k\tau) \tag{24}$$

This gives

$$\tilde{f}(t) = \sum_{k=-\infty}^{\infty} f(t - k\tau) \tag{25}$$

which is completely equivalent to the Fourier series representation for this same function given by Eq. 18.

An interesting situation results if we now sample this periodic time function so as to get

$$\tilde{f}_s(t) = t_0 \tilde{f}(t) \sum_{k=-\infty}^{\infty} u_0(t - kt_0) \tag{26}$$

The transform of this function is $t_0/2\pi$ times the convolution of $F_s(\omega)$ in 23 with $F_u(\omega)$ in Eq. 21, which yields

$$\tilde{F}_s(\omega) = \sum_{\nu=-\infty}^{\infty} F_s(\omega - \nu\Omega) = \sum_{\nu=-\infty}^{\infty} \omega_0 F(\omega - \nu\Omega) \sum_{k=-\infty}^{\infty} u_0(\omega - \nu\Omega - k\omega_0) \tag{27}$$

where we have used different summation indexes in the two sums to emphasize the fact that these are independent.

Now if we choose Ω to be an integer multiple of ω_0 so that (with 15 and 19) we have

$$\frac{\Omega}{\omega_0} = \frac{\tau}{\tau_0} = n = \text{an integer} \tag{28}$$

then

$$\sum_{k=-\infty}^{\infty} u_0(\omega - \nu\Omega - k\omega_0) = \sum_{k=-\infty}^{\infty} u_0(\omega - k\omega_0) \tag{29}$$

and so we get

$$\tilde{F}_s(\omega) = \omega_0 \sum_{\nu=-\infty}^{\infty} F(\omega - \nu\Omega) \sum_{k=-\infty}^{\infty} u_0(\omega - k\omega_0) \tag{30}$$

Suppose we now make use of the periodic function, Eq. 22,

$$\tilde{F}(\omega) = \sum_{\nu=-\infty}^{\infty} F(\omega - \nu\Omega) \tag{31}$$

Then the transform of the sampled periodic time function 26 becomes the sampled periodic frequency function given by

$$\tilde{F}_s(\omega) = \omega_0 \tilde{F}(\omega) \sum_{k=-\infty}^{\infty} u_0(\omega - k\omega_0) \tag{32}$$

It should be noted carefully that whereas $f(t)$ and $F(\omega)$ are a transform pair, the corresponding periodic functions $\tilde{f}(t)$ and $\tilde{F}(\omega)$ defined by Eqs. 25 and 31 are not. However, their respective sampled versions given by Eqs. 26 and 32 are a transform pair. That is to say, with continuous functions, the time- and frequency-domain functions cannot both be periodic; on the other hand, with uniformly sampled functions this situation is possible, but only if the ratio of period to sampling interval is an integer.

4. Relation to Fortescue's Symmetrical Coordinates

Samples of the periodic time function 26 may be related explicitly to samples of the periodic spectrum function 32 in a manner that is interesting and useful both from a theoretical and practical viewpoint. To this end we must evaluate the direct transform of 26 and the inverse transform of 32. These evaluations are aided by the following thoughts.

Strictly speaking the Fourier analysis and synthesis integrals (Eqs. 21 and 22, Ch. XII) are not applicable to periodic functions since they are not absolutely integrable over the infinite domain. What we should do, of course, is use the Fourier series representation (Eqs. 24 and 36, Ch. XI) which is appropriate to periodic functions, but in situations like the present one, it is more convenient to apply the Fourier integrals in a procedure adapted to handle periodic functions.

Let $f_p(t)$ be a periodic time function with the period τ, and let $\hat{f}(t)$ be identical with $f_p(t)$ over one period and zero everywhere else. Let this one period include the point $t = 0$. Then $f_p(t)$ may be regarded as the convolution of $\hat{f}(t)$ with the uniform impulse train

$$\sum_{\nu=-\infty}^{\infty} u_0(t - \nu\tau) \tag{33}$$

having the transform

$$\omega_0 \sum_{\nu=-\infty}^{\infty} u_0(\omega - \nu\omega_0) \tag{34}$$

with

$$\omega_0 = \frac{2\pi}{\tau} \tag{35}$$

If $\hat{F}(\omega)$ is the transform of $\hat{f}(t)$ and $F_p(\omega)$ that of $f_p(t)$, we then have

$$F_p(\omega) = \hat{F}(\omega) \cdot \omega_0 \sum_{\nu=-\infty}^{\infty} u_0(\omega - \nu\omega_0) \tag{36}$$

Since $\hat{f}(t)$ is nonzero only over a finite interval, the evaluation of its transform is straightforward. The desired result is then obtained from Eq. 36 without difficulty.

Analogously we can find the inverse transform of a periodic spectrum function $F_p(\omega)$ with the period Ω. Let $\hat{F}(\omega)$ coincide with $F_p(\omega)$ over one period including $\omega = 0$ and be identically zero otherwise. Then $F_p(\omega)$ is the result of convolving $\hat{F}(\omega)$ with the uniform impulse train

$$\sum_{\nu=-\infty}^{\infty} u_0(\omega - \nu\Omega) \tag{37}$$

having the inverse transform

$$\frac{1}{\Omega} \sum_{\nu=-\infty}^{\infty} u_0(t - \nu t_0) \tag{38}$$

with

$$t_0 = \frac{2\pi}{\Omega} \tag{39}$$

If $\hat{f}(t)$ is the inverse transform of $\hat{F}(\omega)$ and $f_p(t)$ that of $F_p(\omega)$, we have

$$f_p(t) = \hat{f}(t) \cdot t_0 \sum_{\nu=-\infty}^{\infty} u_0(t - \nu t_0) \tag{40}$$

since convolution in the frequency domain corresponds to 2π times multiplication in the time domain.

Let us apply these procedures to the functions 26 and 32. Beginning with the former, we have

$$\hat{F}(\omega) = \int_a^{a+\tau} \tilde{f}_s(t)\, e^{-j\omega t}\, dt = \int_a^{a+\tau} t_0 \tilde{f}(t) \sum_{k=-\infty}^{\infty} u_0(t - kt_0)\, e^{-j\omega t}\, dt \quad (41)$$

Since the integration extends over a single period, the infinite sum is abridged to extend over n impulses for which we can let the index k assume any n consecutive integers. Interchanging the order of summation and integration, and noting that the integrand is zero except at points $t = kt_0$, we get

$$\hat{F}(\omega) = \sum_{k=1}^{n} t_0 \tilde{f}(kt_0)\, e^{-j\omega kt_0} \int_a^{a+\tau} u_0(t - kt_0)\, dt \quad (42)$$

The period over which the integration extends includes one of the k-values in the range 1 to n, and so the integral equals unity. Equation 36 then yields the desired transform of function 26 in the form

$$\tilde{F}_s(\omega) = \frac{2\pi}{n} \sum_{\nu=-\infty}^{\infty} \left[\sum_{k=1}^{n} \tilde{f}(kt_0)\, e^{-j2\pi k\nu/n} \right] u_0(\omega - \nu\omega_0) \quad (43)$$

where we have noted first of all from Eqs. 28 and 35 that $\omega_0 t_0 = 2\pi/n$, and secondly that the summation in 36 limits ω-values in 42 to $\nu\omega_0$.

Comparison with the representation of this same function in Eq. 32 yields the interesting relation

$$\omega_0 \tilde{F}(\nu\omega_0) = \frac{2\pi}{\tau} \sum_{k=1}^{n} t_0 \tilde{f}(kt_0)\, e^{-j2\pi k\nu/n} \quad (44)$$

expressing samples of the periodic frequency function explicitly in terms of samples of the periodic time function.

Going in the opposite direction we find the inverse transform of function 32 by first forming the inverse transform of one of its periods, thus:

$$(t) = \frac{1}{2\pi} \int_a^{a+\Omega} \tilde{F}_s(\omega)\, e^{j\omega t}\, d\omega = \frac{1}{2\pi} \int_a^{a+\Omega} \omega_0\, \tilde{F}(\omega) \sum_{k=-\infty}^{\infty} u_0(\omega - k\omega_0) e^{j\omega t}\, d\omega \quad (45)$$

Interchanging the order of integration and summation, noting that the latter is abridged to n impulses as in the step from Eq. 41 to Eq. 42, we find

$$\hat{f}(t) = \frac{\omega_0}{2\pi} \sum_{k=1}^{n} \tilde{F}(k\omega_0)\, e^{jk\omega_0 t} \int_a^{a+\Omega} u_0(\omega - k\omega_0)\, d\omega \quad (46)$$

The integral again equals unity, and use of Eq. 40 gives the desired result, namely

$$\tilde{f}_s(t) = \frac{1}{n} \sum_{\nu=-\infty}^{\infty} \left[\sum_{k=1}^{n} \tilde{F}(k\omega_0)\, e^{j2\pi k\nu/n} \right] u_0(t - \nu t_0) \quad (47)$$

Comparison with representation 26 for this same function yields the inverse of relation 44, thus:

$$t_0\tilde{f}(\nu t_0) = \frac{1}{\Omega}\sum_{k=1}^{n} \omega_0 \tilde{F}(k\omega_0)\, e^{j2\pi k\nu/n} \tag{48}$$

from which samples of the periodic time function may be calculated directly from samples of the periodic frequency function.

These same results may be obtained in a somewhat different manner that is collaterally interesting. In terms of the samples (Eqs. 5 and 17)

$$\alpha_\nu = \omega_0 \tilde{F}(\nu\omega_0) \quad \text{and} \quad \beta_\nu = t_0\tilde{f}(\nu t_0) \tag{49}$$

Eqs. 4 and 18 yield for the function 26

$$\tilde{f}_s(t) = \sum_{\nu=-\infty}^{\infty} \beta_\nu u_0(t - \nu t_0) = \frac{1}{2\pi}\sum_{\nu=-\infty}^{\infty} \alpha_\nu e^{j\nu\omega_0 t} \tag{50}$$

while 14 and 16 represent function 32 in the form

$$\tilde{F}_s(\omega) = \sum_{\nu=-\infty}^{\infty} \alpha_\nu u_0(\omega - \nu\omega_0) = \sum_{\nu=-\infty}^{\infty} \beta_\nu e^{-j\nu\omega t_0} \tag{51}$$

Suppose we limit the infinite impulse trains in these relations to single periods centered at $t = 0$ and $\omega = 0$. Such truncation of the impulse train in 50 amounts to multiplying it by a rectangular cutoff function of height unity and width τ. In the frequency domain this corresponds to convolution with the function

$$\frac{\tau}{2\pi}\cdot\frac{\sin \omega\tau/2}{\omega\tau/2} \tag{52}$$

giving the spectrum function

$$\sum_{\nu=-\infty}^{\infty} \frac{\tau\alpha_\nu}{2\pi}\cdot\frac{\sin(\omega - \nu\omega_0)\tau/2}{(\omega - \nu\omega_0)\tau/2} = \sum_{\nu=1}^{n} \beta_\nu e^{-j\omega t_0} \tag{53}$$

where it is recognized that this truncation restricts the summation in the second term of 51 to n consecutive ν-values which can begin and end wherever we please since this sum is periodic.

Similarly, truncation of the impulse train in 51 amounts to multiplying it by a rectangular cutoff function of height unity and width Ω. In the time domain this corresponds to convolution with the function

$$\frac{\Omega}{2\pi}\cdot\frac{\sin \Omega t/2}{\Omega t/2} \tag{54}$$

giving the time function

$$\sum_{\nu=-\infty}^{\infty} \frac{\Omega\beta_\nu}{2\pi}\cdot\frac{\sin(t - \nu t_0)\Omega/2}{(t - \nu t_0)\Omega/2} = \frac{1}{2\pi}\sum_{\nu=1}^{n} \alpha_\nu e^{j\nu\omega_0 t} \tag{55}$$

where the truncation now limits summation in the second term of Eq. 50 to n consecutive ν-values.

It is now significant to note that zeros of the $\sin x/x$ function 52 are spaced at intervals $\Delta\omega = 2\pi/\tau = \omega_0$, which is the spacing of samples in the spectrum function. The first sum in 53 is a train of $\sin x/x$ functions with uniform spacing ω_0. For $\omega = k\omega_0$ a typical term involves the $\sin x/x$ function

$$\frac{\sin (k - \nu)\pi}{(k - \nu)\pi} \tag{56}$$

which equals unity for $\nu = k$ and zero for all other ν-values. Although either term in 53 is a continuous function of ω (the effect of truncating the time function renders the spectrum function continuous) it is a function that has the same samples as function 51 at its sampling points, except for the constant factor $\tau/2\pi$.

By the same reasoning we see that the continuous time function 55 has the same samples as function 50 at all of its sampling points, but for the constant factor $\Omega/2\pi$.

Evaluating 53 at points $\omega = k\omega_0$ and 55 at points $t = kt_0$, therefore, yields the results

$$\alpha_k = \frac{2\pi}{\tau} \sum_{\nu=1}^{n} \beta_\nu e^{-j2\pi k\nu/n} \tag{57}$$

and

$$\beta_k = \frac{1}{\Omega} \sum_{\nu=1}^{n} \alpha_\nu e^{j2\pi k\nu/n} \tag{58}$$

respectively, which agree with 44 and 48 as they should.

We can now draw some interesting conclusions from these results. Suppose we consider a pair of functions $f(t)$ and $F(\omega)$ like the ones in the top sketches of Fig. 3. Both are theoretically infinite in extent, but for many practical purposes we can neglect the portions outside of intervals having the widths τ and Ω respectively. If we sample both functions we get a transform pair having the characteristics of functions 26 and 32 which are approximately (and for most practical purposes adequately) represented by their samples throughout any one period. The algebraic relations 57 and 58 are then adequate to transform from the time- to the frequency-domain and back again; and the relationship 51 tells us the incidentally interesting fact that the spectrum function (given in sampled form by the first sum) may be represented as a sum of component spectrum functions (the terms in the second sum) each of which has a constant amplitude and a linear phase. The constant amplitudes of these components (the coefficients β_ν) are simply samples of the associated time function. Equation 58 gives all of these in terms of samples of the pertinent spectrum function. Fourier transformation involving sampled functions is thus rather compactly summarized in convenient algebraic form.

Another more generally useful interpretation of these results, known to the electrical engineering profession as the *theory of symmetrical coordinates*,* may appropriately be pointed out at this time. The basic thought here is that a set of n arbitrary complex numbers may uniquely and reversibly be expressed in terms of n symmetrical sets, each of which is composed of n numbers of equal magnitude and equal angular displacement relative to one another. In practice such symmetrical sets are encountered in balanced polyphase systems. Usefulness of this interpretation lies in the fact that such a system under unbalanced operating conditions may be analyzed by superimposing appropriate balanced (or symmetrical) behavior patterns for which the separate solutions are relatively simple to formulate.

A complex number or Gaussian vector has two degrees of freedom; it can change in length and in direction. A given set of n such vectors possesses $2n$ degrees of freedom. A symmetrical set involving n vectors of the same length with equi-angular spacing between adjacent vectors possesses only 2 degrees of freedom since all amplitudes must change simultaneously and all vectors must rotate together. Thus n symmetrical sets also possess $2n$ degrees of freedom, and hence it should be possible by means of an appropriate linear transformation to transform a set of n arbitrary vectors into n symmetrical sets. In order that this transformation be unique and reversible, it is necessary that the number of vectors in each symmetrical set be equal to the original number of given vectors.

If in the above analysis we regard the quantities

$$a_k = \tilde{F}(k\omega_0) \qquad \text{for } k = 1, 2, \ldots n \tag{59}$$

as a given set of arbitrary Gaussian vectors, and

$$b_\nu \cdot e^{-j2\pi k\nu/n} \qquad \text{for } k = 1, 2, \ldots n \tag{60}$$

as defining the symmetrical set of νth order, then 57 and 58 yield the desired transformation relations, namely

$$a_k = \sum_{\nu=1}^{n} b_\nu e^{-j2\pi k\nu/n} \tag{61}$$

and

$$b_\nu = \frac{1}{n} \sum_{\mu=1}^{n} a_\mu e^{j2\pi\nu\mu/n} \tag{62}$$

* This was first presented in a paper by C. L. Fortescue entitled "Methods of Symmetrical Co-ordinates Applied to the Solution of Polyphase Networks," *Trans. AIEE*, Vol. 37, 1918, Pt. 2, pp. 1027–1140.

The coefficient b_ν represents the lead vector in the νth symmetrical set; its angle determines the orientation of the set, all other members of the set being angularly displaced from the lead vector by uniform increments as is indicated in the set representation 60.

Equations 61 and 62 are inverse sets of linear simultaneous algebraic equations, as is evident from the fact that

$$\frac{1}{n}\sum_{\nu=1}^{n} e^{j2\pi\nu(\mu-k)/n} = \begin{cases} 1 & \text{for } \mu = k \\ 0 & \text{for } \mu \neq k \end{cases} \tag{63}$$

showing that these sets have inverse matrices.

5. Transition from Fourier Series to Fourier Integral Reconsidered

In terms of the concept of a sampled function it is interesting to reconsider the Fourier series representation for a periodic time function and the transition to the Fourier integral representation for an aperiodic function. This transition, as given in the discussion of art. 2, Ch. XII is motivated by the thought that we can regard an aperiodic time function as a periodic one for which the period is arbitrarily large. We can just as directly use a Fourier series for the representation of an aperiodic time function in sampled form, and obtain the desired representation for a continuously defined aperiodic time function by considering an arbitrarily small sampling interval.

Our starting point is the familiar Fourier series and coefficient formula (Eqs. 16 and 17, Ch. XII) which we repeat here:

$$f(t) = \sum_{\nu=-\infty}^{\infty} \alpha_\nu e^{j\nu\omega_0 t} \tag{64}$$

$$\alpha_\nu = \frac{1}{\tau}\int_{-\tau/2}^{\tau/2} f(t)e^{-j\nu\omega_0 t}\,dt \tag{65}$$

in which $\tau = 2\pi/\omega_0$ is the period of $f(t)$ and ω_0 is the fundamental radian frequency in its harmonic series representation. We now consider a change in notation by introducing the discrete frequency variable

$$\omega = \nu\omega_0 \qquad \text{for } \nu = 0, \pm 1, \pm 2, \ldots \tag{66}$$

and the corresponding frequency function

$$F(\omega) = \frac{\alpha_\nu}{\omega_0} \tag{67}$$

Then 64 and 65 assume the forms

$$f(t) = \sum_{\omega=-\infty}^{\infty} F(\omega)e^{j\omega t}\,\Delta\omega \tag{68}$$

and

$$F(\omega) = \frac{1}{2\pi} \int_{-\tau/2}^{\tau/2} f(t) e^{-j\omega t} \, dt \tag{69}$$

where

$$\Delta\omega = \omega_0 \tag{70}$$

is now regarded as a uniform sampling interval. $F(\omega)$ is an aperiodic sampled frequency function; the original Fourier coefficients $\alpha_\nu = \omega_0 F(\nu\omega_0)$, Eq. 67, are its samples; and $f(t)$, as in 64, is a continuously defined but periodic time function.

By merely making some trivial changes in notation, we can interchange the roles of time and frequency functions and obtain representations for a sampled aperiodic time function and its continuously defined but periodic spectrum function. The changes in notation are indicated thus, where the arrow means "is replaced by":

$$\begin{aligned} \omega &\to t \\ t &\to -\omega \\ \omega_0 &\to t_0 \\ \tau &\to -\Omega = \frac{-2\pi}{t_0} \\ \Delta\omega &\to \quad \Delta t = t_0 \end{aligned} \tag{71}$$

The resulting function $F(t)$ we write $h(t)$ and $f(-\omega)$ is denoted by $H(\omega)$. The expressions 69 and 68 then assume the appearance

$$h(t) = \frac{1}{2\pi} \int_{-\Omega/2}^{\Omega/2} H(\omega) e^{j\omega t} \, d\omega \tag{72}$$

and

$$H(\omega) = \sum_{t=-\infty}^{\infty} h(t) e^{-j\omega t} \Delta t \tag{73}$$

respectively.

The frequency variable ω is continuous, and the time

$$t = \nu t_0 \qquad \text{for } \nu = 0, \pm 1, \pm 2, \ldots \tag{74}$$

is now a discrete variable. $h(t)$ is an aperiodic time function in sampled form; the quantities $t_0 h(\nu t_0)$ are its samples; and $H(\omega)$, the spectrum of $h(t)$, is a continuously defined, periodic frequency function.

Observe that these results stem directly from the Fourier series representation 64 and 65. All that we have done is make some changes in notation and in the physical interpretation of what certain symbols stand for. Mathematically there is not the slightest difference between the pair of relations 64 and 65 and the pair 72 and 73 (except that we have interchanged their order). The Fourier series, then, is seen to yield directly the representation for a sampled aperiodic time function.

Transition to the Fourier representation for a continuously defined aperiodic time function is now achieved by allowing the uniform sampling interval t_0 to become arbitrarily small. The period Ω of $H(\omega)$, according to 71, becomes infinite; the time increment Δt is regarded as the differential quantity dt; and the sum 73 by definition becomes replaced by an integral. These relations then read

$$h(t) = \frac{1}{2\pi}\int_{-\infty}^{\infty} H(\omega)e^{j\omega t}\,d\omega \tag{75}$$

and

$$H(\omega) = \int_{-\infty}^{\infty} h(t)e^{-j\omega t}\,dt \tag{76}$$

which are recognized as the Fourier integrals (Eqs. 21 and 22, Ch. XII).

6. Sampling and Lowpass Filtering of a Time Function; the Nyquist Sampling Theorem

When a continuously defined time function $f(t)$ is sampled, a certain amount of the information contained in it is lost, for the samples yield values of the function only at discrete points. An infinity of intermediate values are no longer represented. Intuitively one has the feeling, however, that the amount of information lost will for all practical purposes become negligible if the sampling interval is sufficiently small. The question to be answered is: How can we determine that interval?

The answer is contained in the results illustrated by the sketches of Fig. 3. The center sketch shows that sampling of the time function $f(t)$ causes its spectrum $F(\omega)$ to be repeated at intervals $\Omega = 2\pi/t_0$. If the spectrum drops off for large ω so that frequencies beyond some point $\omega = \omega_c$ contribute negligibly to the interference pattern $f(t)$, and if Ω is at least as large as $2\omega_c$, then the overlapping portions in the periodically repeated version of $F(\omega)$ do not noticeably distort this spectrum throughout any one period, and we can very nearly regain the original spectrum $F(\omega)$ by excising a single period of the periodic spectrum $\tilde{F}(\omega)$. Since regaining $F(\omega)$ from $\tilde{F}(\omega)$ is equivalent to regaining $f(t)$ from its sampled version, we have here the means for telling us: (*a*) whether $f(t)$ can be regained at all with a nonzero sampling interval of reasonable size, and (*b*) how we can determine such an appropriate interval.

The answer to (*a*) depends on how fast the spectrum function drops off for large ω. The discussion in art. 9, Ch. XI is relevant here. It is pointed out there that if $f(t)$ contains discontinuities, it spectrum cannot drop off faster than $1/\omega$ for large ω; if $f(t)$ is continuous but its first derivative contains discontinuities, its spectrum drops off as fast as (but no faster than) $1/\omega^2$; and so forth. Unless $f(t)$ contains impulses, the associated spectrum must drop off at least as fast as $1/\omega$ for large ω. It is, therefore, reasonable to expect in most practical situations that the spectrum will drop off sufficiently fast to allow its truncation without undue distortion of $f(t)$.

The discussion in art. 2 of Ch. XIV provides means for studying the detailed effect upon $f(t)$ for a particular choice of truncation frequency ω_c. Namely, $f(t)$ becomes convolved with the aperiodic Dirichlet window

$$D_a(t) = \frac{\omega_c}{\pi} \cdot \frac{\sin \omega_c t}{\omega_c t} \tag{77}$$

If the aperture of this window, namely $2\pi/\omega_c$, is small enough so that the fastest wiggle in $f(t)$ is still sufficiently well reproduced (see the discussion in art. 2, Ch. XIII, especially Fig. 8), then truncation of the spectrum with this value of ω_c produces a time function $f^*(t)$ which is for all practical purposes the equivalent of $f(t)$.

Strictly speaking $f^*(t)$ is *band-limited* whereas $f(t)$ is not. However, if the difference in the time domain is negligible, then we may regard $f(t)$ to be band-limited also; its spectrum function $F(\omega)$ will essentially be contained within a bandwidth $\Omega = 2\omega_c = 2\pi/t_0$, and the appropriate sampling interval is thus determined.

Now consider the sampled time function $f_s(t)$ for this sampling interval and its periodic spectrum $\tilde{F}(\omega)$. They will have an appearance essentially like that shown in the center sketches in Fig. 3. If we truncate this spectrum with the bandwidth Ω centered at $\omega = 0$ the effect upon $f_s(t)$ is to convolve it with the Dirichlet window 77. Since $f_s(t)$ is given by the impulse train

$$f_s(t) = \sum_{k=-\infty}^{\infty} \beta_k u_0(t - kt_0) \tag{78}$$

this convolution yields the continuous time function

$$f_{sf}(t) = \frac{1}{t_0} \sum_{k=-\infty}^{\infty} \beta_k \frac{\sin \pi[(t/t_0) - k]}{\pi[(t/t_0) - k]} \tag{79}$$

which we regard as a *lowpass-filtered* version of $f_s(t)$ since truncating the spectrum of this sampled time function amounts physically to passing it through an ideal lowpass filter with cutoff frequency ω_c.

Observe that the sampled and lowpass-filtered time function $f_{sf}(t)$ is a *continuous* function of t; the filtering operation (spectrum truncation) converts the discontinuously defined function $f_s(t)$ into one that is continuously defined. However, this continuously defined function is not identical with the original time function $f(t)$ *unless the latter is truly band-limited.* In that event the spectrum $F(\omega)$ in Fig. 3 is wholly contained within an interval of width Ω centered at $\omega = 0$ (it being understood that the sampling interval $t_0 = 2\pi/\Omega$ is properly chosen); the periodic repetition of $F(\omega)$ yielding $\tilde{F}(\omega)$ has no overlapping portions, and hence no distortion in the pertinent time function can result from the lowpass filtering process. The original time function is regained perfectly; or we can say that sampling of such a time function with a properly chosen sampling interval does not represent a loss in the information contained in $f(t)$. This result in essence is known as the *Nyquist sampling theorem.*

When $f(t)$ is not strictly band-limited but its spectrum is more or less completely confined to a finite interval Ω, then it is reasonable to expect that the discrepancy between the continuous function 79 and $f(t)$ is small and for most practical purposes negligible. Herein lies the practical importance of this interesting theorem.

Apropos the question of practical importance, it is significant to notice that although function 79 may not exactly equal $f(t)$ for *all* values of t *it does exactly equal $f(t)$ at the sampling points $t = \nu t_0$* for any integer ν regardless of the choice made for t_0 or how nearly band-limited the function $f(t)$ may be. The truth of this statement may be seen from the fact that for $t = \nu t_0$ the $\sin x/x$ function in 79 reads

$$\frac{\sin \pi(\nu - k)}{\pi(\nu - k)} = \begin{cases} 1 & \text{for } \nu = k \\ 0 & \text{for } \nu \neq k \end{cases} \tag{80}$$

For a fixed ν-value, all terms in the infinite sum are zero except the one for which $k = \nu$ so that, with Eq. 5, we have

$$f_{sf}(\nu t_0) = \frac{\beta_\nu}{t_0} = f(\nu t_0) \tag{81}$$

Thus we conclude that lowpass filtering of a sampled time function with the bandwidth Ω leaves all samples unchanged. To the extent that $f(t)$ is adequately represented by its samples, nothing is lost through the lowpass filtering operation.

7. Spectrum Truncation in the Case of an Infinite Uniform Impulse Train and Results Related Thereto

Some interesting detailed results are obtained if in the sampling and lowpass filtering operation we choose for $f(t)$ a constant. Let the constant equal $1/t_0$ so that samples of $f(t)$ become unity and its corresponding sampled version is the infinite uniform impulse train

$$f_s(t) = f_u(t) = \sum_{k=-\infty}^{\infty} u_0(t - kt_0) \tag{82}$$

having the transform

$$F_u(\omega) = \omega_0 \sum_{k=-\infty}^{\infty} u_0(\omega - k\omega_0) \tag{83}$$

with $\omega_0 = 2\pi/t_0$, as given by Eqs. 6 and 7.

We now consider truncating this spectrum with the rectangular cutoff function which is unity over the range $-\omega_c < \omega < \omega_c$ and otherwise zero. In the time domain this operation corresponds to convolution with the aperiodic Dirichlet kernel given by Eq. 77, and so we get in this case the infinite succession of sin x/x functions

$$f_{uf}(t) = \frac{\omega_c}{\pi} \sum_{k=-\infty}^{\infty} \frac{\sin \omega_c(t - kt_0)}{\omega_c(t - kt_0)} \tag{84}$$

The cutoff frequency ω_c is chosen so that

$$n\omega_0 \le \omega_c < (n + 1)\omega_0 \tag{85}$$

where the equal part of the "less than or equal to" sign on the left is interpreted in such a way that the impulse at $n\omega_0$ is still wholly included; truncation of the infinite train of impulses representing $F_u(\omega)$ in Eq. 83 does not split an impulse. The transform of time function 84 is then given by the finite sum

$$F_{uf}(\omega) = \omega_0 \sum_{k=-n}^{n} u_0(\omega - k\omega_0) \tag{86}$$

Suppose we now transform back to the time domain using the normal synthesis integral. This gives

$$\frac{\omega_0}{2\pi} \int_{-\infty}^{\infty} \sum_{-n}^{n} u_0(\omega - k\omega_0) e^{j\omega t} \, d\omega = \frac{1}{t_0} \sum_{k=-n}^{n} e^{jk\omega_0 t} \tag{87}$$

which (according to Eq. 4 of Ch. XIV) we recognize as being the periodic Dirichlet kernel. We therefore have the interesting result that

$$\begin{aligned} D_n(t) &= \frac{1}{t_0} \sum_{k=-n}^{n} e^{jk\omega_0 t} = \frac{1}{t_0} [1 + 2 \cos \omega_0 t + \cdots + 2 \cos n\omega_0 t] \\ &= \frac{1}{t_0} \left\{ \frac{\sin (n + \frac{1}{2})\omega_0 t}{\sin \omega_0 t/2} \right\} = \frac{\omega_c}{\pi} \sum_{k=-\infty}^{\infty} \frac{\sin \omega_c(t - kt_0)}{\omega_c(t - kt_0)} \end{aligned} \tag{88}$$

where Eqs. 5 and 6 in Ch. XIV are also used.

This result is remarkable for several reasons: (*a*) It yields a simple closed form for the infinite succession of $\sin x/x$ functions. (*b*) An infinite variety of values for ω_c within the range defined by 85 *all yield the same resultant time function*! (*c*) If we let ω_c approach $n\omega_0$ so that the impulse at this point is wholly included, we can identify terms in the infinite sum in Eq. 84 or 88 with the aperiodic Dirichlet kernel $D_a(t)$ defined by Eq. 9, Ch. XIV. Equation 88 then tells us that the periodic Dirichlet kernel is equal to a periodic succession of aperiodic Dirichlet kernels, as expressed by

$$D_n(t) = \sum_{k=-\infty}^{\infty} D_a(t - kt_0) \tag{89}$$

In terms of this result we can prove independently the interesting fact (derived in art. 1 of the preceding chapter) that convolution of a periodic function $f(t)$ with $D_n(t)$ over a single period yields the same result [namely the partial sum $s_n(t)$] as does convolution of $f(t)$ with $D_a(t)$ over the infinite time domain. We begin with the latter and write

$$s_n(t) = \int_{-\infty}^{\infty} f(\xi)\, D_a(t - \xi)\, d\xi = \sum_{\nu=-\infty}^{\infty} \int_{\nu\tau}^{(\nu+1)\tau} f(\xi)\, D_a(t - \xi)\, d\xi \tag{90}$$

where in the second term on the right, the infinite time domain is decomposed into successive periods. Since $f(\xi)$ is periodic we can replace it by $f(\xi - \nu\tau)$, and if we then make the change of variable

$$\xi - \nu\tau = \eta; \qquad \xi = \eta + \nu\tau \tag{91}$$

and observe the effect upon limits, we have

$$s_n(t) = \sum_{\nu=-\infty}^{\infty} \int_0^{\tau} f(\eta)\, D_a(t - \eta - \nu\tau)\, d\eta = \int_0^{\tau} f(\eta) \sum_{\nu=-\infty}^{\infty} D_a(t - \eta - \nu\tau)\, d\eta \tag{92}$$

We now make use of Eq. 89 (noting that t_0 and τ are the same) and have the desired result, namely

$$s_n(t) = \int_0^{\tau} f(\eta)\, D_n(t - \eta)\, d\eta \tag{93}$$

Returning to Eq. 88, we observe that if a value for ω_c is chosen to lie in the center of the range 85 so that

$$\omega_c = (n + \tfrac{1}{2})\omega_0 \tag{94}$$

then the period of $D_n(t)$ is

$$t_0 = \frac{2\pi}{\omega_0} = \frac{\pi}{\omega_c}(2n + 1) \tag{95}$$

Since π/ω_c is the uniform spacing between zeros of any one $\sin x/x$ function in the infinite sum, and successive ones are spaced t_0 seconds apart, we see that all the zeros of these functions coincide. Moreover, the spacing between zeros of the single $\sin x/x$ function on the left of Eq. 88 is likewise π/ω_c. In this case we can visualize rather easily how the periodic succession of $\sin x/x$ functions can add up to the periodic $D_n(t)$.

When $\omega_c = n\omega_0$, leading to the result in Eq. 89, then the zero spacing π/ω_c in the $D_a(t)$ functions is again an integer submultiple of the period t_0 so that all zeros of these functions in the infinite sum coincide, but this zero spacing is not the same as that in the periodic Dirichlet kernel which the infinite sum yields. The result expressed by Eq. 88 is now not so easy to visualize graphically.

Finally if ω_c has any other value in the range 85, then the period t_0 (the spacing between individual $\sin x/x$ functions in the infinite sum) is not an integer multiple of the zero spacing π/ω_c, and overlapping portions of the $\sin x/x$ functions have noncoincident zeros. Moreover, the zero spacing for the periodic Dirichlet kernel which does not depend upon ω_c is still the same, $t_0/(2n + 1)$. The fact that the infinite sum nevertheless yields exactly the same resultant is a point worth careful thought and perhaps calls for some graphical sketching on the part of the reader. We shall not spoil his fun by doing this for him.

Following this same line of thought, if ω_c has any value between zero and ω_0, then the infinite periodic succession of $\sin x/x$ functions in the sum yields the constant value $1/t_0$; if ω_c has any value between ω_0 and $2\omega_0$, the infinite train of $\sin x/x$ functions yields the resultant $(1/t_0)(1 + 2 \cos \omega_0 t)$. We can continue in this manner, but an extremely subtle point is involved here. Compare the situation for which the value of ω_c is negligibly smaller than ω_0 with that for which it is negligibly larger than ω_0. Graphically the appearance of the infinite train of $\sin x/x$ functions is the same in both cases yet one of these yields a constant resultant while the other yields a constant plus a cosine function!

This subtlety becomes even more remarkable if we let $\omega_c = \omega_0$ in such a way as to split the impulse at this point into arbitrary fractional parts. If we split it neatly down the middle, the impulses at $\omega = \pm\omega_0$ have the value $\frac{1}{2}$ and yield $\cos \omega_0 t$ instead of $2 \cos \omega_0 t$ as they do when the impulses are wholly included. However, the infinite train of $\sin x/x$ functions, from a graphical point of view, appears to be the same regardless of how we split the impulse.

We must conclude the following from these considerations: (*a*) The problem involved here is beyond the resolving power of a simple graphical plot since at any point we must consider the superposition of an infinite number of component curves (the separate $\sin x/x$ functions). (*b*) The resultant value at any point is given by the limit of an infinite series. Evidently, when ω_c is chosen so as not to split an impulse, this limit exists and is unique. When an impulse is split, the limit evidently is not unique but depends upon the exact nature of the split. Finally (*c*) resolution of these questions by means of adequate mathematical analysis does not seem warranted here, although our mentioning them in connection with the result 88 is appropriate inasmuch as the question as to what happens when the cutoff frequency ω_c falls upon an impulse quite naturally arises, and hence some consideration must be given to it.

The cases $0 < \omega_c < \omega_0$ and $\omega_0 < \omega_c < 2\omega_0$ may further be interpreted in terms of the process of sampling and lowpass filtering discussed in the preceding article. Suppose the function in question is

$$f(t) = \cos \omega_a t \tag{96}$$

The associated spectrum function, according to Eq. 141, Ch. XII is given by

$$F(\omega) = \pi[u_0(\omega - \omega_a) + u_0(\omega + \omega_a)] \tag{97}$$

and clearly is band-limited. Equation 78 for the samples

$$\beta_k = t_0 \cos \omega_a kt_0 \tag{98}$$

is the sampled version of $f(t)$, and convolution with the Dirichlet kernel 77, equivalent to truncation in the frequency domain with cutoff frequency ω_c, yields the sampled and lowpass filtered version of 96 in the form

$$f_{sf}(t) = \frac{\omega_c t_0}{\pi} \sum_{k=-\infty}^{\infty} \cos \omega_a kt_0 \cdot \frac{\sin \omega_c(t - kt_0)}{\omega_c(t - kt_0)} \tag{99}$$

The bandwidth $\Omega = 2\omega_c$ of spectrum function 97 may be regarded as having any value larger than $2\omega_a$; the cutoff frequency ω_c must be at least as large as ω_a; the sampling interval t_0, therefore, can at most be equal to π/ω_a. Nevertheless, let us begin by considering the sampling interval $t_0 = 2\pi/\omega_0 = 2\pi/\omega_a$ for which function 99 becomes

$$f_{sf}(t) = \frac{\omega_c t_0}{\pi} \sum_{k=-\infty}^{\infty} \frac{\sin \omega_c(t - kt_0)}{\omega_c(t - kt_0)} \tag{100}$$

and see what we get by choosing different values for the truncation frequency ω_c. Thus the choice $0 < \omega_c < \omega_0$, according to 85, corresponds to $n = 0$ and hence Eq. 88 gives

$$f_{sf}(t) = 1 \tag{101}$$

whereas the choice $\omega_0 < \omega_c < 2\omega_0$, corresponding to $n = 1$, yields

$$f_{sf}(t) = 1 + 2 \cos \omega_a t \tag{102}$$

By sampling the cosine function 96 only once per period, we introduce a constant component and hence we cannot claim to have regained the original time function. On the other hand, if we choose $t_0 = 2\pi/\omega_0 = \pi/\omega_a$, corresponding to sampling twice per period, and if we interpret this value as actually less than π/ω_a by an arbitrarily small but nonzero amount so that the truncation frequency ω_c falls just beyond the impulse at $\omega = \omega_a$, then we will exactly regain the cosine function 96 and hence Eq. 99 yields the collaterally interesting formula

$$f_{sf}(t) = \sum_{k=-\infty}^{\infty} (-1)^k \frac{\sin \omega_a(t - kt_0)}{\omega_a(t - kt_0)} = \cos \omega_a t \tag{103}$$

8. Approximation by a Sum of sin x/x Functions and the Lagrangian Interpolation Formula

The Lagrangian interpolation formula enables one to write down at once the expression for a polynomial which will pass through n assigned values at n specified values of the independent variable. If this variable is denoted by x, and $x_1 \cdots x_n$ are points at which the desired polynomial $P(x)$ is to have values $A_1 \cdots A_n$ respectively, and if we define the auxiliary polynomial

$$S(x) = (x - x_1)(x - x_2) \cdots (x - x_n) \tag{104}$$

then we have

$$P(x) = \sum_{\nu=1}^{n} \frac{S(x)}{(x - x_\nu)} \cdot \left[\frac{(x - x_\nu)}{S(x)}\right]_{x=x_\nu} \cdot A_\nu \tag{105}$$

which is known as the *Lagrangian interpolation formula.*

Understanding this formula is aided by observing first that the quantities

$$k_\nu = \left[\frac{(x - x_\nu)}{S(x)}\right]_{x=x_\nu} = \left[\frac{x_k - x_\nu}{S(x_k)}\right]_{k=\nu} \tag{106}$$

are residues of the function $1/S(x)$ in its poles $x_1 \cdots x_n$. Since $S(x_k) = 0$ by definition, we then have

$$\frac{S(x_k)}{(x_k - x_\nu)} = \begin{cases} \dfrac{1}{k_\nu} & \text{for } \nu = k \\ 0 & \text{for } \nu \neq k \end{cases} \tag{107}$$

It should now be fairly evident that Eq. 105 yields

$$P(x_k) = A_k \tag{108}$$

as desired.

The trigonometric sine function may be regarded essentially as a polynomial like $S(x)$ in Eq. 104 having an infinite number of zeros defined by

$$x_\nu = 0, \pm\pi, \pm 2\pi, \ldots \tag{109}$$

Its reciprocal, $1/\sin x$, has simple poles at these points with residues given by

$$k_\nu = \left[\frac{d}{dx}(\sin x)\right]^{-1}_{x=x_\nu} = 1 \tag{110}$$

In the formula 105 we can, therefore, make the identification

$$\frac{S(x)}{(x - x_\nu)} \cdot \left[\frac{(x - x_\nu)}{S(x)}\right]_{x=x_\nu} \rightarrow \frac{\sin(x - x_\nu)}{(x - x_\nu)} \tag{111}$$

and write the Lagrangian formula for an infinite number of uniformly spaced points in the form

$$P(x) = \sum_{\nu=-\infty}^{\infty} \frac{\sin(x - x_\nu)}{(x - x_\nu)} \cdot A_\nu \tag{112}$$

On the other hand, if in Eq. 79 we let

$$\frac{\pi t}{t_0} = \frac{\Omega t}{2} = x \tag{113}$$

we can write this representation for the sampled and lowpass-filtered time function:

$$f_{sf}(t) = \sum_{\nu=-\infty}^{\infty} f(\nu t_0) \frac{\sin(x - x_\nu)}{(x - x_\nu)} \tag{114}$$

with x_ν-values defined by 109. Comparison of Eqs. 112 and 114 then reveals the interesting fact that this representation for a given time function $f(t)$ may be regarded as its *polynomial approximation in terms of the Lagrangian interpolation formula.*

9. Convolution of Impulse Trains Is Like Polynomial Multiplication

Suppose we consider the sampled forms of two time functions that are nonzero over finite intervals. These may be written

$$f_{as}(t) = \sum_{\nu=0}^{m} a_\nu u_0(t - \nu t_0) \tag{115}$$

and

$$f_{bs}(t) = \sum_{k=0}^{n} b_k u_0(t - k t_0) \tag{116}$$

in which

$$a_\nu = t_0 f_a(\nu t_0) \quad \text{and} \quad b_k = t_0 f_b(k t_0) \tag{117}$$

are the samples for a common sampling interval t_0, and m and n are finite integers. Straightforwardly the Fourier analysis integral yields corresponding spectrum functions

$$F_a(\omega) = \sum_{\nu=0}^{m} a_\nu e^{-j\nu\omega t_0} \tag{118}$$

and

$$F_b(\omega) = \sum_{k=0}^{n} b_k e^{-jk\omega t_0} \tag{119}$$

If we let

$$x = e^{-j\omega t_0} \tag{120}$$

these are given by the finite polynomials

$$F_a(\omega) = a_0 + a_1 x + a_2 x^2 + \cdots + a_m x^m \tag{121}$$

and

$$F_b(\omega) = b_0 + b_1 x + b_2 x^2 + \cdots + b_n x^n \tag{122}$$

Significant about each of these spectrum functions is the fact that the polynomial coefficients are time-domain samples. Taking the substitution 120 for granted, we can say that the Fourier transform of a sampled time function is a polynomial in x, in which the coefficients are the pertinent samples. If the polynomial is written in ascending powers of x, then these samples are arranged in the same order in which they appear in the given time function.

The problem that we now wish to discuss is the convolution of the two sampled time functions 115 and 116. In the frequency domain this corresponds to multiplication of the corresponding polynomials 121 and 122. After their product is again written in ascending powers of x, the resultant coefficients are identifiable as successive samples of the desired time-domain solution. Formation of these resultant coefficients involves sorting out terms in the product of 121 and 122 corresponding to like powers of x.

This sorting process is easily done if we visualize the coefficients in these polynomials in the following arrangements. For the coefficient of x^0 we have

$$\begin{array}{cccccccc} & & & & a_0 & a_1 & a_2 & \cdots & a_m \\ b_n & \cdots & b_2 & b_1 & b_0 & & & & \end{array} \tag{123}$$

for the coefficient of x^1,

$$\begin{array}{ccccccccc} & & & a_0 & a_1 & a_2 & \cdots & a_m \\ b_n & \cdots & b_2 & b_1 & b_0 & & & \end{array} \tag{124}$$

for the coefficient of x^2,

$$\begin{array}{ccccccc} & & a_0 & a_1 & a_2 & \cdots & a_m \\ b_n & \cdots & b_2 & b_1 & b_0 & & \end{array} \tag{125}$$

and so forth. Finally for the coefficient of x^{m+n} we have

$$\begin{array}{ccccccccc} a_0 & a_1 & a_2 & \cdots & a_m & & & & \\ & & & & b_n & \cdots & b_2 & b_1 & b_0 \end{array} \tag{126}$$

The resultant coefficients are found by forming and summing products of overlapping coefficients. Thus, if we write the resultant spectrum function as

$$F(\omega) = c_0 + c_1 x + c_2 x^2 + \cdots + c_{m+n} x^{m+n} \tag{127}$$

we have

$$\begin{aligned} c_0 &= a_0 b_0 \\ c_1 &= a_0 b_1 + a_1 b_0 \\ c_2 &= a_0 b_2 + a_1 b_1 + a_2 b_0 \\ &\cdots\cdots\cdots\cdots\cdots \\ c_{m+n} &= a_m b_n \end{aligned} \tag{128}$$

At this point we remind ourselves of the fact that convolution, as discussed in art. 1 of Ch. XIII and illustrated there in Figs. 1 through 5, involves reversal in its left-to-right order of one of the functions, followed by sliding that function like a template over the other one (the scanning operation) and, for any momentary position, forming the product of overlapping portions and integrating (summing) the result. Relations 123 through 126 together with 128 are seen to accomplish precisely this set of operations in terms of samples of the respective time functions and the reminder that the scanning of an impulse of value a by an impulse of value b yields an impulse of value ab.

The convolution of sampled functions is thus reduced to a simple algebraic procedure.

Problems

1. Given:

$$f(t) = u_{-4}(t + 2) - 4u_{-4}(t + 1) + 6u_{-4}(t) - 4u_{-4}(t - 1) + u_{-4}(t - 2)$$

(*a*) Plot this function carefully and accurately, showing all of its essential characteristics.

(*b*) Find and plot with consistent accuracy the corresponding spectrum function $F(\omega)$.

(*c*) If this time function is to be sampled and regained with the same accuracy as above after lowpass filtering, what conditions apply to the uniform sampling interval and the filter bandwidth? Sketch the appropriate filter characteristic. Sketch also the transform of the sampled version of $f(t)$.

2. If $F(\omega)$ in the preceding problem is sampled, what conditions must the uniform sampling interval meet in order that $f(t)$ may still be characterized with the same accuracy as before? Determine and sketch the time function corresponding to the sampled version of $F(\omega)$.

3. In prob. 1, both $f(t)$ and $F(\omega)$ are replaced by their sampled versions. What are respectively the largest sampling intervals for which one may expect consistently accurate representations? From the plots made in prob. 1, measure carefully the corresponding samples of $f(t)$ and $F(\omega)$; then, by means of formulas 44 and 48 or 57 and 58, compute each set of samples in terms of the other and check the resulting accuracy. Comment upon the practical usefulness of this method of Fourier transformation.

4. Sampling of a function $f(t)$ is theoretically carried out through multiplying it by a uniform impulse train like $f_u(t)$ in Eq. 6. In practice one does not have perfect impulses but instead uses rectangular pulses of finite duration (and, let us say, unit enclosed area). This pulse train is amplitude-modulated by the function $f(t)$.

Point out how this circumstance affects the central pictures in Fig. 3 and how an appropriate pulse duration δ can be determined if one expects to regain $f(t)$ within a specified error after appropriate lowpass filtering.

5. Consider the situation described in the previous problem and assume that $\sin x/x$–shaped pulses rather than rectangular ones are used in the approximation to a uniform impulse train. Show that now one can sample $f(t)$ and regain this exact function after appropriate lowpass filtering. Explain how a proper "width" of the $\sin x/x$ pulses is determined, and comment upon the practicality of these results.

6. In prob. 5, could one use $(\sin x/x)^n$–shaped pulses with $n = 2, 3, \ldots$? Consider all aspects of this problem.

7. Convolve

$$\frac{\sin t}{t}$$

with the uniform impulse train given by Eq. 6 and evaluate the result for

$$t_0 = n\pi \qquad \text{with } n = 1, 2, 3, \ldots$$

and again for

$$t_0 = \frac{\pi}{n} \qquad \text{with } n = 1, 2, 3, \ldots$$

Where a question of uniqueness arises, discuss your interpretation adequately.

8. The sinusoidal time function $\cos \omega_a t$ is sampled at regular intervals defined by $t = kt_0$, $k = 0, \pm 1, \pm 2, \ldots$.

Determine and sketch graphically the resulting spectrum function for each of the following conditions:

$$(a)\ \ \omega_a = \frac{\omega_0}{3}; \qquad (b)\ \ \omega_a = \frac{\omega_0}{2}; \qquad (c)\ \ \omega_a = \frac{2\omega_0}{3}; \qquad (d)\ \ \omega_a = \omega_0$$

where

$$\omega_0 = \frac{2\pi}{t_0}$$

If the sampled sinusoid is transmitted through an ideal lowpass filter whose output is expected to be the continuous function $\cos \omega_a t$, point out the particular ω_a-values in the above set for which this expectation is realized, and specify the appropriate cutoff frequency in each case.

CHAPTER XVI

Network Response by Fourier Methods

1. Rational Transforms and Exponential Time Functions

Many of our discussions so far have shown that rational functions play an outstandingly significant role in linear network theory, not only for lumped-parameter systems but for other situations as well, since the additional presence of transcendental and/or multivalued functions (where transmission-line and waveguide sections are involved) must, in most practical instances, be dealt with through the construction of appropriate rational approximants. In the application of Fourier methods to network analysis it is, therefore, of prime importance that we have a rather complete understanding of the time-domain situation for which the associated spectrum function is the j-axis or other contour evaluation of a rational function in the complex frequency domain. Although several aspects of this question have already been touched upon in the discussion of other topics, a more complete and compact presentation will be found appropriate at this time.

From the analysis in Ch. X relative to the time-domain response of linear networks, it is logical to suppose that a linear combination of exponential functions of the form

$$f(t) = u_{-1}(t) \times \sum_{k=1}^{n} A_k e^{s_k t} \tag{1}$$

is the most general time function having a rational transform. Let us consider this question more carefully.

Computing the transform by means of the Fourier analysis integral 22, Ch. XII, we have

$$\begin{aligned} F(j\omega) &= \int_0^\infty \sum A_k e^{(s_k - j\omega)t}\, dt = \sum A_k \int_0^\infty e^{(s_k - j\omega)t}\, dt \\ &= \sum_{k=1}^{n} A_k \left[\frac{e^{(s_k - j\omega)t}}{s_k - j\omega}\right]_0^\infty \end{aligned} \tag{2}$$

If for the moment we assume that all s_k-values have a negative real part so that the time function given by Eq. 1 remains bounded, this result yields

$$F(j\omega) = \sum_{k=1}^{n} \frac{A_k}{j\omega - s_k} \tag{3}$$

and its analytic continuation into the complex s-plane reads

$$F(s) = \sum_{k=1}^{n} \frac{A_k}{s - s_k} \tag{4}$$

which is recognized as the partial fraction expansion of a rational function in the variable s. On the assumption that all s_k-values lie in the left half of the s-plane, its j-axis evaluation is the function $F(j\omega)$ in Eq. 3. Since this function is the Fourier transform of 1, we may regard 4 as the extended Fourier transform that is, as the corresponding Laplace transform of time function 1.

Now we shall show that this time function may be regained from the Fourier transform 3 by applying the synthesis integral 21, Ch. XII. Substitution into this integral followed by interchange of the order of summation and integration gives

$$f(t) = \sum_{k=1}^{n} \frac{A_k}{2\pi} \int_{-\infty}^{\infty} \frac{e^{j\omega t}}{j\omega - s_k}\, d\omega \tag{5}$$

Here we will make the change of variable indicated by

$$j\omega - s_k = \lambda, \quad j\omega = \lambda + s_k \tag{6}$$

and get

$$f(t) = \sum_{k=1}^{n} \frac{A_k e^{s_k t}}{2\pi j} \int_{-j\infty - s_k}^{j\infty - s_k} \frac{e^{\lambda t}}{\lambda}\, d\lambda \tag{7}$$

where the integration now takes place in a complex λ-plane and the effect of substitution 6 upon the limits of integration is taken into account.

If we recognize the complex character of s_k-values by writing

$$s_k = -\sigma_k + j\omega_k \tag{8}$$

with the tacit assumption (made above) that $\sigma_k > 0$, and note that the path of integration in 5 is the j-axis of the s-plane, we see that in 7 it is a line parallel to and σ_k units to the right of the j-axis of the λ-plane. Since the integrand in 7 is analytic in the entire finite λ-plane with the exception of the origin, we see by Cauchy's integral law that the quantity $-s_k$ appearing in the limits may be replaced by an arbitrarily small positive real quantity ϵ. The integral in Eq. 7 may thus be manipulated further as follows:

$$\int_{-j\infty+\epsilon}^{j\infty+\epsilon} \frac{e^{\lambda t}}{\lambda}\, d\lambda = \int_{-j\infty}^{j\infty} \frac{e^{(\lambda+\epsilon)t}}{\lambda+\epsilon}\, d\lambda = \int_{-j\infty}^{j\infty} \frac{\cosh(\lambda+\epsilon)t + \sinh(\lambda+\epsilon)t}{\lambda+\epsilon}\, d\lambda \tag{9}$$

The integration now is along the j-axis of the λ-plane; we can write $\lambda = j\omega$. In anticipation of the limit process $\epsilon \to 0$, we then have

$$j\int_{-\infty}^{\infty} \frac{\cos \omega t}{j\omega + \epsilon}\, d\omega + j\int_{-\infty}^{\infty} \frac{\sin \omega t}{\omega}\, d\omega \tag{10}$$

Thus ϵ is replaced by zero except where this move leads to an improper integral. The first integrand is

$$\frac{\cos \omega t}{\epsilon + j\omega} = \frac{\epsilon \cos \omega t}{\epsilon^2 + \omega^2} - j\,\frac{\omega \cos \omega t}{\epsilon^2 + \omega^2} \tag{11}$$

These terms are respectively even and odd functions of ω. Hence the second contributes nothing, while for the first we let $x = \omega/\epsilon$, $a = \epsilon t$, and have straightforwardly

$$j\int_{-\infty}^{\infty} \frac{\cos ax}{1 + x^2}\, dx = j\pi e^{-a} \to j\pi \qquad \text{for } \epsilon \to 0 \tag{12}$$

For the second integral in 10 we have

$$j\int_{-\infty}^{\infty} \frac{\sin \omega t}{\omega}\, d\omega = \pm j\int_{-\infty}^{\infty} \frac{\sin u}{u}\, du = \pm j\pi \tag{13}$$

where the $\pm$ signs pertain to $t > 0$ and $t < 0$ respectively.

We thus see that the two integrals in 10, equivalent to the integral in 7, yield zero for $t < 0$ and $2\pi j$ for $t > 0$. Equation 7, therefore, evaluates to expression 1 as it should.

We can alternately (and perhaps more simply) evaluate the integral in Eq. 5 by the method of contour integration discussed in art. 3, Ch. XII. To this end we substitute s for $j\omega$ and write 5 in the form

$$f(t) = \sum_{k=1}^{n} \frac{A_k}{2\pi j} \int_{-j\infty}^{j\infty} \frac{e^{st}}{s - s_k}\, ds \tag{14}$$

The integrand is regular in the entire finite s-plane except at the left half-plane points $s = s_k = -\sigma_k + j\omega_k$ where its residues are $e^{s_k t}$. Hence for $t < 0$ the integral equals zero, and for $t > 0$ it equals $2\pi j e^{s_k t}$, whereupon 14 and 1 are seen to be equivalent.

Now let us remove the restriction requiring s_k-values to lie in the left half-plane. Consider the following modification of the time function in Eq. 1:

$$\hat{f}(t) = e^{-at} f(t) = u_{-1}(t) \sum_{k=1}^{n} A_k e^{(s_k - a)t} \tag{15}$$

where the positive real quantity a is so chosen that all

$$s_k - a = \hat{s}_k \tag{16}$$

have negative real parts. This modified time function then has the Fourier transform

$$\hat{F}(j\omega) = \sum_{k=1}^{n} \frac{A_k}{j\omega - \hat{s}_k} \tag{17}$$

and, in the manner just demonstrated above, we can regain $\hat{f}(t)$ by substituting this result into the Fourier synthesis integral. We thus obtain, as in Eq. 5,

$$\hat{f}(t) = \sum_{k=1}^{n} \frac{A_k}{2\pi} \int_{-\infty}^{\infty} \frac{e^{j\omega t}}{j\omega - \hat{s}_k} \, d\omega \tag{18}$$

In view of Eq. 15, however, we now have for $f(t)$ the representation

$$f(t) = \sum \frac{A_k}{2\pi} \int_{-\infty}^{\infty} \frac{e^{(j\omega + a)t}}{j\omega - \hat{s}_k} \, d\omega \tag{19}$$

If we let

$$j\omega + a = s \tag{20}$$

so that, with 16,

$$j\omega - \hat{s}_k = s - (\hat{s}_k + a) = s - s_k \tag{21}$$

then we have

$$f(t) = \sum \frac{A_k}{2\pi j} \int_{-j\omega + a}^{j\omega + a} \frac{e^{st}}{s - s_k} \, ds \tag{22}$$

where the effect of substitution 20 upon the limits of integration is observed.

Except for the difference in these limits of integration, the expression 22 for $f(t)$ is the same as 14. In the latter, all s_k-values are assumed to lie to the left of the j-axis; in 22 they lie to the left of a vertical line parallel to and a units to the right of the j-axis. By contour integration we see at once that 22 yields 1 again, and hence that the appropriate time function is still obtainable from the rational function in Eq. 4, with the important difference, however, that evaluation is now not along the j-axis as in Eq. 3 but along a parallel contour a units to the right of this axis and hence to the right of all poles s_k of this rational function.

By Cauchy's integral law we recognize, of course, that an appropriate contour in any case need not be a straight line. What the above analysis shows is that the contour must pass to the right of every pole of the rational function $F(s)$ in Eq. 4 which is the Laplace transform of the time function in Eq. 1. Other than this, the contour may have any form.

In terms of the concept of interference patterns this contour restriction means simply that if the given time function $f(t)$ contains terms that increase exponentially with time, then it can be obtained from the superposition (interference) of everlasting sinusoids only if among these there are at least some that are likewise exponentially increasing and whose complex frequencies are respectively equal to those of all exponentially increasing terms in $f(t)$ with a slight edge on the real parts (as has already been pointed out in art. 4, Ch. XII and illustrated there by the sketches in Fig. 4).

There is another way of approaching this problem that is of collateral interest. Instead of computing the Fourier transform of time function 1 and then extending it to the complex s-plane, we may attempt to compute the Laplace transform directly by substituting $f(t)$ in Eq. 1 into integral 27, Ch. XII. After interchanging the order of summation and integration, we now get

$$F(s) = \sum_{k=1}^{n} A_k \int_0^{\infty} e^{(s_k - s)t}\, dt = \sum_{k=1}^{n} A_k \left[\frac{e^{(s_k - s)t}}{s_k - s}\right]_0^{\infty} \tag{23}$$

A finite result here is evidently possible only if the real part of $(s_k - s)$ is negative; that is, if

$$\operatorname{Re}(s) > \operatorname{Re}(s_k) \tag{24}$$

In that event we get the representation for $F(s)$ given by Eq. 4; but now we do not know quite how to interpret the condition 24.

A seemingly reasonable conclusion (and one that is quite common*) is to say that $F(s)$ as given by Eq. 4 is defined only over that portion of the s-plane fulfilling condition 24, that this is its "convergence region" or "region of validity". Such statements are misleading, for they misinterpret the real meaning of condition 24. To speak of a convergence region or a region of validity for a rational function is absurd. Equation 4 is a valid representation for the function $F(s)$ throughout the entire s-plane as is manifest from the fact that, in the evaluation of the synthesis integral

$$f(t) = \frac{1}{2\pi j} \int_{-j\infty}^{j\infty} F(s) e^{st}\, ds \tag{25}$$

by contour integration, we are at liberty to allow this contour to violate the restriction 24 if we wish, the only thing that matters being its passage to the right of all poles s_k. This conclusion is not evident in the present

* For example, see S. J. Mason and H. J. Zimmermann, *Electronic Circuits, Signals and Systems*, John Wiley, 1960, pp. 341–342.

derivation, nor is the reason for it as meaningful as when interpreted in terms of the concept of interference patterns which is really the only sound basis for the interpretation of Fourier and Laplace methods of representation.

2. Response of a Lumped Linear Network to a Suddenly Applied Exponential Excitation; the Dominance Requirement

Essentially, this is the problem dealt with in art. 1, Ch. X. There it is attacked and solved by the differential equation method in which the point of departure is an expression of equilibrium for the physical system and the process of constructing a solution is guided by physical and intuitive reasoning although justified rigorously by classical analysis.

Our method of approach here actually differs only in that we will use a Fourier representation for the excitation function. That is to say, instead of considering the suddenly applied exponential excitation as it is, we will represent it as a superposition of everlasting sinusoids. By so doing we obtain an equivalent form of excitation function which no longer suddenly occurs at some finite time but which is composed of sinuosoids (or exponentials) whose time of occurrence lies in the infinitely remote past.

The interference pattern created by these sinusoids, to be sure, has the appearance of suddenly occurring at some finite time (usually chosen as $t = 0$) and of being identically zero before this instant, but the component sinusoids which create this pattern individually have no finite time of origin—they exist throughout the infinite time domain defined by $-\infty < t < \infty$. What do we gain by the use of this artifice?

To answer this question we must review briefly the essential steps in the classical method of solution as outlined in arts. 1 and 2 of Ch. VIII and carried through in art. 1, Ch. X. The net response is found to consist of two parts: the particular integral and the complementary function, alternately designated respectively as the forced and the force-free parts of the response. The particular integral is completely characterized by the pertinent impedance or admittance function (Eq. 9, Ch. VIII) and its determination is theoretically and computationally simple since it involves only the evaluation of this complex function for the given excitation frequency. Evaluation of the complementary function, on the other hand, is computationally far more tedious since it involves solving not only an algebraic equation (the characteristic equation 8, Ch. X) for its roots, but also a set of simultaneous algebraic equations (Eqs. 12, Ch. X) for the constants of integration which give the force-free behavior the ability to meet arbitrary initial conditions.

In any passive network and in any stable one containing active elements, the force-free behavior is not sustained but decays exponentially so that after a relatively short time following the initial instant (referred to as the transient interval) only the forced behavior survives since the forcing function in most cases is a sustained excitation. If the excitation function is represented by a sum of sinusoids (or exponentials) in the Fourier manner, then the "initial instant" is the time of occurrence of each component sinusoid, which is at an infinitely remote time in the past. The "transient interval" then is long past at any finite time and hence the force-free part of the response (the complementary function) need not be evaluated at all because only the particular integral survives. This part of the solution alone represents the total response. Since its evaluation is relatively simple, we stand to gain by using the Fourier artifice, or so it would seem.

Regarding computational simplification, all is not on the credit side of our ledger. First of all, the Fourier artifice replaces a single exponential occurring at $t = 0$ by an infinite sum (or infinite integral) of exponentials occurring at $t = -\infty$. Although we need evaluate only the forced part of the response, this part has an infinite number of components and these must be integrated to form what amounts to the "particular integral" in this situation.

Secondly, the above discussion has made an important tacit assumption which may or may not be fulfilled. This is the assumption that the force-free or natural behavior decays more rapidly than the forced response or particular integral. In most practical situations the latter is a steady sinusoid which neither decays nor grows, while the force-free response involves natural frequencies with negative real parts. Under these circumstances the particular integral is said to *dominate* the response, and the Fourier integral then yields the total solution.

If the excitation is a suddenlv applied exponential, and if the pertinent complex frequency has a real art which may or may not be larger than the real parts of the natural frequencies, whether these are negative or positive, then the Fourier representation of the excitation can still be chosen to involve only sinusoids whose exponential rate of growth (positive or negative) dominates that of the natural behavior. If we group the complex frequency of the excitation with the complex natural frequencies (as we do in Eqs. 10 and 12 of Ch. X) the dominance condition requires simply that the spectrum function for the excitation be evaluated along a contour that passes to the right of all these complex frequencies.

If the network is passive and the excitation a constant-amplitude sinusoid, then the normal Fourier integral with a j-axis path fulfills the dominance requirement. In other situations we may need to choose a path at least partly in the complex plane.

It is significant in this connection to observe that fulfilling the dominance condition is not necessary in order to evaluate the network response by Fourier methods, although these become somewhat more involved when the dominance condition is not fulfilled, for then the particular integral no longer is the complete solution. We must additionally compute the complementary function (involving the evaluation of another Fourier integral) and add these two parts to form the net response as we do in the classical differential equation approach. We will illustrate this process with an example later on.

It is, of course, not sensible to use the Fourier method unless the dominance condition is fulfilled, since one needlessly doubles the computational labor involved. From a theoretical point of view, however, it is important to be aware of the fact that fulfilling the dominance requirement is a convenience, not a necessity, in the use of Fourier or Laplace methods of circuit analysis.

Let us now consider the problem of determining, by Fourier methods, the response of a lumped linear network to a suddenly applied exponential excitation. Let this excitation be a voltage in the form

$$e_i(t) = u_{-1}(t) \cdot E_i e^{s_0 t} \tag{26}$$

as is also defined by Eqs. 2 in Ch. VIII or X. Its transform, according to discussion in the preceding article, is given by the function

$$E_i(s) = \frac{E}{s - s_0} \tag{27}$$

This excitation may be in loop i of some network, and the response, a current in loop j. The pertinent admittance function $y_{ji}(s)$ is the rational function (Eq. 12, Ch. VIII)

$$y_{ji}(s) = \frac{H(s - s_a)(s - s_b) \cdots (s - s_k)}{(s - s_1)(s - s_2) \cdots (s - s_m)} \tag{28}$$

Physically we interpret 27 as being the complex amplitude of a spectral component of the excitation voltage having the complex frequency s. If by $I_j(s)$ we denote the corresponding spectral component of the resulting current in loop j, then we have according to the usual manner of forming the particular integral

$$I_j(s) = \frac{E}{s - s_0} \cdot y_{ji}(s) \tag{29}$$

In the frequency domain this is the desired response; its inverse Laplace transform is the corresponding time-domain response, namely

$$i_j(t) = \frac{1}{2\pi j} \int_{-j\infty}^{j\infty} \frac{E}{s - s_0} \cdot y_{ji}(s) e^{st} \, ds \tag{30}$$

Since the transform 29 is a rational function, the evaluation of this integral is straightforward. We fulfill the dominance requirement of choosing a contour in the s-plane that passes to the right of all poles by $I_j(s)$. By the method of contour integration, we then get zero for $t < 0$ as we should, and for $t > 0$ we have the sum of the residues of the integrand in all of its finite poles.

These residues may be written

$$A_\nu e^{s_\nu t} \tag{31}$$

where

$$A_\nu = \left[(s - s_\nu) \cdot y_{ji}(s) \cdot \frac{E}{s - s_0}\right] \tag{32}$$

are coefficients in the partial fraction expansion of $I_j(s)$ which reads

$$I_j(s) = \frac{A_0}{s - s_0} + \frac{A_1}{s - s_1} + \frac{A_2}{s - s_2} + \cdots + \frac{A_m}{s - s_m} \tag{33}$$

The desired time-domain response is thus given by

$$i_j(t) = A_0 e^{s_0 t} + A_1 e^{s_1 t} + A_2 e^{s_2 t} + \cdots + A_m e^{s_m t} \tag{34}$$

Except for some minor differences in notation, these results agree fully with those found in Ch. X by the differential equation method of approach, and given there by Eqs. 24 and 25.

It may be well to mention here that initial rest conditions are tacitly assumed in these discussions. That is to say, there are no currents in inductances and no charges in capacitances at the initial instant. When this is not the case, additional terms must be added to expression 34 for the current in loop j which result from the initial values of current or charge.

If an inductance carries a current at the initial instant, then we can imagine that it got there either by a voltage impulse applied in series with this element or by a current step in parallel with it. If a capacitance carries a charge at the initial instant, then we can imagine that it is produced by a current impulse in parallel with this capacitance or by a voltage step in series with it. The solution given by Eq. 34 must then be augmented by the addition of responses computed for these voltage or current impulses or steps, also applied at the initial instant, and determined individually for assumed rest conditions.

As pointed out in Ch. X, the step response is immediately obtained from the preceding analysis by setting the complex source frequency s_0 equal to zero, as is evident from Eq. 26. Since impulse response is simply the derivative of step response, it is likewise available from the above analysis.

The only remaining item is concerned with the location of the pertinent initial current or charge relative to the loop j in which the response is to be found. If the location is in loop k and we imagine the current or charge produced by a voltage impulse or step, then the above analysis is carried through with $y_{ji}(s)$ in 28 replaced by $y_{jk}(s)$.

If many initial currents and charges are present, it becomes necessary to determine the same number of transfer admittance functions. However, these differ only in their numerator polynomials since they all have the same poles. In any case, no other method of taking arbitrary initial conditions into account can involve less computational work. In one way or another, the same transfer relationships are involved and need to be calculated.

Regarding the source values which replace initial currents or charges, well-established principles apply. A voltage impulse of value E applied in series with an inductance L establishes instantly a current $i_0 = E/L$; a current impulse of value I applied in parallel with a capacitance C establishes instantly a charge $q_0 = I/C$. Transformation from a voltage to a current source or vice versa, may, if needed, be carried through by well-established principles.*

Now we shall consider a simple example for the purpose of comparing Fourier procedure fulfilling the dominance requirement with Fourier procedure appropriate to a situation where this condition is not fulfilled. Assume a series RL circuit with $R = L = 1$. For a voltage Ee^{st} applied at $t = 0$ the current response is familiarly given by

$$i(t) = \frac{Ee^{st}}{s+1} - \frac{Ee^{-t}}{s+1} \tag{35}$$

where the first term is the particular integral and the second is the complementary function, respectively the forced and the force-free components of the response.

The Laplace representation for a unit impulse occurring at $t = 0$ may be written in the form

$$e(t) = \lim_{T\to\infty} \int_{-j\infty}^{j\infty} \frac{e^{-sT}}{2\pi j} \cdot e^{s(t+T)} u_{-1}(t + T)\, ds \tag{36}$$

Regarding T as some arbitrarily large but finite time, each spectral component is an exponential e^{st} with amplitude

$$E = \frac{e^{-sT}}{2\pi j} \tag{37}$$

suddenly applied at time $t = -T$. The response of our RL circuit to this spectral component is 35 with t replaced by $t + T$ and E by its value in 37.

* See *ICT*, pp. 190–196; also p. 239.

The response of this circuit to a unit impulse occurring at $t = 0$ is obtained by integrating this expression over all spectral components, thus:

$$i(t) = \underbrace{\frac{1}{2\pi j}\int_{-j\infty}^{j\infty} \frac{e^{st}}{s+1}\,ds}_{\text{particular integral}} - \underbrace{\frac{e^{-(t+T)}}{2\pi j}\int_{-j\infty}^{j\infty} \frac{e^{-sT}}{s+1}\,ds}_{\text{complementary function}} \tag{38}$$

These terms correspond respectively to the first and second terms in 35 and hence represent the particular integral (p.i.) and the complementary function (c.f.), as indicated.

If we use contour integration for the evaluation of these integrals it is important to observe that for the first integral the semicircular detour around $s = \infty$ lies in the right half-plane for $t < 0$ and in the left half-plane for $t > 0$, while for the second integral it lies in the right half-plane for all t because only the positive parameter T occurs in the integrand.

The path of integration throughout the fine s-plane must, of course, be the same for both integrals,* but its location we can now choose anywhere we wish. We do not need to consider dominance conditions because 38 contains both the forced and the force-free parts of the response. If the path is so chosen that the particular integral dominates, then the second term in 38 automatically becomes zero; and if the path is chosen so that the particular integral does not dominate, then the second term automatically yields a value that complements the altered particular integral so that the sum of the two terms in 38 remains the same.

Thus *if the path lies to the right of the pole at* $s = -1$, we have by the method of contour integration

$$\begin{aligned} i(t) &= 0 + 0 && \text{for } t < 0 \\ i(t) &= \underset{\text{p.i.}}{e^{-t}} + \underset{\text{c.f.}}{0} && \text{for } t > 0 \end{aligned} \tag{39}$$

If the path lies to the left of the pole at $s = -1$, on the other hand, we see that

$$\begin{aligned} i(t) &= -e^{-t} + e^{-t} && \text{for } t < 0 \\ i(t) &= \underset{\text{p.i.}}{0} + \underset{\text{c.f.}}{e^{-t}} && \text{for } t > 0 \end{aligned} \tag{40}$$

In the first of these two cases the forced response dominates, and the force-free part makes no contribution either for $t < 0$ or for $t > 0$; in the second, for which the forced response does not dominate, the force-free part makes very essential contributions, and indeed is seen to yield the total response for $t > 0$. The net response is the same for either path, for $t < 0$ as well as for $t > 0$.

* Because it is fixed by the synthesis integral 36 characterizing the excitation.

An interesting phenomenon appears when the forced response does not dominate; namely, this part of the response is then nonzero before $t = 0$, the instant at which the excitation is applied. This fact should cause no surprise inasmuch as the Fourier representation of any time function involves sinusoidal components that begin at $t = -\infty$. Hence there is nothing unusual about having these components excite a response before $t = 0$, certainly not so far as a *part* of the net response is concerned. The net response in a physically realizable system, must of course be zero for $t < 0$, and the results in Eqs. 40 bear this out.

3. The Excitation and the Network Are Arbitrary; Time- and Frequency-Domain Characterization of a Linear System

We shall now consider a more general situation in which the excitation is any given function of time and the network is characterized in the frequency domain by a function $H(j\omega)$ not necessarily the j-axis evaluation of a rational function. It may be described by saying that if the input and output are steady sinusoidal time functions then $H(j\omega)$ is the ratio of the complex amplitude of the output to that of the input. For any particular radian frequency ω, this ratio is complex. If input and output are both currents or both voltages, $H(j\omega)$ is dimensionless; if one is a current and the other a voltge, it is an impedance or admittance. Finally, it may be given in graphical or in analytic form, but in either case it is assumed to be defined for the entire j-axis interval $-\infty < \omega < \infty$.

We denote the input and output time functions by $f_1(t)$ and $f_2(t)$ respectively. Their Fourier transforms are given by

$$F_1(j\omega) = \int_{-\infty}^{\infty} f_1(t)e^{-j\omega t}\,dt \tag{41}$$

and

$$F_2(j\omega) = \int_{-\infty}^{\infty} f_2(t)e^{-j\omega t}\,dt \tag{42}$$

Since these may be regarded as the complex amplitudes of steady sinusoidal time functions with radian frequency ω, we can evidently write

$$H(j\omega) = \frac{F_2(j\omega)}{F_1(j\omega)} \tag{43}$$

or

$$F_2(j\omega) = F_1(j\omega) \cdot H(j\omega) \tag{44}$$

Thus the solution to our problem in the frequency domain is given by multiplying the transform of the excitation by the system function or network response function $H(j\omega)$. The response in the time domain is then found by evaluating the inverse Fourier transform of $F_2(j\omega)$. Hence we have

$$f_2(t) = \frac{1}{2\pi}\int_{-\infty}^{\infty} F_1(j\omega)H(j\omega)e^{j\omega t}\,d\omega \tag{45}$$

Compactly stated, the process of computing network response by Fourier methods consists of: (*a*) transformation of the excitation from time to frequency domain, Eq. 41; (*b*) solution of the problem in the frequency domain through multiplication by the system function, Eq. 44; (*c*) transformation back to the time domain by means of the Fourier synthesis integral, Eq. 45.

Dominance of the forced over the force-free response is, of course, a necessary condition for the validity of this procedure. This condition is fulfilled for any stable network whether active elements are involved or not since its natural frequencies are restricted to the left half of the s-plane. The interference pattern for the excitation as defined by the spectrum function 41 involves only constant amplitude sinusoids and hence cannot accommodate functions containing growing exponentials. As pointed out previously, one can accommodate such functions through replacing $j\omega$ by s and extending this transform to the complex plane. In this event it will be necessary similarly to extend the function $H(j\omega)$ since in the evaluation of integral 45 the path of integration will then lie in the plane rather than upon the j-axis.

An interesting special case is that in which the excitation $f_1(t)$ is a unit impulse occurring at $t = 0$. Thus, if

$$f_1(t) = u_0(t) \tag{46}$$

then

$$F_1(\omega) = 1 \tag{47}$$

and Eq. 45 reads

$$f_2(t) = \frac{1}{2\pi}\int_{-\infty}^{\infty} H(j\omega)e^{j\omega t}\,d\omega = h(t) \tag{48}$$

which is the inverse transform of the system function, or *the time-domain characterization of the network itself.* In other words, the impulse response of the network is its time-domain characterization just as the system function or sinusoidal steady-state response function is its frequency-domain characterization; and these two are reversibly related by the synthesis integral 48 and the analysis integral

$$H(j\omega) = \int_{-\infty}^{\infty} h(t)e^{-j\omega t}\,dt \tag{49}$$

Since the network can thus be characterized in either the time or frequency domain, it is logical to ask: Is it possible in terms of the network's time-domain characterization to compute the response $f_2(t)$ from an excitation $f_1(t)$ without transformation to the frequency domain and back again?

The answer is rather obvious if we look at Eq. 44 representing the process of solution in the frequency domain and remind ourselves of the fact (discussed in art. 7, Ch. XIII) that multiplication in the frequency domain corresponds to convolution in the time domain. We thus see at once that the following relationship is true:

$$f_2(t) = \int_{-\infty}^{\infty} h(t - \xi) f_1(\xi) \, d\xi \tag{50}$$

or alternately

$$f_2(t) = \int_{-\infty}^{\infty} h(\xi) f_1(t - \xi) \, d\xi \tag{51}$$

Although we have here derived this formulation for network response with the help of Fourier methods, it is important to be aware of the fact that Fourier theory has nothing whatever to do with this result. The truth of this statement is readily seen if we interpret convolution integral 50 by considering the excitation function $f_1(\xi)$ as represented in terms of its samples $f_1(\xi)\,d\xi$, in the way that the scanning function $K(\xi)$ is represented in the sketch of Fig. 9, Ch. XIII. Regarding each sample as an impulse of value $f_1(\xi)\,d\xi$ occurring at time $t = \xi$, and noting that $h(t - \xi)$ is the response to a unit impulse occurring at $t = \xi$, we find that the integrand in 50 is the response to this typical sample of $f_1(\xi)$ and hence the integral yields the net response.

By means of either integral 50 or 51, network response is formed entirely in the time domain. Fourier transformation is involved neither with the characterization of the network nor with the computation of its time-domain response for a given time-domain exitation. Conditions under which the method is valid are strictly those for which the pertinent convolution integral exists. The question of dominance does not enter here since $h(t)$ is the net response to an applied unit impulse and hence already includes both the forced and the force-free parts of this response. It is not necessary, for evaluation of the convolution integral, that we know whether the forced response dominates or not. The net response as shown by the example in the previous article is the same in either case.

For this reason one is tempted to use the time-domain formulation as a basis for defining the system function or response function of a linear network. The results of this attitude are discussed in the following article.

4. System Function Defined by the Convolution Integral

The convolution integral 50 or 51, which can straightforwardly be formulated in an independent manner (as discussed above), can be used as a basis for a rather general time-domain definition of the function characterizing a linear system. However, one cannot subsequently transform this function to the frequency domain by means of integral 49 without becoming involved in the dominance question in one way or another. In the literature where this method of defining a system function is presented* one finds that this basis aspect of Fourier theory is dealt with by restricting the path of integration in 48 (replacing $j\omega$ by s) and stipulating a "validity region" for the system function 49 in a way that is equivalent to fulfilling the dominance condition but is lacking in justification on these grounds, and hence achieves the desired result at the expense of drawing some false conclusions, such as the one implied by the statement that the synthesis integral is not convergent for other paths.

Our example leading to the results summarized in Eqs. 39 and 40 is relevant here. With regard to the choice of path to which the solutions 40 are pertinent, it is for example concluded† that the integral is divergent and leads to no solution. Actually, as our analysis shows, a Fourier representation for the particular integral does exist; but since it no longer dominates, one must add an appropriate complementary function in order to form the correct solution.

If one wishes to use the time-domain definition of a system function as given by the convolution integral, it is necessary first of all to consider separately the realizable and the nonrealizable cases, the latter corresponding to a fictitious system for which the net response begins *before* the excitation is applied ($t = 0$).

For a realizable system we can use the Laplace transformation (Eq. 49 for $j\omega = s$) to obtain a corresponding frequency-domain function provided in the synthesis integral (Eq. 48 for $j\omega = s$) we restrict the path of integration in such a way that the forced response dominates. This is what we would ordinarily do anyway in order to avoid the necessity of additionally evaluating the complementary function.

For a nonrealizable system, serious inconsistencies appear to be unavoidable. As an example, suppose we assume that the impulse response is given by

$$h(t) = e^{-|t|} \tag{52}$$

Reference to the sketches in Fig. 4, Ch. XII shows that this time function is the sum of time functions for examples (*a*) and (*b*). Luckily the pertinent s-plane paths are identical, and so we can add the respective transforms of these time functions to get the spectrum function for $h(t)$ in Eq. 52, thus:

$$H(s) = \frac{1}{s+1} - \frac{1}{s-1} = \frac{-2}{s^2-1} \tag{53}$$

* See, for example, Mason and Zimmermann, *Electronic Circuits, Signals and Systems*, John Wiley, 1960, p. 340.

† See *op. cit.*, Eq. 7.102, p. 342.

The appropriate path is the j-axis.

But observe that the particular integral evaluated for this path does not dominate. The "system" we are considering here is not allowed to have a natural or force-free behavior. It is not even allowed to possess an equilibrium condition. The forced response is the total response by dictum, and thus we find ourselves in disagreement with the basic principles of classical dynamics.

Such a dilemma is the result of trying to "generalize" a situation beyond its natural capabilities. It just isn't sensible to attempt the definition of a system (or "black box") that responds before the excitation is applied, not even if we label it as being "fictitious." If we merely want to find a spectrum function yielding an interference pattern like any arbitrarily specified time function $h(t)$, this we can do (within limitations discussed in art. 4, Ch. XII). But to designate a fictitious system as being characterized by this pair of functions which, by the basic principles of classical dynamics are violently inconsistent, purports to arouse too much confusion of thought to be justified even on a "make believe" basis.

5. Real-Part Sufficiency

In art. 2, Ch. IX this topic is discussed with reference to the frequency domain. It is shown there that if the j-axis real part of a physically realizable driving-point or transfer impedance (or admittance) of a passive network* is given, inclusive of any impulses that it may contain, then one can uniquely construct that impedance as a function of the complex variable s. We shall now discuss the time-domain aspect of this interesting topic through showing that the time domain response of the implied network may be determined from such a j-axis real part.

The basis for this result is found in the detailed relationship between even and odd parts of a time function $f(t)$ and the real and imaginary parts of its transform $F(j\omega)$, as discussed in art. 5, Ch. XII. Specifically, Eqs. 64 through 68 of that chapter are relevant here. Thus if $F_1(\omega)$ and $F_2(\omega)$ are respectively the real and imaginary parts† of $F(j\omega)$, and $f_e(t)$ and $f_0(t)$ are the even and odd parts of $f(t)$, then we have

$$F_1(\omega) = 2\int_{-\infty}^{\infty} f_e(t) \cos \omega t \, dt \tag{54}$$

$$F_2(\omega) = 2\int_{-\infty}^{\infty} f_0(t) \sin \omega t \, dt \tag{55}$$

* For which integration along the j-axis fulfills dominance requirements.

† Not to be confused with the notation in Eqs. 41 and 42.

and conversely

$$f_e(t) = \frac{1}{\pi} \int_0^\infty F_1(\omega) \cos \omega t \, d\omega \tag{56}$$

$$f_0(t) = -\frac{1}{\pi} \int_0^\infty F_2(\omega) \sin \omega t \, d\omega \tag{57}$$

These are referred to as the sine and cosine transforms.

Suppose we identify $F(j\omega)$ with the frequency-domain response function of a physical network and $f(t)$ with its unit-impulse response. Since

$$f(t) = f_e(t) + f_0(t) \tag{58}$$

and we know that this response must be identically zero before the excitation is applied (i.e. for negative values of t), we can write

$$f(-t) = f_e(t) - f_0(t) \equiv 0 \tag{59}$$

in which the symbol t itself is allowed only positive values. Hence we have

$$f_e(t) \equiv f_0(t) \qquad \text{for } t > 0 \tag{60}$$

From a graphical standpoint this result is easy to understand. Namely, since the even and odd parts of $f(t)$ must cancel for negative t-values, it follows that they must coincide for positive ones. This being true, we have from 58 that

$$f(t) = 2f_e(t) = 2f_0(t) \qquad \text{for } t > 0 \tag{61}$$

For negative t, $f(t)$ is zero, and for positive t it is equal to either twice the even or twice the odd part. Equations 56 and 57 then show that either the real or imaginary part of $F(j\omega)$ alone suffices to determine the net response $f(t)$. We can write (for $t > 0$)

$$f(t) = \frac{2}{\pi} \int_0^\infty F_1(\omega) \cos \omega t \, d\omega = \frac{-2}{\pi} \int_0^\infty F_2(\omega) \sin \omega t \, d\omega \tag{62}$$

and, of course, it follows that $F(j\omega)$ can be constructed from either of its parts, although this result (if desired) is more effectively gotten by the methods discussed in art. 2, Ch. IX, involving operations in the frequency domain alone.

The alternate expression

$$f(t) = \frac{1}{\pi} \int_{-\infty}^\infty F_1(\omega) e^{j\omega t} \, d\omega = \frac{j}{\pi} \int_{-\infty}^\infty F_2(\omega) e^{j\omega t} \, d\omega \tag{63}$$

is seen to be equivalent to Eq. 62, if we note the respective even and odd character of $F_1(\omega)$ and $F_2(\omega)$. It is in many situations preferable since the exponential function is easier to work with than the sines and cosines.

6. Condition for Facsimile Reproduction of Signals

In Eq. 45 for the response of a linear network to an applied signal $f_1(t)$, suppose the system function $H(j\omega)$ is given by

$$H(j\omega) = Ke^{-j\omega t_0} \tag{64}$$

in which K and t_0 are positive constants. We then have

$$f_2(t) = \frac{K}{2\pi}\int_{-\infty}^{\infty} F_1(\omega)e^{j\omega(t-t_0)}\,d\omega \tag{65}$$

and in view of the fact that

$$f_1(t) = \frac{1}{2\pi}\int_{-\infty}^{\infty} F_1(\omega)e^{j\omega t}\,d\omega \tag{66}$$

we see that this result can be written

$$f_2(t) = L \cdot f_1(t - t_0) \tag{67}$$

The output is a delayed replica of the input signal.

The system function 64 which yields this result has a constant amplitude and a linear phase characteristic. More generally we write

$$H(j\omega) = A(\omega)e^{-j\beta(\omega)} \tag{68}$$

and refer to $A(\omega)$ as the amplitude response of our network and to $\beta(\omega)$ as the phase response. For any radian frequency ω and a steady sinusoidal excitation, β is the angle in radians by which the output lags the input; A is the ratio of output amplitude to input amplitude. Facsimile reproduction of the input signal, or distortionless transmission through the pertinent network (viewed as a box with input and output terminal pairs) as expressed in Eq. 67, results if, and only if,

$$\begin{aligned} A(\omega) &= K \qquad \text{a constant} \\ \beta(\omega) &= \omega t_0 \qquad \text{a linear function} \end{aligned} \tag{69}$$

where it is interesting to note that the slope of this linear phase-lag function equals the delay or time of transmission of the signal through the network.

If $A(\omega)$ or $\beta(\omega)$ or both deviate from the conditions 69, the output signal $f_2(t)$ will correspondingly deviate in form from the input signal $f_1(t)$. Since these conditions apply to the infinite frequency domain, it is clear that practical networks always cause some signal distortion. Deviations from conditions 69 on the part of $A(\omega)$ are said to cause *amplitude distortion* while deviations in $\beta(\omega)$ from linearity cause *phase* or *delay distortion.* In most cases both kinds are simultaneously present, and it is difficult if not impossible to separate these effects.

A notable exception to this statement is provided by the so-called all-pass network whose transfer function $H(s)$ has only poles in the left half-plane and only zeros in the right half-plane, these being located at points which are images of the pole positions about the j-axis. From a sketch like the one in Fig. 2 of Ch. IX, we see that this pole-zero pattern yields $A(\omega) = 1$ while $\beta(\omega)$ may vary over wide limits.

In physical optics where we deal with the propagation of light waves through various media the phenomenon of phase distortion is called *dispersion.* This term is also used in network theory in connection with the distortion produced by an all-pass network.

An interesting quasi-physical interpretation of the dispersion phenomenon is had by considering the frequency spectrum decomposed into small intervals or groups $\Delta\omega$ in width. We then observe the propagation of a typical frequency group through the network and visualize the superposition of all such groups at the output end.

For the typical group $\Delta\omega$ centered at $\omega = \omega_0$ we represent the phase function by two terms of a Taylor series, thus:

$$\begin{aligned}\beta(\omega) &= \beta(\omega_0) + \left(\frac{d\beta}{d\omega}\right)_{\omega=\omega_0}(\omega - \omega_0) \\ &= \beta_0 + \omega t_0\end{aligned} \tag{70}$$

where

$$t_0 = \left(\frac{d\beta}{d\omega}\right)_{\omega=\omega_0} \tag{71}$$

and

$$\beta_0 = \beta(\omega_0) - \omega_0 t_0 \tag{72}$$

Representation 68 for the system function then reads

$$H(j\omega) = e^{-j\beta(\omega)} = e^{-j\beta_0}e^{-j\omega t_0} \tag{73}$$

For evaluation of the impulse response $h(t)$, it is convenient to write integral 48 in the form

$$h(t) = \frac{1}{\pi}\operatorname{Re}\int_0^\infty H(j\omega)e^{j\omega t}\,d\omega \tag{74}$$

where Re stand for "real part of." The contribution to $h(t)$ due to a typical frequency group $\Delta\omega$ becomes

$$\Delta h(t) = \frac{1}{\pi}\operatorname{Re}\int_{\omega_0-\Delta\omega/2}^{\omega_0+\Delta\omega/2} e^{-j\beta_0}e^{j\omega(t-t_0)}\,d\omega \tag{75}$$

which evaluates straightforwardly to give

$$\Delta h(t) = \frac{\Delta\omega}{\pi} \cdot \left\{ \frac{\sin (\Delta\omega/2)(t - t_0)}{(\Delta\omega/2)(t - t_0)} \right\} \cos [\omega_0(t - t_0) - \beta_0] \tag{76}$$

The sin x/x function in the curved brackets is slowly varying with respect to the cosine function ($\Delta\omega \ll \omega_0$ by assumption). The wave group (as $\Delta h(t)$ is called in optics) has the appearance of a relatively high-frequency cosine enclosed in an envelope of the form sin x/x. The main hump of this envelope is centered at $t = t_0$; that is to say, the envelope enclosing this typical wave group is delayed in transmission through the network by t_0 seconds, equal to the slope of the phase characteristic at the point $\omega = \omega_0$ at which the group $\Delta\omega$ is centered (Eq. 71). For this reason, t_0 is also called the *envelope delay* or *group delay*, and phase distortion is sometimes also referred to as *envelope distortion.*

The unit impulse function at the input of the network has a uniform spectrum consisting of theoretically an infinite number of frequency groups $\Delta\omega$, all in phase. Like a lineup of marathon runners, they start out together at $t = 0$. If $\beta(\omega)$ is nonlinear, their envelope delays in transmission through the network are not alike, and so their arrival times at the output end have some distribution varying from the minimum to the maximum slope of $\beta(\omega)$ throughout the spectrum. In other words, the various frequency groups suffer dispersion and hence fail to arrive simultaneously at the output end. The impulse that starts out at the input end is literally torn apart and the pieces, arriving at various times at the output, cannot superpose to regain the form of an impulse but yield instead a pulse of finite duration depending upon the degree of nonlinearity of the phase function.

We can apply essentially this same interpretation to the transmission of an impulse through a network having an arbitrary amplitude function $A(\omega)$. Here the wave groups (marathon runners) suffer changes in amplitude as well as dispersion in their arrival times. Obviously, the fit amplitude and linear phase conditions 69 result in all wave groups remaining unaltered in their relative amplitudes and having equal times of arrival, whereupon the form of the input signal is identically regained at the output.

The ideal or distortionless system function 64 substituted into integral 48 yields for the impulse response

$$h(t) = \frac{K}{2\pi} \int_0^\infty e^{j\omega(t-t_0)}\, d\omega = K u_0(t - t_0) \tag{77}$$

(Evaluation of this integral is shown in detail in art. 7, Ch. XII, Eqs. 98 and 99.) The convolution integral 50 or 51 is then seen by inspection to produce a function $f_2(t)$ in agreement with the result expressed by Eq. 67.

The necessary and sufficient condition for facsimile reproduction of signal waveform is, therefore, expressible in the time domain by saying that the impulse response of the pertinent network must be an impulse. Its magnitude and time of occurrence are arbitrary, but its form must be that of the impulse function.

7. The Output Is a Sum of Delayed Replicas of Either the Excitation or the Impulse Response

As Eqs. 50 and 51 show, network response may be expressed as a convolution of the input and the impulse response functions. That is to say, the output can be constructed by scanning the input with the impulse response of the network or by scanning the latter with the input or excitation function. If we apply to this situation the graphical interpretation given to the convolution process in art. 3, Ch. XIII (illustrated there in the sketch of Fig. 10) we obtain a useful and interesting formulation for network response that aids visualizing the effect upon a given signal waveform produced by its transmission through a network with known impulse response.

This formulation is most effectively done by representing either the input or the impulse response in sampled form (art. 1, Ch. XV). Suppose we sample the impulse response function $h(t)$ at uniform intervals t_0, yielding the representation

$$h(t) = \sum_{k=-\infty}^{\infty} a_k u_0(t - kt_0) \tag{78}$$

involving the samples

$$a_k = t_0 h(kt_0) \tag{79}$$

Substitution into the convolution integral 51 then gives

$$f_2(t) = \int_{-\infty}^{\infty} \sum_k a_k u_0(\xi - kt_0) f_1(t - \xi)\, d\xi = \sum_k a_k \int_{-\infty}^{\infty} u_0(\xi - kt_0) f_1(t - \xi)\, d\xi \tag{80}$$

Since the integrand is zero except at the point $\xi = kt_0$, we have

$$f_2(t) = \sum_{k=-\infty}^{\infty} a_k f_1(t - kt_0) \tag{81}$$

Thus the output is given by a sum of delayed replicas of the input, monitored in amplitude by the samples of $h(t)$. If $h(t)$ is compact and hence nonzero over an essentially short interval, this formulation shows that the output signal does not depart too seriously from the form of the input, although the latter becomes somewhat elongated and the fine structure of its waveform is slightly blurred. This "smearing" effect of the input signal evidently becomes worse if $h(t)$ is spread out in time, and we can thus formulate by inspection of the form of $h(t)$ a reasonable estimate of how good or bad the reproduction of an input signal will be if transmitted through the pertinent network.

In an analogous fashion we can begin with a sampled form of the input signal that reads

$$f_1(t) = \sum_{k=-\infty}^{\infty} b_k u_0(t - kt_0) \tag{82}$$

in which the samples are

$$b_k = t_0 f_1(kt_0) \tag{83}$$

Substitution into the convolution integral 50 and evaluation similar to that in Eq. 80 gives

$$f_2(t) = \sum_{k=-\infty}^{\infty} b_k h(t - kt_0) \tag{84}$$

Here the output is expressed as a sum of delayed replicas of the impulse response, monitored in amplitude by samples of $f_1(t)$. If the latter is compact relative to the duration of $h(t)$, the output signal will not look like the input at all, but instead will look like the impulse response. This rather interesting turn of events can be recognized by inspection of the graphical appearance of the input and impulse response functions.

Apropos the smearing effect mentioned above, an extremely interesting and somewhat baffling phenomenon is pertinent here. Consider an input signal $f_1(t)$ whose spectrum $F_1(j\omega)$ is sessentially band-limited within the frequency interval $-\omega_c < \omega < \omega_c$, and a network whose frequency response is the rectangular function

$$H(j\omega) = \begin{cases} 1 & \text{for } -\omega_c < \omega < \omega_c \\ 0 & \text{for } |\omega| > \omega_c \end{cases} \tag{85}$$

The time domain mate of this function is the impulse response

$$h(t) = \frac{\omega_c}{\pi} \cdot \frac{\sin \omega_c t}{\omega_c t} \tag{86}$$

which is nonzero over the infinite interval $-\infty < t < \infty$.

If we sample this impulse response and substitute the samples into expression 81 for the output $f_2(t)$, our graphical interpretation given above would certainly tell us that $f_2(t)$ is a rather badly smeared version of the input signal $f_1(t)$. The truth is, however, that the output is an undistorted version of the input since multiplication of $H(j\omega)$ in 85 by $F_1(j\omega)$ leaves this spectrum function unchanged because its nonzero region is, by assumption, contained within the range $-\omega_c < \omega < \omega_c$. Evidently the sin x/x function has very special properties that defy recognition of this result by inspection of the graphical appearance of the time functions involved in the convolution process.

By the same reasoning, the convolution of two sin x/x functions yields that one whose transform has the narrower bandwidth. Without the pertinent frequency-domain interpretation, this result would be rather difficult to predict. We have here a good example of the principle that Fourier transformation can often be indispensable in aiding the solution of a given problem since situations that are difficult to analyze in one domain can become greatly simplified through transformation to the other domain.

8. The Excitation Function Is Periodic over the Infinite or Semi-Infinite Time Domain

We shall next consider a situation in which a periodic excitation is applied to a passive (or at least a stable) network and a sufficiently long time has elapsed so that all transients have subsided and only the steady periodic response remains. It is this response that we wish to determine.

Both excitation and response are then representable by Fourier series. We write them respectively

$$f_1(t) = \sum_{\nu=-\infty}^{\infty} \alpha_\nu e^{j\nu\omega_0 t} \tag{87}$$

and

$$f_2(t) = \sum_{\nu=-\infty}^{\infty} \beta_\nu e^{j\nu\omega_0 t} \tag{88}$$

Theoretically these functions exist for the doubly infinite interval $-\infty < t < \infty$, notwithstanding the fact that in reality their existence dates back only to a finite time; the important point, however, being that the elapsed time interval is long enough so that the transient terms have, for all practical purposes, died out and the net result is indistinguishable from what it would be if the excitation had always been there.

Linearity of the network allows us to compute each sinusoidal component as though the others were not present. Hence if $H(j\nu\omega_0)$ is the ratio of response to excitation amplitudes (complex, of course) for a single sinusoid of radian frequency $\nu\omega_0$ (the usual definition of a frequency-domain system function), then we have

$$\beta_\nu = \alpha_\nu H(j\nu\omega_0) \tag{89}$$

All Fourier coefficients in the periodic response $f_2(t)$ are thereby determined.

Note that all three spectrum functions $\alpha_\nu = \alpha(\nu\omega_0)$, $\beta_\nu = \beta(\nu\omega_0)$, and $H(j\nu\omega_0)$ are functions of the discrete variable $\nu\omega_0$ for integer values of ν.

The system function here is the same as $H(j\omega)$ defined in Eq. 43, evaluated for the discrete frequencies $\omega = \nu\omega_0$. Only line spectra are involved, and both amplitude and phase relations are taken care of in the familiar manner by the complex character of these functions.

Now let us consider the problem for which the network is at rest before the instant $t = 0$, at which time the periodic excitation $f_1(t)$ given by Eq. 87 is suddenly applied. The net response is desired for all time $t > 0$.

Following the general procedure given in art. 3, we begin by determining the Fourier transform $F_1(j\omega)$ for the pertinent excitation function. In anticipation of the use of complex integration we replace $j\omega$ by s. Then, since the transform of

$$u_{-1}(t)e^{j\nu\omega_0 t} \tag{90}$$

is

$$\frac{1}{s - j\nu\omega_0} \tag{91}$$

we have

$$F_1(s) = \sum_{\nu=-\infty}^{\infty} \frac{\alpha_\nu}{s - j\nu\omega_0} \tag{92}$$

The corresponding transform of the response is

$$F_2(s) = F_1(s) \cdot H(s) \tag{93}$$

and in the time domain we have

$$\begin{aligned} f_2(t) &= \frac{1}{2\pi j}\int_{-j\infty}^{j\infty} \sum \frac{\alpha_\nu H(s)}{s - j\nu\omega_0} e^{st}\, ds \\ &= \sum_{\nu=-\infty}^{\infty} \frac{\alpha_\nu}{2\pi j}\int_{-j\infty}^{j\infty} \frac{H(s)}{s - j\nu\omega_0} e^{st}\, ds \end{aligned} \tag{94}$$

If $H(s)$ is rational, and if its poles are at the points $s = s_a, s_b, \ldots s_n$, we can write the partial fraction expansion

$$\frac{H(s)}{s - j\nu\omega_0} = \frac{A_{\nu 0}}{s - j\nu\omega_0} + \frac{A_{\nu a}}{s - s_a} + \frac{A_{\nu b}}{s - s_b} + \cdots + \frac{A_{\nu n}}{s - s_n} \tag{95}$$

in which

$$A_{\nu 0} = H(j\nu\omega_0); \quad A_{\nu k} = \left[\frac{(s - s_k)H(s)}{s - j\nu\omega_0}\right]_{s=s_k} \quad \text{for } k = a, b, \ldots n \tag{96}$$

The integration in 94 is then readily carried out and we have

$$f_2(t) = \sum_{\nu=-\infty}^{\infty} \alpha_\nu H(j\nu\omega_0)e^{j\nu\omega_0 t} + \sum_{\nu=-\infty}^{\infty} \alpha_\nu\{A_{\nu a}e^{s_a t} + A_{\nu b}e^{s_b t} + \cdots + A_{\nu n}e^{s_n t}\} \tag{97}$$

for $t > 0$ and, of course, $f_2(t) \equiv 0$ for $t < 0$. The first term here coincides with results given by Eqs. 88 and 89, and represents the steady-state part of the response. The second term is the transient response. If we define a set of coefficients

$$B_k = \sum_{\nu=-\infty}^{\infty} \alpha_\nu A_{\nu k} \tag{98}$$

then result 97 can be written in the somewhat simpler form

$$f_2(t) = \sum_{\nu=-\infty}^{\infty} \alpha_\nu H(j\nu\omega_0)e^{j\nu\omega_0 t} + B_a e^{s_a t} + B_b e^{s_b t} + \cdots + B_n e^{s_n t} \tag{99}$$

This same problem can be solved in another way that is collaterally interesting and is, incidentally, applicable whether $H(s)$ is rational or not. As in art. 4, Ch. XV, let $\hat{f}_1(t)$ be identical with the periodic function $f_1(t)$ over one period and zero everywhere else. Let this period begin at $t = 0$. Then we can write

$$f_1(t) = \hat{f}_1(t) + \hat{f}_1(t - \tau) + \hat{f}_1(t - 2\tau) + \cdots \tag{100}$$

where τ denotes the fundamental period.

If $\hat{F}_1(j\omega)$ denotes the Fourier transform of $\hat{f}_1(t)$ and the inverse transform of $H(j\omega)$, $\hat{F}_1(j\omega)$, is denoted by $\hat{f}_2(t)$, then the desired response is given by

$$f_2(t) = \hat{f}_2(t) + \hat{f}_2(t - \tau) + \hat{f}_2(t - 2\tau) + \cdots \tag{101}$$

Note that the separate terms in this expression are nonzero over a semi-infinite range beginning at $t = k\tau$ for $k = 0, 1, 2, \ldots$, whereas those in expression 100 are nonzero over a single period only.

Computation of $f_2(t)$ by Eqs. 98 and 99 for any finite t necessitates the evaluation of infinite summations, which means that we shall have to be content with an approximate result provided by partial sums. Use of the alternate expression 101, on the other hand, while it requires the consideration of a larger number of terms for increasing values of t, yields the desired response in closed form for any finite t.

In this connection the following manipulations are effective in eliminating the infinite summation in Eq. 98. In view of Eq. 100 and the transform $\hat{F}_1(s)$ of $\hat{f}_1(t)$, we can express the transform $F_1(s)$ of the periodic function $f_1(t)$ for $t > 0$ in the form

$$F_1(s) = \hat{F}_1(s) \sum_{k=0}^{\infty} e^{-sk\tau} \tag{102}$$

In the right-half s-plane where the infinite sum appearing here is an absolutely convergent series it may be replaced by the closed form

$$\sum_{k=0}^{\infty} e^{-sk\tau} = \frac{1}{1 - e^{-s\tau}} = \frac{e^{s\tau/2}}{2 \sinh s\tau/2} \tag{103}$$

This function has simple poles on the j-axis where $s\tau/2 = j\nu\pi$ for integer values of ν, or for the s-values

$$s = s_\nu = \frac{j2\pi\nu}{\tau} = j\nu\omega_0 \tag{104}$$

Its residues are simply

$$\rho_\nu = \frac{e^{j\nu\pi}}{\tau \cosh j\nu\pi} = \frac{1}{\tau} \tag{105}$$

Since it is regular everywhere on the j-axis except at the isolated poles, it possesses an analytic continuation into the left half-plane; and the identity theorem shows that this continuation must be given by the same functional expression 103. Hence this closed form is a valid representation for the infinite sum in the entire finite s-plane, notwithstanding the fact that this sum fails to be absolutely convergent in the left half-plane.

The transform $F_1(s)$ in Eq. 102 can, therefore, be expressed in the form

$$F_1(s) = \frac{\hat{F}_1(s)}{1 - e^{-s\tau}} \tag{106}$$

which is seen to be equivalent to the alternate form 92 since its partial fraction expansion (in view of 104 and 105) reads

$$F_1(s) = \frac{1}{\tau} \sum_{\nu=-\infty}^{\infty} \frac{\hat{F}_1(s_\nu)}{s - j\nu\omega_0} \tag{107}$$

and

$$\hat{F}_1(j\omega) = \int_0^\tau f_1(t)\, e^{-j\omega t}\, dt \tag{108}$$

while the Fourier coefficients in 92 are given by

$$\alpha_\nu = \frac{1}{\tau} \int_0^\tau f_1(t)\, e^{-j\nu\omega_0 t}\, dt \tag{109}$$

The expression 98 for the coefficients B_k in Eq. 99 can now also be put into an equivalent closed form. Making use of Eqs. 92 and 96 we have

$$\begin{aligned} B_k &= \sum_{\nu=-\infty}^{\infty} \alpha_\nu A_{\nu k} = \sum_{\nu=-\infty}^{\infty} \frac{\alpha_\nu}{s_k - j\nu\omega_0} [(s - s_k)H(s)]_{s=s_k} \\ &= [(s - s_k)H(s)]_{s=s_k} \sum_{\nu=-\infty}^{\infty} \frac{\alpha_\nu}{s_k - j\nu\omega_0} = [(s - s_k)H(s)]_{s=s_k} F_1(s_k) \end{aligned} \tag{110}$$

With Eq. 106 this yields the desired closed form

$$B_k = [(s - s_k)H(s)]_{s=s_k} \times \frac{\hat{F}_1(s_k)}{1 - e^{-s_k\tau}} \tag{111}$$

It is now possible to obtain a closed-form evaluation* for the first term (the steady-state response) in Eq. 99. Since this term is a periodic function, it is sufficient to evaluate it over one period, let us say, from $t = 0$ to $t = \tau$. Here $f_2(t)$ is given by the first term in Eq. 101, and since the B_k's are given in closed form by Eq. 111, the steady-state term in Eq. 99 may likewise be expressed in closed form.

A collaterally interesting point is the following. If in Eq. 111 we omit dividing $\hat{F}_1(s_k)$ by the factor $1 - e^{-s_k\tau}$, the resulting B_k's nevertheless yield the correct transient term in Eq. 99 for the interval $0 < t < \tau$, and hence the above evaluation of the steady-state term in this equation is seen not to be dependent upon the result 106.

9. Response of an Ideal Filter

The so-called ideal lowpass filter is defined as a network for which the system function $H(j\omega)$ is stipulated to be

$$\begin{aligned} H(j\omega) &= e^{-j\omega t_d} \qquad \text{for } -\omega_c < \omega < \omega_c \\ H(j\omega) &\equiv 0 \qquad \text{otherwise} \end{aligned} \tag{112}$$

According to representation 68, the amplitude $A(\omega)$ is equal to unity for the range $-\omega_c < \omega < \omega_c$, the passband of the filter, and zero everywhere else. The phase function $\beta(\omega)$ is linear over the passband and need not be defined otherwise since the response is zero. The ideal filter, therefore, fulfills the conditions 69 for facsimile reproduction of signal waveforms over its passband.

The unit impulse response is evidently given by the Fourier integral

$$h(t) = \frac{1}{2\pi}\int_{-\omega_c}^{\omega_c} e^{j\omega(t-t_d)}\,d\omega \tag{113}$$

which evaluates to

$$h(t) = \frac{1}{2\pi}\left[\frac{e^{j\omega(t-t_d)}}{j(t-t_d)}\right]_{-\omega_c}^{\omega_c} = \frac{\omega_c}{\pi}\cdot\frac{\sin \omega_c(t-t_d)}{\omega_c(t-t_d)} \tag{114}$$

This result is shown plotted in the upper sketch of Fig. 1 from which we see the distortive effect that the lowpass filter exerts upon the applied signal because of the fact that it transmits only a limited range of frequencies. Apart from the delay which is equal to the slope of the phase function $\beta(\omega)$, the effect of the limited bandwidth is to spread the applied impulse out into a rounded pulse of nonzero duration, preceded and followed by ripples which decrease in amplitude as one recedes from the pulse in either direction. Since the width of the main pulse is $2\pi/\omega_c$ and

* This was first pointed out to the writer by T. G. Stockham.

its height is ω_c/π, the net enclosed area is unity. It is clear, therefore, that as the width of the passband of the filter is increased, the response approaches an impulse as it should. On the other hand, the narrower the passband of the filter, the more its impulse response spreads out in time.

If we now wish to obtain the unit step response of the ideal lowpass filter, we have but to integrate function 114 with respect to t. This process

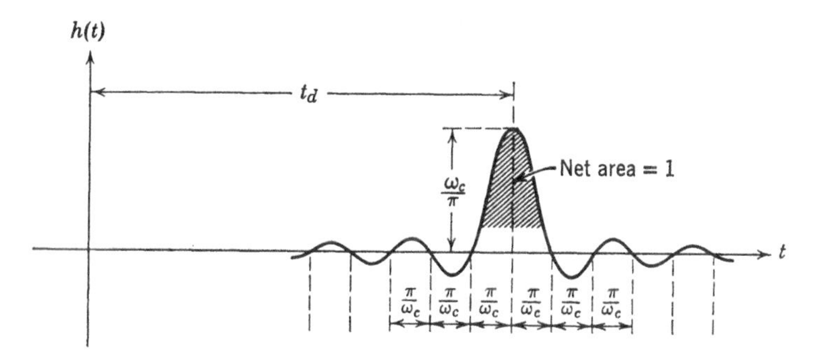

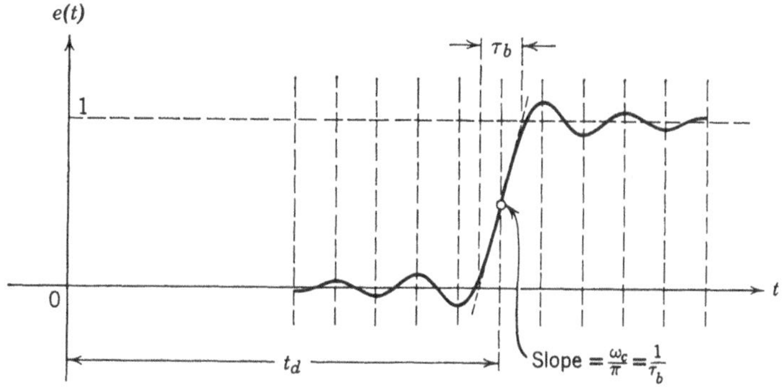

FIG. 1. Impulse and step response of the ideal lowpass filter.

(which is very similar to the manipulations in Eqs. 11 and 12 of Ch. XIV) yields the function

$$e(t) = \int_{-\infty}^{t} h(\xi)\, d\xi = \frac{1}{\pi}\int_{-\infty}^{\omega_c(t-t_d)} \frac{\sin u}{u}\, du = \frac{1}{2} + \frac{1}{\pi}\, Si\omega_c(t - t_d) \quad (115)$$

shown plotted in the lower sketch of Fig. 1.

The point of steepest slope occurs where the impulse response is a maximum and, of course, equals this maximum value. If, as in Eq. 115, we denote this step response by $e(t)$, then

$$\left(\frac{de}{dt}\right)_{t=t_d} = \frac{\omega_c}{\pi} \quad (116)$$

The asymptotic value of $e(t)$ equals the total area under the impulse response $h(t)$ which is unity as it should be since the steady response of the filter to a unit step equals the zero-frequency value of its response function (the dc response) which in this example is unity.

If we define the so-called "rise time" or "buildup time" of the function $e(t)$ as the interval τ_b between the intercepts of the tangent at $t = t_d$ with the zero axis and with the asymptote, then evidently

$$\left(\frac{de}{dt}\right)_{t=t_d} = \frac{1}{\tau_b}, \qquad \text{or} \quad \tau_b = \frac{\pi}{\omega_c} \tag{117}$$

The effect of the limited transmission band of the filter upon the step-function response is that it does not possess the abrupt jump of the applied signal but requires a finite time to follow this abrupt change. Approximately, this buildup time is given by the simple formula 117, according to which it is inversely proportional to the filter bandwidth.

Since, for obvious practical reasons, any communication channel can provide only a finite transmission band, this result is of fundamental importance. It is the variation in amplitude of a signal that contains or conveys the information embodied in it, and hence the speed with which information can be transmitted depends upon the rate at which the output can follow variations in the input. This rate is inversely proportional to τ_b and hence proportional to the bandwidth ω_c.

This result expresses a basic principle in any communication system. To transmit a given amount of information over a frequency channel of given width requires a definite amount of time (subject, of course, to the relation between information and signal; that is, whether the information is in the form of speech or code, whether sentences or whole messages are represented by code symbols, etc.) The time of transmission (not to be confused with the delay t_d) may be shortened by widening the frequency band. A narrower band requires a longer total time. The ultimate conclusion is that a zero bandwidth requires an infinite time, or that no information can be transmitted unless a band of nonzero width is provided.

The results given in Eqs. 114 and 115, and illustrated in Fig. 1 may readily be extended to the corresponding problem involving an ideal bandpass filter through use of the property expressed by Eq. 77 in Ch. XII in which $f(t)$ is identified either with $h(t)$ or with $e(t)$ and in the frequency domain $F(s)$ is identified with the lowpass filter characteristic $H(s)$. Thus the ideal bandpass filter with bandwidth $w = 2\omega_c$ centered at $\omega = \omega_0$ has a system function

$$\tfrac{1}{2}[H(j\omega - j\omega_0) + H(j\omega + j\omega_0)] \tag{118}$$

in which $H(j\omega)$ is the lowpass function 112. Its response to an applied unit impulse is $h(t)$ in Eq. 114 multiplied by $\cos \omega_0 t$, and its response to an excitation

$$u_{-1}(t) \cdot \cos \omega_0 t \tag{119}$$

is the function $e(t)$ in Eq. 115 multiplied by $\cos \omega_0 t$.

In other words, the impulse and step response functions for the lowpass case become the envelope functions for the corresponding bandpass situation. Since the envelope function monitors the response of the bandpass filter, the various remarks made above regarding amplitude distortion caused by the finite bandwidth in the transmission of an impulse or a step function remain essentially unaltered. The buildup time in Eq. 117 is more appropriately written in terms of the bandwidth, as

$$\tau_b = \frac{2\pi}{w} = \frac{1}{\Delta f} \text{ seconds} \tag{120}$$

Here Δf is the bandwidth in cycles per second.

10. The Relative Importance of Amplitude and Phase Distortion in Practical Circuits

The ideal filter has a system function whose amplitude $A(\omega)$ is equal to unity over its transmission ranges and zero over the attenuation ranges. The phase function $\beta(\omega)$ is required to be linear only throughout the transmission ranges because its effect is nil wherever the amplitude function is zero. The ideal filter, therefore, produces no phase distortion. The difference between applied and received signals is due entirely to amplitude distortion. For a given type of signal, the location and width of the passband of the filter is chosen so as to keep the resulting amplitude distortion within tolerable limits. This can usually be done because the spectrum function for most practical signals is essentially confined to a limited band. It should be recognized in this connection that the selection of a suitable bandwidth is solely dependent upon allowable tolerances in the practical application under consideration.

It is clear that an arbitrary function, if replaced by a step approximation, may be regarded as a linear superposition of rectangular pulses. From the preceding discussion it follows that the component pulse of shortest duration is the one which determines the necessary bandwidth, because the buildup time τ_b must at least be equal to this duration; otherwise this shortest pulse will be absent altogether in the reproduced signal. It should be understood in this connection that negative as well as positive pulses must be considered. For example, two pulses separated by a very short zero interval are equivalent to a longer pulse with a short negative one somewhere within this longer interval. It is the duration of the short interval in this case that determines the bandwidth necessary for proper separation of the two pulses.

It is instructive to contrast the distortion caused by an ideal filter with that caused by a network having a flat amplitude function over the entire spectrum but a nonlinear phase. An idealized case may be thought of as having the phase characteristic shown in Fig. 2. This phase has a straight-line variation with slope t_1 from the origin to an angular frequency ω_0, and another straight-line variation but with a larger slope t_2 above this frequency.

Let us consider the transmission of a rectangular pulse of duration δ through this network. Since the corresponding spectrum for this pulse

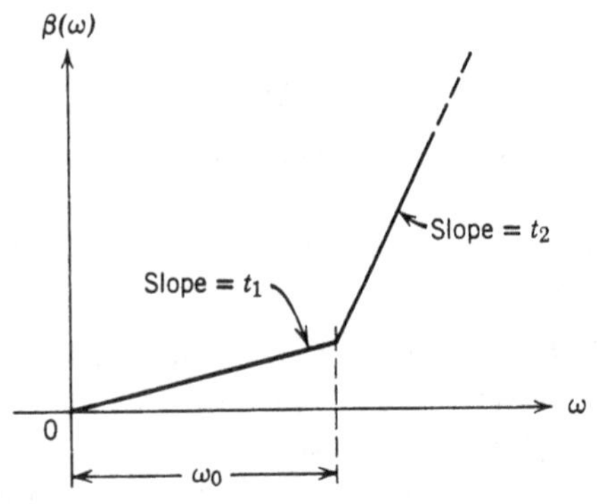

FIG. 2. Filter having its phase represented by a broken line.

is given by a $\sin x/x$ function with zeros where ω equals plus or minus integer multiples of $2\pi/\delta$ (for example, see parts (*a*) and (*b*) of Fig. 5 in Ch. XII), an interesting result might be expected if ω_0 is chosen equal to the relatively low value of π/δ. Utilizing the principle of superposition, the net response may be thought of as consisting of two components. One of these is the response of an ideal lowpass filter with its cutoff at ω_0 and a delay equal to t_1; the other is the response of an ideal highpass filter with its cutoff at ω_0 and a longer delay equal to t_2.

Now the response of the ideal highpass filter is simply equal to the difference between the applied signal delayed by t_2 seconds and the response of an ideal lowpass filter with its cutoff at ω_0 delayed by t_2 seconds. The highpass filter passes that part of the signal which a lowpass filter with the same cutoff and delay would reject. The first of the above mentioned components (the response of an ideal lowpass filter with its cutoff at ω_0 and a delay t_1) has a buildup time τ_b (by Eq. 117) equal to δ seconds; hence it consists of a considerably rounded and elongated pulse. The second of the components consists of some peaks representing the sharp corners of the original signal which are suppressed in the first component, and an oscillatory portion or ripple which is complementary to that in the first component. The first component represents the low-frequency content while the second component represents the high-frequency content contained in the original signal.

If these two contents were delayed by the same amount, the result would be a perfect reproduction of the sign. However, the high-frequency content is delayed by a longer time. If the difference in delay between the low- and high-frequency contents is small compared with the duration δ of the pulse, the net effect is obviously insignificant; but if this difference is of the same order of magnitude as δ, or perhaps larger, then the distortion becomes decidedly noticeable.

It is thus clear that the conditions for which phase distortion becomes noticeable are: first, that the nonlinear portion of the phase characteristic shall lie within the essential part of the frequency spectrum for the given signal, and second, that the variation in delay (slope) shall be at least comparable to the duration of the signal or of a physically perceptable duration in case the signal itself is long.

In the case of speech signals, each spoken syllable may (roughly) be looked upon as an individual signal. The average duration of a spoken syllable is of the order of magnitude of from 0.01 to 0.1 of a second. The networks embodied in most speech amplifiers and similar equipment may have very nonlinear phase characteristics over the audio spectrum, yet no noticeable phase distortion results because the maximum delay and hence the maximum variation in delay in such networks is only a small fraction of a millisecond on account of the relatively few elements involved.

In contrast, some heavily loaded telephone cable circuits have, on account of their length and consequent large number of loading coils, an average delay that is roughly of the same order of magnitude as the spoken syllable. The phase characteristic of such circuits has a slope that increases with frequency, reaching several times the low-frequency value at the upper end of the essential voice spectrum. Unless these circuits are provided with delay equalization, in addition to attenuation equilization, the increased delay of the high-frequency content in speech becomes noticeable in the form of "squeals" at the ends of spoken syllables in addition to a general "hollow barrel" effect.

In circuits intended to transmit television signals the maximum variation in delay must be kept very much smaller on account of the much shorter duration of the separate signal pulses. In the design of networks for such applications the phase requirements are dominant. The effect of amplitude distortion is here evidenced by a partial destruction of the relative half-tone values in the resulting picture. The human eye is comparatively tolerant toward this type of distortion. Phase distortion, on the other hand, causes a shifting in the position of picture elements (smearing) to which the eye is decidedly intolerant.

This situation may roughly be summarized by saying that the human ear is intolerant to amplitude distortion but relatively tolerant toward phase distortion (unless it becomes too severe), while the reverse is true when the eye is the ultimate detector of the received signal. It should be emphasized, however, that in all applications the phase of the system function should be considered as carefully as the amplitude of that function until their relative importance has definitely been established. It is too commonly thought that a flat amplitude characteristic alone is an assurance of quality. A network may have a flat amplitude function over the entire frequency spectrum, and yet distort a given signal beyond recognition because of a nonlinear phase characteristic; and this circumstance can be equally true for audio as well as for high-frequency signals.

11. Compatibility of Amplitude and Phase Functions for Physical Networks

The result illustrated in Fig. 1 for the response of an ideal lowpass filter to the sudden application of an impulse or step places in evidence an additional point which was not specifically emphasized earlier because it has no bearing upon the principles discussed there. A closer examination of this sketch (and of the impulse or step response functions 114 or 115) shows that the output is not zero before $t = 0$, which is impossible in a physical network. Since the mathematical analysis leading to this result is not in error, we must conclude that the assumed system function cannot be that of a physical network.

In Chapter XVIII where we will discuss the detailed aspects of this question more carefully we will find that no physical network can have a response function that is ideal over a limited frequency range and zero otherwise. Relevant here, as the reader may have guessed, is the discussion in art. 5 in which it is shown that, in a physical network, the real or imaginary part of a system function alone suffices to determine the response for $t > 0$ (where the excitation is assumed to be applied at $t = 0$). Equations 54 through 63 are pertinent to this result. By means of Eq. 62 or 63 we can express $F_1(\omega)$ in terms of $F_2(\omega)$ or vice versa. As we shall see, we can similarly relate the phase function $\beta(\omega)$ and the logarithm of $A(\omega)$, the so-called *attenuation function*. Pairs of functions so related are referred to (in the parlance of potential theory*) as *conjugate potential functions*. With reference to the system function of a physical network they are said

* See *MCA*, pp. 330–360.

to be compatible. The amplitude and phase response functions of an ideal filter (as defined above) are incompatible.

The general conclusions drawn from the character of the resulting response for this ideal case are, however, not impaired by this circumstance, as is clear from the fact that the response before $t = 0$ can be made insignificant by assuming a sufficiently large delay time t_d, which, in the preceding discussion, remains arbitrary anyway. From the standpoint of network design or synthesis, on the other hand, the fact that $A(\omega)$ and $\beta(\omega)$ cannot independently be specified, is an important consideration to which we shall return later on.

12. A Filter That Is Ideal with Respect to Its Time-Domain Response

The ideal filter defined in art. 9 is ideal from a frequency-domain standpoint. Its response in the time domain, as shown in Fig. 1, exhibits characteristics that are not desirable for many practical applications. Even though this ideal filter were physically realizable, it would not represent

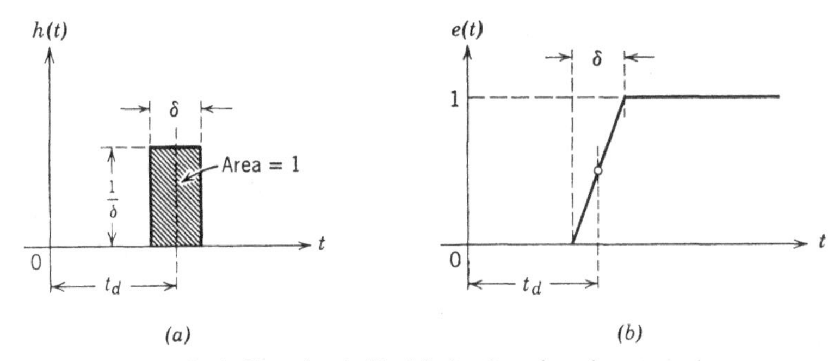

FIG. 3. A filter that is ideal in its time-domain response.

the best or the most acceptable solution to many problems. In other words, it is "ideal" only in a certain sense, and there are numerous situations in practice where these criteria do not apply. They do not apply in most cases where detailed form of the time-domain response is important. For example, the ripply character of the step response function shown in Fig. 1 is often objectionable, the fact that the response overshoots its asymptotic value by the characteristic 9% (as in the Gibbs phenomenon discussed in Ch. XIV) regardless of the filter bandwidth or the delay t_d. Any physical filter whose amplitude characteristic approximates the rectangular form of the ideal filter yields a step response displaying a similar overshoot and ripply approach to its asymptotic final value.

A more desirable form of time-domain response, one that might well be designated as "ideal" in this sense, is shown in the sketches of Fig. 3.

Part (*a*) shows the unit impulse response, part (*b*) the corresponding step response. The latter exhibits a linear rise during an interval δ (the build-up time in this case) followed immediately by the steady final value. The impulse response here may be represented analytically by the relations

$$h(t) = \begin{cases} \dfrac{1}{\delta} & \text{for } t_d - \dfrac{\delta}{2} < t < t_d + \dfrac{\delta}{2} \\ 0 & \text{otherwise} \end{cases} \tag{121}$$

The step response $e(t)$ is the integral of this function as defined in Eq. 115.

The frequency-domain response function yielding this impulse response is found straightforwardly to be

$$H(j\omega) = \frac{\sin \omega\delta/2}{\omega\delta/2} \cdot e^{-j\omega t_d} \tag{122}$$

The amplitude has the form of $\sin x/x$, and the phase is linear throughout the infinite domain with a slope t_d that must be at least as large as $\delta/2$ in order that the response be zero for $t < 0$. If this condition is fulfilled then we may say that the system function 122, unlike the function 112 representing the ideal filter, meets necessary realizability conditions.

In contrast with the rectangular amplitude response of the ideal filter, this one does not possess a clearly definable cutoff, unless we arbitrarily choose the first zero at $\omega = 2\pi/\delta$ as a reasonable cutoff point, and it displays a passband response that drops off continuously as this band limit is approached. We may conclude from this fact that a rounded response over the passband of a filter is preferable to a more rectangular response if overshoot and rippling of the step response is undesirable.

Further detailed consideration of this problem is essentially identical with investigating the effect of spectrum truncation discussed in arts. 2 and 3 of Ch. XIV. The fact that a trapezoidal truncation function, shown there in Fig. 5 and leading to the impulse response in Fig. 6, yields practically no rippling again bears out the fact that a filter with a gradually decreasing response over the passband has more desirable time-domain characteristics and hence is more nearly ideal from this standpoint.

13. Signal Delay in the Presence of Simultaneous Amplitude and Phase Distortion

When the amplitude response is flat over a passband or over the entire frequency domain, and the associated phase function is linear, signal delay is clearly and simply definable in terms of phase slope. In situations where the phase is nonlinear and where the amplitude variation may simultaneously deviate appreciably from a flat response, the concept of signal delay is not as clearly definable nor is its determination as simple. The following ideas and methods are helpful in dealing with many practical problems.

Consider integral 48 for the unit impulse response in terms of a given system function and express the latter in the form 68. We then have

$$h(t) = \frac{1}{2\pi}\int_{-\infty}^{\infty} A(\omega)e^{j(\omega t-\beta)}\, d\omega \tag{123}$$

Since $A(\omega)$ is an even function and $\beta(\omega)$ odd, we can rewrite this integral in the form

$$h(t) = \frac{1}{\pi}\int_{0}^{\infty} A(\omega)\cos(\omega t - \beta)\, d\omega \tag{124}$$

The integrand appearing here is an amplitude-modulated cosine function. Apart from the multiplier $1/\pi$, the value of $h(t)$ for any t equals the net area enclosed by this amplitude-modulated cosine function. This area can at most equal the area enclosed by $A(\omega)$ itself (since, by definition, it is non-negative for all ω) and in general it is less than this area because the oscillatory cosine chops the amplitude function $A(\omega)$ into positive and negative sections. The extent of this chopping effect depends upon the variation of the argument $\omega t - \beta$ with frequency.

If $\beta(\omega)$ has a straight-line variation with slope t_d, then clearly when $t = t_d$ this argument is identically zero; the cosine remains unity for all ω, and $h(t)$ equals $1/\pi$ times the area enclosed by $A(\omega)$. The time t_d at which this maximum of $h(t)$ occurs is regarded as the delay, since the form of $h(t)$, although not like a perfect impulse, will in most cases be that of a fairly compact pulse.

If the phase function $\beta(\omega)$ is nonlinear, the character of the resulting $h(t)$ will depend largely upon where the greatest deviation from linearity occurs in relation to the $A(\omega)$ function. If $\beta(\omega)$ is still approximately linear over that range where $A(\omega)$ has significantly large values (as in the passband of a filter), then the resulting $h(t)$ function is essentially the same as it is for a linear phase since the chopping effect of the cosine function is not important where the values of $A(\omega)$ are relatively small. A reasonably linear phase characteristic over the passband of a filter is, therefore, an assurance that the impulse response will be a compact pulse (subject, of course, to the dependence of its duration upon bandwidth as shown in Fig. 1) with its peak occurring at a time equal to the average slope of the phase in this band.

This definition of signal delay as the time of occurrence of the maximum of $h(t)$ is, of course, an arbitrary one. An alternate definition that is also found useful is formulated as follows. Consider the graphical plot of $h(t)$ versus t and the area enclosed by it. Find the centroid of this area and

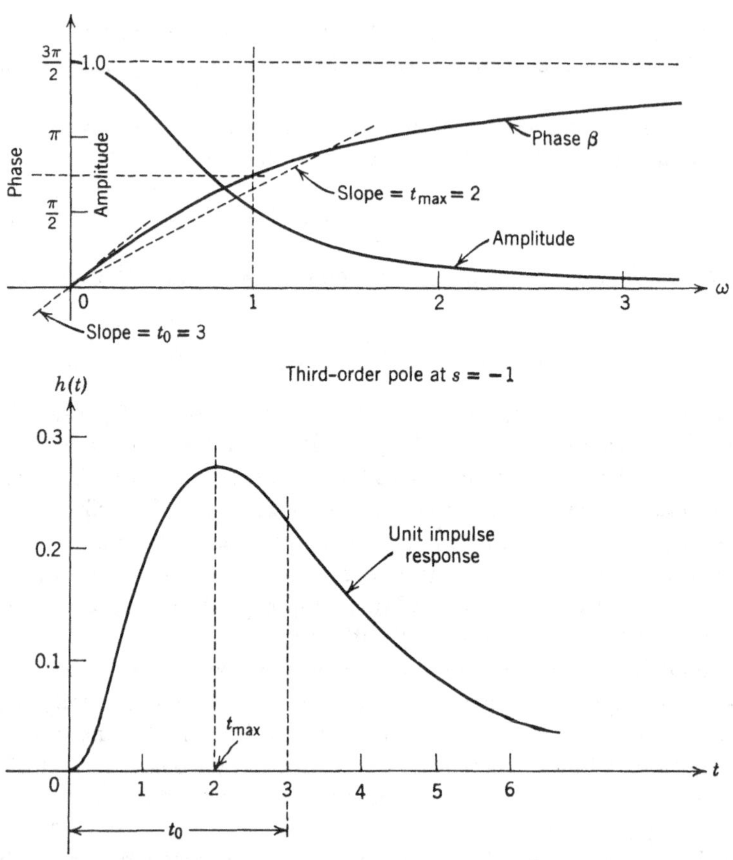

FIG. 4. Example for which the average phase slope yields the better estimate for impulse delay.

define its time of occurrence t_0 as the delay of the pulse $h(t)$. Analytically this formulation reads

$$t_0 = \frac{\int_0^\infty t\, h(t)\, dt}{\int_0^\infty h(t)\, dt} \tag{125}$$

In order to evaluate this expression in terms of frequency-domain characteristics, we consider the system function in logarithmic form by writing

$$H(j\omega) = e^{-(\alpha + j\beta)} \tag{126}$$

Making use of the direct Fourier transformation (Eq. 49)

$$H(j\omega) = \int_0^\infty h(t)\, e^{-j\omega t}\, dt \tag{127}$$

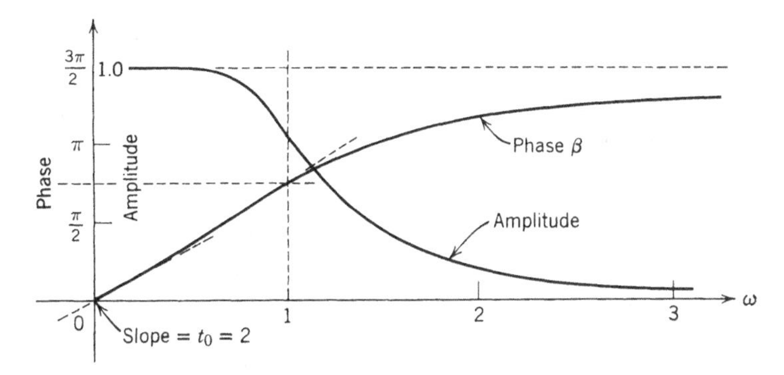

Third-order Butterworth

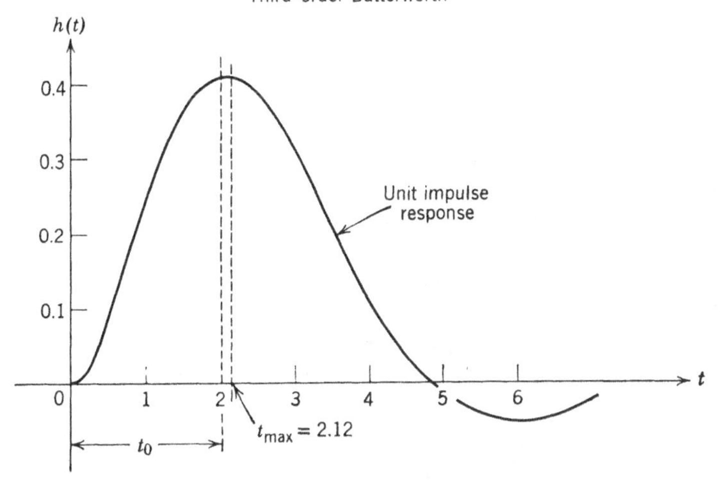

FIG. 5. Example for which either initial or average slope gives good estimate for impulse delay.

and its derivative

$$\frac{dH}{d(j\omega)} = -H(j\omega)\left[\frac{d\alpha}{j\,d\omega} + \frac{d\beta}{d\omega}\right] = -\int_0^\infty t\,h(t)\,e^{-j\omega t}\,dt \tag{128}$$

evaluation for $\omega = 0$, noting that $d\alpha/d\omega$ is zero for $\omega = 0$ because $\alpha(\omega)$ is an even function, gives the remarkably simple result

$$t_0 = \left(\frac{d\beta}{d\omega}\right)_{\omega=0} \tag{129}$$

which states that the slope of the phase at $\omega = 0$ equals the location of the centroid of area enclosed by the $h(t)$-function.

In Figs. 4 and 5 we show examples of how this definition of signal delay turns out in two characteristically different cases. The first (Fig. 4) is a lowpass network whose transfer function has a third-order pole at $s = -1$. The amplitude and phase characteristics $A(\omega)$ and $\beta(\omega)$ are shown in the upper part of this figure; the impulse response in the lower. The phase in this case is entirely convex. Hence its slope at $\omega = 0$ is larger than the slope anywhere else; that is to say, it is larger than the average slope over any region of finite extent. The amplitude characteristic, although lowpass in general character, is such that it is difficult to designate a cutoff frequency. Hence the "average slope of the phase over the passband" is not clearly measurable, although it is less than the slope at $\omega = 0$.

In the plot of $h(t)$ this much is borne out since t_{max} is less than t_0. The "centroid of area" method of defining signal delay does not appear in this example to be a reasonable measure of the actual delay. The situation improves, however, if we consider higher-order poles at $s = -1$. For an nth-order pole we have*

$$t_0 = n \quad \text{and} \quad t_{max} = n - 1 \tag{130}$$

which shows that the centroid of area moves relatively closer to t_{max} as n increases.

Fig 5 is also for a network having a three-pole transfer function, but here the poles have a Butterworth† distribution. The resulting phase function is at first concave and then convex, having a point of inflection at the cutoff $\omega = 1$. The average slope over the passband is quite well defined and is just slightly larger than the slope at $\omega = 0$. The sketch for $h(t)$ bears out these conclusions and shows that either definition for signal delay (average phase slope over the passband or centroid of area) is meaningful and acceptable.

We may conclude that the centroid-of-area definition in general is not as reasonable a measure of signal delay as the average-slope method because it tends to yield too large a value wherever $h(t)$ is a non-negative pulse (as in the example of Fig. 4). The reason the centroid-of-area method is better in the Butterworth case is that here $h(t)$ contains oscillations that contribute negative components to the area at large values of t, thus decreasing the moment integral in the numerator of 125 more than the net area in the denominator and hence preventing the overly large centroid of area which results for a one-signed $h(t)$ like that in Fig. 4.

No simple method of measuring signal delay can, of course, be infallible; and in cases of doubt or in idealized situations where the above methods become degenerate, one must resort to an actual evaluation of the impulse response. In this connection the methods discussed in the next chapter are significantly helpful.

* See *ICT*, p. 442.

† See Ch. IX, art. 4.

14. "Simile" Transforms

Let us consider the time function $f_1(t)$ to be the input to a given transmission network and let us suppose that in the vicinity of some point $t = t_0$ this function has the Taylor series representation

$$f_1(t) = a_0 + a_1(t - t_0) + a_2(t - t_0)^2 + \cdots + a_n(t - t_0)^n + \cdots \quad (131)$$

where

$$a_n = \frac{1}{n!}\left(\frac{d^n f_1}{dt^n}\right)_{t=t_0} \quad (132)$$

At its output, the network is required to produce a function $f_2(t)$ simulating $f_1(t)$ in such a way that the form of $f_1(t)$ for the vicinity of $t = t_0$ is preserved to an extent defined by the Taylor series 131 through a derivative of specified order.

Observe that this requirement is much more severe than the usual one regarding signal waveform preservation. Here the waveform must be so accurately preserved that n successive derivatives as well as the function itself can be correctly evaluated at the point $t = t_0$. Being aware of the fact that differentiation in general accentuates errors in the form of an output function, we recognize the severity of a requirement of this sort.

If the required result is to be achieved at all, we have a feeling that the necessary bandwidth will be huge since the impulse response must be very nearly a perfect impulse in order to produce such a high degree of signal simulation. Nevertheless, a further requirement states that the desired result is to be achieved with a network providing an arbitrarily fixed bandwidth in the frequency domain.

Our first reaction to this problem is that the requirements are impossible of fulfillment, which is true in a literal sense but not true if we interpret the requirements properly and evaluate results in the light of bandwidth limitations imposed by practically realizable networks. Let us clarify this statement by pointing out how, in a world providing only networks with finite bandwidths, we must define what we mean by the concepts of "simulation" as applied to functions and their derivatives.

Suppose the function to be simulated is the impulse. Only an infinite bandwidth can reproduce its form exactly. Any finite bandwidth however large converts the impulse into a pulse of nonzero duration. The sketches in Fig. 1 show that a lowpass bandwidth ω_c yields a pulse duration roughly equal to $\delta = 2\pi/\omega_c$. We can make this duration arbitrarily small by providing a sufficient bandwidth, but in the end we will have to accept some nonzero value δ as a practical definition of what we mean by the duration of a reproducible impulse.

If the impulse has unit value, it must enclose unit area. A rectangular, or roughly rectangular, pulse of duration δ and height $1/\delta$ we shall call an impulse *on a δ-approximate basis.*

To be consistent, we must define successive derivatives of the unit impulse in the same sense. The first derivative of the unit impulse, the unit doublet, is exactly representable by considering a positive impulse of value $1/\delta$ separated by an interval δ from a negative impulse of value $1/\delta$, in the limit as δ becomes zero. On a δ-approximate basis this becomes a quasi-rectangular pulse of width δ and height $1/\delta^2$ followed immediately by a similar negative pulse.

On this same basis the second derivative of a unit impulse (the unit triplet) is given by three quasi-rectangular pulses of width δ immediately adjacent to each other, their heights being respectively $1/\delta^3$, $-2/\delta^3$, $1/\delta^3$; or we may say, their enclosed areas being $1/\delta^2$, $-2/\delta^2$, $1/\delta^2$. The third derivative of the unit impulse (the unit quadruplet) is given by four such adjacent pulses with enclosed areas having a distribution $(1/\delta^3) \times (1, -3, 3, -1)$, and so forth, the relative values being given by the binomial coefficients associated with the pertinent order.*

These δ-approximate versions of the unit impulse and its successive derivatives are observable and measurable with physical apparatus having a finite resolving power (bandwidth). When we speak of measuring or observing the values of a function or its derivatives we can only imply measurement in this sense. Therefore, the problem of reproducing the function in Eq. 131 and its successive derivatives through order n can similarly imply only that this be done in a δ-approximate sense, for nothing else is physically possible.

It should not be concluded that we must sacrifice accuracy in the values of the various derivatives that are measured or observed in this way. The definitions given above for δ-approximate derivatives of the unit impulse are for derivatives of *unit* value. Derivatives of any other values are these same measurable configurations multiplied by suitable amplitude factors. In other words, having the values of successive derivatives of $f_1(t)$ in 131 measured on a δ-approximate basis is entirely equivalent to having them measured on an exact basis. Nothing is lost through conversion to a δ-approximate basis, while the ability to measure with equipment providing a finite bandwidth is assured.

Now let us return to the problem stated at the beginning of this article. The first step is to pass function 131 through a reshaping network that converts it and its successive derivatives through order n to a form measurable on a δ-approximate basis, but that leaves all relative values intact.

* See art. 7, Ch. XII.

This reshaping network clearly is one which does precisely the same job on a perfect unit impulse, namely, converts it to a δ-approximate impulse whose successive derivatives through order n exist and are correctly measurable by apparatus with cutoff $\omega_c = \delta/2\pi$. Let this reshaped version of $f_1(t)$ be $f_1^*(t)$ having a spectrum $F_1^*(j\omega)$. Rectangular truncation of this spectrum at the frequency ω_c (equivalent to passing $f_1^*(t)$ through a lowpass filter with this bandwidth) now leaves the function and its successive derivatives intact because none of these, considered as separate signals, occupies a wider spectrum in the frequency domain and hence is accommodated by the bandwidth ω_c.

As for the system function of the reshaping network, this is none other than function 22 discussed in art. 3 of Ch. XIV shown plotted there for various orders n in the right-hand sketches of Fig. 7; the corresponding time-domain functions are on the left. The latter are reshaped versions of the unit impulse; the one of order n can conservatively be differentiated $n - 1$ times, yielding correctly the values of these derivatives on a δ-approximate basis as desired.

The system functions 22 (in Ch. XIV) which are Fourier transforms of the family of time functions 21 whose convolution with $f_1(t)$ produces the desired reshaping are called "simile transforms" because of their property to convert $f_1(t)$ into a form in which the desired high degree of simulation is made possible.*

A further condition, necessary for the achievement of this result, is that the duration $n\delta$ of a reshaped impulse [$h_n(t)$ in Eq. 21, Ch. XIV] must be no greater than the interval for which n terms of the Taylor series 131 represent an acceptable representation of the time function $f_1(t)$, for the reshaping process expands the "immediate vicinity" of the point $t = t_0$ in this expansion into a δ-approximate vicinity having the nonzero duration $n\delta$. In other words, there can be no additional vicinities where similar response requirements are imposed that are closer to the point $t = t_0$ than $n\delta$ seconds, for overlapping of such successive vicinities would result.

In a sense this condition means that we are trading a bandwidth requirement for time. We are achieving a resolution that would normally require a huge bandwidth by means of a much smaller one, but we are taking a longer time to make this resolution. The interval $n\delta$ may be regarded as a "resolving time." To resolve the character of $f_1(t)$ at the point $t = t_0$ we must expand this vicinity (which theoretically is infinitesimal) into a nonzero interval equal to $n\delta$ seconds. Obviously, points in the function $f_1(t)$ which we can resolve in this manner cannot be spaced more closely than this interval.

* These transformations and their properties were first introduced by Dr. Manuel Cerrillo of the M.I.T. Research Laboratory of Electronics.

Problems

1. A certain network is excited by the voltage function

$$e_1(t) = 1 \qquad \text{for } 0 < t < 10 \text{ seconds}$$
$$e_1(t) \equiv 0 \qquad \text{otherwise}$$

The output is given by

$$e_2(t) = t \qquad \text{for } 0 < t < 1 \text{ second}$$
$$e_2(t) = 1 \qquad \text{for } 1 < t < 10 \text{ seconds}$$
$$e_2(t) = 1 - (t - 10) \qquad \text{for } 10 < t < 11 \text{ seconds}$$
$$e_2(t) \equiv 0 \qquad \text{otherwise}$$

(*a*) What are the magnitude and phase angle of the response function $E_2/E_1 = H(\omega)$?

(*b*) What is the impulse response of this network?

2. The pole-zero pattern of a response function $H(s)$ for a given network is shown in the adjacent sketch. Find the response of this network when the excitation is

$$e^{-t} \cos t$$

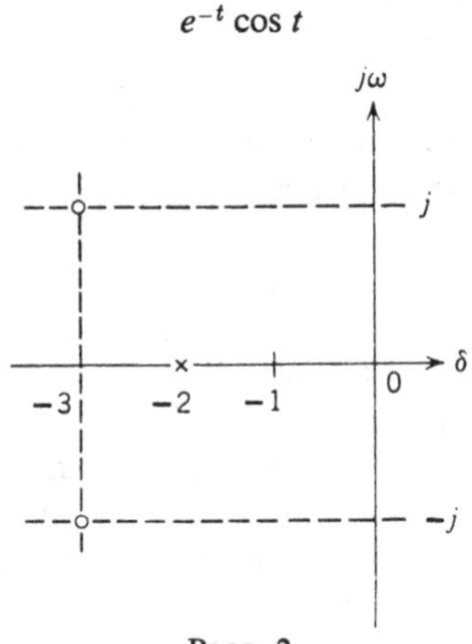

PROB. 2.

3. For the network shown below, obtain the transfer impedance $Z_{12}(s)$ by inspection. Then determine $e_2(t)$ for

$$i_1(t) = e^{-t/2} \cos t$$

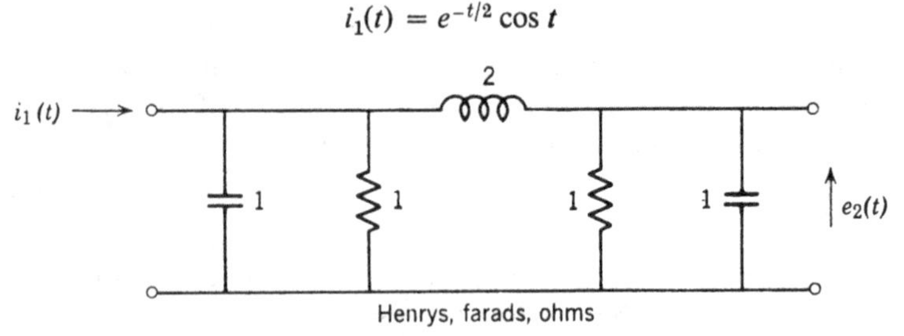

PROB. 3.

4. A given network has the system function

$$H(s) = \frac{s^4 + 7s^3 + 17s^2 + 17s + 7}{(s+1)(s+2)(s+3)}$$

(*a*) Find the impulse response.
(*b*) Evaluate the integral

$$\int_0^\infty e_2(t)\, dt$$

where $e_2(t)$ is the impulse response found in part (*a*).
(*c*) Write down by inspection the first three terms of the Maclaurin expansion $e_2(t) = a_0 + a_1 t + a_2 t^2$.

5. The input to a network is the amplitude-modulated cosine function

$$g_1(t) = f_1(t) \cos \omega_0 t$$

The desired output is

$$g_2(t) = \left[\int_{-\infty}^{t} f_1(\xi)\, d\xi\right] \cos \omega_0 t$$

Find the system function $H(s)$.

6. A periodic current having the form of a succession of identical rectangular pulses of height π and duration $\pi/8$ with a period $\tau = \pi$ is impressed upon an impedance $Z(j\omega)$ whose magnitude is triangular with a peak value of 600 ohms. The peak is centered at $\omega = 28$ radians per second. The base of the triangle is $\Delta\omega = 4$ radians per second, and the impedance has zero magnitude elsewhere—note of course that it is an even function. Its angle may be assumed to pass through zero where the magnitude is maximum. Find the voltage $e(t)$ appearing across this impedance.

7. The accompanying sketches show the magnitude and lag angle for the system function of a bandpass filter. If the input function is given by

$$f_1(t) = u_{-1}(t) \times \cos 100t$$

calculate and neatly plot the output $f_2(t)$.

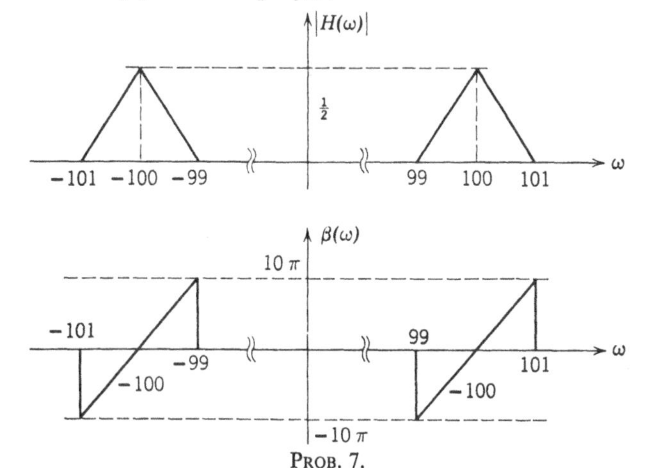

PROB. 7.

8. The unit impulse response of a network is

$$h(t) = u_0(t) - \tfrac{1}{2}u_0(t - 1)$$

(*a*) Find $H(\omega)$.

(*b*) A second network cascaded with this one has an impulse response $g(t)$ such that the overall response in the frequency domain is the constant unity. Find $g(t)$ and check your answer by convolving $g(t)$ with $h(t)$.

9. A parallel *RC* circuit with $R = 1/2$, $C = 1$ is excited by the current source

$$i_s(t) = 0 \qquad \text{for } t < 0$$

$$i_s(t) = e^{-t} \qquad \text{for } t > 0$$

Find the resulting voltage response for $t > 0$. Does the question of dominance arise in this problem? Why or why not? Now consider the same circuit with $R = C = 1$ and

$$i_s(t) = e^{-2t} \qquad \text{for } t > 0$$

Formulate the voltage response in this case for:

(*a*) The dominance condition fulfilled.

(*b*) The dominance condition not fulfilled.

10. The system function of a given network is an entire transcendental function of ω. Does the dominance question need to be considered:

(*a*) If the excitation is a unit impulse?

(*b*) If the excitation is a suddenly applied exponential?

Explain in detail your answers to (*a*) and (*b*).

11. A given network has the system function

$$H(s) = \frac{1}{(s + 2)(s + 3)(s + 4)}$$

It is excited by one or the other of the following input functions:

$$(a)\ f_1(t) = u_{-1}(t) \times e^{-t}, \qquad (b)\ f_1(t) = u_{-1}(t) \times e^{-5t}$$

Does the dominance question need to be considered here? If so in which case and why?

12. A given physical network has the system function

$$H(s) = \frac{1}{(1 - s)(2 + s)(3 + s)}$$

It is excited by the function

$$f_1(t) = u_{-1}(t) \times e^{-t}$$

How must the interference pattern for this excitation function be constructed and why?

13. The functions

$$f_1(t) = u_{-1}(t) \times e^{-t}, \quad f_2(t) = u_{-1}(t) \times e^{-2t}$$

are respectively the input and output of a given network. How would you define the system function of this network in the time domain? Find this time-domain function.

14. In the convolution integral

$$f_2(t) = \int_{-\infty}^{\infty} f_1(x - t)h(x)\,dx$$

f_1 and f_2 are the real input and output functions of a physical network. If each of these has the form of a sum of exponentials (with real and/or complex exponents), show how one can always solve this integral equation in a simple and straightforward way for the time-domain system function $h(t)$ of the network. Does the question of dominance need to be considered in this procedure?

15. The j-axis real part of the system function of a passive network is given by

$$\text{Re}\,[H(j\omega)] = \frac{1}{1 + \omega^2} + \frac{1}{1 + (\omega - 1)^2} + \frac{1}{1 + (\omega + 1)^2}$$

(a) Find the unit impulse response.
(b) Construct the system function $H(s)$.

16. Show how the real-part sufficiency involved in the preceding problem can be extended to realizable active networks by making use of property 74 in Ch. XII as is done in Eq. 15 by introducing a net time function $\hat{f}(t)$.

17. A rectangular pulse of unit height and duration 1 second is transmitted through a network with response function $H(\omega)$ having unit magnitude for all ω and a phase lag that can be approximated by a broken line as in Fig. 2. Assume $\omega_0 = \pi$ radians per second; $t_1 = 4.0$ seconds and $t_2 = 5.5$ seconds. Compute and plot the output pulse.

18. The system function of a given network is defined in the time domain as

$$h(t) = 2 \times \frac{\sin 2\pi(t - 2)}{2\pi(t - 2)}$$

The input is $f(t)$ as given in prob. 1 of Ch. XV. Find and plot the output function.

19. The impulse response of a given network is essentially a rectangular pulse of duration 10^{-4} second and unit enclosed area. Morse code using 5 pulses per letter is transmitted. If we assume an average of 6 letters per word, what is the highest speed in words per minute that can be attained with this facility? Describe and justify any basic assumptions you may make.

20. In connection with the problem characterized by Eqs. 87, 88 and 89 one might consider the following procedure for avoiding the infinite summation in evaluating one period of $f_2(t)$. Let $\hat{f}_1(t)$ and $\hat{f}_2(t)$ be respectively one period of $f_1(t)$ and $f_2(t)$, both including the origin $t = 0$. The Fourier transform of $f_1(t)$ is given by $\hat{F}_1(\omega) \times F_u(\omega)$ and that of $f_2(t)$ is $\hat{F}_2(\omega) \times F_u(\omega)$ in which $\hat{F}_1$ and $\hat{F}_2$ are transforms of $\hat{f}_1$ and $\hat{f}_2$, and F_u is given by Eq. 7, Ch. XV. If $H(\omega)$ is the system function of the network, we have

$$\hat{F}_2(\omega) \times F_u(\omega) = H(\omega) \times \hat{F}_1(\omega) \times F_u(\omega)$$

from which we conclude that $\hat{F}_2(\omega) = H(\omega) \times \hat{F}_1(\omega)$, and get $\hat{f}_2$ from $\hat{F}_2(\omega)$ by Fourier synthesis. What is wrong with this scheme—or is it valid?

CHAPTER XVII

Numerical Methods in Fourier Analysis and Synthesis

1. Introductory Remarks

In many practical problems where Fourier methods can effectively be used, the pertinent functions are specified in graphical rather than in analytic form. In such cases the transformation integrals must be evaluated by some appropriate numerical method. Such methods are frequently useful even if the given functions are in analytic form, for evaluation of the Fourier integrals is not always straightforwardly possible. Moreover, in many situations of this sort, an exact solution valid for all time is not needed; rather, one much prefers an approximate result that can be obtained rapidly and with a moderate amount of computational effort. An effective numerical evaluation procedure is thus recognized to be essential in the practical exploitation of Fourier methods. The present chapter is devoted to a careful discussion of collateral as well as the essential questions that arise in this connection.

Except in situations that are idealized to begin with (such as the problem of the ideal lowpass filter, for example) the use of numerical methods for the evaluation of Fourier integrals involves approximations of one sort or another at various stages in the total procedure. Hence the resulting solution is inexact to some extent and in some characteristic manner. Unless one can determine the nature and extent of the error involved, such an approximate solution has little if any practical value. If one is aware of the fact that a given result is approximate but does not know whether the implied error is 1% or 100%, or what its nature is, then it does not represent a solution at all. Not only must we devise an approximate numerical evaluation procedure, but we must couple with it a method for predicting the nature and extent of the error which will result from its use in any given situation.

Too often, in the discussion of approximate numerical methods, the question of error evaluation is ignored. Instead, the method is applied to some example for which an exact evaluation is also possible so that the two results can be compared. If the comparison is good, a sort of "blind faith" in the method is established which presumably should guarantee its usefulness in all other applications. Unfortunately, this attitude is not consistent with good engineering practice.

The requirement that an error-evaluation procedure must be coupled with any numerical method which we may devise to carry out direct and inverse Fourier transformations might seem offhand to be rather severe in view of the thought that an approximate result plus the error committed equals an exact result, and so we are getting nowhere. This thought, however, overlooks the important fact that we do not need to know the error exactly; we merely need to know its order of magnitude, whether it is of the order of 1 per cent or 10 per cent or 100 per cent, and roughly how it behaves relative to the approximate solution. If we can determine this much with moderate computational effort, then our requirement is adequately fulfilled in almost all practical situations. We shall see that such an objective is rather easily achieved.

Before getting involved in detailed matters, however, it may be well to consider the possible ways in which errors are likely to be committed so that we can be prepared to deal with them more effectively. To be specific, suppose we consider the inverse transformation from a spectrum function to a time function, in other words, Fourier synthesis. The analysis problem involves essentially the same operations, so it doesn't really matter much which one we consider first. However, let us assume that a spectrum function $F(\omega)$ is given and we wish to evaluate its time-domain mate $f(t)$.

The method of approximate integration which we shall discuss here makes use of essentially the same devices or techniques as are found effective in the evaluation of Fourier series coefficients discussed in art. 8, Ch. XI. These are the impulse-train methods which enable one to evaluate integrals by inspection.* We find this technique to be useful not only in dealing with Fourier series but also in the evaluation of convolution integrals as shown in art. 5, Ch. XIII. As is also shown there (Fig. 15), use of this method requires first that the given function, $F(\omega)$ in the present instance, be approximated by a broken line or by confluent parabolic arcs. This process introduces two distinct kinds of error as a result of the following circumstances: (*a*) the confluent straight lines or parabolic arcs do not fit the given function perfectly, and (*b*) the process can extend over only a finite interval of the independent variable whereas the given function may (and usually does) possess an infinite asymptote.

Circumstance (*a*) results in an *approximation error*; circumstance (*b*)

* First published in *Tech. Rept. 268, Mass. Inst. Technol. Res. Lab. Electron.*, Sept. 2, 1953; presented at the National Electronics Conference (Chicago) of that date, and published in their *Transactions*, Vol. 9, Feb. 1954.

requires that the spectrum function be cut off at some finite frequency and hence it causes a *truncation error*.

Chopping the asymptote off a spectrum function causes another kind of improper behavior in the resulting time function which can, however, be corrected by carrying out the impulse-train approximation in a compensating manner. In art. 6, Ch. XII it is shown that the behavior of $f(t)$ near $t = 0$ is linked with the behavior of $F(\omega)$ for $\omega \to \infty$, that is, with the asymptotic behavior of the spectrum function, but that this characteristic is alternately expressible in terms of relations involving the behavior of $F(\omega)$ over the finite domain—the so-called moment conditions. We can, therefore, preserve the proper effect which the asymptotic behavior of $F(\omega)$ has upon $f(t)$ even though this asymptote is removed, if we observe the fulfillment of corresponding moment conditions. Since these involve the detailed nature of the pertinent impulse-train approximation (as is shown below), their fulfillment can be managed while meeting other requirements of the approximation problem as well.

Now that we are aware of the things that must be considered in the formulation of a numerical integration procedure, we are in a proper frame of mind to understand the following detailed discussion.

2. Forms for the Fourier Integrals Appropriate to Impulse-Train Methods of Evaluation

From the discussion in art. 5, Ch. XIII, where impulse-train methods are used in the evaluation of convolution integrals, we note that the impulse train represents some higher derivative of the broken-line or parabolic-arc approximant. For the broken-line approximant it is the second derivative; for the parabolic-arc approximant it is the third derivative. For approximants formed with third- or fourth-degree arcs, the fourth or fifth derivatives yield impulse trains. We have yet to discuss how one decides whether the confluent segments should be arcs of first or second or third degree, etc.; for the moment let us denote the impulse train as representing the νth derivative of the approximant to a given time or frequency function and consider the most effective form in which we can express the desired direct or inverse transform.

To this end consider the Fourier integrals 21 and 22 in Ch. XII which we repeat here for convenience:

$$f(t) = \frac{1}{2\pi}\int_{-\infty}^{\infty} F(\omega)\, e^{j\omega t}\, d\omega \tag{1}$$

$$F(\omega) = \int_{-\infty}^{\infty} f(t)\, e^{-j\omega t}\, dt \tag{2}$$

If we form the derivative of $F(\omega)$ with respect to ω by differentiating under the integral sign in Eq. 2, $f(t)$ becomes multiplied by $-jt$ (as has been pointed out before). After ν successive differentiations, $f(t)$ is multiplied by $(-jt)^\nu$. If we abbreviate the νth derivative of $F(\omega)$ by writing

$$F^{(\nu)}(\omega) = \frac{d^\nu F}{d\omega^\nu} \tag{3}$$

we can say that $(-jt)^\nu f(t)$ and $F^{(\nu)}(\omega)$ are a transform pair and hence that integral 1 yields

$$(-jt)^\nu f(t) = \frac{1}{2\pi}\int_{-\infty}^{\infty} F^{(\nu)}(\omega)\, e^{j\omega t}\, d\omega \tag{4}$$

If we separate $F(\omega)$ into its real and imaginary parts as indicated by

$$F(\omega) = F_1(\omega) + jF_2(\omega) \tag{5}$$

the integral in 4 can be separated into the sum of two integrals, and division by $(-jt)^\nu$ then yields

$$f(t) = \frac{1}{2\pi(-jt)^\nu}\int_{-\infty}^{\infty} F_1^{(\nu)}(\omega)\, e^{j\omega t}\, d\omega + \frac{j}{2\pi(-jt)^\nu}\int_{-\infty}^{\infty} F_2^{(\nu)}(\omega)\, e^{j\omega t}\, d\omega \tag{6}$$

The two terms in this result are respectively the even and odd components of $f(t)$, as may be seen from the following considerations. First of all we recall that $F_1(\omega)$ and $F_2(\omega)$ are even and odd functions of ω. Next we observe that the derivative of an even function is odd, and the derivative of an odd function is even. If we replace the exponential in each integral by its cosine-sine equivalent, noting that the cosine is an even function of both ω and t while the sine is odd, the fact that only the even term in each integrand contributes to the value of the pertinent integral tells us that the first integral in 6 is an even function of t when ν is even and an odd function of t when ν is odd, while the reverse is true for the second integral. Hence the statement made in the opening sentence of this paragraph follows.

We now make use of the real-part sufficiency which applies to physical networks (see art. 5, Ch. XVI) and obtain the time function $f(t)$ in terms of the νth derivative of either the real or imaginary part of the spectrum function. Thus we have for $t > 0$ either

$$f(t) = \frac{1}{\pi(-jt)^\nu}\int_{-\infty}^{\infty} F_1^{(\nu)}(\omega)\, e^{j\omega t}\, d\omega \tag{7}$$

or

$$f(t) = \frac{j}{\pi(-jt)^\nu}\int_{-\infty}^{\infty} F_2^{(\nu)}(\omega)\, e^{j\omega t}\, d\omega \tag{8}$$

When $F_1^{(\nu)}(\omega)$ or $F_2^{(\nu)}(\omega)$ is given in the form of an impulse train, the evaluation of these integrals can be written down by inspection.

Now let us consider the reverse problem of determining the spectrum function $F(\omega)$ from the νth derivative of either the even or odd part of $f(t)$. To this end we observe first that formation of the νth derivative of $f(t)$ in Eq. 1 by differentiating under the integral sign results in $F(\omega)$ being multiplied by $(j\omega)^\nu$. Writing

$$f^{(\nu)}(t) = \frac{d^\nu f}{dt^\nu} \tag{9}$$

and noting that $f^{(\nu)}(t)$ and $(j\omega)^\nu F(\omega)$ are a transform pair, we find that Eq. 2 yields

$$(j\omega)^\nu F(\omega) = \int_{-\infty}^{\infty} f^{(\nu)}(t)\, e^{-j\omega t}\, dt \tag{10}$$

If we separate $f(t)$ into its even and odd parts, thus:

$$f(t) = f_e(t)\ (+ f_0(t) \tag{11}$$

and correspondingly separate the right-hand side of Eq. 10 into two integrals, a line of reasoning similar to that given above shows that these, after division by $(j\omega)^\nu$, are respectively even and odd functions of ω and hence we obtain the results

$$F_1(\omega) = \frac{1}{(j\omega)^\nu} \int_{-\infty}^{\infty} f_e^{(\nu)}(t)\, e^{-j\omega t}\, dt \tag{12}$$

and

$$jF_2(\omega) = \frac{1}{(j\omega)^\nu} \int_{-\infty}^{\infty} f_0^{(\nu)}(t)\, e^{-j\omega t}\, dt \tag{13}$$

Being able to obtain the time function from the real or imaginary part of the spectrum function, as expressed in Eqs. 7 and 8, is a distinct advantage inasmuch as one is relieved of the necessity of dealing with a complex spectrum function. Representations in both the time and frequency domains involve only real functions of real variables. The advantage thus gained is particularly valuable in numerical evaluation since working with real numbers is so much less tedious than working with complex numbers.

In the opposite direction, going from the time to the frequency domain, separation into even and odd parts as is done in Eqs. 12 and 13 does not offer the same advantages as compared with the use of Eq. 10 since $f(t)$ is real anyway. Here the numerical integration procedure can, and often is, just as advantageously applied directly to the integral 10 for the evaluation of $F(\omega)$ from a given time function $f(t)$.

3. The Approximation Problem and Associated Error Criterion

The approximant to a given time or frequency function can take a variety of forms. We can, for example, approximate the function itself by a train of impulses as we do when that function is represented in sampled form.

Such a representation is given by Eq. 4 in Ch. XV and is illustrated there in Fig. 2. Instead we may approximate the function by steps, in which case the first derivative of the approximant is an impulse train. Or we may use a piecewise-linear (broken-line) approximation and obtain an impulse train from its second derivative. A confluent set of second- or higher-degree parabolic arcs may be used to construct the approximant. The questions arising here are: How do we decide which one of these possibilities is best? What are their relative advantages and disadvantages? Can

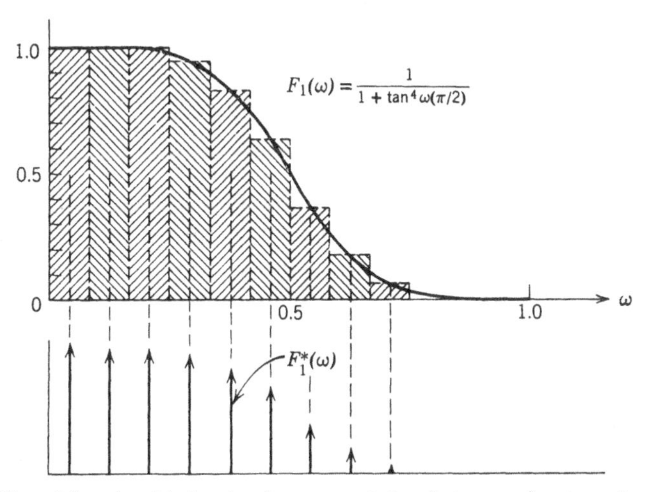

FIG. 1. Plot of function 14 showing its representation in terms of rectangular samples and their approximation by impulses.

we relate the resulting approximation error to the variables involved in this process in a sufficiently simple manner to help us reach a decision in any given situation? We shall try to answer these questions in a logical fashion.

In order to lend concreteness to our discussions we will choose a specific example to illustrate the various possibilities. Suppose we have given the following real part of a spectrum function:

$$F_1(\omega) = \frac{1}{1 + \tan^4 \omega\pi/2} \qquad \text{for } -1 < \omega < 1 \tag{14}$$
$$= 0 \qquad \text{otherwise}$$

and are asked to find the corresponding time function $f(t)$ for $t > 0$.

A plot of this frequency function is shown in Fig. 1 together with its direct approximation (sampled representation) by means of an impulse train. Since this spectrum function is given in analytic form we could, of course, consider evaluation of the time function $f(t)$ by analytic means.

However, evaluation of integral 7 for $\nu = 0$ involving function 14 would seem to pose some difficulty and for this reason we might preferably consider a numerical evaluation process.

Accordingly, we divide the function into rectangular elements as indicated in Fig. 1 by the oppositely cross-hatched areas and assume that $\cos \omega t$ (the real part of the exponential in integral 7) is essentially constant throughout the width of any element and equal to its value at the center of that element. If the areas of the elements are respectively denoted by $a_1, a_2, a_3, \ldots$ and their center frequencies by $\omega_1, \omega_2, \omega_3, \ldots$ then this process of approximation leads to the time function

$$f(t) = \frac{2}{\pi}\{a_1 \cos \omega_1 t + a_2 \cos \omega_2 t + \cdots + a_n \cos \omega_n t\} \tag{15}$$

where n equals the number of elements.

The approximation involved in this result is twofold. First, the rectangular shape of each element leads to a steplike approximation to the function $F_1(\omega)$; and second, the assumption that $\cos \omega t$ is replaceable by $\cos \omega_k t$ leads us to replace each elementary pulse by an impulse whose value equals the respective pulse area, its point of occurrence being at the pulse center. Thus the function $F_1(\omega)$ is replaced by the approximant $F_1^*(\omega)$ consisting of a succession of n impulses as shown in the bottom half of Fig. 1.

Of the two sources of error involved here, the one due to replacing pulses by impulses is the more serious, particularly for large values of t. If the width of each pulse is sufficiently narrow, these errors may be within tolerable limits, but the total number of terms in expression 15 will then become rather large and the computation of $f(t)$ will correspondingly be cumbersome. We shall now consider how we can improve the accuracy in this sort of graphical integration process and actually decrease the number of terms needed to represent $f(t)$.

As illustrated in Fig. 2, we again approximate $F_1(\omega)$ by a step function, which is now denoted by $F_1^*(\omega)$. The derivative of this step-function approximant is the impulse train shown in the bottom half of Fig. 2. If we denote the values of these impulses (inclusive of their algebraic signs) by $a_1, a_2, \ldots$ and their frequencies of occurrence by $\omega_1, \omega_2, \ldots$ we now find from Eq. 7 for $\nu = 1$ the time function

$$f(t) = \frac{-2}{\pi t}\{a_1 \sin \omega_1 t + a_2 \sin \omega_2 t + \cdots + a_n \sin \omega_n t\} \tag{16}$$

where n again is the number of impulses in the representation of $dF^*/d\omega$.

The important difference between this result and that expressed by Eq. 15 is that the only error committed is that due to the step approximation of

$F_1(\omega)$. The impulses are true impulses this time. The expression 16 is, therefore, equally good for all values of the time t.

If we are willing to tolerate the same average error with our present approach as with the previous one, we can probably make a cruder step approximation (which need not, incidentally, involve uniform increments) and obtain an expresson for $f(t)$ involving fewer terms than before.

We can continue to take further advantage of this sort of reasoning. Thus if we construct an approximant $F_1^*(\omega)$ consisting of straight-line segments, the piecewise-linear approximation to $F_1(\omega)$, then its first

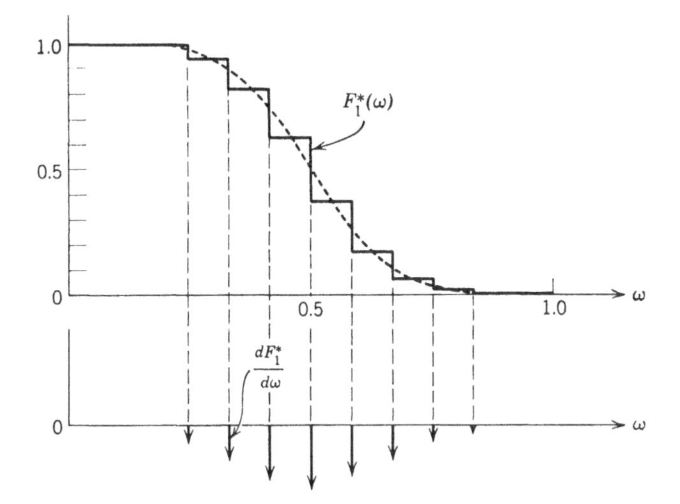

FIG. 2. Step-function approximation of function 14, and its derivative which is an impulse train.

derivative becomes a step approximation to $dF_1/d\omega$ and its second derivative a sum of impulses. For the same number of impulses as before, the piecewise-linear approximation to $F_1(\omega)$ is much better than the step approximation; or for the same approximating tolerance we now have still fewer impulses and hence fewer terms in the resulting expression for $f(t)$ which is readily obtained from Eq. 7 for $\nu = 2$.

The mode of approximation logically following the piecewise-linear procedure is one in which $F_1(\omega)$ is approximated by a set of confluent parabolic arcs. Thus we might try an approximation to $F_1(\omega)$ in Fig. 1 in the form of just two such arcs (after the fashion illustrated in Fig. 15, Ch. XIII), confluent at the point $\omega = 0.5$, one with its apex at $\omega = 0$ and the other with its apex at $\omega = 1$. The first derivative of this approximant is a piecewise-linear approximation to $dF_1/d\omega$ and has just two break points; one at $\omega = 0.5$ and another at $\omega = 1$. The third derivative of this approximant now consists of impulses at the break points of its first derivative, and the corresponding time function is given by just two terms.

This extremely simple expression for $f(t)$ is, however, obtained at the expense of a somewhat large approximating tolerance as further detailed consideration of this problem will show. In the following generalization we will not bother to distinguish the approximant to $F_1(\omega)$ by an asterisk. The νth derivative of this approximant we denote by $F_1^{(\nu)}(\omega)$ and assume that it consists of a sum of impulses.

At the top of Fig. 3 we have plotted the function of Fig. 1 again, and below it the first and second derivatives. We might call attention to the fact that successive differentiation clearly accentuates the variable properties of a function. While the given real part $F_1(\omega)$ is exceptionally smooth and slowly varying, its second derivative is rather jagged in character. This second derivative is shown approximated by straight-line segments (drawn dotted).

If we thus approximate the second derivative of the given real part, then its first derivative becomes approximated by parabolic arcs, and the approximant to the real part itself consists of third-degree parabolic arcs. The fourth derivative of this approximant is an impulse train. In the evaluation of $f(t)$ by integral 7, the integer ν equals 4. Since the broken-line approximation in Fig. 3 has four break points, there will be four terms in the resulting expression for $f(t)$.

The approximation to $F_1(\omega)$ in this case is exceedingly good. Noting how successive integration tends very rapidly to suppress errors in this kind of approximation (as is illustrated by the plots in Fig. 2 of Ch. XVI) it is not difficult to visualize that any residual discrepancy between $F_1(\omega)$ in Fig. 3 and the second integral of the broken-line approximation to $F_1^{(2)}(\omega)$ is so small that it could not possibly be indicated on this graphical plot.

A less good approximation will result if we fit straight-line segments to the first derivative in Fig. 3. Thus a triangular approximation to $F_1^{(1)}(\omega)$ with three break points yields a (second-degree) parabolic-arc approximation to $F_1(\omega)$ and a representation for $f(t)$ consisting of three terms. The approximation mentioned above involving two parabolic arcs (with apexes at $\omega = 0$ and $\omega = 1$), although yielding only two terms in $f(t)$, clearly will be a poorer approximation to $F_1(\omega)$ since the corresponding approximation to $F_1^{(1)}(\omega)$ is a triangle with its base extending from $\omega = 0$ to $\omega = 1$. While we may consider approximating $F_1(\omega)$ in this way, it is apparent by inspection of the first derivative that this is not a good scheme and that the extra break point involved in approximating $F_1^{(1)}(\omega)$ with a narrower triangle is well worth the additional computational effort.

At all events we see the advantage of having sketches for the derivatives

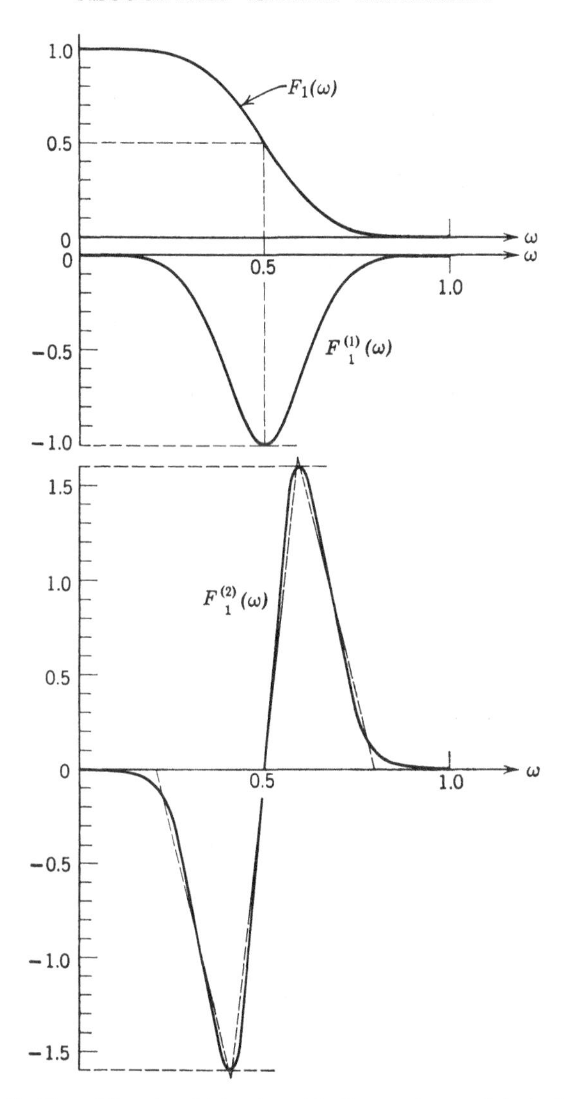

FIG. 3. Plot of function 14 (top) and its first and second derivatives.

of $F_1(\omega)$ as well as the plot of this function itself in deciding upon an appropriate approximation, especially when the use of third- or higher-degree parabolic arcs is contemplated.

We note from the plots in Fig. 3 that each successive derivative has an additional wiggle and hence its broken-line approximation contains an additional break point. Correspondingly, $F_1(\omega)$ is approximated by the next-higher-degree parabolic arcs and with a diminished tolerance. It thus becomes clear how this method of approximation works and how higher-degree arcs produce closer approximations but yield more terms in the resulting expression for $f(t)$. Each additional term potentially achieves a smaller approximation error. Let us see if we can express this correlation in more precise terms.

Fundamental to this discussion is a normalization of the frequency scale so that the interval over which the approximation extends corresponds to the range $0 < \omega < 1$, as in the sketches in Figs. 1, 2 and 3. We wish to interrelate three things: the closeness with which $F_1(\omega)$ is approximated, the order ν of the derivative for which the approximant becomes a sum of impulses, and the number of these impulses or the number of terms needed to represent the associated time function.

Imagine the $(\nu - 2)$th derivative of $F_1(\omega)$ to be approximated by a broken line in such a way that the error (difference between the broken line and the actual curve) changes sign three times within each linear interval, which it normally will do if a line is passed through each inflection point. Within the interval $0 < \omega < 1$ there then are $3(n - 1)$ changes of sign if n equals the total number of break points in the piecewise-linear approximation. Hence there are $3(n - 1)$ half-cycles of error oscillation in the interval $0 < \omega < 1$.

Let $3(n - 1)/2 = \tau/2\pi$, and assume that the error variation can roughly be regarded as sinusoidal with an amplitude ϵ. Then we have an error

$$\begin{aligned} &\text{in } F_1^{(\nu-2)}(\omega) \approx \epsilon \sin \omega\tau \\ &\text{in } F_1^{(\nu-3)}(\omega) \approx -\frac{\epsilon}{\tau} \cos \omega\tau \\ &\text{in } F_1^{(\nu-4)}(\omega) \approx -\frac{\epsilon}{\tau^2} \sin \omega\tau \end{aligned} \tag{17}$$

and so forth. The error in the approximation of $F_1(\omega)$ itself, therefore, has an approximate amplitude

$$\frac{\epsilon}{\tau^{\nu-2}} = \frac{\epsilon}{[3\pi(n - 1)]^{\nu-2}} \tag{18}$$

Although somewhat crude in its derivation, this result is nevertheless useful in estimating the approximation error for particular values of ϵ, n and $\nu \geq 2$. For example, if the first derivative of $F_1(\omega)$ in Fig. 3 is approximated in a piecewise-linear manner involving three break points, the maximum error is readily seen to be about 10 per cent or less. For $\nu = 3$, $n = 3$ and $\epsilon = 0.1$, formula 18 shows that $F_1(\omega)$ is approximated by parabolic arcs with a maximum error of about 0.5 per cent while the associated time function is given by three terms. On the other hand, if the second derivative is approximated by a broken line as indicated, ϵ again is about 0.1, $\nu = 4$ and $n = 4$, for which formula 18 yields an approximation error of the order of 10^{-4}. Such accuracy is not needed and moreover is inconsistent with other data in most practical problems. It is, therefore, seldom necessary to consider parabolic-arc approximations of higher than second order.

4. Direct and Inverse Transforms by Impulse-Train Methods

Consider first the determination of a time function when the νth derivative of the real or imaginary part of its associated spectrum function is given in the form of an impulse train. We have either

$$F_1^{(\nu)}(\omega) = \sum_{k=-n}^{n} a_k u_0(\omega - \omega_k) \tag{19}$$

or

$$F_2^{(\nu)}(\omega) = \sum_{k=-n}^{n} b_k u_0(\omega - \omega_k) \tag{20}$$

The coefficients a_k and b_k are the pertinent impulse values ; ω_k are their frequencies of occurrence, and the integer n equals the number of impulses involved in the approximation range $0 < \omega < 1$ on a normalized basis. Additionally, an impulse may occur at $\omega = 0$, but only if the function in question is even. For both forms 19 and 20 we have

$$\omega_{-k} = -\omega_k \quad \text{and} \quad \omega_0 = 0 \tag{21}$$

Recalling that the derivative of an even function is odd and vice versa, we have for 19

$$\begin{aligned} a_{-k} &= (-1)^\nu a_k \\ a_0 &\neq 0 \qquad \text{for } \nu \text{ even only} \end{aligned} \tag{22}$$

while for 20

$$\begin{aligned} b_{-k} &= -(-1)^\nu b_k \\ b_0 &\neq 0 \qquad \text{for } \nu \text{ odd only} \end{aligned} \tag{23}$$

Substitution into Eqs. 7 and 8 yields $f(t)$ for $t > 0$ in the form

$$\begin{aligned} f(t) &= \frac{1}{\pi(-jt)^\nu} \int_{-\infty}^{\infty} \sum_k a_k u_0(\omega - \omega_k)\, e^{j\omega t}\, d\omega \\ &= \frac{1}{\pi(-jt)^\nu} \sum_{k=-n}^{n} a_k e^{j\omega_k t} \end{aligned} \tag{24}$$

or

$$\begin{aligned} f(t) &= \frac{j}{\pi(-jt)^\nu} \int_{-\infty}^{\infty} \sum_k b_k u_0(\omega - \omega_k) e^{j\omega t}\, d\omega \\ &= \frac{j}{\pi(-jt)^\nu} \sum_{k=-n}^{n} b_k\, e^{j\omega_k t} \end{aligned} \tag{25}$$

If we consider separately the cases for ν even and ν odd, we can put these results into the following more explicit forms.

For ν even:

$$f(t) = \frac{2}{\pi(-jt)^\nu}\left[\frac{a_0}{2} + \sum_{k=1}^{n} a_k \cos \omega_k t\right] = \frac{-2}{\pi(-jt)^\nu}\sum_{k=1}^{n} b_k \sin \omega_k t \tag{26}$$

For ν odd:

$$f(t) = \frac{2j}{\pi(-jt)^\nu}\sum_{k=1}^{n} a_k \sin \omega_k t = \frac{2j}{\pi(-jt)^\nu}\left[\frac{b_0}{2} + \sum_{k=1}^{n} b_k \cos \omega_k t\right] \tag{27}$$

Impulse values at $\omega = 0$, if present, count half in these forms since the summations extend over positive frequencies only.

Corresponding evaluation of the even and odd parts of a spectrum function from the νth derivative of its associated time function is straightforward. The impulse-train representation for the νth derivative of the time function is written

$$f^{(\nu)}(t) = \sum_{k=1}^{n} a_k u_0(t - t_k) \tag{28}$$

where the summation extends over positive integers only since $f(t)$ is assumed zero for $t < 0$. The coefficients a_k are the pertinent impulse values, and their time of occurrence is denoted by t_k. Substitution into Eq. 10 yields for the spectrum function

$$F(\omega) = \frac{1}{(j\omega)^\nu}\int_{-\infty}^{\infty} \sum_k a_k u_0(t - t_k) e^{-j\omega t}\, dt = \frac{1}{(j\omega)^\nu}\sum_{k=1}^{n} a_k e^{-j\omega t_k} \tag{29}$$

Separating into real and imaginary parts, we have *for ν even*

$$F_1(\omega) = \frac{1}{(j\omega)^\nu}\sum_{k=1}^{n} a_k \cos \omega t_k \tag{30}$$

$$F_2(\omega) = \frac{-1}{(j\omega)^\nu}\sum_{k=1}^{n} a_k \sin \omega t_k \tag{31}$$

and *for ν odd*

$$F_1(\omega) = \frac{-j}{(j\omega)^\nu}\sum_{k=1}^{n} a_k \sin \omega t_k \tag{32}$$

$$F_2(\omega) = \frac{-j}{(j\omega)^\nu}\sum_{k=1}^{n} a_k \cos \omega t_k \tag{33}$$

5. The Moment Conditions

As pointed out above, fulfillment of moment conditions over the finite frequency domain is equivalent to providing proper asymptotic behavior and hence gives us the means to control the function $f(t)$ properly near

$t = 0$ even though the asymptote of $F(\omega)$ has been chopped off. Analogous moment conditions in terms of a given $f(t)$ can be used to compensate for truncation of its asymptote and thus provide for proper behavior of the associated spectrum function near $\omega = 0$.

From the relations 24 through 33 it is quite evident that this aspect of our problem needs clarification since the form of these results might lead one to conclude that they "blow up" for $t = 0$ or $\omega = 0$.

If $f(t)$ has a zero of order μ at $t = 0$ then $t^{\nu}f(t)$ has one of order $\mu + \nu$ at that point. The moment conditions expressed by Eqs. 89 and 90, Ch. XII, in terms of the impulse-train representations 19 and 20, then yield

$$\int_{-\infty}^{\infty} \omega^p F^{(\nu)}(\omega)\, d\omega = \int_{-\infty}^{\infty} \omega^p \sum_{k=-n}^{n} (a_k + jb_k)u_0(\omega - \omega_k)\, d\omega = 0$$
$$\text{for } p = 0, 1, 2, \ldots \mu + \nu - 1 \quad (34)$$

We thus have the following specific conditions upon the coefficients a_k and b_k:

$$\sum_{k=-n}^{n} a_k\omega_k^p = \sum_{k=-n}^{n} b_k\omega_k^p = 0 \qquad \text{for } p = 0, 1, 2, \ldots \mu + \nu - 1 \quad (35)$$

In expression 24 for $f(t)$ we replace the exponential function by its Maclaurin expansion

$$e^{j\omega_k t} = \sum_{r=0}^{\infty} \frac{(j\omega_k t)^r}{r!} \quad (36)$$

and have, in view of 35,

$$f(t) = \frac{(-1)^{\nu}}{\pi} \sum_{r=0}^{\infty} \left[\sum_{k=-n}^{n} a_k\omega_k^r\right] \frac{(jt)^{r-\nu}}{r!}$$
$$= \frac{(-1)^{\nu}}{\pi} \sum_{r=\mu+\nu}^{\infty} \left[\sum_{k=-n}^{n} a_k\omega_k^r\right] \frac{(jt)^{r-\nu}}{r!} \quad (37)$$

or, if we let $r - \mu - \nu = m$, this result can be written

$$f(t) = \frac{(-1)^{\nu}}{\pi} \sum_{m=0}^{\infty} \left[\sum_{k=-n}^{n} a_k\omega_k^{m+\mu+\nu}\right] \frac{(jt)^{m+\mu}}{(m+\mu+\nu)!} \quad (38)$$

which places in evidence the μth-order zero of $f(t)$ at $t = 0$. Hence the moment conditions 34 yielding the results 35 are seen to insure that $f(t)$ in formula 24 does not "blow up" for $t = 0$ but exhibits its proper behavior for the vicinity of this point.

Analogously the conditions 35 and Maclaurin expansion 36 substituted into Eq. 25 for $f(t)$ yield the companion result to 38 which reads

$$f(t) = \frac{j(-1)^{\nu}}{\pi} \sum_{m=0}^{\infty} \left[\sum_{k=-n}^{n} b_k\omega_k^{m+\mu+\nu}\right] \frac{(jt)^{m+\mu}}{(m+\mu+\nu)!} \quad (39)$$

In view of relations 21, 22 and 23, having to do with the even or odd character of functions 19 and 20, we observe that some of the conditions 35 are automatically fulfilled. Thus if p is even and ν odd or if p is odd and ν even, then 21 and 22 show that conditions 35 involving the coefficients a_k are fulfilled regardless of their specific values. Similarly, if p and ν are both either even or odd, then 21 and 23 show that conditions 35 involving coefficients b_k are fulfilled regardless of their values. Taking the relations 21, 22 and 23 for granted, therefore, we can regard conditions 35 as being trivial for the a_k's if $p \pm \nu$ is odd, for the b_k's if $p \pm \nu$ is even. In the nontrivial cases it is more effective to extend the sums in 35 over positive integers only. We then have

$$\sum_{k=0}^{n} a_k \omega_k^p = 0 \qquad \text{for } p = 0, 1, 2, \ldots \leq \mu + \nu - 1 \text{ and } p \pm \nu \text{ even} \tag{40}$$

$$\sum_{k=0}^{n} b_k \omega_k^p = 0 \qquad \text{for } p = 0, 1, 2, \ldots \leq \mu + \nu - 1 \text{ and } p \pm \nu \text{ odd} \tag{41}$$

If coefficients a_0 or b_0 are present, one-half of their values should be substituted in these sums since the other half belongs with terms for negative k-values which are omitted.

These conditions on the impulse values insure some other rather interesting and important properties which we shall now point out. Return to the impulse-train representations 19 and 20, and integrate r times so as to get

$$F_1^{(\nu-r)}(\omega) = \sum_{k=-n}^{n} a_k u_{-r}(\omega - \omega_k) \tag{42}$$

and

$$F_2^{(\nu-r)}(\omega) = \sum_{k=-n}^{n} b_k u_{-r}(\omega - \omega_k) \tag{43}$$

Now consider $\omega = 0$ and note that only terms for $k \leq 0$ contribute and that the term for $k = 0$ counts half. Moreover the singularity function of order $-r$ is then replaceable by

$$u_{-r}(\omega_k) = \frac{\omega_k^{r-1}}{(r-1)!} \tag{44}$$

so that 42 and 43 (21, 22 and 23 being noted) yield

$$F_1^{(\nu-r)}(0) = \frac{(-1)^\nu}{(r-1)!} \sum_{k=0}^{n} a_k \omega_k^{r-1} \tag{45}$$

$$F_2^{(\nu-r)}(0) = \frac{-(-1)^\nu}{(r-1)!} \sum_{k=0}^{n} b_k \omega_k^{r-1} \tag{46}$$

Identifying the integer $r - 1$ in these expressions with p in 40 and 41, we find that 40 and 45 yield

$$F_1^{(\nu-r)}(0) = 0 \qquad \text{for } \nu - r \text{ odd and } r \leq \mu + \nu \tag{47}$$

while from 41 and 46 we have

$$F_2^{(\nu-r)}(0) = 0 \qquad \text{for } \nu - r \text{ even and } r \leq \mu + \nu \tag{48}$$

For $\nu - r$ odd, function 42 is odd; and for $\nu - r$ even, function 43 is odd also. An odd function is zero for $\omega = 0$ and so the results 47 and 48 seem merely to state what we already know. However, they are by no means trivial, for in effect they state that the moment conditions 34 and their consequence as stated by conditions 35 or by 40 and 41 *insure* among other things that those functions 42 and 43 that are supposed to be odd actually turn out that way, which does not necessarily follow for any values of the coefficients a_k and b_k even though they have the properties 22 and 23. The latter merely insure that the impulse trains 19 and 20 themselves are even or odd as required by a particular ν-value; there is no assurance that the respective even or odd character of each function will result after successive integration so that after ν such integrations $F_1(\omega)$ turns out to be even and $F_2(\omega)$ odd.

The point at issue here is the following. Nonzero values of the impulse trains 19 and 20 are confined to the approximation range which, by normalization, we assume to be the interval $-1 < \omega < 1$. By the integral of either of these functions we mean one that represents the accumulated area under either function from $\omega = -1$ to some variable point within the approximation range or beyond it. For $\omega > 1$ this integral function must be zero, and the next must be zero for $\omega > 1$ also, through to the function $F_1(\omega)$ or $F_2(\omega)$ itself. Only a very special distribution of impulse values a_k and b_k will yield such a result. The moment conditions, if fulfilled, insure that the coefficients a_k and b_k have these special values.

What would happen if these conditions were not fulfilled? The property of being nonzero only over the range $-1 < \omega < 1$ means that when we integrate the impulse trains 19 and 20 once we get step functions that start at $\omega = -1$ with the value zero and return to the value zero at $\omega = 1$. At $\omega = 0$ the first integral of 19, being odd, must pass through zero. If we integrate again, the resulting broken line must stop with the value zero at $\omega = 1$ again, and in the case of function 20 it must pass through zero at $\omega = 0$. We may express this peculiarity of the functions 19 and 20, and of their successive integrals, by saying that they remain *attached* to the ω-axis, that they do not drift away from it, as they might if conditions 35 were not fulfilled.

If, through insufficient accuracy in our calculations, these conditions are not fulfilled closely enough, then functions formed by successive integration of 19 and 20 will show a "drift error" that becomes rapidly larger with each successive integration and hence can completely swamp the approximation error discussed above. The importance of fulfilling moment conditions with considerable care thus becomes evident.

It may be mentioned in passing that conditions 40 and 41 include all necessary relations 47 and 48, even for $\mu = 0$ for which the asymptotic behavior of $F(\omega)$ is such that $f(t)$ has a step at $t = 0$. In cases for which $f(t)$ has a simple or higher-order zero at $t = 0$, the moment conditions 40 and 41 include relations beyond those necessary to insure zero drift on the part of $F_1(\omega)$ and $F_2(\omega)$.

Now let us turn our attention to moment conditions replacing asymptotic behavior of the time function. These determine the character of $F(\omega)$ near $\omega = 0$. They are essential in the numerical evaluation of direct transforms through use of formulas 28 and 29 (or the more specific real- and imaginary-part formulas 30 through 33).

If $F(\omega)$ has a zero of order μ at $\omega = 0$ then $(j\omega)^\nu F(\omega)$ has one of order $\mu + \nu$ at that point, and the moment conditions expressed by Eqs. 94 and 95 of Ch. XII yield for the time function 28

$$\int_0^\infty t^p f^{(\nu)}(t)\,dt = \int_0^\infty t^p \sum_{k=1}^n a_k u_0(t - t_k)\,dt = \sum_{k=1}^n a_k t_k^p = 0$$

$$\text{for } p = 0, 1, 2, \ldots \mu + \nu - 1 \quad (49)$$

In Eq. 29 we now replace the exponential by its Maclaurin series

$$e^{-j\omega t_k} = \sum_{r=0}^\infty \frac{(-j\omega t_k)^r}{r!} \quad (50)$$

and get in view of 49

$$F(\omega) = (-1)^\nu \sum_{r=0}^\infty \left[\sum_{k=1}^n a_k t_k^r\right] \frac{(-j\omega)^{r-\nu}}{r!} = (-1)^\nu \sum_{r=\mu+\nu}^\infty \left[\sum_{k=1}^n a_k t_k^r\right] \frac{(-j\omega)^{r-\nu}}{r!} \quad (51)$$

If we again let $r - \mu - \nu = m$, this result can be written

$$F(\omega) = (-1)^\nu \sum_{m=0}^\infty \left[\sum_{k=1}^n a_k t_k^{m+\mu+\nu}\right] \frac{(-j\omega)^{m+\mu}}{(m + \mu + \nu)!} \quad (52)$$

thus substantiating the fact that $F(\omega)$ has the required μth-order zero at $\omega = 0$, and that formula 29, or the equivalent result expressed by Eqs. 30 through 33, does not "blow-up" for $\omega = 0$.

We can also show that the moment conditions 49 assure zero drift when $f^{(\nu)}(t)$ in Eq. 28 is integrated ν successive times. What we need to show is that the impulse train given by Eq. 28 is zero for any t-value larger than the largest t_k, and that the same is true for each successive integral formed from this impulse train between limits zero and t, through to the function $f(t)$ itself.

In this sense, the rth successive integral of impulse train 28 reads simply

$$f^{(\nu-r)}(t) = \sum_{k=1}^{n} a_k u_{-r}(t - t_k) \tag{53}$$

For $t - t_k > 0$ we have according to the definition of this family of singularity functions

$$u_{-r}(t - t_k) = \frac{(t - t_k)^{r-1}}{(r - 1)!} \tag{54}$$

By the binomial theorem

$$(t - t_k)^{r-1} = \sum_{q=0}^{r-1} \binom{r-1}{q} t^{r-q-1} t_k^q \tag{55}$$

so that 53 can be written

$$f^{(\nu-r)}(t) = \sum_{q=0}^{r-1} \binom{r-1}{q} \frac{t^{r-q-1}}{(r-1)!} \sum_{k=1}^{n} a_k t_k^q \tag{56}$$

where it is tacitly understood that this expression is valid only for values of t larger than the largest t_k, for which it is identically zero if

$$\sum_{k=1}^{n} a_k t_k^q = 0 \qquad \text{for } q = 0, 1, \ldots r - 1 \tag{57}$$

The moment conditions 49 assure this result for integer q-values through $\mu + \nu - 1$. Hence function 56 is identically zero for $r \leq \mu + \nu$, and our objective is achieved even if $\mu = 0$, for which $F(\omega)$ does not have a zero at $\omega = 0$.

6. Truncation Error Evaluation

We will confine our discussion here to the determination of a time function from a given frequency spectrum since essentially the same techniques apply also to the reverse situation.

As shown in Ch. XIV, truncation of the spectrum with a rectangular truncation function having a cutoff frequency ω_c is equivalent to convolution in the time domain with the aperiodic Dirichlet kernel

$$D_a(t) = \frac{\omega_c}{\pi} \cdot \frac{\sin \omega_c t}{\omega_c t} \tag{58}$$

The nature and extent of the error which is thus produced in the time function $f(t)$ depends largely upon the character of this function. If it contains discontinuities, then a Gibbs phenomenon type of error is produced at each of these, the extent of which depends solely upon the magnitude of the pertinent discontinuity and not upon the truncation frequency. If $f(t)$ contains no discontinuities but its first derivative does, then a less severe type of error is produced which (as we shall see) is inversely proportional to the truncation frequency and hence can be reduced to a tolerable value by choosing this frequency sufficiently large. If discontinuities do not appear until one forms the second derivative of $f(t)$, then the error is inversely proportional to the square of the truncation frequency, and in most cases is unimportant.

Whether $f(t)$ or its derivatives contain discontinuities can, as we have seen, be recognized from the asymptotic behavior of the associated spectrum function $F(\omega)$. We recall (art. 9, Ch. XI and arts. 5, 6, 7, Ch. XII) that if $f(t)$ has discontinuities, then $F(\omega)$ can converge no faster than $1/\omega$ for large ω; conversely, if this is the asymptotic character of $F(\omega)$, then we can conclude that $f(t)$ does have discontinuities (at least one). If $F(\omega)$ converges as fast but no faster than $1/\omega^2$ for larger ω then $f(t)$ is continuous but its first derivative is not. If $F(\omega)$ behaves like $1/\omega^n$ for $\omega \to \infty$, then discontinuities do not appear until the $(n - 1)$th-order derivative of $f(t)$ is formed.

From the asymptotic behavior of a given spectrum function (which, as discussion in the previous article shows, can be inferred from moment conditions in case the asymptote has already been chopped off) we can, therefore, predict at once the nature of the truncation error that will result in any given situation. The details of this process as well as a means for estimating the extent of such error will now be discussed in greater detail.

Let us begin by assuming that the given spectrum function $F(\omega)$ has an asymptote that does not approach zero, indicating that the associated $f(t)$ contains impulses. In the vicinity of one of these, having unit value, the error for a truncation frequency ω_c is the difference between the Dirichlet kernel 58 and the unit impulse itself, namely an error function

$$\epsilon_0(t) = \frac{\omega_c}{\pi} \cdot \frac{\sin \omega_c t}{\omega_c t} - u_0(t) \tag{59}$$

where the point at which this impulse occurs is chosen as a time origin.

If instead, $F(\omega)$ behaves like $1/\omega$ for $\omega \to \infty$ and hence $f(t)$ contains discontinuities, we obtain the error function in the vicinity of one of these (having unit value) simply by integrating $\epsilon_0(t)$, since integration in the time domain multiplies $F(\omega)$ by $1/j\omega$ and vice versa. The pertinent error function then reads

$$\epsilon_{-1}(t) = \int_{-\infty}^{t} \epsilon_0(\xi)\, d\xi = \frac{1}{2} + \frac{1}{\pi} Si(\omega_c t) - u_{-1}(t) \tag{60}$$

in which $u_{-1}(t)$ is the unit step and $Si(x)$ is the sine-integral function discussed in art. 1, Ch. XIV and again in art. 9, Ch. XVI.

Next if $F(\omega)$ behaves like $1/\omega^2$ for $\omega \to \infty$ and the associated $f(t)$ is continuous but its first derivative is not, then the error in the vicinity of a break point involving unit change in slope is obtained by integrating the

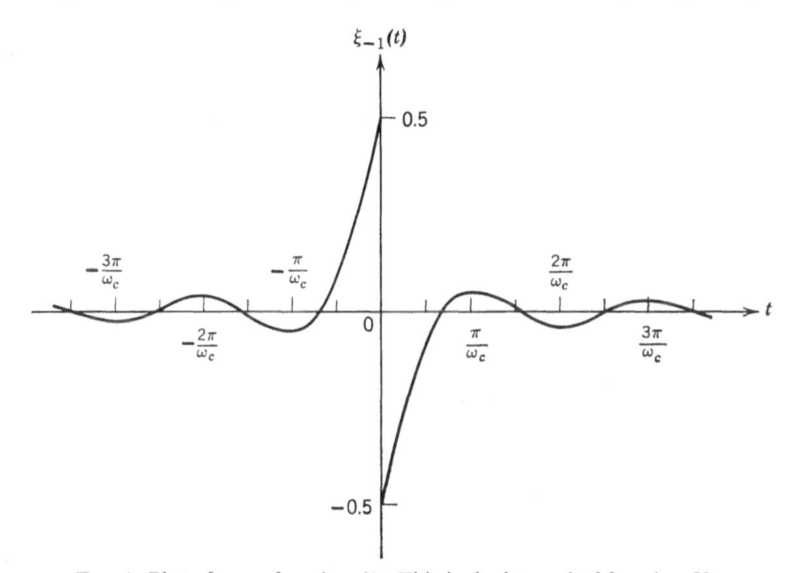

FIG. 4. Plot of error function 60. This is the integral of function 59.

error 59 a second time, or the error function 60 once. Although its integral is not tabulated, we can compute and plot the result as indicated by

$$\epsilon_{-2}(t) = \int_{-\infty}^{t} \epsilon_{-1}(\xi)\, d\xi \tag{61}$$

Functions 60 and 61 are shown plotted in Figs. 4 and 5. Figure 4 shows the error in the vicinity of a unit step (the familiar Gibbs phenomenon); Fig. 5, the error with which the unit ramp function is approximated. Except for some minor ripples, the essential error here consists of the appearance of a fillet at the break point having a value $1/\pi\omega_c$.

One can readily obtain this value by observing that the transform of $\epsilon_0(t)$ in Eq. 59 is $H(j\omega) - 1$ where $H(j\omega)$ is the rectangular cutoff function of unit height and width $2\omega_c$ centered at $\omega = 0$. The transform of $\epsilon_{-2}(t)$ is obtained through division by $(j\omega)^2$, and hence we have

$$\epsilon_{-2}(0) = \frac{1}{2\pi}\int_{-\infty}^{\infty} \frac{1 - H(j\omega)}{\omega_2}\, d\omega = \frac{1}{\pi}\int_{\omega_c}^{\infty} \frac{d\omega}{\omega^2} = \frac{1}{\pi\omega_c} \tag{62}$$

If the spectrum function $F(\omega)$ converges like $1/\omega^3$ or $1/\omega^4$ for large ω, then the pertinent error functions $\epsilon_{-3}(t)$ and $\epsilon_{-4}(t)$, per unit discontinuity in the second and third derivatives of $f(t)$, are the next two successive integrals of function 61. $\epsilon_{-3}(t)$ thus turns out to be an odd function with maximum ripples adjacent to $t = 0$ of approximate magnitude $1/2\omega_c^2$, and $\epsilon_{-4}(t)$ is even about $t = 0$ with a maximum at this point equal to $1/3\omega_c^3$. It seems hardly worth working these things out in detail since the values of these errors will in most cases be negligible anyway.

In situations where we begin with a time function, compute its spectrum, and then use numerical methods to regain the time function, these results

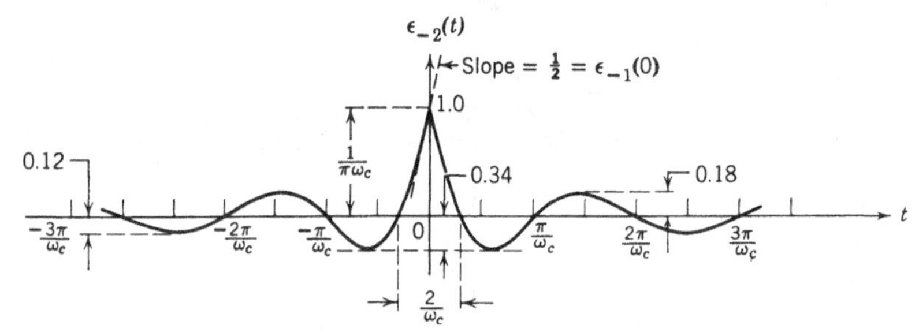

FIG. 5. Plot of function 61 which is the second integral of 59.

enable us to estimate the truncation error for any chosen cutoff frequency or determine the latter for a specified maximum error because, in such a problem, we know the values of discontinuities in the function or its derivatives. For example, if $f(t)$ consists of confluent straight-line segments, then the essential error due to truncation amounts to the insertion of fillets at the break points. Since Fig. 5 shows the fillet size per unit change in slope, this type of error is readily computed at each break point from the change in slope that occurs there.

On the other hand, if we are merely given a spectrum function and do not know what the time function is supposed to be, then although the asymptotic behavior of $F(\omega)$ tells us whether to expect discontinuities in $f(t)$ or in df/dt or in d^2f/dt^2, etc., we do not know where these occur or how large they are. In such a situation we may have to begin by choosing ω_c at random. Knowing, for example, that $f(t)$ has discontinuities in its first derivative, we can expect to recognize where possible break points lie if the chosen cutoff is adequately large. Although we may need to make a second choice for ω_c, a knowledge of the nature of the expected error, and the dependence of its magnitude upon ω_c, gives us a means to estimate whether the chosen value of ω_c is large enough in view of a stated upper bound on the allowable error.

7. Determination of Impulse Response from Specified Input and Output Time Functions

Input and output, or excitation and response functions as well as the network are characterized in the time domain, the network characterization being its unit impulse response. In the present problem, input and output are given; the impulse response of the network is to be found. As we have seen (Ch. XVI, art. 3, Eqs. 50 and 51) the output time function is obtained by convolving the input with the impulse response. If we denote the input by $a(t)$, the impulse response by $b(t)$, and the output by $c(t)$, then the pertinent analytic relationship reads

$$c(t) = \int_{-\infty}^{\infty} a(\xi)b(t - \xi)\,d\xi = \int_{-\infty}^{\infty} a(t - \xi)b(\xi)\,d\xi \tag{63}$$

In the usual network problem, $a(t)$ and $b(t)$ are given; the response $c(t)$ is to be found. In the problem that we wish to consider, $a(t)$ and $c(t)$ are given and $b(t)$ is the unknown. Mathematically we are dealing with a particular sort of *integral equation.*

If the Fourier transforms of these functions are denoted by $A(\omega)$, $B(\omega)$ and $C(\omega)$, then the frequency-domain equivalent of Eq. 63 reads

$$C(\omega) = A(\omega) \cdot B(\omega) \tag{64}$$

and if $B(\omega)$ is the unknown, we have simply

$$B(\omega) = \frac{C(\omega)}{A(\omega)} \tag{65}$$

In the frequency domain it is equally as simple to find the system function when input and output are given as it is to find the output from a given input and system function. We simply divide instead of multiplying.

Thus the solution to integral equation 63 for given functions $a(t)$ and $c(t)$ can straightforwardly be found by using the preceding numerical methods to transform to the frequency domain and back again. However, this process is a rather long one, and it is, therefore, logical to try and find a method whereby this integral equation can be solved in the time domain directly.

The clue to finding such a method is given by the results of art. 9, Ch. XV, where we show that convolution of time functions represented by impulse trains (sampled form) is equivalent to polynomial multiplication. In terms of the preceding discussions of this chapter, such impulse trains may represent a first or higher derivative of the pertinent time functions. Since at the moment we are not concerned with numerical accuracy, let us suppose that $a(t)$ and $b(t)$ are step-function approximants, whereupon $c(t)$ in 63 is a broken-line or piecewise-linear approximation to the pertinent smooth function. If, for simplicity, we denote differentiation with respect to t by a prime, we have the following uniform impulse train representations to deal with:

$$a'(t) = \sum_{k=0}^{m} a_k u_0(t - kt_0) \tag{66}$$

$$b'(t) = \sum_{k=0}^{n} b_k u_0(t - kt_0) \tag{67}$$

$$c''(t) = \sum_{k=0}^{q} c_k u_0(t - kt_0) \tag{68}$$

in which t_0 is the uniform sampling interval, and

$$q = m + n \tag{69}$$

If (as in art. 9, Ch. XV) we use the notation

$$x = e^{-j\omega t} \tag{70}$$

then the Fourier transforms of functions $a'(t)$, $b'(t)$ and $c''(t)$ are respectively given by the polynomials

$$j\omega A(\omega) = a_0 + a_1 x + a_2 x^2 + \cdots + a_m x^m \tag{71}$$

$$j\omega B(\omega) = b_0 + b_1 x + b_2 x^2 + \cdots + b_n x^n \tag{72}$$

$$(j\omega)^2 C(\omega) = c_0 + c_1 + c_2 x^2 + \cdots + c_q x^q \tag{73}$$

where the coefficients are impulse values (samples) of these differentiated time functions. The polynomials are finite in extent for finite integers m and n.

Assuming that areas enclosed by the time functions $a(t)$ and $b(t)$, equal respectively to $A(0)$ and $B(0)$, are finite and nonzero, the transforms 71 and 72 of their derivatives have simple zeros at $\omega = 0$. These derivative functions 66 and 67, therefore, enclose zero area (according to moment conditions 94 and 95, Ch. XII) and hence the impulse values a_k and b_k satisfy conditions

$$\sum_{k=0}^{m} a_k = 0 \quad \text{and} \quad \sum_{k=0}^{n} b_k = 0 \tag{74}$$

It follows that the transform of $c''(t)$ given by Eq. 73 has a second-order zero at $\omega = 0$, so that moment conditions here yield

$$\sum_{k=0}^{q} c_k = 0 \quad \text{and} \quad \sum_{k=0}^{q} kt_0 c_k = 0 \tag{75}$$

If conditions 74 are fulfilled then both conditions 75 must automatically be satisfied since 73 is equal to the product of 71 and 72.

Our problem involving the solution of integral equation 63 is now straightforward. From a step approximation to $a(t)$ and a piecewise-linear approximation to $c(t)$ we read off by inspection the values of coefficients a_k and c_k, thus determining the polynomials 71 and 73. According to Eq. 65, the polynomial 72 is then found by the familiar process of long division, and the step approximation to the desired function $b(t)$ is constructible from the resulting coefficients b_k.

Formally this process is simple. Unless we better understand some of its subtleties, however, we may have difficulty interpreting some of the results. For example, we can expect a function $b(t)$ of finite duration (i.e., a polynomial 72 of finite degree) only if the polynomial 71 is a factor of the polynomial 73; otherwise the long-division process will continue indefinitely, the function $f(t)$ will extend over the infinite time domain, and, incidentally, it would seem that the relation 69 among the integers m, n and q would not be fulfilled since we start with finite values for m and q and obtain an infinite value for n.

This conclusion is false, since continuation of the long-division process implies the insertion of zero coefficients c_k beyond the last nonzero value, thus in effect extending the integer q indefinitely also. In other words, indefinite continuation of the division process is equivalent in Eq. 68 to regarding the region $t > qt_0$ where $c(t)$ is identically zero as part of its legitimate interval of definition.

Another difficulty may arise, however, if 71 is not a factor of 73. Indefinite continuation of the division process may yield a divergent series for the representation 72, thus indicating that the specified input and output time functions can result only from an unstable network. This result is actually not a failure of our method for the solution of integral equation 63; rather it shows that a solution interpretable in terms of a stable network does not necessarily exist.

We can understand this situation rather easily by reference to Eq. 65 where the frequency-response function of the network is expressed as a ratio of the transforms of output and input time functions. Poles of this response function $B(\omega)$ are poles of $C(\omega)$ and any zeros of $A(\omega)$ not also contained in $C(\omega)$. Since such zeros can lie in the right as well as in the left half-plane, it is possible to obtain a function $B(\omega)$ having some right half-plane poles, whereupon the implied network is unstable and its impulse response is divergent.

In order to anticipate this situation it would be advantageous to tell whether a given time function has a transform with right half-plane zeros. This we can do in the following manner. Consider the step function $a(t)$ and the transform of its first derivative as given by Eq. 71. In view of the substitution 70 which maps the right half of the s-plane upon the interior of a unit circle in the x-plane, we can say that the transform of $a(t)$ has no right half-plane zeros if polynomial 71 has no zeros within the unit circle of the x-plane.

If we make the further change of variable

$$x = \frac{1 - \lambda}{1 + \lambda} \quad \text{or} \quad \lambda = \frac{1 - x}{1 + x} \tag{76}$$

which maps the interior of a unit circle in the x-plane upon the right half of the λ-plane, the polynomial 71 is converted into

$$\lambda A^*(\lambda) = a_0(1 + \lambda)^m + a_1(1 - \lambda)(1 + \lambda)^{m-1} + \cdots + a_m(1 - \lambda)^m \tag{77}$$

Since the origin of the s-plane maps upon the origin of the λ-plane, it is clear that polynomial 77 must contain the factor λ as indicated. This fact also follows from 74 which states that the coefficients a_k add to zero, hence 77 can have no constant term. Given the step approximation to a time function $a(t)$ we can form the polynomial $A^*(\lambda)$ directly. If this polynomial is Hurwitz (see art. 5, Ch. VIII) then the transform of $a(t)$ has no right half-plane zeros.

The method will, of course, fail if the step approximation is too crude. It can readily be extended to a piecewise-linear or parabolic arc approximation and thus be made to yield good accuracy without an unduly high degree in the polynomial $A^*(\lambda)$.

Returning to the process of dividing polynomial 71 into 73, an obvious a priori condition is that q be larger than m since by 69, $n = q - m$ and n must be larger than nothing. If this condition seems not to be fulfilled we can, of course, simply add as many zero coefficients c_k as we wish, bearing in mind the pertinent things that are pointed out above. However, let us consider a situation in which q is considerably larger than m to start with, and suppose we are looking for a polynomial 72 of finite degree n. We said before that this could result only if 71 were a factor of 73, which is true, but let us see whether somehow we can bring about this fortuitous circumstance and at what cost.

The cost is easily recognizable. If the last $m + 1$ coefficients c_k in polynomial 73 are left unspecified, then obviously we can choose their values such that the long-division process terminates after $n = q - m$ steps. Thus, if the desired output time function is specified except for the last m of its sampling intervals (leaving the last $m + 1$ samples arbitrary) then the polynomial 73 can be made to contain 71, and a stable impulse response $b(t)$ is straightforwardly determined regardless of whether the polynomial $A^*(\lambda)$ derived from the input time function is Hurwitz or not.

It is interesting to note in this connection that we can carry out the long division with polynomials 71 and 73 oriented in either ascending or descending powers of x. If oriented in ascending powers, the unspecified portion of the output $c(t)$ is at the tail end of this function, in the so-called *coda region*; if oriented in descending powers, the unspecified portion of $c(t)$ is at the head end or in the *precursor region.*

If, in this process, one augments the polynomial 73 by inserting zero coefficients so as to prolong the long division, the precursor or coda regions can be detached from the specified output function $c(t)$. Thus one can specify this complete function, and then let the termination process in the long division determine a completely separate precursor or coda portion which must be attached to the specified $c(t)$ in order to achieve stability. We will now illustrate these matters with several examples.

In Fig. 6 are shown a pair of time functions $a(t)$ and $c(t)$ that are to be the input and output respectively of some network whose impulse response is to be found. By inspection, the pertinent a_k and c_k coefficients are

$$a_k = 1, 1, -2 \qquad \text{for } k = 0, 1, 2 \tag{78}$$

$$c_k = 1, -\tfrac{1}{2}, -\tfrac{1}{2}, -\tfrac{1}{2}, -\tfrac{1}{2}, 1 \qquad \text{for } k = 0, 1, 2, 3, 4, 5 \tag{79}$$

Suppose we check $a(t)$ to see if its transform has right half-plane zeros. By Eq. 77 and the values 78 we have

$$\begin{aligned} \lambda A^*(\lambda) &= (1 + \lambda)^2 + (1 - \lambda)(1 + \lambda) - 2(1 - \lambda)^2 \\ &= 2\lambda(3 - \lambda) \end{aligned} \tag{80}$$

Obviously $A^*(\lambda)$ is not Hurwitz, so we expect stability difficulties. The long-division process takes the form

$$\begin{array}{ccc|ccccccc}
 & & & 1 & -\frac{3}{2} & 3 & -\frac{13}{2} & 12 & -24 & \\ \hline
1 & 1 & -2 & 1 & -\frac{1}{2} & -\frac{1}{2} & -\frac{1}{2} & -\frac{1}{2} & 1 & \\
 & & & 1 & 1 & -2 & & & & \\ \cline{4-6}
 & & & & -\frac{3}{2} & \frac{3}{2} & -\frac{1}{2} & & & \\
 & & & & -\frac{3}{2} & -\frac{3}{2} & 3 & & & \\ \cline{5-7}
 & & & & & 3 & -\frac{7}{2} & -\frac{1}{2} & & \\
 & & & & & 3 & 3 & -6 & & \\ \cline{6-8}
 & & & & & & -\frac{13}{2} & \frac{11}{2} & 1 & \\
 & & & & & & -\frac{13}{2} & -\frac{13}{2} & 13 & \\ \cline{7-9}
 & & & & & & & 12 & -12 & \\
 & & & & & & & 12 & 12 & -24 \\ \cline{8-10}
 & & & & & & & & -24 & 24
\end{array} \tag{81}$$

from which it becomes clear that the process does not terminate, and is obviously divergent.

We can continue the function $c(t)$ in such a way as to terminate the division in two additional steps. Picking up the process 81 at this point takes the form

$$
\begin{array}{r|rrrrrr}
 & 12 & -24 & (c_6 + 48) & & & \\
\hline
1 \quad 1 \quad -2 & \cdot\ \cdot\ \cdot & \cdot\ \cdot & c_6 & c_7 & c_8 & \\
 & 12 & -12 & c_6 & \cdot & \cdot & \\
 & 12 & 12 & -24 & \cdot & \cdot & \\
\cline{2-4}
 & & -24 & (c_6 + 24) & c_7 & \cdot & \\
 & & -24 & -24 & 48 & \cdot & \\
\cline{3-5}
 & & & (c_6 + 48) & (c_7 - 48) & c_8 & \\
 & & & (c_6 + 48) & (c_6 + 48) & -2c_6 & -96 \\
\cline{4-7}
\end{array}
\tag{82}
$$

Since the coefficients b_k must sum to zero we see that $c_6 + 48$ in the quotient must equal 16, or c_6 must equal -32. Termination of the division process

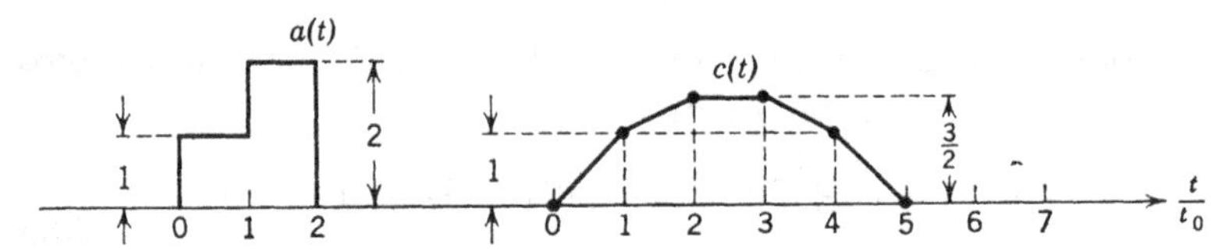

FIG. 6. Pertinent to the illustrative example relating to the data given by Eqs. 78 and 79.

then requires c_7 to be 64 and c_8 to be -32. In the sketch of Fig. 6, these values introduce a huge negative spike in the interval between $t = 5t_0$ and $t = 7t_0$.

This is the cost of demanding a stable system in spite of a right half-plane zero in the transform of $a(t)$. Adding the huge negative spike to $c(t)$ has the effect of introducing a factor $3 - \lambda$ into the corresponding polynomial $C^*(\lambda)$ obtained from Eq. 77, thus canceling this factor in $A^*(\lambda)$ and preventing the system function $B(\omega)$ from having a right half-plane pole.

Now let us consider the same problem with the input function $a(t)$ reversed in orientation so that the a_k-values 78 are replaced by

$$
a_k = 2 \quad -1 \quad -1 \qquad \text{for } k = 0, 1, 2 \tag{83}
$$

This process replaces t by $-t$ and hence s by $-s$. Incidentally, x in Eq. 70 is replaced by $1/x$ and hence λ in Eq. 76 is replaced by $-\lambda$. The essential point of all this is that the factor $3 - \lambda$ in Eq. 80 becomes $3 + \lambda$. Reversing the orientation of $a(t)$ in Fig. 6 should, therefore, lead to a stable impulse response. The long-division process now reads

$$
\begin{array}{ccc|ccccccccc}
 & & & \frac{1}{2} & 0 & 0 & -\frac{1}{4} & -\frac{3}{8} & \frac{3}{16} & -\frac{3}{32} & \frac{3}{64} & \\
\hline
2 & -1 & -1 & 1 & -\frac{1}{2} & -\frac{1}{2} & -\frac{1}{2} & -\frac{1}{2} & 1 & 0 & 0 & 0 \\
 & & & 1 & -\frac{1}{2} & -\frac{1}{2} & & & & & & \\
\cline{4-6}
 & & & & 0 & 0 & -\frac{1}{2} & -\frac{1}{2} & 1 & & & \\
 & & & & & & -\frac{1}{2} & \frac{1}{4} & \frac{1}{4} & & & \\
\cline{7-9}
 & & & & & & & -\frac{3}{4} & \frac{3}{4} & 0 & & \\
 & & & & & & & -\frac{3}{4} & \frac{3}{8} & \frac{3}{8} & & \\
\cline{8-10}
 & & & & & & & & \frac{3}{8} & -\frac{3}{8} & 0 & \\
 & & & & & & & & \frac{3}{8} & -\frac{3}{16} & -\frac{3}{16} & \\
\cline{9-11}
 & & & & & & & & & -\frac{3}{16} & \frac{3}{16} & 0 \\
 & & & & & & & & & -\frac{3}{16} & \frac{3}{32} & \frac{3}{32} \\
\cline{10-12}
 & & & & & & & & & & \frac{3}{32} & -\frac{3}{32}
\end{array}
\tag{84}
$$

From the pattern thus established we can see that the succeeding terms in the quotient vary like $1/2^{n-1}$ and hence the impulse response, although not finite, is convergent. If we introduce coefficients c_6, c_7 and c_8 as in 82 the process can be terminated in the following manner:

$$
\begin{array}{ccc|ccccc}
 & & & -\frac{3}{8} & \frac{3}{16} & (c_6 - \frac{3}{16})/2 & & \\
\hline
2 & -1 & -1 & \cdots & \cdot & c_6 & c_7 & c_8 \\
 & & & -\frac{3}{4} & \frac{3}{4} & c_6 & & \\
 & & & -\frac{3}{4} & \frac{3}{8} & \frac{3}{8} & & \\
\cline{4-6}
 & & & & \frac{3}{8} & (c_6 - \frac{3}{8}) & c_7 & \\
 & & & & \frac{3}{8} & -\frac{3}{16} & -\frac{3}{16} & \\
\cline{5-7}
 & & & & & (c_6 - \frac{3}{16}) & (c_7 + \frac{3}{16}) & c_8 \\
 & & & & & (c_6 - \frac{3}{16}) & \dfrac{(c_6 - \frac{3}{16})}{-2} & \dfrac{(c_6 - \frac{3}{16})}{-2} \\
\cline{6-8}
\end{array}
\tag{85}
$$

In order that the coefficients b_k in the quotient may sum to zero, c_6 must have the value $\frac{1}{16}$; and termination of the division process then yields $c_7 = -\frac{1}{8}$ and $c_8 = \frac{1}{16}$. These values represent a triangle $\frac{1}{16}$ unit tall filling the interval between $t = 5t_0$ and $t = 7t_0$ in the sketch of Fig. 6. We can make this triangle even more insignificant by first adding some zero coefficients to the dividend before applying the termination process.

If we return to the problem as originally stated in Fig. 6, but consider introducing a precursor in the function $c(t)$, the procedure is to divide as in 81, but with the coefficient order reversed in both dividend and divisor.

Since $c(t)$ is a symmetrical function, the dividend coefficients remain the same. The net result is like the division 84 except for a reversed algebraic sign in the divisor. Using the termination process 85 (except that we will denote c_6, c_7, c_8 respectively by c_{-1}, c_{-2}, c_{-3}) we find a precursor in the form of a triangle $\frac{1}{16}$th of a unit tall. The fact that this very small precursor triangle has the same effect in canceling the right half-plane zero in the transform of $a(t)$ as does a huge spike in the coda region is quite remarkable and illustrates the subtle nature of a problem of this sort.

8. Examples of Numerical Transform Evaluation and Error Control

As our first example let us choose a spectrum function and compute the associated time function. Specifically we will draw a graph appropriate to the real part of a function $F(\omega)$ that behaves like $1/(j\omega)^3$ for large ω. If we draw this graph in the form of confluent parabolic arcs then this is a situation for which $\nu = 3$ and* $\mu = 2$. The nontrivial moment conditions 40 must then hold for the integers $p = 1$ and 3. Additionally it is convenient to normalize the amplitude of the real part $F_1(\omega)$ so that its zero-frequency value is unity. Equation 45 can be used for this purpose, and together with conditions 40, one then has the relations

$$\sum a_k\omega_k = 0 \tag{86}$$

$$\sum a_k\omega_k^2 = -2 \tag{87}$$

$$\sum a_k\omega_k^3 = 0 \tag{88}$$

The simplest graph we can draw will involve at least three parabolic arcs, and in that event the a_k values must fulfill the equations

$$\begin{aligned} a_1\omega_1 + a_2\omega_2 + a_3\omega_3 &= 0 \\ a_1\omega_1^2 + a_2\omega_2^2 + a_3\omega_3^2 &= -2 \\ a_1\omega_1^3 + a_2\omega_2^3 + a_3\omega_3^3 &= 0 \end{aligned} \tag{89}$$

Normalization of the frequency scale amounts to choosing $\omega_3 = 1$. The only freedom left is the choice of ω_1 and ω_2. Suppose we let $\omega_1 = \frac{1}{3}$ and $\omega_2 = \frac{2}{3}$. Equations 89 then yield

$$a_1 = 45, \quad a_2 = -36, \quad a_3 = 9 \tag{90}$$

The real-part function which these a_k values imply is found to be given by the following three confluent parabolic arcs:

$$F_1(\omega) = 1 - 9\omega^2 \qquad \text{for } 0 < \omega < \tfrac{1}{3} \tag{91}$$

$$F_1(\omega) = \tfrac{7}{2} - 15\omega + \tfrac{27}{2}\omega^2 \qquad \text{for } \tfrac{1}{3} < \omega < \tfrac{2}{3} \tag{92}$$

$$F_1(\omega) = -\tfrac{9}{2} + 9\omega - \tfrac{9}{2}\omega^2 \qquad \text{for } \tfrac{2}{3} < \omega < 1 \tag{93}$$

* The time function $f(t)$ has a second-order zero at $t = 0$.

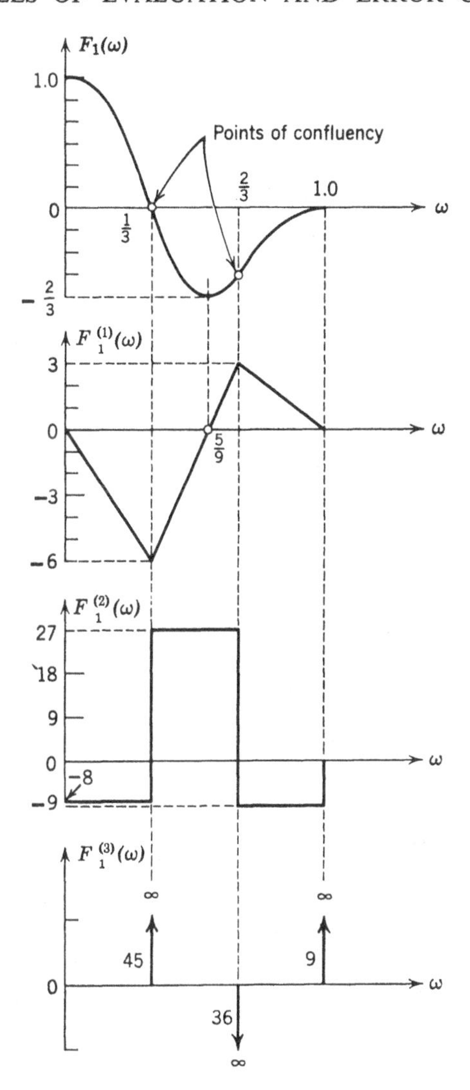

FIG. 7. Illustrative example in which the real-part function $F_1(\omega)$ is defined by the parabolic arcs 91, 92 and 93.

Plots of this function and its first three derivatives are shown in Fig. 7. As required by the assumed asymptotic behavior (implying a total phase shift of $3\pi/2$ radians for the range $0 < \omega < \infty$) the real part has one change of sign, and encloses zero area, which (according to relation 47) is the meaning of moment condition 88.

The time function is obtained from Eq. 27 as

$$\begin{aligned} f(t) &= \frac{2}{\pi t^3}[45 \sin(t/3) - 36 \sin(2t/3) + 9 \sin t] \\ &= \frac{2}{\pi}\left[\frac{t^2}{27} - \frac{t^4}{729} + \cdots\right] \end{aligned} \tag{94}$$

from which we see that the proper behavior at $t = 0$ is obtained.

This example shows, incidentally, that fulfillment of moment conditions as required by an assumed or given asymptotic behavior curtails to some extent the ability to approximate arbitrary functions with a limited number

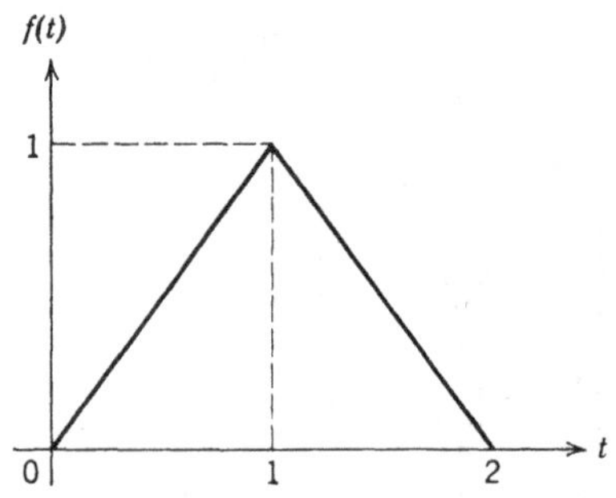

FIG. 8. Triangular time function to which the numerical integration process is applied.

of terms. However, this fact is entirely logical since the added requirement of meeting certain asymptotic conditions can be met only through relinquishing some other desired features or by increasing correspondingly the total number of adjustable parameters.

In our next example we will illustrate the approximation and error control process by assuming a time function, calculating its spectrum, and then regaining the time function through numerical integration.

Let us consider the triangular time function shown in Fig. 8. Its second derivative consists of a train of three impulses at the points $t = 0, 1, 2$. The first and last of these have the value $+1$, the second has the value -2. Making use of elementary properties (art. 5, Ch. XII) we can write down the Laplace transform at once. It reads

$$\begin{aligned} F(s) &= \frac{1 - 2e^{-s} + e^{-2s}}{s^2} \\ &= \left[\frac{1 - e^{-s}}{s}\right]^2 \end{aligned} \tag{95}$$

For $s = j\omega$, the real part is given by

$$F_1(\omega) = \left[\frac{\sin \omega/2}{\omega/2}\right]^2 \cdot \cos \omega \tag{96}$$

It is shown plotted in the top part of Fig. 9 over the range from $\omega = 0$ to $\omega = 2\pi$. Beyond this point the values of $F_1(\omega)$ become small rather rapidly and so we shall truncate the spectrum here and see how closely we regain the triangular time function.

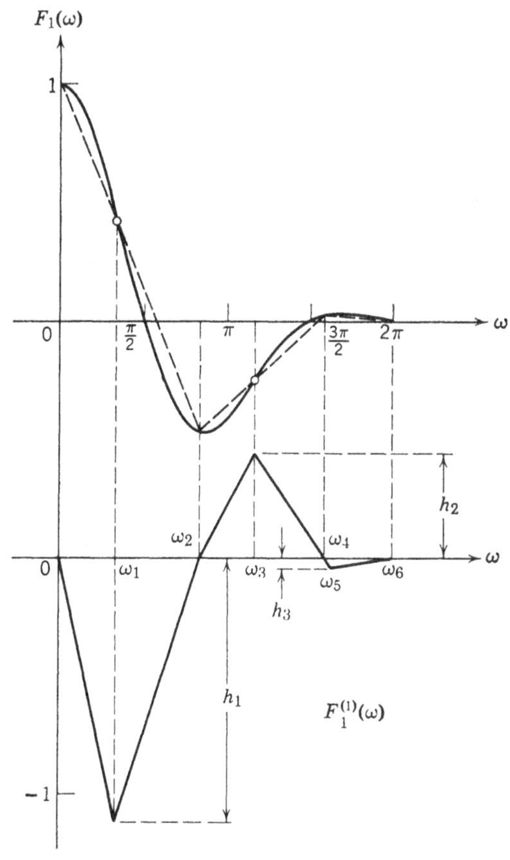

FIG. 9. Filling parabolic arcs to the transform of the time function shown in Fig. 8.

From the form of $F_1(\omega)$ it seems reasonable that a rather close approximation may be had by use of confluent parabolic arcs. The first derivative $F_1^{(1)}(\omega)$ is then a broken line, and the third derivative is an impulse train.

In fitting parabolic arcs to the curve of $F_1(\omega)$, we will follow the procedure outlined in art. 5, Ch. XIII and illustrated there in Fig. 15.

Consider the first triangular portion of $F_1^{(1)}(\omega)$ which is sketched in the bottom part of Fig. 9. The base of this triangle extends from $\omega = 0$ to a point $\omega = \omega_2$ where the first minimum of $F_1(\omega)$ occurs, and the negative area contributed by this triangle must equal the ordinate difference between $F_1(0)$ and $F_1(\omega_2)$. From calculations yielding the plot of $F_1(\omega)$, this ordinate difference is found to have the value 1.470 and $\omega_2 = 5\pi/6$ with reasonable accuracy. If we denote the altitude of this first triangle by h_1, then we have

$$\frac{h_1}{2} \cdot \frac{5\pi}{6} = 1.47 \quad \text{or} \quad h_1 = 1.123 \tag{97}$$

Next we must decide where to place the peak of this first triangle. In this regard we are guided by the fact that the location of this peak becomes the point of confluency of the parabolic arcs and that the ordinate value common to the two arcs at this point (in Fig. 9 the point $\omega = \omega_1$) varies linearly from $F_1(0) = 1$ to $F_1(\omega_2) = -0.470$ as a function of the location of ω_1 between $\omega = 0$ and $\omega = \omega_2$. Hence if we draw a dotted straight line from the maximum of $F_1(\omega)$ at $\omega = 0$ to its first minimum at $\omega = \omega_2$, as is done in the sketch of Fig. 9, the intermediate ordinates of this line are values of the arcs at any chosen point of confluency. Since we want this value to coincide with that of $F_1(\omega_1)$, the point $\omega = \omega_1$ is seen to be determined by the intersection of the dotted straight line and the curve for $F_1(\omega)$. In our example this point falls at $\omega_1 = \pi/3$ with consistent accuracy.

The next triangular portion of $F_1^{(1)}(\omega)$ is similarly determined. A dotted line from the minimum of $F_1(\omega)$ at $\omega = \omega_2$ to its next maximum at $\omega = \omega_4 = 19\pi/12$ intersects $F_1(\omega)$ at $\omega = \omega_3 = 7\pi/6$. The ordinate difference between ω_2 and ω_4 (from data used to plot F_1) is calculated to be 0.490. Hence the altitude h_2 of the second triangle is given by

$$\frac{h_2}{2} \cdot \frac{3\pi}{4} = 0.490 \quad \text{or} \quad h_2 = 0.416 \tag{98}$$

The altitude of the third and last triangle is found from the calculations

$$\frac{h_3}{2} \cdot \frac{5\pi}{12} = 0.02 \quad \text{or} \quad h_3 = 0.0306 \tag{99}$$

The point of confluency $\omega = \omega_5$ for this last pair of arcs is found by demanding fulfillment of the pertinent moment condition rather than from a graphical construction like that used previously, since the moment condition must be fulfilled in order to insure the proper behavior of $f(t)$ at $t = 0$, and values of $F_1(\omega)$ in this final interval are so small anyway that the exact character of the approximating arcs is relatively unimportant.

We therefore proceed at this point to calculate the slopes of the successive straight-line segments in the broken-line approximation to $F_1^{(1)}(\omega)$, as follows:

$$
\begin{aligned}
s_1 &= \frac{-h_1}{\omega_1} = \frac{-1.123}{\pi/3} = -1.072 \\
s_2 &= \frac{h_1}{\omega_2 - \omega_1} = \frac{1.123}{\pi/2} = 0.715 \\
s_3 &= \frac{h_2}{\omega_3 - \omega_2} = \frac{0.416}{\pi/3} = 0.3975 \\
s_4 &= \frac{-h_2}{\omega_4 - \omega_3} = \frac{-0.416}{5\pi/12} = -0.318 \\
s_5 &= \frac{h_3}{\omega_6 - \omega_5} = \frac{0.0306}{2\pi - \omega_5}
\end{aligned} \tag{100}
$$

The last of these calculations cannot be finished because we do not know the value of ω_5. However, we proceed with the calculation of the impulse values a_k in the impulse-train representation 19 for $F_1^{(3)}(\omega)$. We have

$$
\begin{aligned}
a_1 &= s_2 - s_1 = 1.787 \\
a_2 &= s_3 - s_2 = -0.3175 \\
a_3 &= s_4 - s_3 = -0.7155 \\
a_4 &= s_5 - s_4 = s_5 + 0.318 \\
a_5 &= -s_5
\end{aligned} \tag{101}
$$

Regarding moment conditions we observe in our present problem that the integers μ and ν are 1 and 3 respectively. Conditions 40 should, therefore, hold for the integers $p = 1$ and 3. The one for $p = 3$ (as in the previous problem) assures that the net area enclosed by the parabolic-arc approximation to $F_1(\omega)$ be zero, so that $f(t)$ will begin with a zero value for $t = 0$. Even if the approximation were perfect over the interval $0 < \omega < 2\pi$ its enclosed area would not be zero because the spectrum is chopped off at $\omega = 2\pi$. It is logical, therefore, to ignore the moment condition for $p = 3$ and recognize the resulting discrepancy in $f(t)$ at $t = 0$ as a truncation error. We have, then, to consider only the one condition

$$a_1\omega_1 + a_2\omega_2 + a_3\omega_3 + a_4\omega_5 + a_5\omega_6 = 0 \tag{102}$$

Noting the values of ω_k's given above and the results 101 for the a_k, we have

$$1.787 \cdot \frac{\pi}{3} - 0.3175 \cdot \frac{5\pi}{6} - 0.7155 \cdot \frac{7\pi}{6} + 0.318\omega_5 - s_5(\omega_6 - \omega_5) = 0$$

From the expression for s_5 in 100 we note that the last term here is $-h_3$, and so 102 becomes

$$1.872 - 0.831 - 2.621 + 0.318\omega_5 - 0.0306 = 0$$

whereupon

$$\omega_5 = \frac{1.6106}{0.318} = 5.07 \tag{103}$$

We can now complete the calculations 100 and 101, and get

$$\begin{aligned} s_5 &= \frac{0.0306}{1.21} = 0.0253 \\ a_4 &= 0.0253 + 0.318 = 0.3433 \\ a_5 &= -0.0253 \end{aligned} \tag{104}$$

Substitution into Eq. 27 yields the time function

$$f(t) = \frac{2}{\pi t^3} 1.787 \sin \frac{\pi t}{3} - 0.3175 \sin \frac{5\pi t}{6} - 0.7155 \sin \frac{7\pi t}{6} + 0.3433 \sin 1.614\pi t - 0.0253 \sin 2\pi t \tag{105}$$

from which the tabulated values and the plot in Fig. 10 are obtained.

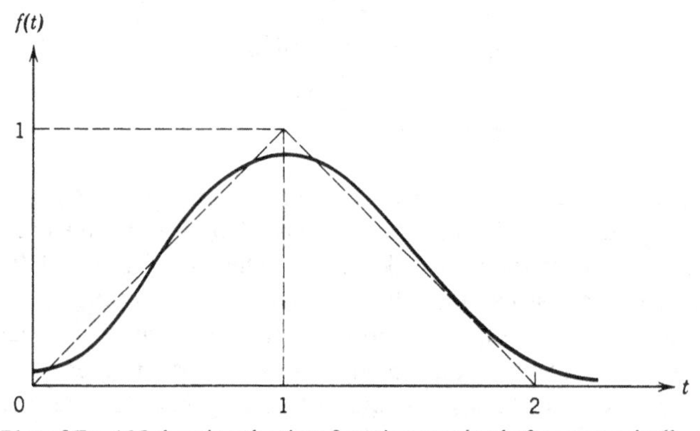

FIG. 10. Plot of Eq. 105 showing the time function regained after numerically evaluating the direct and inverse Fourier transforms starting with the function in Fig. 8.

According to Fig. 5 and the truncation frequency $\omega_c = 2\pi$, the approximate fillet size per unit change in slope is predicted to be $\frac{1}{2}\pi^2$ or about 5%. At the points $t = 0$ and $t = 2$ in Fig. 10 the fillets are about 5 and 8% respectively, while at $t = 1$ where the change in slope is 2, the resulting time function exhibits a 10% fillet as it should. It would seem, therefore, that the truncation error prediction is quite well in agreement with actual results obtained in this example, in which only reasonable graphical accuracy was maintained in determining the significant values entering into the calculation.

Had we been given only the spectrum function in Fig. 9 together with the fact that the asymptote varies as $1/\omega^2$ for large ω, which tells us that $f(t)$ has breaks in its slope, our truncation error analysis in conjunction with the numerical result 105 would give us a strong clue as to the true form of $f(t)$ and hence an estimate of the truncation error without any a priori knowledge about the time function.

Problems

1. Calculate and plot the time function $f(t)$ corresponding to the real-part function $F_1(\omega)$ given by Eq. 14:

(*a*) By approximating its first derivative in Fig. 3 by straight-line segments with three break points.

(*b*) By approximating its second derivative by a broken line as indicated in Fig. 3.

Estimate the approximation error in each case by means of formula 18. Compare the time functions obtained by these two approximations.

2. The time function shown in the adjacent sketch is transmitted through a network with the response function $H(j\omega)$ whose magnitude is unity from $-\omega_c$ to ω_c and otherwise zero. The phase may be assumed linear.

If the output is to approximate $f(t)$ with a maximum error not in excess of 2%, what should be the cutoff frequency ω_c? What is the nature of the error function?

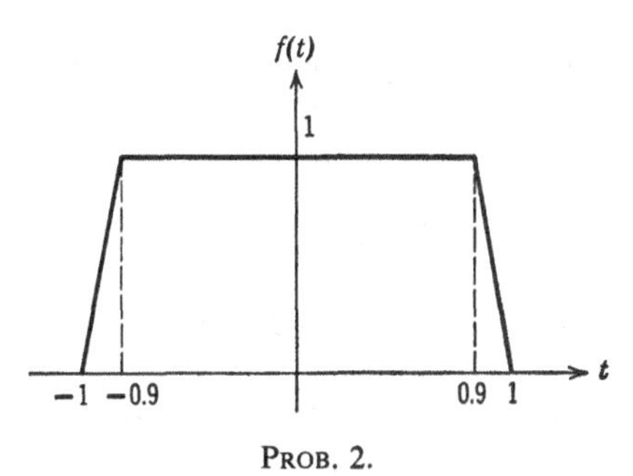

PROB. 2.

3. A time function in the form of a rectangular pulse of height unity and duration δ is to be transmitted through a lowpass network with response function $H(j\omega)$ which is essentially band-limited but whose exact form is left unspecified. The reproduced pulse should be an approximate rectangle not less than 0.8 in average height, having no deviation from a true rectangle greater than 2% from this height or 2% from the width of the input pulse. Determine an appropriate form for $H(j\omega)$ with minimum cutoff. Calculate and plot the reproduced pulse.

4. The system function of a network is to be determined such that it will delay a step function with acceptable tolerances. To this end, the real part of this function is assumed in the form

$$F_1(\omega) = H(j\omega) \times \cos \omega t_0$$

where $H(j\omega)$ is a cutoff function with truncation frequency ω_c. If we write $\omega_c t_0 = n\pi$, n is some integer. The form of H may be rectangular or triangular as discussed in art. 2 of Ch. XIV. It may also be trapezoidal as shown in Fig. 5, art. 3 of that chapter.

Using the methods given there together with techniques discussed in the present chapter, compute and plot $f(t)$ for some simple cases and recommend what you regard as an adequate solution.

5. Given:

$$F_1(\omega) = 1 + \cos \omega t_0 \qquad \text{for } -\frac{\pi}{t_0} < \omega < \frac{\pi}{t_0}$$
$$= 0 \qquad \text{otherwise}$$

(*a*) Find $f(t)$ by analytic evaluation of the synthesis integral.
(*b*) Approximate $F_1(\omega)$ by parabolic arcs and calculate $f(t)$.
Compare results.

6. Consider the real-part function

$$F_1(\omega) = \frac{\sin \omega\delta/2}{\omega\delta/2}$$

Approximate it by parabolic arcs for the interval $-4\pi/\delta < \omega < 4\pi/\delta$ and calculate the corresponding $f(t)$.

Alternately get $f(t)$ by truncating $F_1(\omega)$ in its given form at $\omega = \pm 4\pi/\delta$. Plot both time functions and compare.

7. Approximate the real-part function

$$F_1(\omega) = \frac{1}{1 + \omega^4}$$

with parabolic arcs and truncate at $\omega = 2$ or 3. Carry out the approximation so as to fulfill the appropriate moment conditions.

Compute and plot $f(t)$. Then determine $f(t)$ analytically and compare.

8. Given the following input and output functions:

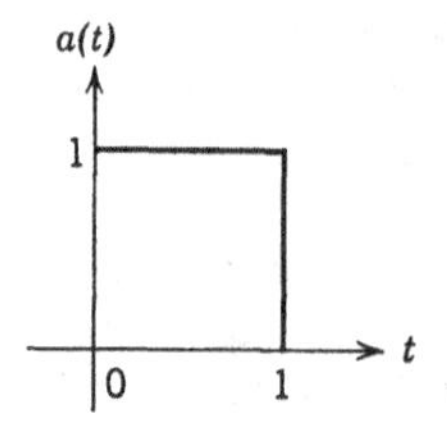

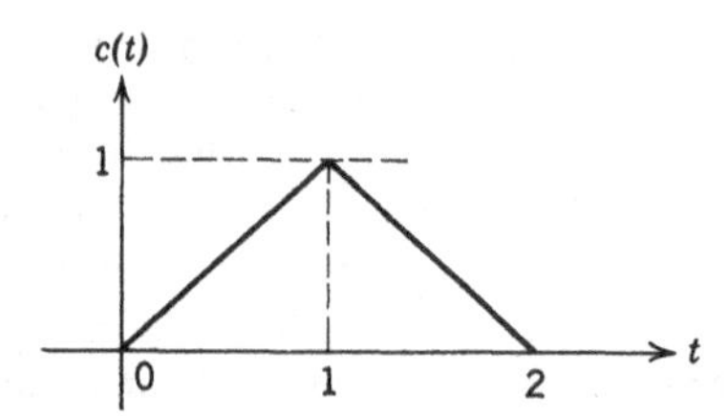

PROB. 8.

By the method of art. 7, find the impulse response and note that the result checks exactly with that found by analytic means.

9. Keeping the same $a(t)$ as in the previous problem but changing the output to

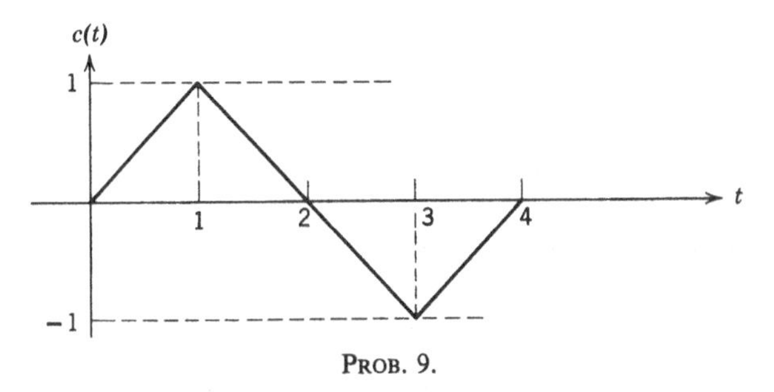

PROB. 9.

again find the impulse response $b(t)$ and plot.

Check your answer by convolving $a(t)$ with $b(t)$.

10. The adjacent sketch illustrates a function $a(t)$ that rises linearly in the

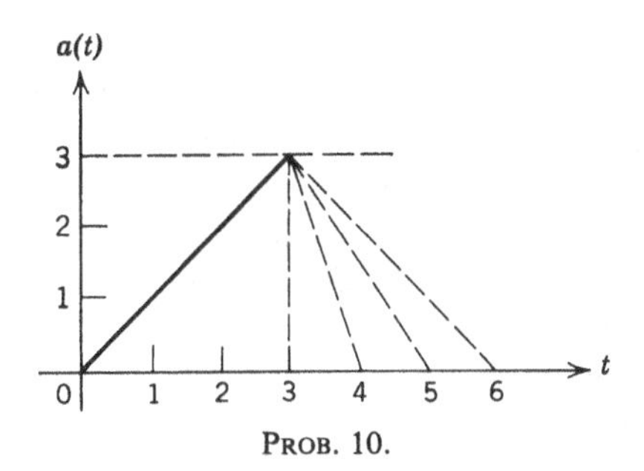

PROB. 10.

interval $0 < t < 3$ and then decreases linearly to zero at four different rates as indicated by the dotted lines. Determine which of these functions have transforms with right half-plane zeros.

CHAPTER XVIII

Hilbert Transforms

1. Potential Theory and Conjugate Functions

A *potential function* is a particular one in a broader class known as the *state function.** The latter is a scalar function whose value depends solely upon the values of its independent variables—not in any case also upon collateral conditions or how these values come about. Functions representing the electric or magnetic stored energy in a linear system (the quadratic forms discussed in art. 1, Ch. V), for example, are state functions since they are completely determined by values of the voltage across capacitances or currents through inductances. In electrostatics or magnetostatics, the electric or magnetic scalar potentials are functions of position only; for every point in space characterized by values of the coordinates x, y, z, the pertinent function has a corresponding scalar value that depends upon nothing else besides the location in space of the point in question. In a mountainous terrain, the altitude of any point in terms of its geographical coordinates has the same property.

All such functions have one basic characteristic in common, namely, they possess a complete differential, like

$$dU = \frac{\partial U}{\partial x}\,dx + \frac{\partial U}{\partial y}\,dy + \frac{\partial U}{\partial z}\,dz + \cdots \tag{1}$$

Although the term "potential function" is sometimes loosely used to denote a state function ("thermodynamic potential" is an example of such use), it more accurately should be reserved for that subclass which, in addition to having the property expressed in Eq. 1, also satisfies Laplace's equation

$$\frac{\partial^2 U}{\partial x^2} + \frac{\partial^2 U}{\partial y^2} + \frac{\partial^2 U}{\partial z^2} + \cdots = 0 \tag{2}$$

A three-dimensional *Newtonian* potential function is the familiar one pertinent to gravitational theory; a two-dimensional or *logarithmic*

* The German term is *Ortsfunktion*.

potential function is the one relevant to our present discussions because of its close relationship to functions of a complex variable. Thus the Cauchy-Riemann equations* that assure the uniqueness of the derivative of a complex function show that the real and imaginary parts of that function satisfy Laplace's equation in two dimensions and hence may be regarded as logarithmic potential functions.

If we write the complex function in conventional notation as

$$w = f(z) = u(x, y) + jv(x, y) \tag{3}$$

where the independent variable is

$$z = x + jy \tag{4}$$

The Cauchy-Riemann equations are the expressions

$$\frac{\partial u}{\partial x} = \frac{\partial v}{\partial y} \tag{5}$$

$$\frac{\partial u}{\partial y} = -\frac{\partial v}{\partial x} \tag{6}$$

If we form the partial derivative of 5 with respect to x and that of 6 with respect to y, addition yields

$$\frac{\partial^2 u}{\partial x^2} + \frac{\partial^2 u}{\partial y^2} = 0 \tag{7}$$

In a similar manner we get from 5 and 6 the corresponding Laplace equation for $v(x, y)$:

$$\frac{\partial^2 v}{\partial x^2} + \frac{\partial^2 v}{\partial y^2} = 0 \tag{8}$$

In this connection it is significant to note that identities of real and imaginary parts u and v in 5 and 6 become interchanged by merely reversing the algebraic sign of one of them. For example, we may replace u by v and v by $-u$ in 5 and 6 without altering these equations in any way. This simple procedure for interchanging identities of real and imaginary parts will be made use of later on. In the theory of two-dimensional potential functions, any two that bear this relationship to one another are referred to as *conjugate potential functions* (the term "conjugate" here having no relation to the use of this same adjective in designating conjugate complex numbers).

The real and imaginary parts of network impedance functions or of the logarithm of such functions (the loss and phase-lag or gain and phase-lead

* For a more detailed discussion see *MCA*, pp. 254–257.

functions—see art. 4, Ch. III) are related in this manner. The implications of this relationship show that the real part alone is sufficient to characterize the pertinent complex function, as is demonstrated for the frequency domain in art. 2, Ch. IX, for the time domain in art. 5, Ch. XVI, and again in art. 11, Ch. XVI where this same relationship is referred to as the *compatibility* of amplitude and phase functions.

Essentially these discussions show that we can construct the pertinent imaginary part from a given real part (or vice versa) or, what amounts to the same thing, we can construct the complex function from either the real or imaginary part. In the frequency domain this result is accomplished by purely algebraic means based upon stability or right half-plane analyticity requirements; time-domain methods, which are analytic in character, make use of Fourier transformation and the property of physical networks that proclaims the response to be identically zero before the excitation is applied. The algebraic methods require that a given real or imaginary part be in rational algebraic form, while the method based upon Fourier analysis can also be used when a given function is piecewise analytic or in graphical form, although its implementation may then require use of numerical integration procedures as given in the last chapter.

In the present chapter we will introduce an analytic procedure that is particularly adapted to handle situations in which the given function is in graphical or in piecewise analytic form, and for which an adaptation of the numerical integration methods enables one to write down the solution to simple examples almost by inspection. Where the given function is in rational algebraic form, this method will be seen to coincide (as we might expect) with the purely algebraic one discussed earlier.

Validity conditions are, of course, the same as before, namely that the pertinent complex function be analytic in the right half-plane. If this function is an impedance, an admittance, or a dimensionless transfer function, then the j-axis need not be included in the analyticity region provided j-axis impulses of the real part are taken into account (as pointed out in the discussion of art. 2, Ch. IX). Zeros of the complex function may lie in both right and left half-planes; in the case of a transfer function there is thus no restriction to a minimum-phase property. The real (or imaginary) part of a transfer function determines that function completely regardless of whether it is minimum phase or not.

However, when we deal with the logarithm of a transfer function (the complex propagation function $\gamma = \alpha + j\beta$ where α is loss in nepers and β is phase lag in radians), then the transfer function is restricted to have neither zeros nor poles in the right half-plane, for the logarithm function takes zeros as seriously as poles—it becomes infinite for both. Since a transfer function having no right half-plane zeros is minimum-phase, we see that if we use the proposed method to determine $\beta(\omega)$ from $\alpha(\omega)$ or vice versa, we will automatically get a minimum-phase propagation function $\gamma(\omega)$.

Formally the present method for computing a real from an imaginary part, or vice versa, involves a pair of infinite integrals called the *Hilbert transforms*. In the following articles we give first, two derivations that are somewhat oversimplified and hence do not allow us to see their restrictions or conditions of validity but do instead provide some physical understanding of the modus operandi in terms of well-known network properties. After developing some effective evaluation techniques and illustrating their use with several examples, we return (in art. 7) to the derivation of these transforms in a more adequate manner that not only shows clearly the conditions under which their use is justified but also yields a number of collaterally interesting and useful by-products.

2. Derivation of Hilbert Transforms Using the Conjugate of an Impulse

The basic thought involved in the present method of determining the conjugate of a given potential function is that if the conjugate of a unit impulse (with ω as the independent variable) is known, then the conjugate of any function may be found from its representation as a superposition of impulse-like elements (samples), the representation given in Eq. 3, Ch. XV and illustrated there in Fig. 1. Justification for this method lies in the fact that the Cauchy-Riemann Eqs. 5 and 6 are linear; hence the conjugate of a sum of functions is equal to the sum of their individual conjugates. In other words, the additive property holds; we can find the conjugate of a function by decomposing it into elements, finding the conjugate of each, and adding these.

The conjugate of an impulse is readily found by considering the function

$$F(s) = \frac{1}{s} \tag{9}$$

having the j-axis evaluation (Eq. 130, Ch. XII)

$$F(j\omega) = F_1(\omega) + jF_2(\omega) = \pi u_0(\omega) + \frac{1}{j\omega} \tag{10}$$

We thus see that the conjugate of a unit impulse

$$u_0(\omega - \xi) \tag{11}$$

occurring at the point $\omega = \xi$ is given by

$$\frac{-1}{\pi(\omega - \xi)} \tag{12}$$

If an arbitrary real-part function $F_1(\omega)$ is represented in the form (see Eq. 3, Ch. XV)

$$F_1(\omega) = \int_{-\infty}^{\infty} F_1(\xi)\, u_0(\omega - \xi)\, d\xi \tag{13}$$

it follows at once that the corresponding imaginary-part function is given by

$$F_2(\omega) = -\frac{1}{\pi}\int_{-\infty}^{\infty} \frac{F_1(\xi)\, d\xi}{\omega - \xi} \tag{14}$$

Transformation in the opposite direction is immediately obtained by the simple expedient of interchanging $F_1(\omega)$ and $F_2(\omega)$ in 14 and reversing the algebraic sign of one of these functions. Thus we have

$$F_1(\omega) = \frac{1}{\pi}\int_{-\infty}^{\infty} \frac{F_2(\xi)\, d\xi}{\omega - \xi} \tag{15}$$

The pair of integrals 14 and 15 are known as *Hilbert transforms.*

3. Derivation Using Convolution with the Transform of a Unit Step

This method of deriving the Hilbert transform is based upon the fact that the time-domain response of a physical system may be expressed as twice the even part of this response for $t > 0$ (Eq. 61, Ch. XVI). In terms of the unit step function this result can be written

$$f(t) = 2f_e(t) \cdot u_{-1}(t) \tag{16}$$

Recalling that multiplication in the time domain is equivalent to $\frac{1}{2}\pi$ times convolution in the frequency domain (art. 8, Ch. XIII), and for convenience indicating the convolution process by placing an asterisk between the functions in question (a commonly used notation) we have

$$F(j\omega) = \frac{1}{\pi} F_1(\omega) * \left[\pi u_0(\omega) + \frac{1}{j\omega}\right] \tag{17}$$

since $F_1(\omega)$ is the transform of $f_e(t)$ and, according to Eqs. 9 and 10, the bracket expression is the Fourier transform of a unit step (its Laplace transform being $1/s$).

Rewriting the indicated convolution 17 in conventional form, we find

$$F(j\omega) = F_1(\omega) - \frac{j}{\pi}\int_{-\infty}^{\infty} \frac{F_1(\xi)\, d\xi}{\omega - \xi} \tag{18}$$

where we make use of the fact that convolution of any function with a unit impulse yields that same function. In view of the first two terms in Eq. 10, we then have

$$F_2(\omega) = \frac{-1}{\pi}\int_{-\infty}^{\infty} \frac{F_1(\xi)\,d\xi}{\omega - \xi} \tag{19}$$

in agreement with Eq. 14.

4. Alternative Forms

The last method of derivation makes us aware of the fact that the Hilbert transform is a convolution integral. Essentially it represents the convolution of $1/\omega$ with the function $F_1(\omega)$ or $F_2(\omega)$ in Eqs. 14 and 15 respectively. Hence we can apply the generalizations discussed in art. 4, Ch. XIII, according to which the result remains unaltered if one function is differentiated and the other integrated. For example, Eq. 14 may thus be altered to read

$$F_2(\omega) = -\frac{1}{\pi}\int_{-\infty}^{\infty} F_1'(\xi)\ln|\omega - \xi|\,d\xi \tag{20}$$

where differentiation is denoted by a prime, and the absolute value sign in the argument of the logarithm function is needed because only real functions are involved here. A similar manipulation of 15 yields

$$F_1(\omega) = \frac{1}{\pi}\int_{-\infty}^{\infty} F_2'(\xi)\ln|\omega - \xi|\,d\xi \tag{21}$$

We can further modify these forms by recognizing the even and odd character of $F_1(\omega)$ and $F_2(\omega)$ respectively. Thus Eqs. 14 and 15 may be written

$$\begin{aligned} F_2(\omega) &= -\frac{1}{\pi}\int_0^{\infty} F_1(\xi)\left(\frac{1}{\omega - \xi} + \frac{1}{\omega + \xi}\right)d\xi \\ &= -\frac{2\omega}{\pi}\int_0^{\infty}\frac{F_1(\xi)\,d\xi}{\omega^2 - \xi^2} \end{aligned} \tag{22}$$

and

$$\begin{aligned} F_1(\omega) &= \frac{1}{\pi}\int_0^{\infty} F_2(\xi)\left(\frac{1}{\omega - \xi} - \frac{1}{\omega + \xi}\right)d\xi \\ &= \frac{2}{\pi}\int_0^{\infty}\frac{\xi F_2(\xi)\,d\xi}{\omega^2 - \xi^2} \end{aligned} \tag{23}$$

Relations 20 and 21 may similarly be rewritten. Noting that $F_1'(\xi)$ in 20 is odd, one has

$$\begin{aligned} F_2(\omega) &= -\frac{1}{\pi}\int_0^{\infty} F_1'(\xi)(\ln|\omega - \xi| - \ln|\omega + \xi|)\,d\xi \\ &= \frac{1}{\pi}\int_0^{\infty} F_1'(\xi)\ln\left|\frac{\omega + \xi}{\omega - \xi}\right| d\xi \end{aligned} \tag{24}$$

Analogously, 21 may be rewritten in the form

$$F_1(\omega) = \frac{1}{\pi}\int_0^\infty F_2'(\xi)(\ln|\omega - \xi| + \ln|\omega + \xi|)\, d\xi$$
$$= \frac{1}{\pi}\int_0^\infty F_2'(\xi)\ln|\omega^2 - \xi^2|\, d\xi \tag{25}$$

A large variety of modified forms are obtainable by further manipulations of this sort, or by applying the method of integration by parts, or by introducing a logarithmic frequency variable.* Whereas some of the special forms obtained in this way are effective in dealing with particular problems, we prefer to discuss the use of impulse-train techniques for the numerical evaluation of the ones we already have.

5. Use of Impulse-Train Techniques

A logical continuation of the process leading to forms 20 and 21 is to differentiate $F_1(\omega)$ and $F_2(\omega)$ in 14 and 15 twice and convolve with the second integral of $1/\omega$, or convolve their third derivatives with the third repeated integral of $1/\omega$, and so forth. If $F_1(\omega)$ or $F_2(\omega)$ is approximated by a broken line or by parabolic arcs so that the νth derivative is a sum of impulses, then essentially the same numerical methods as those discussed in the last chapter can be used to evaluate Hilbert transforms. If the successive integrals of $1/\omega$ are available in tabulated form, one can thus write down the desired conjugate function by inspection.

In order to implement this scheme it is convenient to denote the function $1/\omega$ by $v_0(\omega)$, and its successive integrals by $v_{-1}(\omega)$, $v_{-2}(\omega)$, and so forth. Functions in this sequence are respectively the conjugates of $u_0(\omega)$, $u_{-1}(\omega)$, $u_{-2}(\omega)$, etc. They are given by the expressions

$$v_0(\omega) = -\frac{1}{\pi\omega} \tag{26}$$

$$v_{-1}(\omega) = -\frac{1}{2\pi}\ln\omega^2 \tag{27}$$

$$v_{-2}(\omega) = -\frac{1}{2\pi}(\omega\ln\omega^2 - 2\omega) \tag{28}$$

$$v_{-3}(\omega) = -\frac{1}{2\pi}\left(\frac{\omega^2}{2}\ln\omega^2 - \frac{3\omega^2}{2}\right) \tag{29}$$

$$v_{-4}(\omega) = -\frac{1}{2\pi}\left(\frac{\omega^3}{6}\ln\omega^2 - \frac{11\omega^3}{18}\right) \tag{30}$$

and so forth.

* For a detailed discussion see *MCA*, pp. 339–349.

If the νth derivative of a given real-part function is a sum of impulses, then that function itself may be written

$$F_1(\omega) = \sum_{k=1}^{n} a_k[u_{-\nu}(\omega - \omega_k) + (-1)^\nu u_{-\nu}(\omega + \omega_k)] \tag{31}$$

and the conjugate function (corresponding imaginary part) is then

$$F_2(\omega) = \sum_{k=1}^{n} a_k[v_{-\nu}(\omega - \omega_k) + (-1)^\nu v_{-\nu}(\omega + \omega_k)] \tag{32}$$

where the coefficients a_k are pertinent impulse values and the ω_k are their frequencies of occurrence. The integer n equals the total number of impulses for $\omega > 0$.

Evaluation of $F_2(\omega)$ through substitution of the appropriate expression for $v_{-\nu}(\omega)$, as given in Eqs. 26 through 30, is simplified if we observe that the second terms in functions $v_{-\nu}(\omega)$ for $\nu \geq 2$ contribute nothing to the sum 32 because of moment conditions as expressed in Eq. 35, Ch. XVII. In this connection we observe that the p-values to which these moment conditions apply are adequate even for $\mu = 0$, since the highest-power term in Eqs. 28, 29, 30, . . . is just $\nu - 1$ as required.

In situations where $F_1(\omega)$ is not the transform of a time function (for example, if we apply this computational method to the real and imaginary parts of a propagation function), the moment conditions through the same p-values still apply because they must be fulfilled in order to assure zero drift, as is shown by Eqs. 53 through 57 in Ch. XVII. Although the demonstration there is for a time function, it applies with minor changes in notation to the frequency function 31 as well.

We may thus write for the imaginary-part function 32

$$F_2(\omega) = \frac{-1}{\pi} \sum_{k=1}^{n} a_k[\ln|\omega - \omega_k| - \ln|\omega + \omega_k|] \qquad \text{for } \nu = 1 \tag{33}$$

$$F_2(\omega) = \frac{-1}{\pi} \sum_{k=1}^{n} a_k[(\omega - \omega_k)\ln|\omega - \omega_k| + (\omega + \omega_k)\ln|\omega + \omega_k|] \qquad \text{for } \nu = 2 \tag{34}$$

$$F_2(\omega) = \frac{-1}{\pi} \sum_{k=1}^{n} a_k[(\omega - \omega_k)^2\ln|\omega - \omega_k| - (\omega + \omega_k)^2\ln|\omega + \omega_k|] \qquad \text{for } \nu = 3 \tag{35}$$

$$F_2(\omega) = \frac{-1}{6\pi} \sum_{k=1}^{n} a_k[(\omega - \omega_k)^3\ln|\omega - \omega_k| + (\omega + \omega_k)^3\ln|\omega + \omega_k|] \qquad \text{for } \nu = 4 \tag{36}$$

and so forth.

It is usually helpful to have an approximate expression for the phase near $\omega = 0$ from which we can readily evaluate the phase slope at the origin. In this regard the following linear terms for Maclaurin expansions of the functions 33 through 36 are pertinent:

$$F_2(\omega) = \frac{2\omega}{\pi} \sum_{k=1}^{n} \frac{a_k}{\omega_k} + \cdots \qquad \text{for } \nu = 1 \tag{37}$$

$$F_2(\omega) = \frac{-2\omega}{\pi} \sum_{k=1}^{n} a_k \ln \omega_k + \cdots \qquad \text{for } \nu = 2 \tag{38}$$

$$F_2(\omega) = \frac{2\omega}{\pi} \sum_{k=1}^{n} a_k \omega_k \ln \omega_k + \cdots \qquad \text{for } \nu = 3 \tag{39}$$

$$F_2(\omega) = \frac{-2\omega}{\pi} \sum_{k=1}^{n} \frac{a_k \omega_k^2}{2} \ln \omega_k + \cdots \qquad \text{for } \nu = 4 \tag{40}$$

6. Examples

Consider the real-part function shown in part (*a*) of Fig. 1. Its first derivative is a sum of two impulses, a positive one at $\omega = -\omega_1$ and an

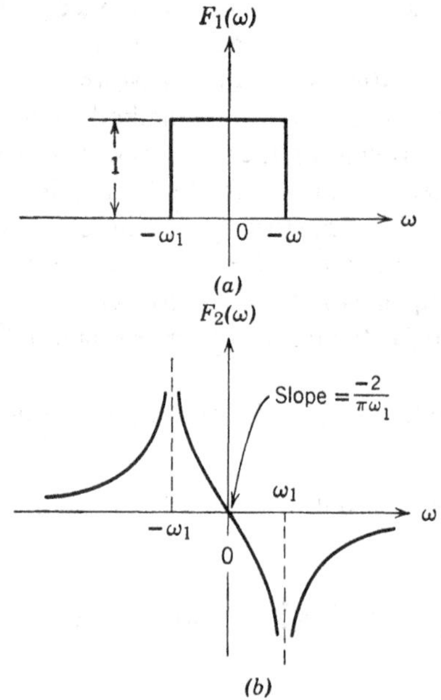

FIG. 1. Example of the use of Eq. 37 for evaluation of $F_2(\omega)$ for the given $F_1(\omega)$.

equal negative one at $\omega = \omega_1$, their absolute values being unity. The sum 31 consists of a single term with $a_1 = -1$. Hence Eq. 33 yields

$$F_2(\omega) = \frac{1}{\pi} \ln \left| \frac{\omega - \omega_1}{\omega + \omega_1} \right| \tag{41}$$

and Eq. 37 shows that

$$\left(\frac{dF_2}{d\omega}\right)_{\omega=0} = \frac{-2}{\pi\omega_1} \tag{42}$$

This result is shown in part (*b*) of Fig. 1.

The function $F_2(\omega)$ exhibits logarithmic infinities at the discontinuities of $F_1(\omega)$, and its slope at $\omega = 0$ is inversely proportional to the width of the pulse-like F_1 function. As ω_1 becomes larger and larger, F_1 degenerates into a constant and F_2 becomes identically zero. The conjugate of a constant is zero.

This fact enables us to derive an interesting collateral result from the example in Fig. 1. We can change the algebraic signs of both F_1 and F_2, add unity to the resulting real part, and then multiply both by a constant A so as to form the new pair of conjugate potential functions

$$\alpha(\omega) = A(1 - F_1) \tag{43}$$

and

$$\beta(\omega) = \frac{A}{\pi} \ln \left| \frac{\omega + \omega_1}{\omega - \omega_1} \right| \tag{44}$$

which we regard as the real and imaginary parts of a propagation function $\gamma = \alpha + j\beta$. Thus $\alpha(\omega)$ represents the loss (in nepers) and $\beta(\omega)$ the phase lag (in radians) of some transmission network.

From the plots in Fig. 2 we see that they are appropriate to a lowpass filter for which the region $-\omega_1 < \omega < \omega_1$ represents a passband. In fact, the amplitude characteristic becomes identical with that assumed for the ideal lowpass filter in art. 9 of Ch. XVI if the constant A is infinite. We can now appreciate that this ideal lowpass filter is not realizable since the phase function 44 does not exist for an infinitely large A-value. On the other hand, for a sufficiently large finite A-value we may regard the characteristics shown in Fig. 2 as approximating those of an ideal filter, and in this sense the results given in art. 9, Ch. XVI do have practical significance.

Consider next the real-part function shown in part (*a*) of Fig. 3. An attenuation function $\alpha(\omega)$ related to it in the manner that $\alpha(\omega)$ in Fig. 2 is related to $F_1(\omega)$ in Fig. 1 is sketched in part (*b*) of the same figure. To apply Eq. 31 here, we regard the triangular function $F_1(\omega)$ as the degenerate form of a trapezoid and have

$$a_1 = \frac{-1}{\omega_2}, \quad \omega_1 = 0, \quad a_2 = \frac{1}{\omega_2} \tag{45}$$

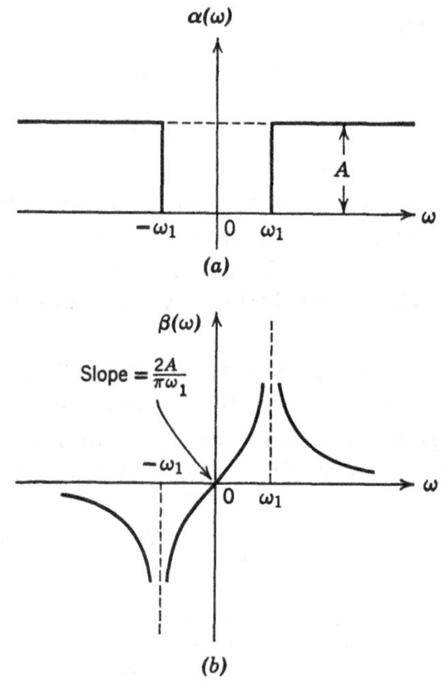

FIG. 2. Attenuation $\alpha(\omega)$ and associated phase $\beta(\omega)$ derived from an adaptation of the results in Fig. 1.

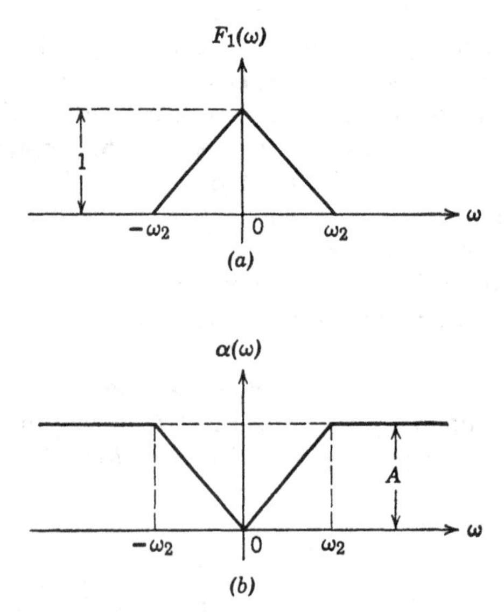

FIG. 3. Triangular real part and related attenuation function.

The integer $\nu = 2$, and so Eq. 32 is evaluated by substituting the values 45 into formula 34. This gives

$$F_2(\omega) = \frac{1}{\pi\omega_2}\,[\omega \ln \omega^2 - (\omega - \omega_2) \ln |\omega - \omega_2| - (\omega + \omega_2) \ln |\omega + \omega_2|] \tag{46}$$

The loss function in Fig. 3, part (*b*), according to Eq. 43, is thus seen to have a minimum associated phase lag given by

$$\beta(\omega) = \frac{A}{\pi\omega_2}\,[(\omega + \omega_2) \ln |\omega + \omega_2| + (\omega - \omega_2) \ln |\omega - \omega_2| - \omega \ln \omega^2] \tag{47}$$

which is shown, plotted in Fig. 4.

A number of detailed characteristics are of interest. First, the phase slope is infinite wherever the attenuation function has a breakpoint,

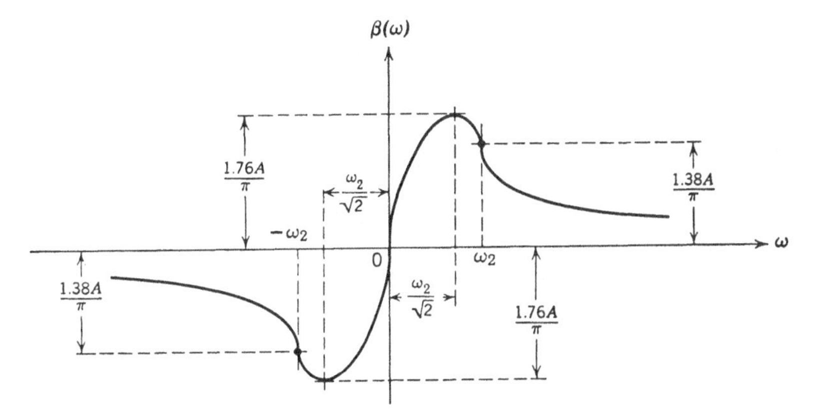

FIG. 4. Imaginary part associated with the loss function shown in Fig. 3.

namely at $\omega = 0$ and at $\omega = \pm\omega_2$. Second, the phase maxima are now finite as opposed to the logarithmically infinite values occurring in the example of Fig. 2. We have

$$\beta_{\max} = \frac{A}{\pi} \ln \frac{\sqrt{2} + 1}{\sqrt{2} - 1} \tag{48}$$

and

$$\beta(\pm\omega_2) = \pm\,\frac{2A}{\pi} \ln 2 \tag{49}$$

This example is collaterally interesting in that the centroid-of-area method for predicting signal delay as discussed in art. 13, Ch. XVI here cannot be applied at all since the relation $t_0 = (d\beta/d\omega)_{\omega=0}$ for the time at which the centroid of area occurs yields an infinite result. The average-slope method, although also somewhat difficult to apply, is more meaningful.

In our next example we make use of the fact that the conjugate property of a pair of functions is not destroyed by integration or differentiation, although an even function becomes odd and an odd one, even. We can in this way obtain another interesting pair of conjugate potential functions through integrating those in Fig. 1. The integral of $F_1(\omega)$ shown there yields a linear ramp in the region $-\omega_1 < \omega < \omega_1$ and constant values below and above this range. By adding a suitable constant (which we can do because its conjugate is zero) we can make this result look like an odd function and regard it as an imaginary part. Its conjugate mate (the associated real-part function) then is the negative integral of $F_2(\omega)$ in Fig. 1 or in Eq. 41, negative because we interchange identities of real and imaginary parts. This integration yields the function

$$\frac{\omega_1}{\pi}\left[\ln\left|\frac{\omega^2}{\omega_1^2}-1\right|-\frac{\omega}{\omega_1}\ln\left|\frac{\omega-\omega_1}{\omega+\omega_1}\right|-2\right] \tag{50}$$

Now the linear-ramp function looks like the phase characteristic of an ideal lowpass filter, and so the present results are most meaningful if we interpret them as the real and imaginary parts α and β of a propagation function. In this regard we are reminded of the fact that asymptotic values of phase shift in a physical network are integer multiples of π radians; hence it is appropriate to multiply both real- and imaginary-part functions by a factor $n\pi/\omega_1$. After dropping the constant term in 50 we then have

$$\alpha(\omega) = n\ln\left|\frac{\omega^2}{\omega_1^2}-1\right| - n\frac{\omega}{\omega_1}\ln\left|\frac{\omega-\omega_1}{\omega+\omega_1}\right| \tag{51}$$

$$\begin{aligned}\beta(\omega) &= n\pi\frac{\omega}{\omega_1} \qquad \text{for } -\omega_1 < \omega < \omega_1 \\ &= \pm n\pi \qquad \text{for } |\omega| > \omega_1\end{aligned} \tag{52}$$

These results are illustrated in Fig. 5.

It is interesting to compare these and the pair in Fig. 2 with the characteristics of the so-called ideal lowpass filter whose response is calculated in art. 9, Ch. XVI. The latter characteristics, as we pointed out, are not compatible; the impulse or step response is not zero for $t < 0$ and hence no physical network can have exactly such characteristics. Those in Fig. 2 or Fig. 5, on the other hand, are compatible.

In Fig. 2, the attenuation characteristic is that of the ideal filter for a sufficiently large A-value; in Fig. 5, the phase characteristic is ideal. We thus see that if the amplitude characteristic of an actual filter is ideal then the phase cannot be, and vice versa.*

The pictures in Figs. 2 and 5 show what the conjugate function looks like when its mate is ideal, and a better appreciation is thereby had of both the nature and the degree of idealization involved in the characteristics of a so-called ideal filter.

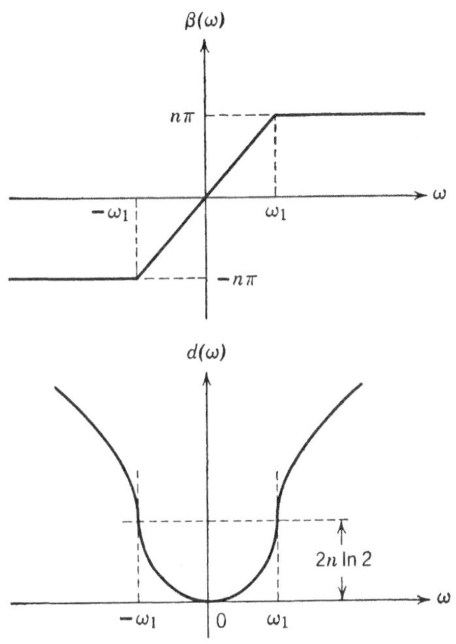

FIG. 5. By integrating the conjugate pair in Fig. 1 (adding a suitable constant to the first) we get the new pair shown here. The phase β is that of an ideal bandpass filter; therefore $\alpha(\omega)$ is the associated attenuation of such a filter.

A further application of the results summarized in Eqs. 31 through 40 is illustrated in Fig. 6 where the pertinent functions again are attenuation and phase. The piecewise-linear α-function has two breakpoints on the positive frequency scale at

$$\omega_1 = 1.0 \quad \text{and} \quad \omega_2 = 1.25 \tag{53}$$

where the changes in slope are respectively

$$a_1 = 4 \quad \text{and} \quad a_2 = -4 \tag{54}$$

* This statement assumes a minimum-phase network. Through cascading an all-pass network, the phase of a given filter can be made more nearly linear without affecting its attenuation function.

With these values Eq. 34 gives

$$\beta(\omega) = \frac{-4}{\pi}[(\omega - 1)\ln|\omega - 1| + (\omega + 1)\ln|\omega + 1| - (\omega - \tfrac{5}{4})\ln|\omega - \tfrac{5}{4}| - (\omega + \tfrac{5}{4})\ln|\omega + \tfrac{5}{4}|] \quad (55)$$

and Eq. 38 yields the slope

$$\left(\frac{d\beta}{d\omega}\right)_{\omega=0} = -\frac{2}{\pi}[4\ln 1 - 4\ln\tfrac{5}{4}] = \frac{8}{\pi}\ln\tfrac{5}{4} = 0.568 \quad (56)$$

From Eq. 55 we readily calculate the values for

$$\begin{aligned} \omega &= 0.5, 1.0, 1.125, 1.25, 2.0 \\ \beta &= 0.312, 1.003, 1.230, 1.036, 0.411 \end{aligned} \quad (57)$$

These yield the plots in Fig. 6. Comparison with those in Fig. 2 is interesting, the essential difference being that the attenuation in the present example has a nonzero rise interval instead of a discontinuity at the cutoff;

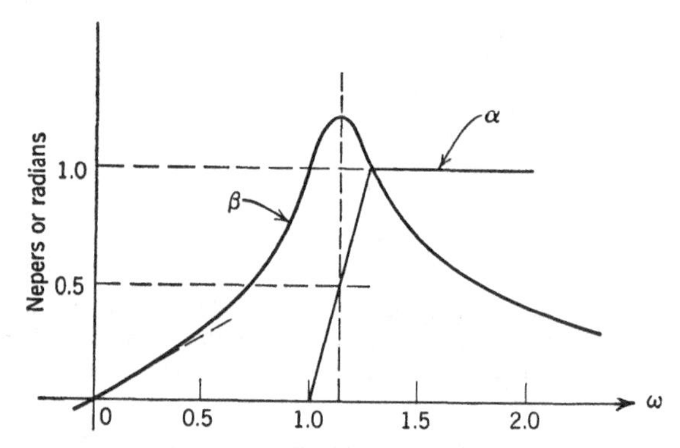

FIG. 6. Phase and attenuation pair simultating the phase and attenuation of a lowpass filter (Eqs. 31 through 40).

the effect of this change upon the phase is to prevent it from having an infinite maximum. The attenuation characteristic in Fig. 2 is too strongly idealized; the one in Fig. 6 is more realistic.

Another example of this sort is shown in Fig. 7. Here $\alpha(\omega)$ is assumed to have the form of two confluent parabolic arcs. One apex is at $\omega = 1$, the other is at $\omega = 3$, and the point of confluency is at $\omega = 2$. The amplitude of α is again normalized at 1 neper. Since we can multiply the scales for α and β by any desired constant, the results in Fig. 7 are applicable to any asymptotic α-value.

By inspection one has

$$a_1 = 1, \quad a_2 = -2, \quad a_3 = 1 \tag{58}$$

and

$$\omega_1 = 1, \quad \omega_2 = 2, \quad a_3 = 3 \tag{59}$$

The integer $\nu = 3$; the relevant moment conditions read

$$\sum_{k=-n}^{n} a_k \omega_k^p = 0 \qquad \text{for } p = 0, 1, 2 \tag{60}$$

These are readily seen to be fulfilled in view of the fact that the third derivative of $\alpha(\omega)$ is odd and hence $a_{-k} = -a_k$.

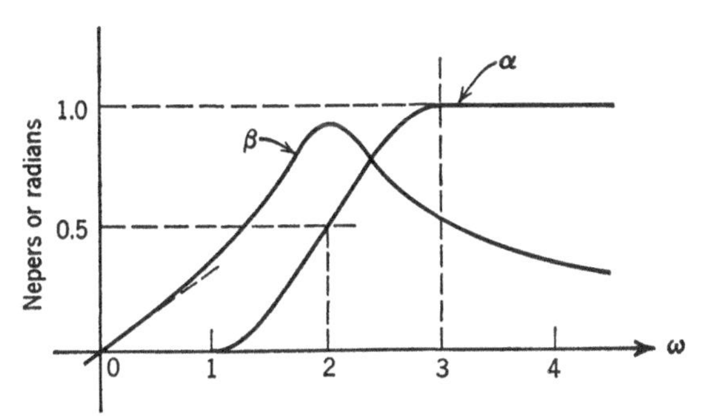

FIG. 7. This pair (also characteristic of a lowpass filter) is derived from Eqs. 58 through 64.

With the values 58 and 59, Eq. 35 yields the following expression for the phase:

$$\begin{aligned}\beta(\omega) = \frac{-1}{2\pi} [&(\omega - 1)^2 \ln |\omega - 1| - (\omega + 1)^2 \ln |\omega + 1| \\ &- 2(\omega - 2)^2 \ln |\omega - 2| + 2(\omega + 2)^2 \ln |\omega + 2| \\ &+ (\omega - 3)^2 \ln |\omega - 3| - (\omega + 3)^2 \ln |\omega + 3|]\end{aligned} \tag{61}$$

By choosing the values

$$\omega = 1, 2, 3, 4 \tag{62}$$

several terms in 61 become zero and we can easily calculate the corresponding β-values:

$$\beta = 0.39, 0.91, 0.54, 0.37 \tag{63}$$

Together with

$$\left(\frac{d\beta}{d\omega}\right)_{\omega=0} = 0.335 \tag{64}$$

calculated from Eq. 39, we have sufficient data to plot the curve in Fig. 7 with reasonable accuracy. Comparison with Fig. 6 shows that the more gradual rise in the attenuation function results in a significant decrease in the phase maximum.

In applying these methods to practical problems, the artifice of frequency scaling is essential in avoiding awkward numbers. Availability of tables or plots for the functions $x \ln x$, $x^2 \ln x$, $x^3 \ln x$, . . . over a range $0 < x < 10$ is then all that is needed to expediate the calculations.

7. Derivation and Generalization of the Hilbert Transform Using Cauchy's Integral Formula

The foregoing derivations of the Hilbert transform, although useful in the solution of many practical problems, do not provide an understanding of its limitations or the precise conditions for which it can lead to valid results. A derivation based upon Cauchy's integral formula,* for which the validity conditions are well known, fills in the needed mathematical background.

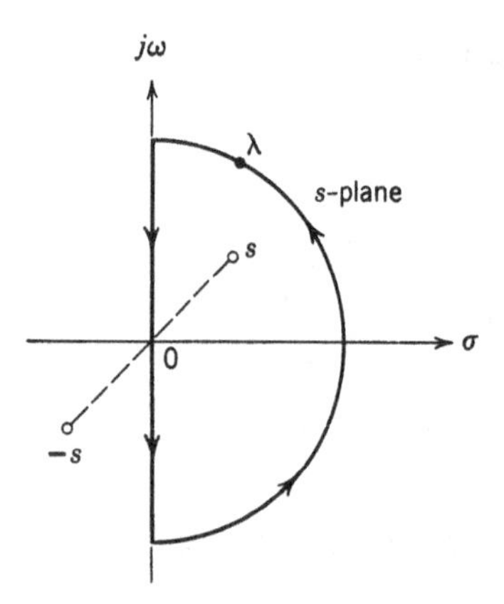

FIG. 8. Pertinent to applying Cauchy's integral formula, Eq. 65, to the derivation of Hilbert transforms.

Let the function $\gamma(s)$ be analytic in the right half of the s-plane, inclusive of the j-axis. Physically, $\gamma(s)$ may be a propagation function $\alpha + j\beta$, or it may be a driving-point or transfer impedance, or any other of the familiar functions characterizing a network in the frequency domain. In any case, we can apply Cauchy's integral formula to a region bounded by the contour shown in Fig. 8, which consists of the j-axis and a semicircular arc of arbitrarily large but finite radius that lies entirely in the right half-plane. If s denotes an internal point, and λ a point on the boundary, Cauchy's formula reads

$$\gamma(s) = \frac{1}{2\pi j} \oint \frac{\gamma(\lambda)\, d\lambda}{\lambda - s} \tag{65}$$

while for the external point $-s$, Cauchy's integral law yields

$$0 = \frac{1}{2\pi j} \oint \frac{\gamma(\lambda)\, d\lambda}{\lambda + s} \tag{66}$$

* See *MCA*, pp. 272–275.

By addition or subtraction of Eqs. 65 and 66, we obtain

$$\gamma(s) = \frac{1}{2\pi j} \oint \left(\frac{1}{\lambda - s} \pm \frac{1}{\lambda + s} \right) \gamma(\lambda)\, d\lambda \tag{67}$$

and hence we have either

$$\gamma(s) = \frac{s}{j\pi} \oint \frac{\gamma(\lambda)\, d\lambda}{\lambda^2 - s^2} \tag{68}$$

or

$$\gamma(s) = \frac{1}{j\pi} \oint \frac{\lambda\gamma(\lambda)\, d\lambda}{\lambda^2 - s^2} \tag{69}$$

In either of these integrals, the path of integration may be abridged to the j-axis if the contribution from the arc is zero, which it is if the integrand for large values of λ varies inversely as λ to some power larger than unity. In that event the contribution from the arc is proportional to

$$\int_{\text{arc}} \frac{d\lambda}{\lambda^{1+a}} \tag{70}$$

where $a > 0$. If we let

$$\lambda = \rho e^{j\theta} \tag{71}$$

this integral becomes

$$j \int_{-\pi/2}^{\pi/2} \frac{e^{ja\theta}\, d\theta}{\rho^a} \tag{72}$$

which is smaller than an arbitrarily small but nonzero value for a sufficiently large but still finite value of ρ (the radius of the arc).

If the asymptotic behavior of $\gamma(\lambda)$ is described by

$$\gamma(\lambda) \to \lambda^n \qquad \text{for } \lambda \to \infty \tag{73}$$

then in Eq. 68 the contribution from the arc is zero if $n < 1$, and in Eq. 69 if $n < 0$. Under these conditions, we can write integrals 68 and 69 in the forms

$$\gamma(s) = \frac{s}{j\pi} \int_{-j\infty}^{j\infty} \frac{\gamma(\lambda)\, d\lambda}{s^2 - \lambda^2} \tag{74}$$

and

$$\gamma(s) = \frac{1}{j\pi} \int_{-j\infty}^{j\infty} \frac{\lambda\gamma(\lambda)\, d\lambda}{s^2 - \lambda^2} \tag{75}$$

where the change in algebraic sign of the integrand is the result of traversing the j-axis in the opposite direction from that shown in Fig. 8. Since λ is now restricted to the j-axis, we shall write $\lambda = j\xi$ and

$$\gamma(\lambda) = \gamma(j\xi) = \alpha(\xi) + j\beta(\xi) \tag{76}$$

where $\alpha(\xi)$ and $\beta(\xi)$ may be regarded as attenuation and phase-lag functions along the j-axis in the normal manner. Since these are respectively even and odd, integrals 74 and 75 become

$$\gamma(s) = \frac{s}{\pi}\int_{-\infty}^{\infty}\frac{\alpha(\xi)\,d\xi}{s^2+\xi^2} \tag{77}$$

and

$$\gamma(s) = \frac{-1}{\pi}\int_{-\infty}^{\infty}\frac{\xi\beta(\xi)\,d\xi}{s^2+\xi^2} \tag{78}$$

These results are slightly generalized versions of the Hilbert transforms that permit computation of the complex propagation function (or system function, if α and β are its j-axis real and imaginary parts) for any right half-plane point, as well as for points on the j-axis, since $s = \sigma + j\omega$ and σ may have any finite positive value, as well as the value zero.

To interpret 77 and 78 on the j-axis we recognize that

$$\begin{aligned}\left(\frac{s}{s^2+\xi^2}\right)_{s=j\omega} &= \frac{1}{2}\left(\frac{1}{s-j\xi}+\frac{1}{s+j\xi}\right)_{s=j\omega}\\ &= \frac{\pi}{2}[u_0(\omega-\xi)+u_0(\omega+\xi)] - \frac{j\omega}{\omega^2-\xi^2}\end{aligned} \tag{79}$$

and

$$\begin{aligned}\left(\frac{\xi}{s^2+\xi^2}\right)_{s=j\omega} &= \frac{1}{2j}\left(\frac{1}{s-j\xi}-\frac{1}{s+j\xi}\right)_{s=j\omega}\\ &= \frac{-j\pi}{2}[u_0(\omega-\xi)-u_0(\omega+\xi)] - \frac{\xi}{\omega^2-\xi^2}\end{aligned} \tag{80}$$

In other words, we must not overlook the j-axis impulses that result when the integrands in 77 and 78 are evaluated along the j-axis. For integral 77 we thus obtain

$$\alpha(\omega)+j\beta(\omega) = \frac{1}{2}\int_{-\infty}^{\infty}[u_0(\omega-\xi)+u_0(\omega+\xi)]\alpha(\xi)\,d\xi + \frac{j\omega}{\pi}\int_{-\infty}^{\infty}\frac{\alpha(\xi)\,d\xi}{\xi^2-\omega^2} \tag{81}$$

The first of these integrals yields $\alpha(\omega)$, and, recognizing that the integrand in the second integral is even, we have

$$\beta(\omega) = -\frac{2\omega}{\pi}\int_{0}^{\infty}\frac{\alpha(\xi)\,d\xi}{\omega^2-\xi^2} \tag{82}$$

in agreement with Eq. 22.

Similarly, integral 78 with Eq. 80 substituted gives

$$\alpha(\omega)+j\beta(\omega) = \frac{j}{2}\int_{-\infty}^{\infty}[u_0(\omega-\xi)-u_0(\omega+\xi)]\beta(\xi)\,d\xi - \frac{1}{\pi}\int_{-\infty}^{\infty}\frac{\xi\beta(\xi)\,d\xi}{\xi^2-\omega^2} \tag{83}$$

Since the first of these integrals yields $j\beta(\omega)$ and the integrand in the second is again even, we get from 83

$$\alpha(\omega) = \frac{2}{\pi}\int_0^\infty \frac{\xi\beta(\xi)\,d\xi}{\omega^2 - \xi^2} \tag{84}$$

which is the same as Eq. 23.

Significant about this method of deriving the Hilbert transforms is the fact that it yields the pertinent validity conditions because these are

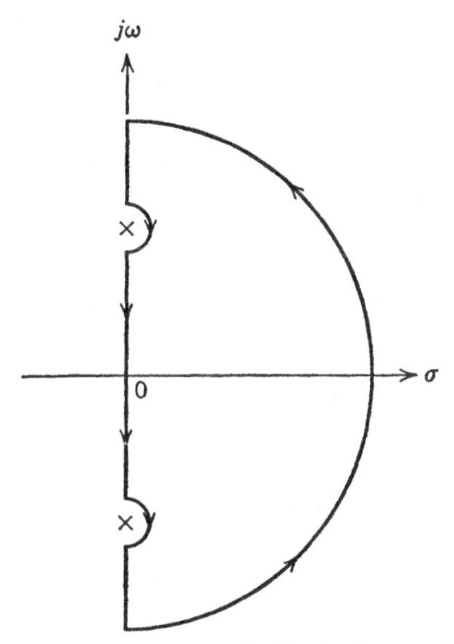

FIG. 9. Modification of closed contour in Fig. 8 when j-axis poles are present.

provided by the restrictions upon Cauchy's integral formula. Since we want the Hilbert transform to apply to the entire finite j-axis, it is necessary that the contour enclose either the entire right or entire left half-plane. In other words, the essential basis for the Hilbert transform is that the function, whose j-axis real and imaginary parts are involved, be analytic in a half-plane. Which half-plane this is doesn't matter so far as validity of the Hilbert transform is concerned; however, if the function in question characterizes a network, then clearly it must be the right half-plane.

Cauchy's integral formula requires analyticity of the pertinent function not only within the enclosed region but also upon the boundary, and this means that we cannot permit j-axis poles. So far as a propagation function is concerned, we are willing to accept this restriction, but for any other frequency-domain response function the exclusion of j-axis poles would seem to imply that we cannot permit impulses in the real- or imaginary-part functions. This, however, is not the case if we interpret the boundary appropriately.

As shown in Fig. 9, the presence of j-axis poles can be tolerated if we modify the contour in their immediate vicinities by inserting semicircular detours that bypass them. The Hilbert transform as derived from Cauchy's formula then strictly speaking applies to this modified boundary and is valid even if we make the radii of the semicircular detours arbitrarily small but nonzero. We then interpret the results as applying to the j-axis exclusive of the detours themselves where the relationship between real and imaginary parts is separately provided by the interpretation given in art. 4, Ch. VIII. Indeed, all our dealings with impulse functions require their separate treatment by a technique that is not included in the normal analytic procedure. The Hilbert transforms are not an exception to this rule, and the inclusion of impulse functions in their evaluation is justified solely on grounds provided by the conformal representation given in the Ch. VIII article referred to above.

8. The Paley-Wiener Criterion for an Attenuation Function and a Corresponding One for Phase

In Eq. 82 we observe that if the asymptotic behavior of $\alpha(\xi)$ is described by

$$\alpha(\xi) \to \xi^n \qquad \text{for } \xi \to \infty \tag{85}$$

then a finite value for this integral can result only if $n < 1$. Essentially, this condition is the so-called *Paley-Wiener criterion.* If we consider integral 77 whence Eq. 82 stems, we can say that the asymptotic character of $\alpha(\xi)$ must be such that this integral has a finite value for all finite values of s in the right half-plane. For this to be so, it is sufficient that the integral have a finite value for $s = 1$, or that the integral

$$\int_0^\infty \frac{\alpha(\xi)\, d\xi}{1 + \xi^2} \tag{86}$$

have a finite value. This is the form in which the Paley-Wiener criterion is stated.*

Since derivations in the present article are implicitly restricted to minimum-phase networks, one might question the general validity of this result on the basis of our method of obtaining it. That is to say, we may conclude that it applies only to minimum-phase networks and not to networks generally. That, indeed, it is generally true may be seen from the fact that any network can always be decomposed into a minimum-phase network in cascade with an all-pass network (art. 3, Ch. IX), since any right half-plane zeros of the given network can be assigned to the all-pass

* R. E. A. C. Paley and Norbert Wiener, "Fourier Transforms in the Complex Domain," *Am. Math. Soc. Colloq. Pub.*, Vol. 19 (1934), Ch. I, "Quasi-analytic Functions," Theorem XII, pp. 16–17.

constituent and left half-plane poles of the latter can be canceled by zeros of the minimum-phase constituent. The net attenuation being that of the minimum phase constituent, any limitations upon its asymptotic behavior are clearly also binding upon the attenuation function of the total or more general network.

We might ask: What is the significance of the Paley-Wiener criterion? Essentially it states that a linear increase with frequency (involving any finite slope) is an upper bound on the asymptotic behavior of any physically realizable attenuation function. No physical network can provide an attenuation which asymptotically increases faster than a constant times ω. For example no attenuation can increase like $\omega^{3/2}$ or ω^2 for $\omega \to \infty$.

More than this, an attenuation function cannot be infinite over a continuous band of frequencies whether finite or infinite in extent, although it can be infinite at discrete points so long as these infinities are integrable. That is to say, the integral 82 fails to yield a finite result if $\alpha(\xi)$ is infinite over a continuous frequency range even if this range has finite width.

A case in point is the ideal lowpass filter as defined in art. 9, Ch. XVI, where its response is assumed to be unity over the range $-\omega_c < \omega < \omega_c$ and zero otherwise. Since zero response means infinite attenuation, this definition clearly violates the Paley-Wiener criterion. As pointed out in connection with the example of Fig. 2, the phase function becomes divergent everywhere if the attenuation A is allowed to be infinite. The ideal filter, as ordinarily defined, is not realizable.

Another commonly encountered example of a non-physically-realizable characteristic is that for which the amplitude response [$A(\omega)$ in Eq. 68, Ch. XVI] is given by the Gaussian error function $e^{-\omega^2}$. Here too, any attempt to compute the associated phase leads to a divergent integral.

On the other hand, the lowpass filter discussed in art. 12, Ch. XVI, whose response is ideal in the time domain (as shown there in Fig. 3), can be approximated arbitrarily closely by a physical network for a finite delay equal to or larger than one-half the rise interval δ. For $t_d = \delta/2$, the response function (Eq. 122, Ch. XVI) reads

$$H(j\omega) = \frac{1 - e^{-j\omega\delta}}{j\omega\delta} \tag{87}$$

Although zero at an infinite number ofdiscrete points on the j-axis, it maybe approximated arbitrarily closely by a rational function fulfilling realizability conditions (see art. 4, Ch. X).

A related question is the following: How must the behavior of $\beta(\xi)$ in Eq. 84 be restricted if this integral is to have a finite value? Here we find that if we write

$$\beta(\xi) \to \xi^n \qquad \text{for } \xi \to \infty \tag{88}$$

we must require $n < 0$. In fact, the conditions $n < 1$ for $\alpha(\xi)$ in Eq. 82 and $n < 0$ for $\beta(\xi)$ in Eq. 84 are just the conditions assumed initially for the asymptotic behavior of $\gamma(\lambda)$ in Eqs. 68 and 69 in order to render contributions from the semicircular arcs zero, so that the subsequent derivation of Hilbert transforms can be carried out. Since the resulting restriction upon $\alpha(\xi)$ agrees with the Paley-Wiener criterion, we do not question it; but the restriction just stated with regard to $\beta(\xi)$ is rather puzzling because we know that phase functions of minimum-phase networks are not so restricted.

The condition $n < 0$ states that $\beta(\xi)$ must become zero for $\xi \to \infty$, whereas we know that in networks whose transfer functions have zeros at $s = \infty$, the phase approaches a constant nonzero asymptote. In fact we can conceive of minimum-phase networks with continuously increasing phase, for which, therefore, $\beta(\xi)$ becomes proportional to ξ for large ξ. What is the explanation of this seeming inconsistency?

It is simply that $\alpha(\omega)$ in Eq. 84 (unlike $\beta(\omega)$ in Eq. 82) is determined only within an arbitrary additive constant; and the value of this constant is sometimes infinite. It is infinite, for example, in any situation in which the phase $\beta(\omega)$ approaches a constant asymptote. In such a case we should not try to compute $\alpha(\omega)$ from $\beta(\omega)$ but instead find

$$\alpha(\omega) - \alpha(0) = \frac{2\omega^2}{\pi} \int_0^\infty \frac{\beta(\xi)\, d\xi}{\xi(\omega^2 - \xi^2)} \tag{89}$$

in which any additive constant in the function $\alpha(\omega)$ drops out. In other words, integral 89 yields the functional variation in $\alpha(\omega)$, which is really all that we are interested in. For its determination according to integral 89, we see that the asymptotic character described by 88 allows n to be larger than unity as long as $n < 2$. This then, is the true restriction upon asymptotic phase behavior. *The phase of a physically realizable network can at most increase at an asymptotic rate that is limited by a quadratic parabola.*

The generality of this statement, like that made in connection with the Paley-Wiener criterion, is not difficult to establish although we might argue that a non-minimum-phase network can possibly provide a phase that has a faster asymptotic rate of increase. If this were possible it could come about only because of the cascading of an all-pass constituent that breaks through the asymptotic restriction just stated. It is, however, easy to see that the phase of an all-pass network is limited by the same criterion. Consider an all-pass pole-zero configuration. If we remove all the zeros, it becomes a minimum-phase configuration but the phase function remains the same except for a factor $\frac{1}{2}$.

As an example of the use of Eq. 89 we readily compute for the specification

$$\beta(\xi) = n\pi\xi \qquad \text{for } -1 < \xi < 1$$

$$\beta(\xi) = n\pi \qquad \text{for } |\xi| > 1 \tag{90}$$

that

$$\alpha(\omega) - \alpha(0) = n \ln |\omega^2 - 1| - n\omega \ln \left| \frac{\omega - 1}{\omega + 1} \right| \tag{91}$$

in agreement with Eq. 51 for the phase specification 52.

Again for $\beta(\xi) = \xi$, Eq. 89 gives* $\alpha(\omega) - \alpha(0) \equiv 0$, which agrees with the well-known physical fact that a linear phase shift over the infinite spectrum is associated with a constant attenuation, albeit an infinite attenuation if we conceive of obtaining this result with a minimum-phase network.

9. Evaluation of the Hilbert Transform for a Rational Function

It is generally thought that usefulness of the Hilbert transform is restricted to situations in which the given real or imaginary part is either graphically or piecewise-analytically specified over intervals of the j-axis, as, for example, in specification 90 or in the computation of phase associated with an attenuation function defined by confluent straight-line segments or arcs. Computation of the imaginary part from Eq. 82, when the real part is given as a rational function of the frequency variable ω, is regarded as leading to an integral that is difficult to evaluate; and so such problems are usually solved by other methods (like those discussed in art. 2, Ch. IX). However, through use of Eq. 74 or 77, equivalent to 82, together with methods of complex integration, we can use the Hilbert transforms for problems of this sort with the same facility as the algebraic methods. In fact, we shall show that use of the Hilbert transform, Eq. 82, or Eq. 77 leads directly to the algebraic formulation given in art. 2, Ch. IX.

Let us do this for an impedance $Z(s)$, for which we write $Z(j\omega) = R(\omega) + jX(\omega)$, and have Eq. 82 in the form

$$X(\omega) = -\frac{2\omega}{\pi} \int_0^\infty \frac{R(\xi)\, d\xi}{\omega^2 - \xi^2} \tag{92}$$

* For $\beta = \xi$ this integral becomes

$$\frac{\omega}{\pi} \int_0^\infty \left(\frac{1}{\xi + \omega} - \frac{1}{\xi - \omega} \right) d\xi = \frac{\omega}{\pi} \ln \left| \frac{\xi + \omega}{\xi - \omega} \right| \Bigg]_0^\infty = 0$$

The j-axis real part of the impedance $Z(s)$ is expressible in the familiar form

$$R(\omega) = \tfrac{1}{2}[Z(s) + Z(-s)]_{s=j\omega} = [R(-s^2)]_{s=j\omega} \tag{93}$$

in which $R(-s^2)$ is the even part of $Z(s)$.

Since we want to use methods of complex integration for the evaluation of integral 92, it must be converted into a contour integral. This process is just the reverse of the above derivation of Eq. 82 from Eqs. 74 and 77. If we use Eq. 74, and close the path of integration by adding the semicircular arc shown in Fig. 8, we obtain

$$Z(s) = \frac{1}{j\pi}\oint \frac{sR(-\lambda^2)\,d\lambda}{\lambda^2 - s^2} \tag{94}$$

where traversal of the path is again counterclockwise. Since the even part of any physical impedance must be regular at infinity, the conditions for a zero contribution from the semicircular arc are obviously fulfilled.

If the left half-plane poles of $Z(s)$ are denoted by λ_ν, and the residues of $Z(s)$ in these poles are k_ν, then a partial fraction expansion of $R(-\lambda^2)$ reads

$$R(-\lambda^2) = \frac{1}{2}\sum_\nu \left(\frac{k_\nu}{\lambda - \lambda_\nu} - \frac{k_\nu}{\lambda + \lambda_\nu}\right) + R \tag{95}$$

because the residues of $R(-\lambda^2)$ in its left half-plane poles (according to Eq. 93) are equal to $k_\nu/2$, while those in the right half-plane poles are the negatives of these values, as is clear from the quadrantal symmetry of the pole-zero pattern of $R(-\lambda^2)$. R_0 is the asymptotic value of $R(-\lambda^2)$.

The partial-fraction expansion of the integrand in Eq. 94 is then seen to be

$$\frac{sR(-\lambda^2)}{\lambda^2 - s^2} = \frac{R(-s^2)}{2(\lambda - s)} - \frac{R(-s^2)}{2(\lambda + s)} + \frac{1}{2}\sum_\nu \frac{sk_\nu}{\lambda_\nu^2 - s^2}\left(\frac{1}{\lambda - \lambda_\nu} - \frac{1}{\lambda + \lambda_\nu}\right) \tag{96}$$

By Cauchy's residue theorem the value of integral 94 is equal to $2\pi j$ multiplied by the sum of residues of the integrand in those poles enclosed by the contour of Fig. 8. These poles are at $\lambda = s$ and at $\lambda = \lambda_\nu$. The residues are evident in Eq. 96, and so the evaluation of Eq. 94 yields

$$Z(s) = R(-s^2) - \sum_\nu \frac{sk_\nu}{\lambda_\nu^2 - s^2} \tag{97}$$

which, with Eq. 95, becomes

$$Z(s) = \frac{1}{2}\sum_\nu \left(\frac{k_\nu}{s - \lambda_\nu} + \frac{k_\nu}{s + \lambda_\nu}\right) + \frac{1}{2}\sum_\nu \left(\frac{k_\nu}{s - \lambda_\nu} + \frac{k_\nu}{s + \lambda_\nu}\right) + R_0 \tag{98}$$

or

$$Z(s) = \sum_{\nu} \frac{k_\nu}{s - \lambda_\nu} + R_0 \tag{99}$$

Hence the impedance generated by the rational even-part function $R(-\lambda^2)$ is found as readily through use of the Hilbert transform and complex integration as it is by the algebraic method given in art. 2, Ch. IX. In fact, the computational work is exactly the same in the two methods.

10. The Resistance-Integral Theorem; Its Variants and Applications

An interesting and useful result may be obtained by inspection of the Hilbert transform, Eq. 92 (or any of the equivalent forms derived earlier). The behavior of the imaginary-part function $X(\omega)$ for $\omega \to \infty$ is evidently that of a capacitive reactance since we have

$$X(\omega) \to -\frac{2}{\pi\omega}\int_0^\infty R(\xi)\, d\xi \qquad \text{for } \omega \to \infty \tag{100}$$

If we write this as

$$X(\omega) \to \frac{-1}{C\omega} \qquad \text{for } \omega \to \infty \tag{101}$$

then we have

$$\int_0^\infty R(\xi)\, d\xi = \frac{\pi}{2C} \tag{102}$$

The total area enclosed by the real part of an impedance determines its asymptotic behavior, which is that of a capacitive reactance for any finite nonzero real-part area. Physically this result means that if this area is known to be finite and nonzero, then a realization for the pertinent impedance must involve a shunt capacitance across the input terminals having a value determined by Eq. 102. This result is known as *Bode's resistance-integral theorem.*

Formula 102 is particularly useful in situations where the shunt capacitance is parasitic rather than part of a network schematic. If we can in some way determine the total real-part area, then this parasitic capacitance can be calculated, or in cases where we can estimate the parasitic shunt capacitance, the net resistance area can be calculated by this formula.

From Eq. 92 we can alternately obtain the behavior of the imaginary-part function $X(\omega)$ for $\omega \to 0$. This yields an inductive reactance since we have

$$X(\omega) \to \frac{2\omega}{\pi}\int_0^\infty \frac{R(\xi)}{\xi^2}\, d\xi \qquad \text{for } \omega \to 0 \tag{103}$$

If we write this as

$$X(\omega) \to L\omega \qquad \text{for } \omega \to 0$$

then we have

$$L = \frac{2}{\pi} \int_0^\infty \frac{R(\xi)}{\xi^2}\, d\xi \tag{104}$$

from which the inherent shunt inductance associated with an impedance may be calculated from its known real-part function.

Duals of the results stated by Eqs. 102 and 104 read respectively

$$\int_0^\infty G(\xi)\, d\xi = \frac{\pi}{2L} \tag{105}$$

and

$$C = \frac{2}{\pi} \int_0^\infty \frac{G(\xi)}{\xi^2}\, d\xi \tag{106}$$

in which $G(\xi)$ is the conductance (real part) of an admittance function. For $\omega \to \infty$ it degenerates into a series inductance having the value L given by 105 and for $\omega \to 0$ it degenerates into a series capacitance having the value C given by 106.

We can obtain more detailed relations from Eq. 92 by making use of the asymptotic expansion

$$\frac{1}{\omega^2 - \xi^2} = \frac{1}{\omega^2} + \frac{\xi^2}{\omega^4} + \frac{\xi^4}{\omega^6} + \cdots \tag{107}$$

which is valid for large ω. Integral 92 may thus be written

$$X(\omega) = \frac{2}{\pi\omega} \int_0^\infty \left(1 + \frac{\xi^2}{\omega^2} + \frac{\xi^4}{\omega^4} + \cdots\right) R(\xi)\, d\xi \tag{108}$$

from which we can draw the following specific conclusions.

If $Z(s) \to 1/s$ for $s \to \infty$, then

$$\int_0^\infty R(\xi)\, d\xi \text{ is finite and nonzero} \tag{109}$$

If $Z(s) \to 1/s^2$ for $s \to \infty$, then

$$\int_0^\infty R(\xi)\, d\xi = 0 \tag{110}$$

If $Z(s) \to 1/s^3$ for $s \to \infty$, then

$$\int_0^\infty R(\xi)\, d\xi = 0 \quad \text{but} \quad \int_0^\infty \xi^2 R(\xi)\, d\xi \text{ is finite and nonzero} \tag{111}$$

and so forth.

It is obvious that the reverse of each of these statements is also true. Thus the asymptotic behavior of $Z(s)$ determines its real-part properties, and conversely, the real-part properties imply a corresponding asymptotic behavior for the function $Z(s)$.

These results are rather closely related to moment conditions discussed in art. 3, Ch. X and in art. 6, Ch. XII. For example, from Eqs. 90 and 91 in Ch. XII we have

$$\int_{-j\infty}^{j\infty} s^n Z(s)\, ds = (j)^{n+1}\int_{-\infty}^{\infty} \omega^n R(\omega)\, d\omega + (j)^{n+2}\int_{-\infty}^{\infty} \omega^n X(\omega)\, d\omega = 0 \tag{112}$$

for $n = 0, 1, 2, \ldots, k$, and the tacit assumption that

$$Z(s) \to \frac{1}{s^{k+2}} \qquad \text{for } s \to \infty \tag{113}$$

Since $R(\omega)$ and $X(\omega)$ are respectively even and odd, we have

$$\int_0^{\infty} \omega^n R(\omega)\, d\omega = 0 \qquad \text{for even } n \leq k \tag{114}$$

which is in agreement with results given by Eqs. 109 through 111 and their continuation.

Correlation with asymptotic behavior as discussed in art. 3, Ch. X is provided by the following. If $Z(s)$ has the partial-fraction expansion

$$Z(s) = \sum_{\nu=1}^{n} \frac{k_\nu}{s - s_\nu} \tag{115}$$

then its Maclaurin expansion about $s = \infty$ (as given by Eqs. 62, 63 of Ch. X) reads

$$Z(s) = \sum_{\nu=1}^{n} k_\nu\left(\frac{1}{s} + \frac{s_\nu}{s^2} + \frac{s_\nu^2}{s^3} + \cdots\right) \tag{116}$$

while the corresponding impulse response for $t > 0$, given by

$$f(t) = \sum_{\nu=1}^{n} k_\nu e^{s_\nu t} \tag{117}$$

has a Maclaurin expansion about $t = 0$ of the form

$$f(t) = \sum_{\nu=1}^{n} k_\nu\left(1 + s_\nu t + \frac{s_\nu^2 t^2}{2!} + \cdots\right) \tag{118}$$

With the help of singularity functions (Eq. 107, Ch. XII), this may be written

$$(t) = \sum_{\nu=1}^{n} k_\nu[u_{-1}(t) + s_\nu u_{-2}(t) + s_\nu^2 u_{-3}(t) + \cdots] \tag{119}$$

Comparison of 116 and 119 shows that if

$$Z(s) \to \frac{A}{s^m} \qquad \text{for } s \to \infty \tag{120}$$

then

$$f(t) \to Au_{-m}(t) \qquad \text{for } t \to 0 \tag{121}$$

This interesting correlation between asymptotic behavior in the frequency and time domains shows for $m = 1$ that $f(0) = A$ and for $m > 1$ that $f(0) = 0$. Since, by Eq. 62, Ch. XVI we have

$$f(0) = \frac{2}{\pi} \int_0^\infty R(\omega)\, d\omega \tag{122}$$

it follows for $m = 1$ (writing $A = 1/C$) that

$$\int_0^\infty R(\omega)\, d\omega = \frac{\pi}{2C} \tag{123}$$

in agreement with Eq. 102, and for $m > 1$ that

$$\int_0^\infty R(\omega)\, d\omega = 0 \tag{124}$$

which also agrees with the above results.

The most useful practical application of Bode's resistance-integral theorem is in the design of very-wide-band amplifiers where the limit on obtainable gain-bandwidth product is imposed by parasitic shunt capacitance. The pertinent $Z(s)$ here is the transfer impedance of an interstage coupling network between successive stages of a vacuum tube amplifier.

If a two-terminal network is involved, $Z(s)$ is its driving-point impedance. In the simplest situation of this kind, $Z(s)$ is normally a resistance R and the parasitic capacitance in parallel with it is represented by the sum of plate-cathode and grid-cathode capacitances respectively of the preceding and succeeding vacuum tubes. Inclusive of this net shunt capacitance (which we can denote by C), $Z(s)$ is a parallel RC combination. Since the magnitude of $Z(s)$ drops to $1/\sqrt{2}$ of its maximum zero-frequency value where $RC\omega = 1$, we have for this simplest situation the gain-bandwidth product (apart from a transconductance factor of the vacuum tube)

$$R\omega_c = \frac{1}{C} \tag{125}$$

where ω_c denotes the cutoff frequency (in this case the half-power point).

The maximum gain obtainable per stage is determined by the value R, the bandwidth by ω_c. For a fixed amount of parasitic capacitance C

we can have a high gain over a narrow band or a lower gain over a wider band. The product of gain and bandwidth is unalterably fixed by the tube capacitances.

By using a more elaborate network for $Z(s)$, we can obtain a larger gain-bandwidth product for the same C-value. Optimally the impedance $Z(s)$ should have a constant magnitude over the band $-\omega_c < \omega < \omega_c$ because we want the gain or the amplifier response to be constant in this region, and its real part should be zero or nearly zero outside of this band because the impedance level for a fixed amount of real-part area (fixed parasitic capacitance C) is maximum if none of this area is wasted by lying outside the band.

An impedance fulfilling these specifications is given by the function

$$Z(s) = \frac{K}{s + \sqrt{1 + s^2}} = Ke^{-(\alpha + j\beta)} \tag{126}$$

which is discussed in art. 4, Ch. IX and illustrated there in Fig. 9. The cutoff frequency is normalized at 1 radian per second. Over the passband, the attenuation α is zero and hence the impedance has a constant magnitude K, while its phase angle varies from zero to $\pi/2$. Outside the passband this angle remains constant at the value $\pi/2$, thus yielding an identically zero real part. Since the function 126 is irrational its realization will require approximation by a rational function. We will see that this can very easily be managed. In the meantime let us see what gain-bandwidth product can be obtained with it.

To begin with we will un-normalize the cutoff frequency by letting

$$s = \frac{j\omega}{\omega_c} \tag{127}$$

If we then choose the multiplier

$$K = \frac{2}{C\omega_c} \tag{128}$$

the impedance becomes

$$Z(j\omega) = \frac{2}{(j\omega + \sqrt{\omega_c^2 - \omega^2})C} \tag{129}$$

which has the asymptotic behavior

$$Z(j\omega) \to \frac{1}{j\omega C} \qquad \text{for } \omega \to \infty \tag{130}$$

as it should.

Next let us see whether the resistance-integral relation 102 is fulfilled. In the passband where $\alpha = 0$ we have

$$R(\omega) = K \cos \beta \tag{131}$$

and from 126 and 127

$$\sqrt{1 - (\omega/\omega_c)^2} + j(\omega/\omega_c) = e^{j\beta} \tag{132}$$

Hence we have

$$\sqrt{1 - (\omega/\omega_c)^2} = \cos\beta \tag{133}$$

$$\omega/\omega_c = \sin\beta \tag{134}$$

and

$$d\omega = \omega_c \cos\beta \, d\beta \tag{135}$$

so that

$$\int_0^\infty R(\omega)\, d\omega = K\omega_c \int_0^{\pi/2} \cos^2\beta \, d\beta = \frac{\pi K\omega_c}{4} \tag{136}$$

For the K-value given by Eq. 128 this is the result 102 as it should be. Equation 128, incidentally, gives the gain-bandwidth product

$$K\omega_c = \frac{2}{C} \tag{137}$$

which shows that a two-terminal coupling network having the impedance 126 or 129 is twice as good as a pure resistance R for which the gain-bandwidth product is given by Eq. 125. Since this impedance is optimum, the result 137 represents the best that can be done by a two-terminal interstage network.

Let us now see how we can realize this impedance. Writing for the sake of abbreviation

$$x = \frac{\omega}{\omega_c} \tag{138}$$

the key to a realization procedure lies in the fact that

$$jx + \sqrt{1 - x^2} = \frac{1}{-jx + \sqrt{1 - x^2}} \tag{139}$$

which is the same as saying that the impedance 129 has constant magnitude over the passband. We thus have

$$jx + \sqrt{1 - x^2} = 2jx - jx + \sqrt{1 - x^2} = 2jx + \frac{1}{jx + \sqrt{1 - x^2}} \tag{140}$$

If we apply the same relation to the denominator of the second term, and continue to do so, we get the infinite continued fraction development

$$jx + \sqrt{1 - x^2} = 2jx + \cfrac{1}{2jx + \cfrac{1}{2jx + \cdots}} \tag{141}$$

which suggests a uniform ladder network in which the shunt admittances and series impedances are given by successive terms in this continued fraction. For identification of the shunt branches we have from Eqs. 128 and 138

$$2jx = j\omega KC \tag{142}$$

This is a capacitive susceptance normalized by the impedance level factor K. If we define an inductive reactance such that

$$j\omega L \times \frac{1}{j\omega C} = K^2 \tag{143}$$

which is reciprocal to the reactance of the shunt branch with respect to the constant K^2, then we have for the identification of series branches

$$2jx = \frac{j\omega L}{K} \tag{144}$$

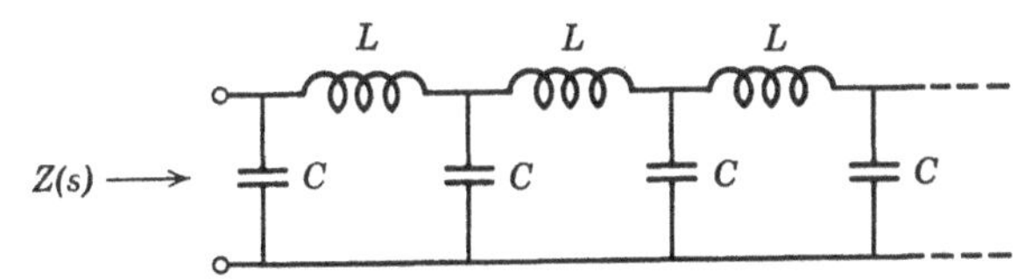

FIG. 10. Realization of the impedance given by Eq. 145.

Since the normalized impedance 126 or 129 is thus seen to have the continued-fraction representation

$$\frac{Z}{K} = \cfrac{1}{2jx + \cfrac{1}{2jx + \cfrac{1}{2jx + \ddots}}} \tag{145}$$

its realization is recognized to be the infinite uniform ladder network shown in Fig. 10.

Now we must find a way of terminating this ladder in order to get a finite and hence realizable approximation. The answer to this question is readily obtained by a glance at relation 140 which tells us that we can terminate the continued fraction 145 at any point by writing $jx + \sqrt{1 - x^2}$ in place of the repetitive term $2jx$. If this final term is a series branch then it involves an inductance $L/2$ in series with a terminating resistance $R = K\sqrt{1 - x^2}$; if it is a shunt branch then it consists of a capacitance $C/2$ in parallel with a terminating conductance $G = \sqrt{1 - x^2}/K$. These terminations are shown in Fig. 11. Although the desired terminating resistance or conductance is frequency-dependent, we obtain an approximate result by replacing it with a constant nominal value. The zero-frequency value is a common choice although a slightly smaller value is somewhat better in an average sense.

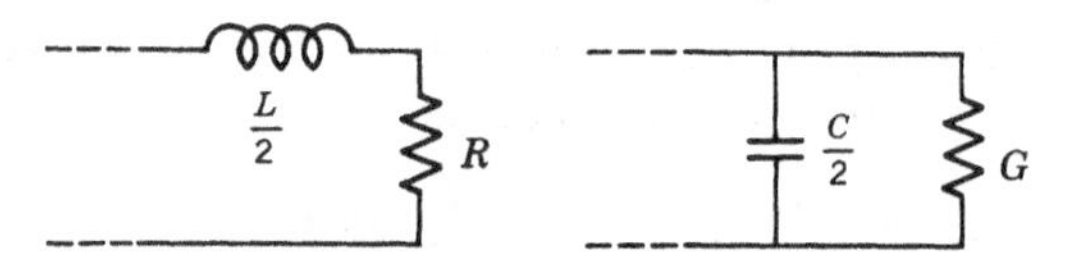

FIG. 11. Possible terminations for the network of Fig. 10.

On a normalized basis, a specific example of this mode of approximating the irrational impedance 126 by a lumped network is shown in part (*a*) of Fig. 12. Part (*b*) shows a plot of the resulting real part. Its desired theoretical behavior is represented by the dotted curve (which, incidentally is the

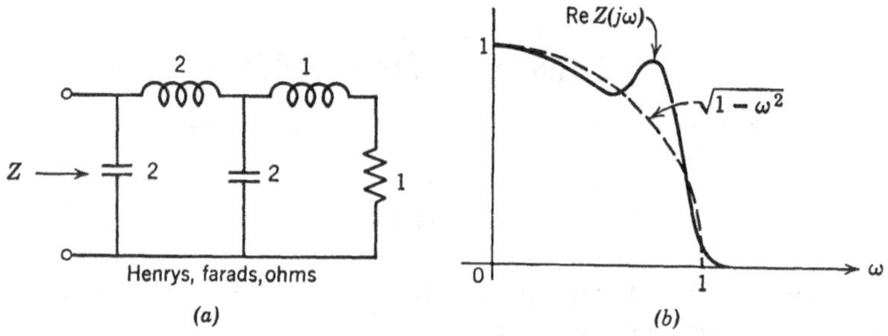

FIG. 12. Network which is an approximate realization of the infinite continuous fraction according to the methods illustrated in Figs. 10 and 11.

arc of a circle). It is surprising to note that the approximation of the resulting real-part function to its desired behavior, even in this simple case, is reasonably good. It can be made arbitrarily good by adding more branches to the ladder network. A more economical way to improve the

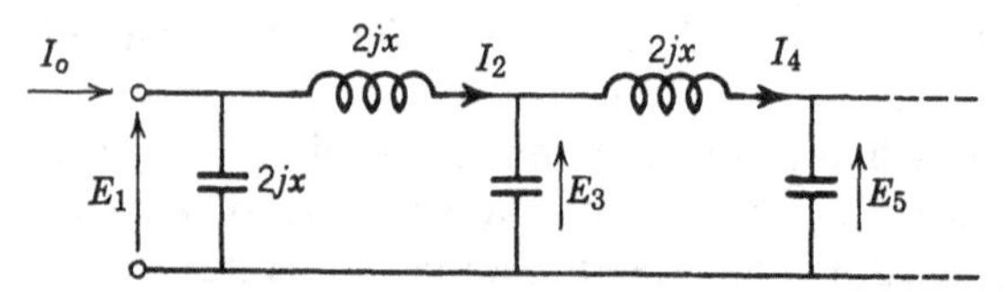

FIG. 13. Four-terminal interstage to increase the gain-bandwidth product over that obtainable with the two-terminal network in Fig. 12.

convergence of this realization makes use of more elaborate terminating networks than the simple ones shown in Fig. 11. Further detailed discussion of this item, however, does not seem appropriate here.

Instead we shall turn our attention to showing how we can further increase the gain-bandwidth product by using a four-terminal interstage network (or two terminal-pair network) to couple successive vacuum-tube stages. With reference to notation given in the network sketch of Fig. 13 and the preceding analysis, we have

$$E_1 = I_0\left(\frac{1}{jx + \sqrt{1 - x^2}}\right) \tag{146}$$

$$\begin{aligned} I_2 = I_0 - 2jxE_1 &= I_0\left(1 - \frac{2jx}{jx + \sqrt{1 - x^2}}\right) \\ &= \frac{-jx + \sqrt{1 - x^2}}{jx + \sqrt{1 - x^2}} I_0 = \frac{I_0}{(jx + \sqrt{1 - x^2})^2} \end{aligned} \tag{147}$$

$$E_3 = I_2\left(\frac{1}{jx + \sqrt{1 - x^2}}\right) = \frac{I_0}{(jx + \sqrt{1 - x^2})^3} \tag{148}$$

For the ladder network of Fig. 10 we can, therefore, define a normalized transfer impedance

$$\frac{Z_{n0}}{K} = \frac{E_n}{I_0} = \frac{1}{(jx + \sqrt{1 - x^2})^n} \tag{149}$$

which is the input impedance raised to an *odd* power. Outside the passband, the angle of this impedance is an odd multiple of $\pi/2$; hence the real-part function is identically zero there, as it is for the driving-point impedance. The magnitude of the transfer impedance, like that of the driving-point impedance, is constant over the passband. An appropriate four-terminal interstage coupling network is, therefore, obtained from the circuit of Fig. 10 by simply attaching a pair of output terminals across any of the succeeding capacitances.

This expedient makes available a total of twice as much shunt capacitance to absorb parasitic capacitances of the associated tubes. On the assumption that grid-cathode and plate-cathode capacitances are equal, we thus get the gain-bandwidth product

$$K\omega_c = \frac{4}{C_{\text{total}}} \tag{150}$$

which results from substituting the relation

$$C_{\text{total}} = 2C \tag{151}$$

into Eq. 137. The simple network of part (*a*) in Fig. 12 (with output terminals across the second shunt capacitance) can thus be used as a reasonably good interstage coupling means.

Although one is tempted on heuristic grounds to conclude that this scheme represents the largest possible gain-bandwidth product obtainable, we can show by means of the resistance-integral theorem that a still larger value can be achieved if we are willing to resort to a more elaborate network for its realization. Our point of departure in this argument is still the impedance function 149 which for $n = 3, 5, \ldots$ has an asymptotic behavior like $1/\omega^3$, $1/\omega^5$, ... Its real part, as pointed out above, is nonzero only over the passband, and the net enclosed area is now zero for reasons expressed by Eqs. 120 through 124. It is interesting to check this fact independently.

The real part of this transfer impedance (we denote it by r_{12}) is given by

$$r_{12}(\omega) = K \cos n\beta \tag{152}$$

which reverts to Eq. 131 for $n = 1$. Equation 135 still applies, and so we have

$$\begin{aligned}
&\int_0^\infty r_{12}(\omega)\, d\omega \\
&\quad = K\omega_c \int_0^{\pi/2} \cos n\beta \cos \beta \, d\beta \\
&\quad = \frac{K\omega_c}{2} \int_0^{\pi/2} [\cos (n+1)\beta + \cos (n-1)\beta]\, d\beta \\
&\quad = \frac{K\omega_c}{2} \left\{ \frac{\sin (n+1)\beta}{n+1} + \frac{\sin (n-1)\beta}{n-1} \right\}_0^{\pi/2} \\
&\quad = \frac{K\omega_c}{2} \left\{ \frac{n \sin n\beta \cos \beta - \cos n\beta \sin \beta}{n^2 - 1} \right\}_0^{\pi/2} = 0 \qquad \text{for } n = 3, 5, \ldots
\end{aligned} \tag{153}$$

Since the net resistance area is zero, we must determine the maximum obtainable area enclosed by $|r_{12}(\omega)|$. The resistance-integral theorem relates to driving-point impedances, and so we must express r_{12} in terms of the real parts of the driving-point impedances of the coupling network. To this end we make use of the relation 56 in Ch. VIII which we repeat here for convenience:

$$Z(s) = z_{11}x_1^2 + 2z_{12}x_1x_2 + z_{22}x_2^2 \tag{154}$$

Since this must be a p.r. function, its j-axis real part

$$\text{Re}\,[Z(j\omega)] = r_{11}(\omega)x_1^2 + 2r_{12}(\omega)x_1x_2 + r_{22}(\omega)x_2^2 \tag{155}$$

is not allowed to become negative for any real values of x_1 and x_2. In other words, the quadratic form 155 must be positive or semidefinite. According to discussion in art. 4, Ch. V, this condition requires

$$r_{11}r_{22} - r_{12}^2 \geq 0 \tag{156}$$

in addition to $r_{11} \geq 0$ and $r_{22} \geq 0$ which are fulfilled anyway since these are real parts of driving-point impedances.

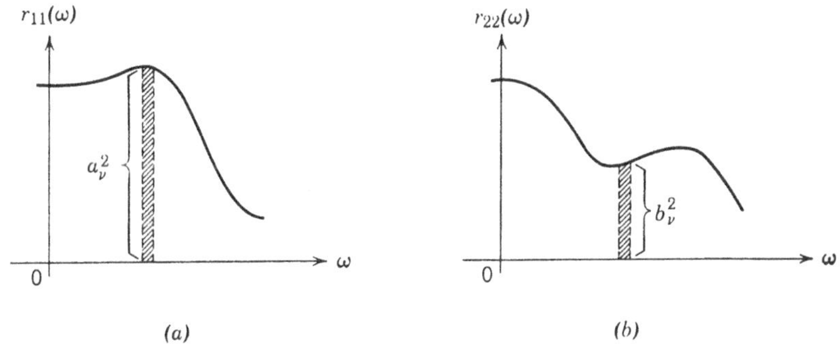

Fig. 14. Elements a_ν^2 and b_ν^2 of the input and transfer impedances of a two-terminal interstage network.

In the interest of maximizing the area enclosed by $|r_{12}|$, we fulfill condition 156 by making $r_{12}^2 = r_{11}r_{22}$ and have

$$\int_0^\infty |r_{12}|\, d\omega = \int_0^\infty \sqrt{r_{11}}\sqrt{r_{22}}\, d\omega \tag{157}$$

For the driving-point functions we have

$$\int_0^\infty r_{11}(\omega)\, d\omega = \frac{\pi}{2C_1} \quad \text{and} \quad \int_0^\infty r_{22}(\omega)\, d\omega = \frac{\pi}{2C_2} \tag{158}$$

where C_1 and C_2 are the shunt capacitances associated with terminal pairs 1 and 2 of the pertinent coupling network. These resistance-integral relations do not help us evaluate the maximum of 157, however, unless we know how to apportion areas between $r_{11}(\omega)$ and $r_{22}(\omega)$. The following reasoning answers this question.

Refer to the sketches of $r_{11}(\omega)$ and $r_{22}(\omega)$ shown in Fig. 14 where the amplitudes of typical area elements are denoted by a_ν^2 and b_ν^2. Relations 157 and 158 can thus be represented by the expressions

$$\int_0^\infty |r_{12}|\, d\omega = \sum_{\nu=0}^{\infty} a_\nu b_\nu \tag{159}$$

$$\sum_{\nu=0}^{\infty} a_\nu^2 = \frac{\pi}{2C_1} \quad \text{and} \quad \sum_{\nu=0}^{\infty} b_\nu^2 = \frac{\pi}{2C_2} \tag{160}$$

The problem of making 159 a maximum subject to constraints 160 can now be given a simple geometrical interpretation. If we regard the quantities a_ν and b_ν as components of two vectors, then 159 is their scalar product and 160 states that each vector has a fixed length. Since the scalar product is a maximum when the two vectors coincide in direction, we see that we must make a_ν/b_ν constant or choose

$$r_{22}(\omega) = \rho^2 r_{11}(\omega) \tag{161}$$

where we may regard ρ as the transformation ratio of our coupling network. Choosing $\rho = 1$ is equivalent to setting $C_1 = C_2$, but this need not be done unless desired.

We now have

$$\int_0^\infty |r_{12}|\, d\omega = \rho \int_0^\infty r_{11}(\omega)\, d\omega = \frac{\rho\pi}{2C_1} = \frac{\pi}{2\sqrt{C_1 C_2}} \tag{162}$$

since

$$\frac{C_1}{C_2} = \rho^2 \tag{163}$$

because capacitances are inversely proportional to impedance levels. The fact that the net area enclosed by $r_{12}(\omega)$ is zero enables us to write

$$\int_0^\infty |r_{12}|\, d\omega = 2\int_0^\infty (\text{pos. } r_{12})\, d\omega \tag{164}$$

where (pos. r_{12}) denotes the positive portions of $r_{12}(\omega)$. According to Eq. 152 these occur in the β-intervals 0 to $\pi/2n$, $3\pi/2n$ to $5\pi/2n$, $7\pi/2n$ to $9\pi/2n$, . . . $(n-2)\pi/2n$ to $\pi/2$. Integral 153 evaluated for these intervals gives

$$\int_0^\infty |r_{12}|\, d\omega = 2K\omega_c\left[\int_0^{\pi/2n} \cos n\beta \cos\beta\, d\beta + \int_{3\pi/2n}^{5\pi/2n} \cos n\beta \cos\beta\, d\beta + \cdots + \int_{(n-2)\pi/2n}^{\pi/2} \cos n\beta \cos\beta\, d\beta\right] \tag{165}$$

Evaluation as in Eq. 153 yields straightforwardly

$$\int_0^\infty |r_{12}|\, d\omega = \frac{2nK\omega_c}{n^2-1}\left[\cos\frac{\pi}{2n} + \cos\frac{3\pi}{2n} + \cos\frac{5\pi}{2n} + \cdots + \cos\frac{(n-2)\pi}{2n}\right] \tag{166}$$

This trigonometric polynomial can be put into a closed form,* which gives

$$\int_0^\infty |r_{12}|\, d\omega = \frac{nK\omega_c}{n^2-1}\,\frac{\sin(n-1)\pi/2n}{\sin\pi/2n} = \frac{nK\omega_c}{n^2-1}\cot\frac{\pi}{2n} \tag{167}$$

* See *MCA*, Eq. 15, p. 438.

where the fact that n is an odd integer (required by the condition that the real part be zero outside the band) is made use of.

From Eqs. 162 and 167 we now get

$$K\omega_c = \frac{\pi(n^2 - 1)}{2n\sqrt{C_1C_2}} \tan \frac{\pi}{2n} \tag{168}$$

For large values of n this expression approaches the value

$$K\omega_c = \frac{\pi^2}{4\sqrt{C_1C_2}} \tag{169}$$

Here

$$C_{\text{total}} = C_1 + C_2 \tag{170}$$

so that we can write the result 169

$$K\omega_c = \frac{\pi^2/2}{C_{\text{total}}} \times \tfrac{1}{2}(\sqrt{C_1/C_2} + \sqrt{C_2/C_1}) \tag{171}$$

The second factor in this expression is unity for $C_1 = C_2$ and very nearly unity even for a ratio $C_1/C_2 = 2$ or $\frac{1}{2}$. For comparison with Eqs. 137 and 150, we can write the present result as

$$K\omega_c = \frac{\pi^2/2}{C_{\text{total}}} \tag{172}$$

which shows that the network of Fig. 10 or 12 is not optimum. However, the result 172 is only about 25% larger than 150, and to realize it requires a rather elaborate network. Practical expediency will probably favor the simple network although from a theoretical standpoint it is useful to know the value of the ultimate gain-bandwidth product obtainable in an interstage network design.

The resistance-integral theorem is similarly useful in enabling one to determine the ultimate gain-dependence upon bandwidth obtainable in the design of an input or output network. Here the network is used either to couple a source to a vacuum tube or to couple the latter to a load. Again, the parasitic grid-cathode or plate-cathode capacitance of the tube limits the obtainable gain and its dependence upon bandwidth.

Two situations are distinguished here according to whether the coupling network requires resistive loading at only one end or at both ends. These are shown in Figs. 15 and 16 respectively. Figure 15, part (*a*) shows an input network situation where the capacitance C is absorbed by the grid-cathode capacitance of a vacuum tube; part (*b*) of the same figure is drawn for an output network situation. Here C is absorbed by the plate-cathode capacitance of a vacuum tube which is assumed to be a pentode having an essentially infinite plate resistance. The same situation in the case of a triode vacuum tube is shown in Fig. 16 where the finite plate resistance is represented by R_1. Here the coupling network which must provide for both C and R_1 as well as R_2, is required to accommodate

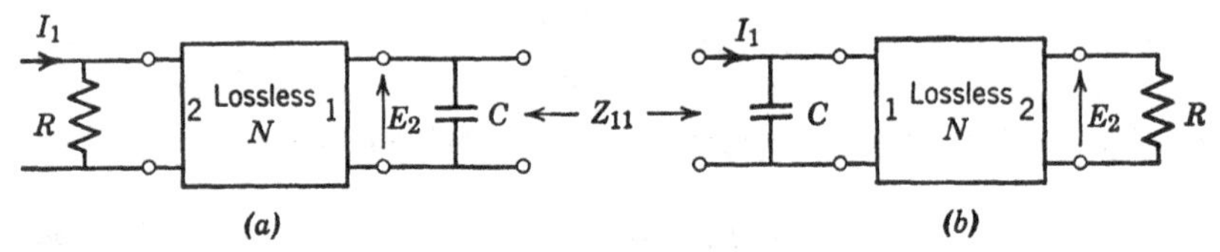

FIG. 15. The interstage network requires resistive loading at one end only.

resistive loading at both ends. In all cases the network within the box is assumed lossless.

The input and output situations in Fig. 15 are accommodated by essentially the same network which is oppositely oriented in these two applications, the resistance R representing either the internal resistance of a

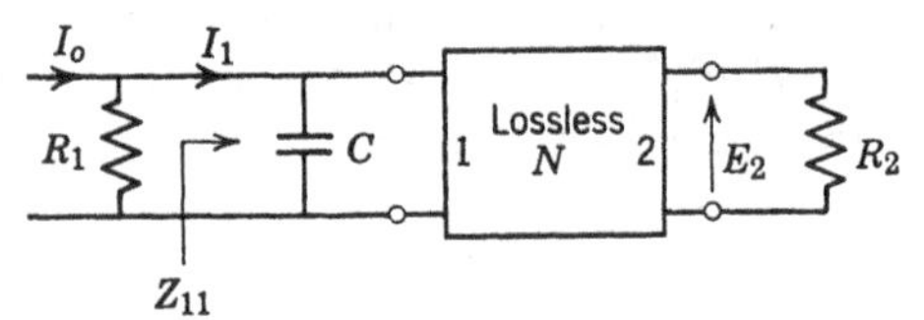

FIG. 16. The interstage requires resistive loading at both ends.

source or the resistance of a load. By the reciprocity theorem the transfer impedance

$$Z_{12}(j\omega) = \frac{E_2}{I_1} \tag{173}$$

is the same for either orientation. If we write for the input impedance

$$Z_{11}(j\omega) = R_{11}(\omega) + jX_{11}(\omega) \tag{174}$$

then the fact that the input power in situation (b) must equal the power delivered to the load yields

$$|I_1|^2 R_{11}(\omega) = \frac{|E_2|^2}{R} \tag{175}$$

whereupon, with Eq. 173, we have the interesting result

$$R \times R_{11}(\omega) = |Z_{12}(j\omega)|^2 \tag{176}$$

If the coupling network is to be designed for a specified transfer impedance magnitude we immediately have $R_{11}(\omega)$ which (by methods discussed in this chapter) determines the input impedance $Z_{11}(j\omega)$. Its development into a lossless structure terminated in a resistance, which can always be managed, and in some useful cases very easily, leads directly to a solution of this synthesis problem. At the moment we are more interested in the following relation:

$$\int_0^\infty R_{11}(\omega)\, d\omega = \frac{\pi}{2C} = \int_0^\infty \frac{|Z_{12}(j\omega)|^2}{R}\, d\omega \tag{177}$$

The best we can do in a coupling network design is to have $|Z_{12}|$ equal a constant K over the passband and be zero otherwise. For a given parasitic capacitance C and bandwidth ω_c, this choice for the behavior of $|Z_{12}|$ will evidently yield the largest gain factor K. Equation 177 then gives

$$\frac{K^2\omega_c}{R} = \frac{\pi}{2C} \tag{178}$$

or

$$K = \sqrt{\frac{\pi R}{2C\omega_c}} \tag{179}$$

which is the maximum flat response obtainable over a band in either situation shown in Fig. 15.

For a double-loaded output network as shown in Fig. 16, the situation is a bit more complicated. Here we define a transfer function

$$T(j\omega) = \frac{2E_2}{I_0} = \frac{2R_1Z_{12}}{R_1 + Z_{11}} \tag{180}$$

where Z_{12} is given by Eq. 173 as before and we recognize that

$$\frac{I_1}{I_0} = \frac{R_1}{R_1 + Z_{11}} \tag{181}$$

The power delivered to the lossless network or to the load is

$$P_1 = |I_1|^2R_{11} = P_2 = \frac{|E_2|^2}{R_2} \tag{182}$$

while the maximum power deliverable by the source with internal resistance R_1 is

$$P_{\text{max}} = \frac{|I_0|^2R_1}{4} \tag{183}$$

The squared magnitude of the transfer function $T(j\omega)$ is thus seen to be

$$|T(j\omega)|^2 = 4\left|\frac{E_2}{I_0}\right|^2 = \frac{P_1}{P_{\text{max}}}(R_1R_2) \tag{184}$$

which suggests that we introduce a normalized transfer function

$$\frac{T(j\omega)}{\sqrt{R_1R_2}} = t(j\omega) = \frac{2E_2/I_0}{\sqrt{R_1R_2}} \tag{185}$$

for which Eq. 184 reads

$$|t(j\omega)|^2 = \frac{P_1}{P_{\max}} \tag{186}$$

Since its maximum value obviously is unity, the quantity

$$1 - |t(j\omega)|^2 = |\rho(j\omega)|^2 \tag{187}$$

may be regarded as a measure of the per unit power that is undeliverable to the load. By Eqs. 181, 182 and 183 we have

$$|\rho(j\omega)|^2 = 1 - \frac{P_1}{P_{\max}} = 1 - \frac{4R_{11}|I_1|^2}{R_1|I_0|^2} \tag{188}$$

$$= 1 - \frac{4R_1R_{11}}{|R_1 + Z_{11}|^2} = \left|\frac{R_1 - Z_{11}}{R_1 + Z_{11}}\right|^2 \tag{189}$$

as may readily be verified by substituting expression 174 for Z_{11} and working out the algebra. We thus recognize that

$$\rho(j\omega) = \frac{R_1 - Z_{11}}{R_1 + Z_{11}} \tag{190}$$

which is commonly referred to as the *reflection coefficient* pertinent to the input end of the network in Fig. 16.

The quantity*

$$10 \log \frac{1}{|\rho(j\omega)|^2} = 20 \log \frac{1}{|\rho(j\omega)|} \tag{191}$$

is called the *return loss* measured in *decibels*, and

$$\ln \frac{1}{|\rho(j\omega)|} \tag{192}$$

is the return loss in *nepers*. It is this return loss that we wish to maximize for a given parasitic capacitance C and passband width ω_c.

In the design of a coupling network for the situation shown in Fig. 16. with a specified $|t(j\omega)|$, one begins by determining $|\rho(j\omega)|^2$ from Eq. 187 Since $\rho(s)$ must be a rational function we can write

$$\rho(s) = \frac{p(s)}{q(s)} \tag{193}$$

* log designates a logarithm to the base 10.

and have

$$|\rho(j\omega)|^2 = \left|\frac{p(s)p(-s)}{q(s)q(-s)}\right|_{s=j\omega} \tag{194}$$

To construct $\rho(s)$ from $|\rho(j\omega)|^2$ we must form $p(s)$ from $p(s)p(-s)$ and $q(s)$ from $q(s)q(-s)$. The latter polynomial has zeros in quadrantal symmetry since those of $q(-s)$ are images of the zeros of $q(s)$ about $s = 0$. From Eq. 190 it is clear that the poles of $\rho(s)$ must lie in the left half-plane; $q(s)$ must be a Hurwitz polynomial. Hence we can form $q(s)$ simply by assigning to it all zeros of $q(s)q(-s)$ that lie in the left half-plane. In similarly forming $p(s)$ from the zeros of $p(s)p(-s)$, however, we do not have to restrict our choice to left half-plane zeros because the numerator in Eq. 190 allows zeros to be in either half of the s-plane. These must now be so chosen that the return loss 192 is maximized over the passband.

We can construct a large variety of ρ-functions according to how we distribute the zeros between the two half-planes, and correspondingly we get a variety of input impedances

$$\frac{Z_{11}}{R_1} = \frac{1 - \rho(s)}{1 + \rho(s)} \tag{195}$$

These determine different lossless network realizations for the box N, all of which, however, have the same transfer-function magnitude $|t(j\omega)|$. Let us denote a ρ-function having all left half-plane zeros by ρ_{left}, and one having all right half-plane zeros by ρ_{right}. If we denote the zeros of ρ_{left} by $s_1, s_2, \ldots s_n$, where $s_\nu = -\sigma_\nu + j\omega_\nu$ with $\sigma_\nu > 0$, then we can write

$$\rho_{\text{right}} = \frac{(s + s_1)\cdots(s + s_n)}{(s - s_1)\cdots(s - s_n)} \cdot \rho_{\text{left}} \tag{196}$$

The asymptotic behavior of $Z_{11}(s)$ differs according to whether it is generated by ρ_{left} or ρ_{right}. We can write

$$[Z_{11}(s)]_{\text{left}} \to \frac{1}{C_l s} \qquad \text{for } s \to \infty \tag{197}$$

and

$$[Z_{11}(s)]_{\text{right}} \to \frac{1}{C_r s} \qquad \text{for } s \to \infty \tag{198}$$

For the reflection coefficients Eq. 190 then gives

$$\rho_{\text{left}} \to 1 - \frac{2}{R_1 C_l s} \qquad \text{for } s \to \infty \tag{199}$$

and

$$\rho_{\text{right}} \to 1 - \frac{2}{R_1 C_r s} \qquad \text{for } s \to \infty \tag{200}$$

Since for large s we have

$$\frac{(s+s_1)\cdots(s+s_n)}{(s-s_1)\cdots(s-s_n)} \to 1 + \frac{2}{s}\sum_{\nu=1}^{n} s_\nu = 1 - \frac{2}{s}\sum_{\nu=1}^{n} \sigma_\nu \tag{201}$$

Eqs. 196, 199 and 200 yield

$$\frac{1}{R_1C_r} = \frac{1}{R_1C_l} + \sum_{\nu=1}^{n} \sigma_\nu \tag{202}$$

We thus see that the resulting real-part area for the input impedance $Z_{11}(j\omega)$ and hence its asymptotic behavior and corresponding shunt capacitance depend upon the distribution of zeros of $\rho(s)$ between the right and left half-planes. Specifically we see that the resulting shunt capacitance is largest if all zeros are chosen to lie in the left half-plane.

The maximum return loss may now readily be calculated. To this end we consider the integral of $\ln 1/\rho_{\text{left}}$ for the closed contour shown in Fig. 8 consisting of the j-axis and an infinite semicircle lying in the right half-plane. Since the integrand is analytic in this half-plane, the integral around the contour is zero. However, let us consider separately the contributions from the j-axis and the semicircular arc. On the j-axis the imaginary part of the integrand (the angle of $1/\rho$) is an odd function and hence contributes nothing, while the real part $\ln 1/|\rho|$ is the same for ρ_{left} and ρ_{right} as is evident from Eq. 196. On the semicircular arc we can use the representation 199 giving

$$\oint \ln\left(1 - \frac{2}{R_1C_ls}\right) ds \to \frac{2}{R_1C_l}\oint \frac{ds}{s} \tag{203}$$

If we let $s = e^{j\theta}$, $ds/s = j\,d\theta$, and this integral becomes

$$\frac{2j}{R_1C_l}\int_{\pi/2}^{-\pi/2} d\theta \to \frac{-2\pi j}{R_1C_l} \tag{204}$$

where a clockwise traversal is assumed since this coincides with a traversal of the j-axis from $-j\infty$ to $j\infty$.

Putting these parts together, we thus have

$$j\int_{-\infty}^{\infty} \ln\left|\frac{1}{\rho}\right| d\omega - \frac{2\pi j}{R_1C_l} = 0 \tag{205}$$

or

$$\int_0^{\infty} \ln\left|\frac{1}{\rho}\right| d\omega = \frac{\pi}{R_1C_l} \tag{206}$$

An optimum situation results if $|\rho|$ equals unity outside the passband and a constant, say ρ_0, over this band. Equation 206 then yields for *the maximum return loss over the band*

$$\ln 1/\rho_0 = \frac{\pi}{R_1C_l\omega_c} \text{ nepers} \tag{207}$$

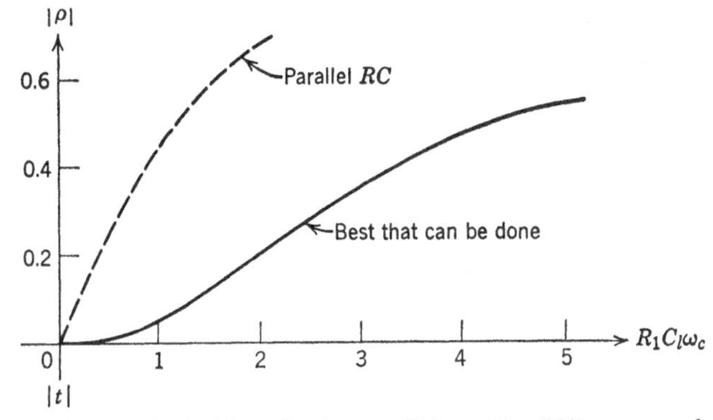

FIG. 17. Maximum obtainable reflection coefficient (Eq. 208) compared with that resulting with a simple parallel RC interstage.

The *minimum* average reflection coefficient over the band is given by

$$\rho_0 = e^{-(\pi/R_1 C_l \omega_c)} \tag{208}$$

By way of comparison consider the simple parallel RC circuit for which

$$Z_1(j\omega) = \frac{R}{1 + jRC\omega} \tag{209}$$

and

$$\rho = \frac{jRC\omega}{2 + jRC\omega}, \quad |\rho| = \frac{RC\omega}{\sqrt{4 + R^2C^2\omega^2}} \tag{210}$$

The reflection coefficient given by Eq. 208 is shown plotted in Fig. 17 in which the dotted curve is for the RC circuit. The absolute value of normalized response, related to ρ_0 by Eq. 187, is shown in Fig. 18.

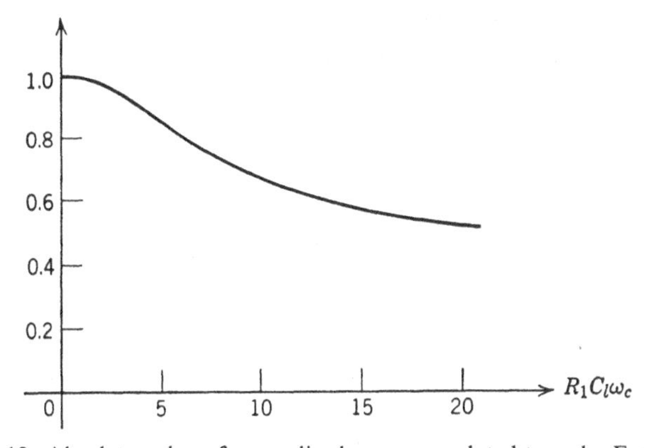

FIG. 18. Absolute value of normalized response related to ρ_0 by Eq. 187.

Although limitations imposed by the resistance-integral theorem upon the design of input or output networks take quite a different form from those imposed on the design of an interstage coupling network, particularly when resistance loading must be provided for at both ends, it is significant to note that the same asymptotic properties implied by this theorem enable one to determine upper bounds on behavior patterns obtainable with passive networks. Being aware of the best that can be done in a given design problem enables one to assess the relative merit of a specific result and thus gain a better perspective upon either the need for or the futility of searching for further possible improvement.

Problems

1. Given the real part of a spectrum function

$$F_1(\omega) = \frac{\sin \omega\delta}{\omega\delta}$$

(*a*) Construct the transform $F(j\omega)$.
(*b*) Calculate $f(t)$.
(*c*) Find the step response of a network whose frequency response is $F(j\omega)$.

2. Given the real-part function

$$F_1(\omega) = \left(\frac{\sin \omega\delta/2}{\omega\delta/2}\right)^2$$

(*a*) Find $F(j\omega)$ and the corresponding $f(t)$.
(*b*) Find and plot the step response of a network characterized by the frequency function $F(j\omega)$.

3. A network is to be designed so that it has a maximally fast step response with minimal overshoot and ripple combined with a frequency response that is lowpass in character with maximum attenuation beyond its cutoff which preferably is an infinite loss point.

Formulate an approach to this problem.

4. The j-axis real part of an impedance $Z(s)$ is shown on the left below. The sketch on the right indicates that the realization of $Z(s)$ involves a shunt capacitance C. Find the minimum value of C.

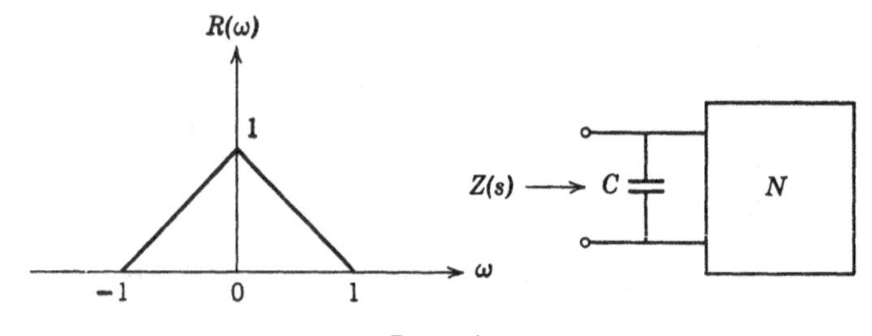

PROB. 4.

5. The adjacent sketch shows the desired magnitude of a driving-point impedance as a function of frequency in rectangular form. Prove that a passive network realizing this impedance does not exist.

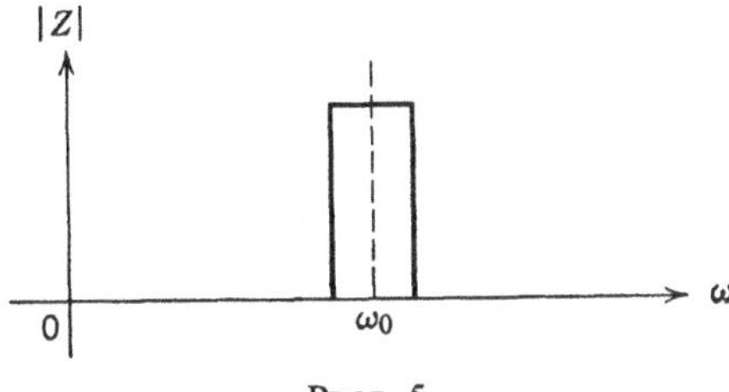

PROB. 5.

6. Relative to the adjacent sketch of the magnitude of a driving-point impedance, and using results plotted in Fig. 6, determine the smallest $\Delta\omega$ possible for a fixed value of w if $Z(s)$ is to be a p.r. function.

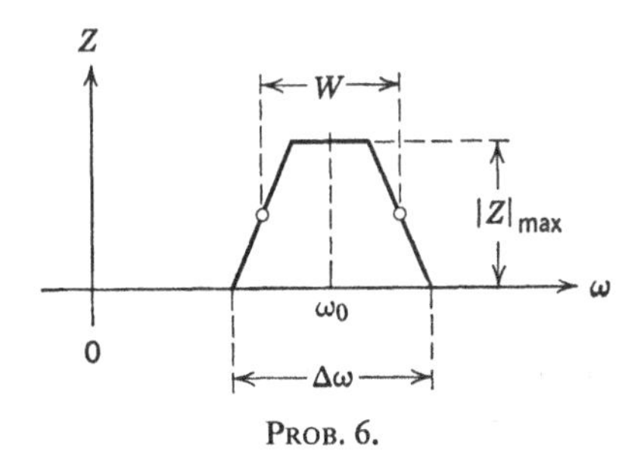

PROB. 6.

7. The real part $R(\omega)$ of an impedance $Z(j\omega)$ is a lowpass trapezoidal cutoff function of unit height and cutoff frequency $\omega_c = 1$. The realization of $Z(j\omega)$ is shown in the adjacent sketch. If the capacitance $C = \frac{3}{2}\pi$, compute the minimum value of L.

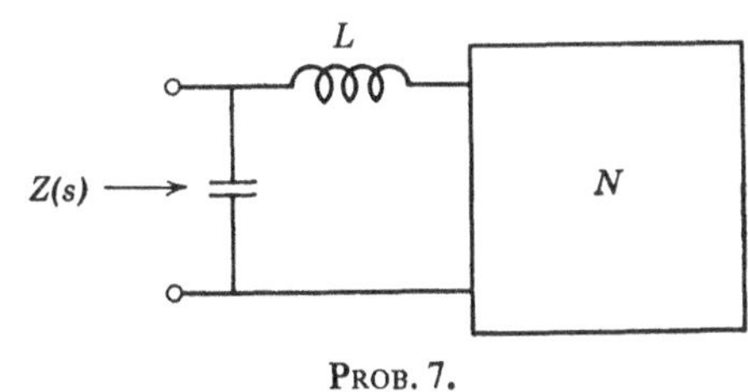

PROB. 7.

8. Given:

$$R(\omega) = \frac{2\alpha\omega^2}{(1 - \omega^2)^2 + 4\alpha^2\omega^2}$$

Evaluate:

$$\int_0^\infty R(\omega)\, d\omega \quad \text{and} \quad \int_0^\infty \frac{R(\omega)}{\omega^2}\, d\omega$$

9. Given:

$$\operatorname{Re}[H(j\omega)] = \frac{1}{(1+\omega^2)(\omega^4 - \omega^2 + 1)}$$

Construct $H(j\omega)$.

10. Given:

$$F_1(\omega) = \frac{\sin \omega\delta}{\omega\delta} \cos \omega t_0 \qquad \text{with } t_0 = 3\delta$$

(*a*) Find $F(j\omega)$ and plot its magnitude.

(*b*) If $F(j\omega)$ is approximated by a rational function over the interval $0 < \omega < 3\pi/\delta$, what is the smallest number of poles that will be needed?

11. Given:

$$|F(j\omega)| = \frac{1}{\sqrt{1+\omega^{10}}}$$

(*a*) Calculate and plot the associated phase angle.

(*b*) Calculate and plot $F_1(\omega)$.

(*c*) Calculate and plot the impulse response.

12. Given:

$$|F(j\omega)| = \cos \omega\delta \qquad \text{for } -\frac{\pi}{2} < \omega\delta < \frac{\pi}{2}$$

$$\equiv 0 \qquad \text{otherwise}$$

(*a*) Compute the impulse response assuming that the associated phase is linear with a slope t_d.

(*b*) Compute the actual associated phase angle.

(*c*) Compute the actual impulse response. Plot it and compare it with that found in part (*a*).

Index